AF677377

Coastal Watershed Management

WITPRESS

WIT Press publishes leading books in Science and Technology.
Visit our website for the current list of titles.
www.witpress.com

WITeLibrary

Home of the Transactions of the Wessex Institute, the WIT electronic-library provides the international scientific community with immediate and permanent access to individual papers presented at WIT conferences. Visit the WIT eLibrary at http://library.witpress.com

International Series on Progress In Water Resources

Series Editor

A.H-D. Cheng
Department of Civil Engineering
University of Mississippi
P O Box 1848
University, MS 38677-1848
USA

Honorary Editors

J. Bear
Technion Institute of Technology
Israel

G. Pinder
University of Vermont
USA

Associate Editors

Y. N. Abousleiman
University of Oklahoma
USA

A. Aldama
Mexican Inst of Water Technology
Mexico

P. Anagnostopoulos
Aristotle University of Thessaloniki
Greece

B. Bobee
Universite du Quebec
Canada

C.A. Brebbia
Wessex Institute of Technology
UK

C-L. Chiu
University of Pittsburgh
USA

M. da Conceicao Cunha
University of Coimbra
Portugal

A.B. de Almeida
Instituto Superior Tecnico
Portugal

J.P. du Plessis
University of Stellenbosch
South Africa

R.A. Falconer
Cardiff University of Wales
UK

G. Gambolati
Universita di Padova
Italy

P. Holz
Brandenburg Technical University
Germany

K.L. Katsifarakis
Aristotle University of Thessaloniki
Greece

D. Koga
Saga University
Japan

L.F. Konikow
U S Geological Survey
USA

M.A. Losada
Universidad de Granada
Spain

K. Onishi
Ibaraki University
Japan

A.C. Rodrigues
Universidade Nova de Lisboa
Portugal

W.W-G. Yeh
University of California at Los Angeles
USA

B.C. Yen
University of Illinois at Urbana-Champaign
USA

Coastal Watershed Management

Edited by:

Ali Fares and Aly I. El-Kadi
University of Hawaii-Manoa, Hawaii

WIT*PRESS* Southampton, Boston

A. Fares & A.I. El-Kadi
University of Hawaii-Manoa, Hawaii

Published by

WIT Press
Ashurst Lodge, Ashurst, Southampton, SO40 7AA, UK
Tel: 44 (0) 238 029 3223; Fax: 44 (0) 238 029 2853
E-Mail: witpress@witpress.com
http://www.witpress.com

For USA, Canada and Mexico

WIT Press
25 Bridge Street, Billerica, MA 01821, USA
Tel: 978 667 5841; Fax: 978 667 7582
E-Mail: infousa@witpress.com
http://www.witpress.com

British Library Cataloguing-in-Publication Data

A Catalogue record for this book is available
from the British Library

ISBN: 978-1-84564-091-0
ISSN: 1461-6513

Library of Congress Catalog Card Number: 2007942009

The texts of the papers in this volume were set individually by the authors or under their supervision.

No responsibility is assumed by the Publisher, the Editors and Authors for any injury and/or damage to persons or property as a matter of products liability, negligence or otherwise, or from any use or operation of any methods, products, instructions or ideas contained in the material herein. The Publisher does not necessarily endorse the ideas held, or views expressed by the Editors or Authors of the material contained in its publications.

© WIT Press 2008

Printed in Great Britain by Cambridge Printing

All rights reserved. No part of this publication may be reproduced, stored in a retrieval system, or transmitted in any form or by any means, electronic, mechanical, photocopying, recording, or otherwise, without the prior written permission of the Publisher.

The authors are grateful to their families for their understanding, encouragment and assistance.

Fares dedicates this work to his wife Samira, daughters Amna and Sara, sons Othman and Ayoub, and parents Ahmed, Hassna and Yougouta.

El-Kadi dedicates this book to his wife Faten and children Shereen, Aladdin and Enjy.

Contents

Preface

Coastal watersheds differ from others by their unique features, including proximity to the ocean, weather and rainfall patterns, subsurface features, and land covers. Land use changes and competing needs for valuable water and land resources are especially more distinctive to such watersheds. Surface water is a valued resource of significant economic, ecologic, cultural, and aesthetic importance. Streams supply irrigation water and can be the main source of drinking water in some places. Streams also provide important habitats for many unique native species. Water quality of receiving waters, such as estuaries, bays, and nearshore waters, are negatively impacted by stream chemical, biological, and sediment pollutants. Coastal groundwater aquifers are negatively affected by land use changes, with associated reduction in recharge and increase in chemical use, and are subjected to the threat of saltwater intrusion. Limited water resources and concerns regarding water quality necessitate the need for best management practices. Watershed problems and pertinent management practices are site specific with conditions that drastically change based on the watershed nature. Hence, there is need for a better understanding of the various physical, chemical, and biological processes involved.

This book covers recent research relevant to coastal watersheds. It addresses the impact of stream chemical, biological, and sediment pollutants on the quality of receiving waters, such as estuaries, bays, and near-shore waters. The contents of the book can be divided into three sections; a) overview of hydrological modeling, b) water quality assessment, and c) watershed management. Chapter 1 presents a general overview of hydrological modeling with emphasis on tropical watershed hydrology. Water quality of coastal watersheds is discussed in chapters 2 through 5. Nutrient bioavailability via runoff from agricultural soils in a watershed in Australia is presented in chapter 2. Chapter 3 explores sediment tracing techniques including artificial and cosmogenic radionuclides, exotic particles, fingerprinting, and rare earth elements. Chapter 4 discusses the importance of and threats to coastal wetlands. Chapter 5 reviews four components of fine particle dynamics: sources and supply mechanisms; in-stream transport and deposition; biological impacts; and spatial and temporal scales of study and variability. Watershed management issues include effect of nitrogen best management practices on water quality (Chapter 6); effects of changing land use on nutrient loads and water quality (Chapter 7), effects of land use changes and groundwater pumping on salt water intrusion

(Chapter 8); a restoration and protection plan for a coastal watershed (Chapter 9); estimation of benefits from restoring coastal ecosystems (Chapter 10), economic value of watershed conservation (Chapter 11); and impact of best management practices in coastal watershed (Chapter 12). Two case studies are also presented in this book. Chapter 13 explores the link between watershed development, hydrologic response and increased risk of waterborne disease as a result of flooding and presence of commensal rodents chronically infected with leptospirosis in a Hawaii watershed. Chapter 14 presents a protection and restoration plan for a watershed in Hawaii which can serve as a model for many similar areas.

This book differs from other hydrology books by dealing with coastal watersheds which are characterized by their unique features concerning weather and rainfall patterns, subsurface characteristics, and land use and cover. In addition to academia, the book should be of interest to organizations concerned with watershed management, such as local and federal governments and environmental groups. Although the book covers coastal regions, it should be of importance to wide range of readers working in other environments. Most contents in the book require minimum background in hydrology, but some chapters require familiarity with hydrological processes, modeling, and watershed management. Overall, the book is expected to satisfy a great need toward understanding and managing critical areas in many parts of the world.

A. Fares & A.I. El-Kadi
University of Hawaii-Manoa, Hawaii

Acknowledgements

Many people cooperated and assisted in completing this work. Their vital suggestions and critical reviews have improved the clarity and contents of this book. The authors are grateful and thankful to these colleagues for their contribution to the success of this work. Following is list of the names of these colleagues arranged alphabetically:

- Younes Alila, Associate Professor, Hydrology and Watershed Management, Department of Forest Resources Management, Faculty of Forestry, The University of British Columbia, Vancouver, British Columbia Canada.
- Mark Brinson, Professor, Biology Department, East Carolina University.
- Williamson B.C Chang, Professor, William S. Richardson School of Law, University of Hawaii at Manoa.
- Chris Craft, Associate Professor, School of Public and Environmental Affairs Indiana University.
- Roger S. Fujioka, Professor, Water Resources Research Center, University of Hawaii at Manoa.
- Stephen B. Gingerich, Research Hydrologist, United States Geological Survey, Honolulu, Hawaii.
- Mary Kentula, Wetland Ecologist, EPA Western Ecology Division, Corvallis, Oregon.
- Stephen Lau, Emeritus Professor, University of Hawaii at Manoa.
- Greg Noe, Scientist USGS, 430 National Center, Reston, VA USA.
- Paul F. Pedone, USDA-NRCS Oregon State Geologist, Oregon NRCS State Office, Portland, OR.
- Joy Zedler, Professor of Botany and Aldo Leopold Chair in Restoration Ecology, Botany Department, University of Wisconsin-Madison.

Special thanks are extended to Farhat Abbas and Ahmet Dogan for their assistance in organizing the material presented in this book. The authors are also thankful to Alan Mair, Amjad Ahmad, Nghia D. Tran, Mohammad Safeeq, and Chui Cheng for their help during editing process of this work.

CHAPTER 1

Overview of the hydrological modeling of small coastal watersheds on tropical islands

A. Fares
College of Tropical Agriculture and Human Resources,
University of Hawaii-Manoa, Honolulu, HI, USA.

Abstract

Increased population growth especially in coastal areas has resulted in substantial land use and land covers changes that in turn have generated concerns about the effects of such activities on their natural resources and especially on the quality and quantity of water resources. Watershed models based upon sound physical theory and well calibrated can provide useful tools for assisting hydrologists and natural-resources managers to choose the best management practices for these sites. This chapter presents an overview of coastal-watershed modeling. It depicts the basic hydrological components of coastal watersheds; it also discusses the different governing equations implemented in the different models to describe the surface and subsurface water flow processes simulated by these models. In addition, governing equations for erosion and contaminant transport mechanisms were also presented for physically based and empirical modeling approaches. The chapter discusses the two main approaches (numerical and analytical) of solving the water flow and sediment transport governing equations models. Salt water intrusion as a result of natural disasters (Tsunami and hurricanes, e.g. Katrina) was also discussed. This chapter provides an overview of a few coastal-watershed hydrology case studies using different watershed models. By addressing various issues of coastal watershed modeling, this work is intended to assist resource managers, researchers, consultant groups and government agencies to select, use and evaluate different watershed models to be able to adopt sustainable watershed-management practices.

1 Introduction

Rapid growth of global population and changes in economic environment have triggered land-use change that can be linked to changes in climate, biodiversity, and water quantity and quality. The impacts of these changes have more pronounced effects on coastal watersheds, especially those of small islands, i.e. Caribbean Islands, Hawaiian Islands, and Pacific Islands. A watershed is defined as a geographic area of land that drains water to a shared destination such as a river system or any other water body. The size of a watershed can be small, representing a single tributary within a larger system, or quite large and cover thousands of square kilometers. Small islands are characterized by a large number of small and steep watersheds with highly permeable volcanic rocks and soils. Rainfall is spatially and temporally variable resulting from a combination of both the location within the island and altitude. Tropical rainfall comprises more than two-thirds of the global rainfall [1]. Great variations of rainfall occur within small distances on tropical islands. For example, on the island of Kaua'i, Hawaii annual rainfall increases from 500 mm near Kekaha to over 11,000 mm at Mt. Wai'ale'ale, an average gradient of 0.42 mm/m [2]. This is caused mainly by orographic characteristics of rains, which are formed by humid air above oceans carried by trade winds from the sea over the steep and high terrain of the islands. These coastal watersheds contain some of the most productive and diverse natural systems. They comprise complex and highly specialized ecosystems, which extend from the mountains to the adjacent coastal areas that include estuaries, coral reefs, and stream delta, which are vital natural resources for different stakeholders. Intensive management practices in these relatively sensitive environments have generated concerns about the effects of land use/cover changes on the quality and quantity of surface water in adjacent coastal areas and groundwater of the whole system.

Hydrologists are often requested to describe, interpret the behavior of these complex systems. Although some conclusions can be made using best physical and biological science judgments, in many instances human reasoning alone is inadequate to synthesize the collection of factors involved in analyzing complex hydrological problems. Intensive field experiments can be conducted to answer many of these practical management questions; however, such investigations are commonly site specific, dependent upon climatological and edaphic conditions, and costly in time and resources.

Hydrological watershed models based upon sound physical theory can provide practical management tools to assist natural-resources managers meet the challenge of description and interpretation. Such management tools combine the subtlety of human judgment with the power of personal computers to allow more effective use of available data and account for more complexity. Watershed models have been successfully used to perform complex analyses and to make informed predictions concerning the consequences of proposed actions. They also increased the accuracy of estimates for alternative practices to a level beyond the best human judgment decisions.

1.1 Characteristics of small coastal watersheds on tropical islands

Many unique characteristics of coastal islands result from their isolation, small size and exposure to the marine environment. Most of the tropical islands are the results of volcanic activities, which make them mountainous in nature, e.g. Hawaii. These islands are continuously exposed to winds, waves, tides, salts, animals, and human activities making them vulnerable to natural and man-made stresses. Generally, the larger the island, the more diverse is its ecosystem, the more varied and numerous are its plants and animals life, and the more tolerant it is to disturbance. The tropical island climate is strongly moderated by the ocean. Island soils are acidic, infertile, and shallow, with a thin organic layer. Larger islands often contain marshes and bogs. Vegetative cover varies, depending on local conditions, soil type, and past clearing practices. Most of the larger islands are forested and mature softwood stands predominant on their landscapes.

Groundwater is the main source of freshwater on islands, but its depletion and contamination is limiting its use. In tropical islands, groundwater is generated entirely by rain on the island, which percolates into the aquifer. Most of the islands are highly rocky and have impervious soil layers that reduce water infiltration, causing more surface runoff. Sometimes high groundwater demand under limited source causes saltwater intrusions into the groundwater supply [3]. A methodical understanding of hydrologic cycle components and characteristics of coastal watersheds on tropical islands is needed to select a hydrological model suitable for a particular scenario. This chapter covers the following aims: 1) to describe the main characteristics of hydrological models; 2) to give an overview of available hydrological models applicable to small island coastal watersheds; 3) to review major environmental problems in coastal watersheds; and 4) to present case studies on the application of hydrological models to coastal watersheds.

2 Classification of models

Models are simplified representation of real systems and are often used to predict the response of the modeled system under the influence of different management scenarios. Models are classified based on process description (deterministic vs. stochastic), timescale (single event vs. continuous), space scale (distribute vs. lumped), techniques of solution (analytical vs. numerical), and their use (watershed, groundwater) (Table 1).

Physical models are based on the mathematical-physics equations of mass and energy transfer intended to avoid and/or minimize the need for calibration. The physical models are physical representations of a smaller- or larger-scale real system. A physical model is used to simulate some phenomenon on a large-scale by using a small-scale experiment either in a field or a laboratory. Geometric and dynamic scales of physical models are important characteristics. Models can be also classified as linear or nonlinear, deterministic or stochastic, steady state or transient, and lumped or distributed. A linear model is the one in which objective functions are expressed by linear equations. A steady-state model does not account for the element of time, while a transient model is one with an explicit time dimension. A deterministic model is one in which its variables do not vary randomly. Stochastic models have some randomness and uncertainty that are described by statistical properties, such as trend, seasonality, mean, variance, skewness, covariance, correlation, and variance function.

Table 1: Characteristics of some watershed models.

Model	Simulation type	Runoff generation	Overland flow	Channel flow	Watershed representation	Use
HSPF	Continuous	Soil moisture accounting	Kinematic wave	Kinematic wave	Lumped	Watershed hydrology and water quality
PSRM	Continuous and event based	SCS curve number and soil moisture	Cascade	Kinematic wave	Distributed	Runoff and sediment yield simulation
MIKE-SHE	Continuous	Richards' equation	Saint-Venant equations	Saint-Venant equations	Distributed	Hydrologic and hydraulic simulation
DHSVM	Continuous	Saturation Excess	Kinematic wave	Muskingum	Distributed	Hydrologic simulation
HEC-1	Event based	SCS curve number	Unit hydrograph	Muskingum	Lumped	Rainfall runoff process
TOPMODEL	Continuous	Green–Ampt	Saint-Venant equations	Saint-Venant equations	Distributed	Stream flow and water quality
GLEAMS	Event based	SCS curve number	Kinematic wave	No channel routing	Lumped	Water quality and quantity
SWAT	Continuous	SCS curve number or Green–Ampt	Kinematic wave	Muskingum	Distributed	Runoff, nonpoint-source pollution
WEPP	Continuous and event based	Hortonian flow	Kinematic wave	Kinematic wave	Distributed	Erosion
AnnAGNPS	Continuous	SCS curve number	Kinematic wave	Kinematic wave	Distributed	Water quality and quantity

Some deterministic models may include stochastic processes to add the dimension of spatial and temporal variability to some of the subprocesses, such as infiltration. A lumped model does not account for the spatial variability of inputs and outputs parameters, while a distributed model does.

3 Mathematical description of the components of hydrologic cycle

Hydrological models represent one or many components of the hydrological cycle, such as precipitation, infiltration, evapotranspiration, and runoff. The main components of the watershed hydrological cycle are briefly discussed in the following sections.

3.1 Precipitation

Precipitation (rain or snow) is generally one of the most important components of the hydrological cycle. In this text, precipitation and rainfall will be used interchangeably. Rainfall is characterized by its total amount, duration, intensity and spatial distribution. Under tropical conditions, rainfall is the main form of precipitation and causes most of the water-related disasters. Rainfall is modeled to estimate annual and seasonal water yield, design water-harvesting structures, and predict flood peaks, erosion and chemical transport from a given watershed. In most of the tropical islands, the rainfall is spatially and temporally variable, posing complications and challenges for modeling exercises. A stochastic approach has been used to analyze rainfall spatially and temporally. Details on stochastic rainfall model are provided by Loukas *et al.* [4].

Osborn and Lane [5] identified three major directions in rainfall analysis: (a) determining the optimum sampling in time and space to answer specific questions, (b) determining the accuracy of rainfall estimates based on existing sampling systems, and (c) simulating precipitation patterns in varying degree of complexity based on existing sampling system for input to hydrologic models. Loukas and Quick [6] developed an event based watershed response model that uses a linear reservoir-routing technique and simulates the fast runoff. The whole process is infiltration controlled and they reported good simulation results of the watershed response [6]. Assuming a linear routing, Nash [7] related the storage factor, KF, to the lag time of the watershed as follows:

$$t_1 = nKF, \tag{1}$$

where Nash's n is the number of the linear reservoirs or the shape parameter of the Nash unit hydrograph. The time lag, t_1, is defined as the time between the centroid of rainfall excess hydrograph and the hydrograph peak. Chuptha and Dooge [8] and Rosso [9] have shown that n is a function of only the geomorphology of the watershed and KF is a function of the geomorphology and precipitation characteristic of the watershed. Yang *et al.* [10] and Sarino and Serrano [11] reported that *KF* is the most uncertain parameter of Nash's model.

3.2 Evapotranspiration

Evapotranspiration is responsible for significant water losses from a watershed. Types of vegetation and land use significantly affect ET. Factors that affect ET include plant type, the plant's growth stage or level of maturity, rooting depth, per cent soil cover, solar radiation, humidity, temperature, and wind speed. The amount of water transpired depends on the rooting depth of plants because water transpired through leaves is extracted by the roots from the soil in the root zone. Plants with deep-reaching roots can transpire more water than a similar plant with a shallow root system. Solar radiation is the major source of energy for ET and usually contributes from 80 to 100 per cent of the total ET.

Vapor pressure at saturation as a function of air temperature is described by the following equation:

$$e_s = \exp\left[\frac{16.78T - 116.9}{T + 237.3}\right] \text{ for } 0 < T < 50\ ^\circ\text{C}, \qquad (2)$$

where e_s is saturation vapor pressure (kPa) and T is air temperature (°C).

Actual vapor pressure of the air (e_a) is calculated by the following equation:

$$e_a = \frac{e_s RH}{100}, \qquad (3)$$

where RH is relative humidity.

Advancements in the field ET measurement have been significant during the past three decades. Now, there is a choice of models based on data type and quality, and suitability of field conditions. Watershed models use different ET submodels, i.e. Penman [12], Priestly–Taylor [13], Thornthwaite [14]. Penman [12] mathematical model combines the vertical energy budget with horizontal wind effects. ET calculation/measurement has been determined using one of the following: (i) water budget, e.g. Fares and Alva, [15], (ii) mass transfer, e.g. Harbeck, [16], (iii) combination, e.g. Penman, [12], (iv) radiation, e.g. Priestley and Taylor, [13], and (v) temperature based, e.g. Thornthwaite, [14]. Detailed information on many of these methods is available in the literature, e.g. Jensen *et al.* [17]; and Morton, [18]. Penman model improvements and adaptations were made by many researchers by including the direct net radiation estimates, improved wind profile theory and effect of plants [19, 20]. The Penman–Monteith model is probably the most suitable ET model for watershed studies, particularly in tropical islands where high intensity winds have significant effect on ET.

The Penman–Monteith [12] approach includes all parameters that govern energy exchange and the corresponding latent heat flux (evapotranspiration) from uniform expansion of vegetation. It calculates evapotranspiration (mh^{-1}) as follows:

$$\lambda ET = \frac{\Delta(R_n - G) + \rho_a C_p((e_s - e_a)/r_a)}{\Delta + \gamma(1 + (r_s / r_a))}, \tag{4}$$

where R_n is net radiation (MJ m^{-2} h^{-1}), G is soil heat flux (MJ m^{-2} h^{-1}), ($e_s - e_a$) is vapor pressure deficit of the air (kPa), ρ_a is mean air density at constant pressure (kg m^{-3}), C_p is specific heat capacity of the air (MJ m^{-3} °C^{-1}), Δ is the slope of saturation vapor pressure–temperature relationship times air pressure (kPa °C^{-1}), γ is the psychometric constant (kPa °C^{-1}), λ is the latent heat of vaporization (MJ m^{-3}), and r_s and r_a are the surface and aerodynamic resistances (s m^{-1}).

3.3 Infiltration and subsurface flow

Infiltration is the rate of the downward entry of water into soil; it is one of the most important hydrological processes of the water cycle. Infiltration is the process that partition water input, e.g. rainfall, irrigation, between the subsurface flow and the runoff. It is driven by matric and gravitational forces; thus, factors affecting infiltration include soil physical properties, initial water content, rainfall intensity, and soil surface sealing or crust. The infiltration rate is usually expressed in units of length per unit time. Several efforts have been made to characterize infiltration for field application including a model based on a storage concept [21] that was later modified by Holtan and Lopez [22]. An approximate model utilizing Darcy's law was proposed by Green and Ampt [23] that was later modified by several researchers mainly Bouwer [24], and Chu [25] who applied the Green–Ampt equation for unsteady-state cases. Some of these efforts involved a simple concept that permits the infiltration rate or cumulative infiltration rate to be expressed mathematically in terms of time and some soil physical properties. Parameters in such models can be determined from soil water properties based on initial and boundary conditions. Below are a few of the infiltration models that have been implemented in different watershed models.

Horton [21, 26] developed the following infiltration model:

$$f_p = f_c + (f_o - f_c)e^{-\beta t}, \tag{5}$$

where f_p is infiltration capacity (LT^{-1}), f_c is final constant infiltration rate (LT^{-1}), f_o is initial ($t = 0$) infiltration rate (LT^{-1}), β is a soil parameter that describes the rate of decrease of infiltration, and t is time (T). The parameters of Horton's model, f_o, f_c, and β are derived based on infiltration tests.

The Green–Ampt model [23] was based upon a very simple physical model of the soil; it considers that the total saturation is behind the wetting front and the saturated water content is constant but not necessarily total porosity. The original equation was derived for infiltration from a ponded surface into a deep homogeneous soil with uniform water content. Water is assumed to enter the soil as piston flow resulting in a sharply defined wetting front that separates a zone that has been wetted from totally unwetted zone. Infiltration capacity (f_p) is calculated as follows:

$$f_p = K_s \frac{S_{cw} + L_w}{L_w}, \tag{6}$$

where f_p is the infiltration capacity (LT^{-1}), K_s is the saturated hydraulic conductivity (LT^{-1}), S_{cw} is the soil suction at wetting front (L), and L_w is the depth of the wetting front from ground surface.

The depth of the wetting front (L_w) can be related to the cumulative amount of infiltration, F(L) as follows:

$$F = (\theta_s - \theta_i)L_w, \tag{7}$$

where θ_s and θ_i are the saturated and initial soil-water content, respectively.

The infiltration rate $f(t)$ becomes:

$$f(t) = K_s(1 + S_{cw}(\theta_s - \theta_i)/F) \text{for } t > t_p \tag{8a}$$

$$f(t) = P \qquad \text{for } t > t_p, \tag{8b}$$

where t_p is the time the water begins to pond at the soil surface.

3.4 Surface flow

Surface runoff also known as surface flow is that portion of precipitation that, during and immediately following a storm event, ultimately appears as flowing water in the drainage network of a watershed [27]. Surface flow is a major component of water cycle in coastal area and small-island watersheds where excess water gets much less time to infiltrate and runs out quickly through streams into the sea. Surface runoff is influenced by soil type, rainfall intensity, topography of the watershed, and vegetation type. The theoretical hydrodynamic equations governing the overland flow are generally attributed to Barre de St. Venant and were formulated in the late 19th century [27]. The St. Venant equations are based on conservation of mass and conservation of momentum for a control volume. The basic continuity equation is given by:

$$\iint \rho\bar{v}\, d\bar{A} = \frac{\partial}{\partial t}\iiint v\, dV, \tag{9}$$

where ρ is the fluid density, $\bar{v}$ is the velocity vector, $\bar{A}$ is the area vector, t is the time, and V is the volume.

The law of conservation of linear momentum may be expressed as:

$$\bar{F} + \iiint \bar{B}\rho\; dV = \iint \bar{v}(\rho\bar{v}\; d\bar{A}) + \frac{\partial}{\partial t}\iiint \bar{v}\rho\; dV, \tag{10}$$

where $\overline{F}$ is the sum of all surface forces on the control volume, and $\overline{B}$ is the sum of all internal forces per unit mass.

3.4.1 St. Venant equations

The St. Venant equations are commonly used for prediction and control design for irrigation and drainage channels. This is one of the most commonly used physical-based models for predicting overland flow. The St. Venant continuity equation is given by:

$$v\frac{\partial A}{\partial x}+A\frac{\partial v}{\partial x}b\frac{\partial h}{\partial t}=0. \tag{11}$$

The dynamic, or momentum, equation is:

$$g\frac{\partial h}{\partial x}v\frac{\partial v}{\partial x}+\frac{\partial v}{\partial t}=g(i-j), \tag{12}$$

where, A is the cross-sectional area of the section, h is the depth of flow at the section, v is the mean velocity at the section, b is the width of the top of the section, x is the position of the section measured from the upstream end, t is the time, g is the acceleration due to gravity, and j is the energy loss/unit length of the channel/unit weight of fluid.

The St. Venant equations cannot be solved explicitly except by making some unrealistic assumptions. Therefore, numerical techniques have to be used. The St. Venant equations work under following assumptions:

- Flow is one-dimensional
- Hydrostatic pressure prevails and vertical accelerations are negligible
- Streamline curvature and the bottom slope of the channel are small
- Manning's equation is used to describe resistance effects
- The fluid is incompressible

3.4.2 Kinematic equation

Lighthill and Whitham [28] proposed a quasi-steady approach known as the kinematic wave approximation. The discharge Q after the replacement of the St. Venant equation by a much simpler kinematics wave equation is given by

$$Q=ay^{m}, \tag{13}$$

where, α and m are parameters, and y is the depth of flow.

The dynamic term in the momentum equation was ignored since it has negligible affect especially in cases where backwater effects were absent. Woolhiser and Liggett [29] showed that the effect of neglecting dynamic terms in the momentum equation could be assessed by the value defined as,

$$k=\frac{S_oL}{HF^2}, \tag{14}$$

where, k is the dimensionless parameter is the length of the bed slope, H is the equilibrium flow depth at outlet, and F is the equilibrium Froude number for flow at the outlet.

3.4.3 SCS method

The SCS curve number equation is an empirical equation that estimates runoff from small agricultural watersheds by a 24-h rainfall event. The curve number method [27] has been widely used to estimate direct runoff. Runoff (Q) is calculated using the following equation:

$$\begin{aligned} Q &= \frac{(P-I_a)^2}{(P-I_a)+S} \quad &\text{if } P > I_a \\ Q &= 0 \quad &\text{if } P \leq I_a, \end{aligned} \tag{15}$$

where P is the rainfall (in), S is the potential maximum retention after runoff begins (in), and I_a is the initial abstraction (in).

The initial abstraction (I_a) quantifies the water losses before runoff begins. It is defined as a percentage of potential maximum retention (S):

$$I_a = 0.2S. \tag{16}$$

The potential maximum retention is a function of curve number:

$$S = \frac{1000}{CN} - 10, \tag{17}$$

where CN is the curve number, which ranges from 0 for completely permeable surface to 100 for an impermeable surface but practically ranges between 40 and 98.

The curve number is determined by the hydrologic soil group, cover type, hydrologic condition, and antecedent moisture condition. Although the method is designed for a single storm event, it can be scaled to predict average annual runoff values. For designing flood-control structures, the rational method is most commonly used.

3.4.4 Rational method

Several empirical methods of similar form have been developed that require input of rainfall estimates for storms of given frequencies. Possibly the best known and widely used is the simple and aptly named rational formula [30]. The rational equation is an empirical equation that has been used for predicting the peak discharge from a small watershed and for design of flood-control structures. The peak discharge ($ft^3\ h^{-1}$) in rational equation is described as:

$$q = CiA, \tag{18}$$

where C is a runoff coefficient, i is rainfall intensity in in h^{-1} for a given frequency and A is the area of the watershed in acres.

3.5 Subsurface and groundwater flow

Darcy [31] found that soil water movement in porous media (q) is directly proportional to the hydraulic gradient (i) as follows:

$$q = -Ki, \tag{19}$$

where q is flux or volume of water moving through the soil per unit area per unit time (LT^{-1}), K is the hydraulic conductivity (LT^{-1}), which is dependent on the properties of the fluid and porous medium, i is hydraulic gradient (LL^{-1}) expressed in the x-direction as follows:

$$i = \partial H / \partial x, \tag{20}$$

where H is the hydraulic head, which is the sum of the pressure head (h) and elevation head (z).

For saturated soils, the hydraulic conductivity is constant with respect to h; whereas for unsaturated conditions, hydraulic conductivity can vary with time and space if the soil is heterogeneous or anisotropic. In unsaturated conditions, K becomes a function of pressure head (h), then the water flux is expressed as follows:

$$q = K(h) \partial H / \partial x. \tag{21}$$

Water flow in variably saturated porous media is described by Richard's equation that combines the mass balance for an element volume of porous media with Darcy's law. The 1D form of this equation for flow in the vertical direction is as follows:

$$C_w(h)\frac{\partial h}{\partial t} = \frac{\partial}{\partial z}\left[K(h)\left(\frac{\partial h}{\partial z}+1\right)\right] \pm S, \tag{22}$$

where $C_w(h)$ is the water capacity function which is equal to the inverse slope of $h(\theta)$, θ is water content, and S is the source/sink term. This form of the Richard's equation has been used to simulate both saturated and unsaturated subsurface flow for different initial and boundary conditions.

4 Contaminant transport

Water quality is important for sustainable development in watersheds. Water is the transport agent of energy, nutrient chemicals, and sediments. Increasing amounts of potentially hazardous chemicals released from various agricultural operations have been polluting soil–water ecosystems. Understanding the transport of these chemicals through surface and subsurface water flow is essential for the management of our natural resources to ensure sustainable crop production and minimize pollution of water resources. Farming and ranching have also allowed an excess of nutrients, sediment and chemicals to runoff [32]. Leaching of agrochemicals through the root zone of agricultural crops continues to endanger the long-term groundwater quality in agricultural areas. Hubbard and Sheredan [33] documented that in many agricultural areas, nitrate-nitrogen (NO_3–N) levels in drinking water were significantly higher than the maximum contaminant level of 10 mg L^{-1} set by the US Environmental Protection Agency.

The fate of a pollutant, in soil, is determined by advection, diffusion and dispersion processes. In this section different transport processes in saturated (groundwater) and unsaturated (vadose) zones are discussed.

4.1 Surface-water contamination

Surface-water contamination occurs when hazardous substances coming from different sources dissolve or mix with receiving water bodies, e.g. streams, lakes, and oceans. Because of the close relationship between sediments and surface water, contaminated sediments are often considered part of surface-water contamination. Sediments not only contaminate the water but also threat wetlands and streams by depositing pollutants on the bottom of streams, lakes, and oceans. Surface water can be contaminated by hazardous substances either coming from agricultural fields or flowing from an outfall pipe or channel or by mixing with contaminated storm water runoff. Effluent coming from industrial sources or from some older sewage systems that overflow during wet weather to streams can cause substantial amounts of water contamination. Stormwater runoff becomes contaminated when rain water mixes with contaminated soil and either dissolves the contamination held in the soil or carries contaminated soil particles. Surface water can also be contaminated when contaminated groundwater reaches the surface through a rising groundwater table in the rainy season or via a spring.

4.2 Soil erosion

Soil water erosion is the processes of soil detachment, deposition, and transport through a watershed. Erosion is a natural process that can be induced by human activities. There are three main types of soil water erosion: sheet and rill, gully and channel, and mass wasting. Sheet and rill erosion is caused primarily by the action of raindrops and surface-water movement. Raindrops have high energy and initially start the erosion process by splashing and loosening surface soil particles. Gully erosion occurs in well-defined channels. Mass wasting occurs when large masses of soil move at once as a result of a landslide, or more slowly over time. Human activities, such as building construction, road construction, timber harvest, grazing, and agriculture activities can accelerate soil-erosion processes.

Soil erosion is a two-stage process. First, sediment is detached, then it is transported. Soil-particle detachment by rainfall is a function of the kinetic energy of the rainfall. After its detachment, sufficient overland flow energy must be available for a soil particle's transport or it will be deposited. Sediment transport occurs in two associated forms a suspended and a bedload. A suspended load is much more uniformly distributed throughout the flow depth than a bedload. The transport capacity stays mostly in the vicinity of the deposition of suspended sediment due to the small fall velocities. Bedload is that portion of the load that moves along the bottom of the flow by rolling, sliding and saltation. It is generally composed of the larger soil particles, and consequently is highly transport dependent. As such, a decrease in transport capacity causes instantaneous deposition of the excess bedload.

4.3 Modeling soil erosion

Modeling soil erosion has been achieved using physically based models, e.g. rill inter-rill erosion model [34] and empirical models, e.g. USLE [35] and its revised version RUSLE [36].

4.3.1 Empirical erosion models

The RUSLE is an empirical model that predicts annual soil water erosion (tons/acre/yr) resulting from sheet and rill erosion in croplands. It is the official tool used for conservation planning in the US. Many other countries have also adapted this model. It is defined as follows:

$$A = R * K * L * S * C * P, \tag{23}$$

where,
A = Annual soil loss (tons acre^{-1} yr^{-1}) resulting from sheet and rills.
R = Rainfall – runoff erosivity factor; it has been mapped for the entire USA.
K = Soil erodibility factor; it is a function of the inherent soil properties, including organic matter content, particle size, permeability, etc.
L = Slope length factor. This factor accounts for the effects of slope length on the rate of erosion.
S = Slope steepness factor; it accounts for the effects of slope angle on erosion rates.
C = Cover management factor; it accounts for the influence of soil and cover management, such as tillage practices, cropping types, crop rotation, and leaving areas fallow, on soil erosion rates.
P = Supporting practices factor; it accounts for the influence of conservation practices, e.g. contouring, strip cropping, and terracing.

Despite their wide use in many watershed models, USLE and RUSLE have some theoretical problems, such as interaction among the variables and water flow, on which soil loss is closely dependent, is underestimated in the models [37]. It is difficult to identify the events that most likely result in large-scale erosion because USLE/RUSLE are not event-responsive equations. They ignore the processes of rainfall-runoff as well as the heterogeneities in input such as vegetation cover and soil types [38]. They do not account for gully erosion, mass movement and sediment deposition [39].

Erosion estimated with these empirical models, e.g. USLE and RUSLE, is often higher than that measured at watershed outlets. The sediment-delivery ratio (SDR) is used to correct for this reduction effect. SDR is defined as the fraction of gross erosion that is transported for a given time interval. It is a measure of the sediment transport efficiency, which accounts for the amount of sediment that is actually transported from the eroding sources to a measurement point or watershed outlet compared to the total amount of soil that is detached over the same area above that point. In relatively large watersheds, most sediment is deposited within the watershed and only a fraction of the soil that is eroded from the hillslope reaches the stream network or the watershed outlet.

Physically, SDR stands as a mechanism for compensating for areas of sediment deposition that becomes increasingly important with increasing watershed area. There are many factors that must be addressed when calculating the sediment-delivery ratio in any watershed. Some of the factors that influence the SDR include: hydrological inputs (mainly rainfall), landscape properties (e.g. vegetation, topography and soil properties) and their complex interactions at the land surface.

4.3.2 Physically based erosion models

Water erosion prediction by physically based erosion models, e.g. WEPP [40] uses the physically based rill interrill concept to predict soil erosion [34]. A physically based model computes detachment and transport by raindrop impact, and detachment, transport and deposition by flowing water. It also predicts sheet and rill erosion from the top of the hillslope to receiving channel; it also considers sediment deposition. The sediment continuity equation for overland flow used is as follows:

$$\frac{\partial(ch)}{\partial t} + \frac{\partial(cq)}{\partial x} = e_i + e_r, \tag{24}$$

where c is total sediment concentration (kg m^{-3}), h is the average, local overland flow depth (m), q is discharge per unit width (m^2 s^{-1}), x is distance in the direction of flow (m), e_i is interrill erosion rate per unit area (kg s^{-1} m^{-2}), and e_r is net rill erosion or deposition rate per unit area (kg s^{-1} m^{-2}).

The sediment yield equation assumes constant rainfall [41] for a runoff event and is as follows:

$$Q_s(x) = QC_b = Q\{B/K+(K_i - B/K)[1-\exp(-K_r x)]/K_r x\}, \tag{25}$$

where Q_s is total sediment yield for the entire amount of runoff per unit width of the plane (kg m^{-1}), Q is the total storm runoff volume per unit width (m^3 m^{-1}), C_b is mean sediment concentration over the entire hydrograph (kg m^{-3}), K_r and B are rill coefficients, K_i is an interrill coefficient, K is a slope resistance coefficient, x is distance in the direction of flow (m), and the other variables are described earlier.

Lane *et al.* [42] extended this sediment-yield equation for a single plane to irregular slopes approximated by a cascade of planes. From the input data, parameter estimation procedures derived from calibrating WEPP erosion model using rainfall simulator data were used to compute the depth-discharge coefficient, interrill erodibility, rill erodibility, and sediment-transport coefficient [43].

4.4 Subsurface-water contamination

Subsurface-water contamination occurs when hazardous substances such as chemical fertilizer and pesticides from landfill, factory affluent and agricultural farm leach to groundwater. Several reviews of solute-transport modeling have been written, such as those by Mercer and Faust [44], Anderson and Woessner [45] and Zheng and Bennett [46]. Freeze and Cherry [47] cover many of the transport equations and offer clear descriptions of many transport mechanisms. Diffusion, dispersion, and advection are the basic processes by which solute moves from one place to another. Diffusion is a molecular-scale process, which causes the spreading of the solute due to concentration gradients and random motion. Diffusion causes a solute in water to move from an area of higher concentration to an area of lower concentration. This process continues as long as a concentration gradient exists. The mass of fluid diffusing is proportional to the concentration gradient, which can be expressed using Fick's first law.

Dispersion is caused by heterogeneities in the medium that create variation in flow velocities and flow paths. This variation may occur due to a velocity difference from one channel to another, or due to variable path lengths. Dispersion is a function of average linear velocity and dispersivity of the medium. Dispersivity in a soil column is on the order of centimeters, while in the field it is on the order of one to one thousand of meters. Mass transport due to dispersion can occur in both longitudinal (parallel to flow direction) as well as transverse (perpendicular to flow direction) directions. In most cases, transverse dispersivity is much smaller than the longitudinal dispersivity. Hydrodynamic dispersion is the process by which solutes spread out and are diluted compared to simple advection alone. It is defined as the sum of the molecular diffusion and mechanical dispersion.

4.5 Solution techniques

4.5.1 Analytical techniques

Several analytical models have been developed to solve the water flow and solute transport equations for specific boundary and initial conditions [48–50]. Analytical solutions are conceptually limited and so does their application to real problems. The geometry of the problem must be regular and simple, e.g. circular, rectangular; as such, they are not applicable to complex boundary conditions and are also limited to idealized conditions. Conceptually, analytical solutions are limited by several simplifying assumptions that were used to develop the solution. To overcome these limitations of analytical solution, a numerical approximating technique has been used to solve the transport equations.

4.5.2 Numerical techniques

These techniques are more flexible than analytical solutions because they can describe complex systems with proper arrangements of grid cells. In general, these solution techniques break up the study field into small grid cells of different shapes that best describe the system. These techniques have some limitations. The common numerical methods used to implement mathematical formulation of partial differential equations of flow and solute transport are finite-difference, finite-volume and finite-element, method of characteristics, collocation methods, and boundary-element methods as explained by Bedient *et al.* [51].

The finite element and finite-difference methods are the most common methods for simulating water flow and solute transport. Finite-difference methods are more simple, straightforward and easy to understand. A variety of algorithms were developed to solve finite-difference equations. Finite-difference methods represent the simulated system with a grid of square or rectangular shape cells. Partial differential equations governing water flow and solute transport can be approximated by differences and solved by iteration [44]. This approximation leads to errors that can be significant [52]. The finite-element method operates by breaking the space in elements of different shapes and sizes that gives more flexibility to describe irregular simulated systems and variable boundary conditions. The major disadvantages of the finite-element method are its high computing requirement [53] and its difficult formulation process.

5 Integrating GIS with watershed models

Geographic information systems play a significant role in facilitating spatial data preparation and analysis because of its ability to store, retrieve, manipulate, analyze, and map geographic data. Using GIS, hydrologists were able to readily produce high-quality maps incorporating model output and geographic entities, further enabling visual support during decision-making processes. Advanced analyses and interpretations were possible using several spatial analysis capabilities of the GIS.

Lumped watershed models simplify most of their input parameters and use spatial averaged values for them over the entire simulated watershed. Similarly their outputs are also spatially averaged. These types of models have been used as great teaching tools; however, they were not embraced as research and management tools to evaluate real management scenarios and nonpoint-source pollution problems. They are unable to determine critical areas of the watershed that are contributing substantially to pollutant loads generated from the watershed of interest. In many nonpoint-source pollution problems, there is a lack of time and resources to conduct intensive field work to identify the spatial contribution of different parts of watersheds to the sediment and pollutant loads leaving a watershed. Thus, use of distributed watershed models is the only viable option that can help manage many of these watersheds with reasonable investment of time and resource.

The use of distributed watershed models has been gaining momentum for the last few decades because of their capabilities in depicting the spatial distribution of water flow and erosion processes. However, from the start, their major obstacle was their requirements for large amounts of time and resources needed to assemble and manipulate the input and output data sets even for small watersheds. The amount of data increases substantial and consequently so does the time to analyze it as the size of the watershed increases and more heterogeneity is introduced. A logical step in helping watershed hydrologists use distributed watershed models is to interface these models with a practical data management scheme such as geographic information systems (GIS) that would manage, help analyze and display spatially distributed data.

Distributed models create grid/mesh of the simulate watershed domains. These meshes are composed of cells, also know as units. The mesh is generated based on topographic characteristics from the digital elevation model (DEM) data. The water flow and sediment transport equations, are solved within each cell at each time step during the duration of simulations. The impact of grid size on the performance of watershed models is well reported in the literature. A sensitivity analysis that used different grid cell sizes (2, 4, 10, 30 and 90 m) reported significant effects of the grid cell size on the computed topographic parameters and hydrographs [54]. Moore and Thompson [55] found that the slope and topographic index values varied with grid cell size for scales ranging from 20 m to 680 m in three 100 km^2 study areas in southeastern Australia.

6 Performance of hydrologic model

The performance and behavior evaluation of hydrological models is commonly made through comparison of different efficiency criteria. To achieve adequate reliability of the simulation models, it is important that they are rigorously calibrated and validated before any analysis and/or management scenario analysis are conducted. It is highly recommended to do the sensitivity analysis of model parameters before starting the calibration process. Model calibration and evaluation efforts are performed to achieve a reasonable correspondence between measured field data and the output of the model.

6.1 Sensitivity analysis and model evaluation

Sensitivity analysis is the study of how the variation in the output of a model can be apportioned, qualitatively or quantitatively, to different sources of variation in input. It is the technique of identifying the parameters with little and high impact on the performance of the tested model. Parametric sensitivity is a vital part of most optimization techniques [56]. This modeling tool, if properly used, can provide a better understanding of the correspondence between the model and physical process being modeled. McCuen [56] explained the sensitivity in mathematical form using the Taylor series expansion of the explicit function; thus, from the definition, sensitivity S can be given by:

$$S = \frac{\partial F_0}{\partial F_i} = \left[x(F_i + \Delta F_i,\ F_{j/j \neq i}) - x(F_1,\ F_2,\ ....,\ F_n) \right] / \Delta F_i. \tag{26}$$

For parametric and component sensitivity, the factor F_0 replaced by an output function (ϕ) and F_i with a parameter under consideration (p_i). Thus, the parametric sensitivity, S_{pi}, can be given by:

$$S_{pi} = \frac{\partial \phi}{\partial p_i}. \tag{27}$$

Currently, there are several available methods for sensitivity analysis [57, 58]. The new Morris method, in addition to the overall sensitivity, offers estimates of the two factor interaction effects [59]. Several studies have addressed the problem of sensitivity analysis in land-surface schemes using different approaches. Bastidas *et al.* [60], using the BATS (biosphere-atmosphere transfer scheme) in two different climatic regions of the US showed that a sensitivity analysis performed before the calibration process reduces the number of parameters prompted for calibration. Their findings suggest that the sensitivity analysis is efficient in reducing the computational time needed in the calibration.

Model evaluation is intimately related to model development. No matter whether models are physically based or conceptually based, they all have some empirical constraints, which could be due to lack of sufficient observational evidence on some processes and/or limitations set by available computing resources [61]. The model evaluation is an essential process to evaluate the model performance and to assess how well the model represents the real physical system. The purpose of model evaluation is to lead the modeling system toward better results [61]. Model evaluation could be based on anything from accessibility of the model to the real data testing. In modeling terms, the goodness of fit after calibration between the observed data and simulated data is one way to represent it. There are several ways to express the error between model prediction and real data; i.e. mean absolute error, root mean square error, average relative error and the coefficient of efficiency given by Nash and Sutcliffe [62].

6.2 Calibration and validation of models

An important part of any modeling exercise is the model calibration. Calibration is a process wherein certain parameters of the model are altered in a systematic fashion and the model is repeatedly run until the simulated results match field-observed values within an acceptable level of accuracy. The process of model calibration is quite complex and limited by the model itself, input, and output data. Imperfect knowledge of watershed characteristics, mathematical structures of the hydrological processes and model limitations can cause error in calibration process. Before starting model calibration, field conditions at the site should be properly characterized. Lack of proper site characterization may lead to a wrong representation of the simulated system. There are two primary parts in the model-calibration process [63]. The first is to decide how to judge whether one set of parameters is preferred over another; second is to find the preferred set of parameters. Model calibration can be performed either by trial and error or by automated techniques. Automated calibration can be performed by means of specifying an objective or a set of objective functions [63]. Uncertainty in models and data leads to uncertainty in model parameters and model predictions. To avoid these uncertainties, Bevin and Binley [64] proposed generalized likelihood uncertainty estimation (GLUE) that uses prior distributions of parameter sets and a method for updating these estimates as new calibration data becomes available. Automated parameter-estimation techniques for model calibration are accurate and rapid. Validation of hydrologic models is a process of matching the simulated results with observed values without altering the calibrated parameters. General methodologies related to model calibration and validation has been considerably discussed [65]. However, as noted by Hassanizadeh and Carrera [66] no consensus on methodology exists. Some efforts were made during the past three decades to develop methods for calibration and validation of lumped models, but limited attention has been devoted to distributed models that are relatively more complicated [65]. Refsgaard and Storm [67] emphasized that a rigorous parameterization procedure is crucial in order to avoid methodological problems in the subsequent phases of model calibration and validation.

7 Overview of available hydrologic models

***Soil Water Assessment Tool* (*SWAT*):** The Soil Water Assessment Tool [68] is a watershed-scale, distributed, conceptual and continuous simulation model, used as a soil and water assessment tool. It can also be used as a field scale model too. There are several versions of SWAT available, and the recent one is SWAT2000 that includes bacteria transport, Green–Ampt infiltration, the Muskingum routing method, a weather generator, and the SCS curve number for runoff estimation. For potential evapotranspiration calculations, users have options between Penman–Monteith, Priestley–Taylor, and Hargreaves methods. Event-based erosion caused by rainfall and runoff is modeled using a modified universal soil loss equation (MUSLE).

***Distributed Hydrology Soil Vegetation Model* (*DHSVM*):** This is a distributed, physically based, and continuous simulation watershed and field-scale model. DHSVM was developed by Wigmosta *et al.* [69] at the University of Washington, Seattle. This model accounts for topographic effects on soil moisture, groundwater, and surface-water relocation in a complex topography. It includes canopy interception, evaporation, transpiration, and snow accumulation and melt, as well as runoff generation via the saturation excess mechanisms. Canopy evapotranspiration is represented via a two-layer Penman–Monteith formulation that incorporates local net solar radiation, surface meteorology, soil characteristics and moisture status, and a species-dependent leaf-area index and stomatal resistance. Snow accumulation and ablation are modeled using an energy-balance approach that includes the effects of local topography and vegetation cover. Saturated subsurface flow is modeled using a quasi-three-dimensional routing scheme.

***System Hydrologique Européen* (*MIKE SHE*):** The original MIKE SHE [70] model was developed and became operational in 1982, under the name Système Hydrologique Européen (SHE). The model was sponsored and developed by three European organizations: the Danish Hydraulic Institute (DHI), the British Institute of Hydrology, and the French consulting company SOGREAH. MIKE SHE is an integrated, physically based, distributed model that simulates hydrological and water-quality processes on a basin scale. This model is able to simulate both surface and groundwater with precision equal to that of models focused separately on either surface water or groundwater. The MIKE SHE modeling system simulates most major hydrological processes of water movement, including canopy and land-surface interception after precipitation, snowmelt, evapotranspiration, overland flow, channel flow, unsaturated subsurface flow, and saturated groundwater flow. It also simulates major water-quality components. A grid network represents spatial distributions of the model parameters, inputs, and results with vertical layers for each grid. MIKE SHE uses the Kristensen and Jensen [71] method for calculating actual evapotranspiration. It includes Muskingum and Muskingum–Cunge methods for simplified channel routing.

***Annualized Agricultural Nonpoint-Source Model (AnnAGNPS)*:** Annualized Agricultural Nonpoint Source designed by the US Department of Agriculture (USDA ARS and NRCS), is a continuous distributed simulation model widely used for watershed assessment. It expands the capabilities of its predecessor AGNPS [72] which is a single-event model. Runoff is calculated using the SCS curve number equation [73], but is modified if a shallow frozen surface soil layer exists. Curve numbers are modified daily based upon tillage operations, soil moisture, and crop stage. Actual evapotranspiration is a function of potential evapotranspiration calculated using the Penman–Monteith equation [12] and soil-water content. Soil water erosion is estimated using RUSLE [36] that was modified to be implemented at the watershed scale in AnnAGNPS [74]. AnnAGNPS uses a GIS interface for processing input and output data. However, selecting the proper grid size was identified as a major factor influencing sediment yield calculations [75]. The border conditions before a rainfall-runoff event are calculated by the model rather than by individual user input. Additionally, long-term simulations are possible using AnnAGNPS as compared to event-based AGNPS model.

***Nonpoint-Source Pollution and Erosion Comparison Tool (N-SPECT)*:** The coastal services center of the National Oceanic and Atmospheric Administration (NOAA) developed the Nonpoint-Source Pollution and Erosion Comparison Tool (N-SPECT) to examine the relationships between land cover, soil characteristics, topography, and precipitation in order to assess spatial and temporal patterns of surface-water runoff, nonpoint-source pollution, and erosion. N-SPECT was developed as a decision-support tool for coastal watersheds. N-SPECT uses the SCS curve number method for runoff estimates and generates a curve number grid based on the combination of land cover and hydrological soil group at each cell within a given study area. Soil erosion is calculated either using RUSLE or MUSLE equations when the model is used to simulate annual or single event, respectively.

***Physically Based Runoff Prediction Model (TOPMODEL)*:** This is a physically based distributed, continuous simulation watershed model. TOPMODEL was developed by Beven and Kirkby [76], it predicts watershed discharge and a spatial soil-water saturation pattern based on precipitation and evapotranspiration time series and topographic information. TOPMODEL is a set of conceptual tools that can be used to reproduce the hydrological behavior of watersheds in a distributed or semidistributed way. The Penman–Monteith method is implemented in the model to estimate ET. Runoff is computed according to the infiltration excess mechanism, thus, TOPMODEL uses the exponential Green–Ampt equation of Beven [77]. Detailed background information of the model and some of its applications can be found in Beven [78]. TOPMODEL assumes that whole basin is homogeneous, which could be unrealistic and applicable for only smaller basins. The model is very sensitive to parameters like soil hydraulic conductivity decay, the soil transmissivity at saturation, the root zone storage capacity and the channel routing velocity in larger watersheds [79]. The calibrated values of parameters are also related to the grid size used in the digital terrain analysis [80–82]. The time step and the grid size have also been shown to influence TOPMODEL simulations [83].

Hydrological Simulation Program – FORTRAN (HSPF): The Hydrological Simulation Program – FORTRAN (HSPF) was developed by the EPA-Athens laboratory [84]. HSPF is a comprehensive, conceptual, continuous watershed simulation model that simulates the water quantity and quality processes that occur in a watershed, including sediment transport and movement of contaminants. It is an analytical tool that has application in planning, design, and operation of water-resources systems. The model enables the use of probabilistic analysis in the fields of hydrology and water-quality management through its continuous simulation capability. This model is classified as a lumped model, but it can reproduce spatial variability by dividing the basin in hydrologically homogeneous land segments and it can simulate runoff for each subbasin independently, using different meteorological input data and watershed parameters. Runoff flow rate, sediment loads, nutrients, pesticides, toxic chemicals, and other water-quality constituent concentrations can be predicted. The model can simulate continuous, dynamic, or steady-state behavior of both hydrologic/hydraulic and water-quality processes in a watershed. HSPF also may be applied to urban watersheds through its impervious-land module. A large number of parameter requirements increases the problem associated with parameter selectivity and physical meaningfulness of model parameters. The model relies heavily on calibration against field data for parameterization [85]. HSPF does not explicitly model agricultural management practices and their effects on runoff or water quality.

Water-Erosion Prediction Project (WEPP) Model: The WEPP erosion model, developed by USDA-ARS is a continuous simulation computer program that predicts soil loss and sediment deposition from overland flow on hill slopes, soil loss and sediment deposition from concentrated flow in small channels, and sediment deposition in impoundments. In addition to the erosion components, it also includes a climate component that uses a stochastic generator to provide daily weather information, a hydrology component that is based on a modified Green–Ampt infiltration equation and solutions of the kinematic wave equations, a daily water-balance component, a plant growth and residue decomposition component based on the erosion productivity impact calculator (EPIC) model, and an irrigation component. The WEPP model computes spatial and temporal distributions of soil loss and deposition, and provides explicit estimates of when and where in a watershed or on a hill slope erosion might occur so that appropriate conservation measures can be selected to best control soil loss and sediment yield. Theoretically, it can exactly predict how rainfall will interact with the soil on a site during a particular rainstorm or during the course of an entire year [86]. The model uses the soil–water-balance component based on the corresponding component of the simulator for water resources in rural basins (SWRRB) model [87]. The infiltration component of the hill slope model is based on the Green and Ampt equation as modified by Mein and Larson [88], with the ponding-time calculation for an unsteady rainfall [25]. The water-balance and percolation components of the hill-slope model are based on the water-balance component of the SWRRB [87], with some modifications for improving estimation of percolation and soil evaporation parameters. WEPP considers only Hortonian flow or flow that occurs when the rainfall rate exceeds the infiltration rate. The model uses two methods of computing the peak discharge: a semianalytical solution of the kinematic-wave model and an approximation of the kinematic-wave model. The first method is used when WEPP is run in single-event mode, while the second is used when WEPP is run in continuous simulation mode [89, 90]. WEPP requires large number of data sets that may limit model use in watersheds where relatively less data is available. Many of the model parameters need to be calibrated to avoid problems with model identifiablity and the physical interpretability of model parameter [38]. The WEPP model does not include gully erosion and the rill-interrill concept of erosion that may limit its application for all types of soil and field conditions [38]. WEPP does not model nitrate or phosphorus losses from agricultural landscapes.

***CREAMS/GLEAMS*:** Chemicals, runoff, and erosion from agricultural management systems (CREAMS) model [91] was developed by the US Department of Agriculture-Agricultural Research Service to aid in the assessment of agricultural best management practices for pollution control. CREAMS is commonly used for evaluation of agricultural best management practices (BMPs) for pollution control. Daily erosion, sediment yield, and associated nutrient and pollutant loads are estimated at the boundary of the agricultural area. Runoff estimates are based on the SCS curve number method. CREAMS calculates runoff volume, peak flow, infiltration, evapotranspiration, soil-water content, and percolation on a daily basis. Daily erosion and sediment yield are also estimated and average concentrations of sediment associated and solute chemicals are calculated for the runoff, sediment, and percolating water [91]. By incorporating a component for vertical flux of pesticides in the root zone, the groundwater loading effects of agricultural management systems (GLEAMS) model [92] was established. GLEAMS is partitioned into three components, namely hydrology, erosion/sediment yield, and pesticides. Surface runoff is estimated using the SCS Curve Number Method [93]. Soils are divided into multiple layers of varying thickness for water and pesticide routing [92]. Both CREAMS and GLEAMS are maintained by the USDA Agricultural Research Service. The major limitation of the model is that it is a lumped model, it assumes the whole watershed is uniform in soil topography and land use, a highly unrealistic assumption.

8 Specific environmental problems in coastal watersheds

Saltwater intrusion is a natural process influenced by humans; it occurs in almost all coastal aquifers. Saltwater intrusion is the movement of salt water into freshwater resources, such as a groundwater aquifer or a freshwater marsh. This intrusion may occur as the result of a natural process like a storm surge from a hurricane. For freshwater, more often it results from human activities such as construction of navigation channels or oil field canals. Climate change has led to a rise in sea level with loss of coastal wetlands and increased saltwater intrusion [94]. The December, 2005 tsunami in the Indian Ocean and Hurricane Katrina in New Orleans and southern Louisiana (August, 2005) resulted in salt-water intrusion into surface and subsurface freshwater sources. Salt water intrusion into water bodies such as rivers, wells, inland lakes, and groundwater aquifers has occurred in many of the affected countries. A post-tsunami study conducted by the Indian Agricultural Research Institute [95] showed that in deep brown coastal soil zones, the quality of shallow groundwater has deteriorated. The electrical conductivity of shallow groundwater (25 m below ground level) changed from the pretsunami value of 0.5 dS m^{-1} to the post tsunami value of 4.8 dS m^{-1}. An estimated 62,000 groundwater wells were contaminated by seawater in Sri Lanka alone. However, in the Maldives islands saltwater intrusion from the tsunami has rendered many of the reservoirs useless. The extent of damage caused by these natural disasters to groundwater resources is still unknown and needs to be assessed.

The coastal areas of the world accommodate high populations and overexploitation of the groundwater has become a common issue along the coast where good-quality groundwater is available. Consequently, many coastal regions in the world experience extensive saltwater intrusion in aquifers resulting in severe deterioration of the quality of groundwater resources. The extent of this saltwater intrusion depends on climatic conditions, aquifer characteristics and groundwater use. In Australia, serious problems of saltwater intrusion exist in the coastal plain of Queensland [96–98]. Many coastal areas in the United States have experienced sea-water intrusion due to both increased groundwater withdrawal and increased urbanization [99].

Saltwater-intrusion problems in coastal aquifers are not new and different researchers have used different numerical and physical techniques to simulate the problem. The initial model was developed independently by Ghyben in 1888 and by Herzberg in 1901. This simple model is known as the Ghyben–Herzberg model and is based on the hydrostatic balance between fresh and saline water in a U-shaped tube. They showed that the saltwater occurs at a depth h below sea level represented by:

$$h = \frac{\rho_s}{\rho_s - \rho_f} h_f, \tag{28}$$

where, ρ_f and ρ_s are, respectively, the density of fresh and saline water, and h_f is the elevation of fresh water level above mean sea level. More detailed information on the subject is covered in this book by Dogan and Fares (Chapter 8)

9 Applications of hydrologic models to coastal watersheds: case studies

Earlier in the chapter, we talked about different types of watershed modeling approaches of rainfall runoff and sediment transport. This section focuses on overviewing some of the watershed hydrology studies that use some of the watershed models described in previous sections of the chapter.

A study of nonpoint-source modeling was published by Corbett *et al.* [100] on a forested and urban watershed in South Carolina coast. The two selected watersheds were 27 km apart and were adjacent to high-salinity salt marshes. Storm-water runoff volumes, flow rates, and sediment loads from both watersheds were compared based on 10 rainfall events using the agricultural nonpoint-source (AGNPS) model. Their results show that although AGNPS was intended for agricultural watersheds, it can also simulate forested and urban watershed reasonably well. Simulation results reported significantly higher runoff volume (14.5%) and sediment loads from the urban watershed than from the forested watershed. In the AGNPS model, runoff volumes were governed by the total impervious area and ignoring the spatial characteristics of watershed, i.e. size, shape, location, and contiguity. Adding simulated impervious surface area increased runoff volumes linearly and peak flow rates exponentially. Flow rates and sediment loads were controlled by impervious surface spatial characteristics. The authors reported maximum sediment loads from the urban watershed when disconnected patches of impervious surface covered 35% of the watershed. Maximum differences between the forested and urban watersheds occurred at low rainfall depths [100]. They recommended the incorporation of groundwater dynamics, the spatial and temporal variability of rainfall, and accumulation and wash-off of specific pollutants [100].

Vieux and Needham [101] studied the sensitivity of AGNPS to variations of grid-cell sizes in an agricultural and forested watershed near Morris, Minnesota. By varying the grid cells between one hectare and 16 hectares, simulated flow path lengths were seen to decrease with increasing grid-cell size. A corresponding variability in AGNPS sediment yield was also observed due to change in flow path length. It was observed that the sediment-delivery ratio using the one-hectare grid cells, was 71% greater than the 16-hectare grid-cells. This research showed that cell-size selection for a discrete watershed analysis should be based on the spatial variability of parameters in the watershed.

The Texas Natural Resource Conservation Commission (TNRCC) published a study of water quality in the Nueces Coastal Basins in 1994. TNRCC used GIS techniques for the establishment of a nonpoint-source pollution-potential index (NSPPI) in an effort to identify areas with high potential risk of nonpoint-source loadings. Components of the NSPPI are based on the RUSLE equation [36]. In addition to the elements from the RUSLE, the NSPPI also includes nonsediment-related hazardous pollutants, such as pesticides or heavy metals. For each of the input parameters to the RUSLE equation and independent related hazardous pollutant factors in the pollution-potential index, a separate GIS layer, was created with component values assigned to the reclassified polygons from the original source map. Through application of this index to the study areas of the San Antonio–Nueces and Nueces–Rio-Grande coastal basins, Texas, the TNRCC concluded that the region generally had a moderate potential for nonpoint pollutant sources, but that areas of higher potential are the agricultural land in regions of maximum slope and erodible soils [102].

Baird *et al.* [101] compared the effectiveness of SWAT [67] and HSPF [84] to assess nonpoint-source pollution. They found that average annual predicted streamflow was approximately 10% less than the average observed streamflow over the period between 1987 and 1992. Predicted streamflow values for each year between 1986 and 1993 showed errors in excess of 68%, when compared with observed annual streamflow values [103]. However, they also reported that the average annual predicted streamflow calculated by HSPF was within 0.4% of the average observed value over the period from 1987 to 1992.

Nutrient and sediment loadings were predicted using HSPF by applying expected mean concentration values to land uses in the Oso Creek watershed, Austin, Texas. They presented sets of land-use-based loads for each month in the eight-year modeling period. Summation of the land-use-based loads resulted in a total load of pollutant from the watershed. Variability of the loadings from year to year naturally corresponded to the observed variability of stream flows from year to year [104]. Overall, the HSPF model was seen to be more robust and to provide more accurate results than the SWAT model. Cuo *et al.* [104] used the DHVSM model to simulate the soil moisture, net radiation and stream flow in a tropical mountainous watershed in Pang Khum, Chang Mai, Thailand. They reported that the model performed reasonably well despite being applied in a region and at a scale that contrasted strongly with those in which it was developed. DHSVM computes the channel discharge for each channel segment using a linear reservoir routing scheme. It incorporates lateral inflow via both overland flow and intercepted subsurface flow [69, 105]. Doten *et al.* [106] evaluated the road-removal scenario and a basin-wide fire scenario in a mountainous forested watershed. Their study under forest fire, showed an increase in all erosion components due to decreases in root cohesion and increases in surface runoff and thus transport capacity. Also, road erosion rate decreased with decreasing road density. Cuo *et al.* [104] reported that road significantly alters the runoff and they attributed the effect to Horton Overland Flow (HOF) generated on the road surface. Ziegler *et al.* [107] reported that the use of a HOF-based model to simulate runoff and sediment transport on unpaved roads provides not only lower-bound estimates of these processes, but also realistic approximations for typical events.

A numerical modeling exercise was carried out [108] using a modified version of the SHARP model to study the groundwater withdrawal in Lihue basin, Kauai, Hawaii. Izuka and Gingerich [109] studied the effects of groundwater withdrawals proposed for Hanamaulu and Puhi, Kauai, Hawaii. The Lihue Basin is a large semicircular depression in southeastern Kauai, the fourth-largest island (553 miles^2) in the tropical, north-Pacific archipelago of Hawaii. The simulations were carried out in both steady and transient states at different pumping rates. Simulated groundwater withdrawals in the model were based on water-use data obtained in 1993 from the Hawaii State Commission on Water Resources Management.

Numerical simulations indicate that groundwater withdrawals from the Hanamaulu and Puhi areas of the southern Lihue Basin will result in depression of water levels and reductions in stream base flows in and near proposed new water-supply wells. Except for areas such as Puhi and Kilohana, which have unique hydraulic characteristics that are of limited extent, the freshwater lens in most inland areas of the southern Lihue Basin is thick and hydraulic conductivities are low. Effects of the projected withdrawals on streams depend on the withdrawal rate and proximity of the pumped wells to streams. However, shifting ground-water withdrawals away from streams with small base flow and toward streams with large base flow can reduce the relative effect on individual streams.

Mair *et al.* [110] evaluated streamflow, rainfall, and ground-water pumping data for the upper part of the Makaha valley coastal watershed on the island of Oahu, Hawaii to identify corresponding trends and relationships. They found that streamflow declined over the 46-year period of record during the ground-water pumping period. Mean and annual streamflow declined by 42% (135 mm) and 56% (175 mm), respectively, and the mean number of dry stream days per year increased from 8 to 125. Rainfall across the study area appeared to have also declined though it was not clear whether the reduction in rainfall was responsible for all or part of the observed streamflow decline. Mean annual rainfall at one location in their study area declined by 14% (179 mm) and increased by 2% (48 mm) at the watershed head water.

Fares [111] evaluated the performance of AnnAGNPS watershed model, in simulating runoff and soil erosion in a 50-km^2 watershed located on the Island of Kauai, Hawaii. The model was calibrated and validated using 2 years of observed stream flow and sediment load data. Alternative scenarios of spatial rainfall distribution and canopy interception were evaluated. They reported that initially, the model produced high CN values, which resulted in increased simulated runoff. To overcome this problem the initial CN values were reduced to their lower limit values for the corresponding land-cover types. Simulations showed that in order to account for the canopy-interception effect, a site-specific canopy-interception model was preferable over the algorithm provided in AnnAGNPS. Accurate representation of the spatial distribution of precipitation is critical for accurate model performance. It was demonstrated that even with a limited number of climate stations within the watershed, an adequate representation of spatial rainfall distribution can be achieved using an accurate annual precipitation map. Monthly runoff volumes predicted by AnnAGNPS compared well with the measured data ($R^2 = 0.90$), however, up to 60% difference between the actual and simulated runoff were observed during the driest months (May and July). Prediction of daily runoff was less accurate ($R^2 = 0.55$).

During sensitivity analysis it was found that sediment yield from the watershed was closely related to: vegetation root mass, average canopy fall height, soil erodibility, percentage of ground residue cover, and canopy cover ratio. The latter two parameters had the greatest influence on sediment yield. The entire watershed was covered by dense vegetation, which protects the soil from direct rainfall impact. Under these conditions high sediment yield was observed on areas with low clay content and on steep slopes. The RUSLE erosion factor K, which is directly related to soil properties, was the single most important parameter, which influenced the spatial variability of sediment losses. Predicted and observed sediment yields on a daily basis were moderately correlated ($R^2 = 0.5$). For the events of small magnitude, the model generally overestimated sediment yield, while the opposite was true for larger events. Total monthly sediment yield varied within 50% of the observed values, except for May 2004. It was found that approximately one third of the watershed area had low sediment yield (0–1 t ha^{-1} y^{-1}), and presented limited erosion threat. However, five per cent of the area had sediment yields in excess of 5 t ha^{-1} y^{-1}. Fifty one per cent of the total area of the watershed contributed with less than 10% of total sediment generated; however, 49% of watershed generated over 90% of the total sediment. The results are based on the use of original NRCS soil classification and USGS land-cover maps with baseline curve numbers.

The model was recalibrated due to the availability of an updated NRCS soil classification and a higher resolution and species-specific land-cover map developed by USGS [111, 112]. Predicted runoff and sediment load using the new parameters were more accurate compared to those estimated with the original soil classification and Landsat land-cover map. For 2003, runoff and sediment were overpredicted by 99% of the measured values. The recalibrated input parameters were used to predict runoff and sediment for 2004 as well. The USGS land-cover map with the updated soil classification produced slight overestimates of runoff and sediment load.

In Hanalei, Feral pigs are one of the major causes of pollution. The soil disturbance due to their activities in the watershed results in increased sedimentation in the bay. The implementation of feral pig damage estimates resulted in a substantial increase of sedimentation due to the high sensitivity of the model to the surface residue cover parameter. With nearly 90% of the study area affected by feral pig activity, as a result, the predicted sedimentation was almost 2.5 times larger than that without pig damage. This substantial increase in sedimentation was expected due to sensitivity of the model to the surface residue cover parameter.

10 Summary

Considerable concern has arisen over potential ecological and environmental impacts of nonpoint-source pollution originating from different parts of coastal watersheds as a result of different management practices and land-use changes. A number of experimental investigations have been reported in the literature for acquiring information essential to optimum watershed management, conservation, and regulatory decision. Impacts of different management practices may range from a few days to several years. Unfortunately, field investigations are typically site and weather specific. Thus, total reliance upon results from field experiments requires a very large resource base acquired over a long time span.

Watershed models offer practical tools optimizing two finite management assets, time and money. Modeling endeavors may be used to lessen the number of field experiments required, and underscore important parameters and variables that most influence this system. A combination of carefully planned field investigations and physically based distributed watershed models offers an effective means to make informed analyses and/or predictions concerning sensitive coastal watersheds.

The hydrology of most coastal watersheds is very similar; however, the hydrology of small islands watersheds has many unique features due to the strong dominance of the surrounding ocean, the continuous effect of prevalent winds, steep topography and their relatively small size. Although these islands have substantial amounts of surface water, groundwater is the main source of their freshwater and as such its contamination is limiting its use. Most of these islands are highly rocky and have impervious soil layers that reduce water infiltration, causing more surface runoff.

After an overview of the some of the major criteria on classification of watershed models the chapter gives an overview of some of the watershed models currently used and featured in the literature. Because this overview information was tabulated, it allows for a practical comparison between these different models using the following criteria: simulation type, runoff generation, overland flow, channel flow, watershed representation, and their use. This information will be useful for the users to help them select the appropriate model based on their modeling needs.

Many of the mathematical equations implemented in many watershed models were presented to describe the major components of the hydrologic cycle and processes, e.g. surface and subsurface water flow, erosion prediction and sediment delivered, and evapotranspiration. More than one equation has been used to describe the same process or hydrological components because of differences between the different models, i.e. deterministic or stochastic, lumped or distributed. The two main approaches used to solve the mathematical equations implemented in watershed models are analytical and numerical solutions. The solution techniques section of this chapter gave a general overview of these two techniques, with emphasis on their advantages and shortcomings.

The chapter also discussed the close connection between GIS and watershed models and benefits of integrating them. Distributed watershed models take advantages of GIS's significant role in facilitating spatial-data preparation and analysis because of its ability to store, retrieve, manipulate, analyze, and map geographic data. As a result, watershed hydrologists were able to generate high-quality maps incorporating model output and geographic entities further enabling visual support during decision-making processes. Advanced analyses and interpretations were possible using several spatial analysis capabilities of the GIS. GIS allowed the identification of the critical areas of the watershed that are contributing substantially to the pollutant loads generated from the watershed of interest.

A brief overview was given to the main steps in any watershed modeling exercises that included model calibration and validation, and sensitivity analysis. This was followed by brief description of seven watershed models SWAT, MIKE SHE, AnnAGNPS, N-SPECT, TOPMODEL, HSPF, WEPP and CREAMS/GLEAMS. A few cases studies were then discussed with special emphasis on tropical coastal watersheds.

References

[1] NASA;http://earthobservatory.nasa.gov/Library/TRMM. Accessed on March 15, 2007.

[2] Giambelluca, T.W., Nullet, M.A. & Schroeder, T.A., *Rainfall Atlas of Hawaii* Report R-76. Honolulu: Hawaii State Department of Land and Natural Resources, 1986.

[3] Atkinson, S.F., Miller, G.D., Curry, D.S. & Lee, S.B., *Salt-water Intrusion: Status and Potential in the Contiguous United States*, Lewis Publishers, Chelsea, Michigan, pp. 390, 1986.

[4] Loukas, A., Quick, M.C. & Russell, S.O., A physically based stochastic-deterministic procedure for the estimation of flood frequency. *Water Resources Management*, **10(6)**, pp. 415–437, 1996.

[5] Osborn, H.B. & Lane, L.J., Precipitation. *Hydrologic modeling of small watersheds*, eds. C.T. Haan, H.P. Johnson & D.L. Brakensiek, ASAE, Michigan, USA, pp. 81–118, 1982.

[6] Loukas, A. & Quick, M.C., Hydrologic behavior of a mountainous watershed. *Canadian Journal of Civil Engineering*, **20(1)**, pp. 1–8, 1993.

[7] NASH, J.E., Systematic determination of unit hydrograph parameters. *Journal of Geophysics Research*, **64(1)**, 1959.

[8] Chuptha, P. & Dooge, J.C.I., The shape parameter of the geomorphologic unit hydrograph. *Journal of Hydrology*, 117, pp. 81–97, 1990.

[9] Rosso, R., Nash model relation to Horton order ratios. *Water Resources Research*, **20(7)**, pp. 914–920, 1984.

[10] Yang, J.C., Tarng, S.Y. & Tung, Y.K., Analyzing uncertainty of IUH of Nash. *Hydraulic Engineering*, eds. H.W. Shen, S.T. Su & F. Wen, ASCE, San Francisco, pp. 1927–1932, 1993.

[11] Sarino & Serrano, S.E., Development of the instantaneous unit hydrograph using stochastic differential equations. *Stochastic Environmental Research and Risk Assessment*, **4**, pp. 151–160, 1990.

[12] Penman, H.L., Natural evaporation from open water, Bare soil and grass. *Proceedings of the Royol Society of London*, **193**, pp. 120–145, 1948.

[13] Priestley, C.H.B. & Taylor, R.J., On the assessment of the surface heat flux and evaporation using large-scale parameters. *Monthly Weather Review*, **100**, pp. 81–92, 1972.

[14] Thornthwaite, C.W., An approach toward a rational classification of climate. *Geographical Review*, **38**, pp. 55–94, 1948.

[15] Fares, A. & Alva. A.K., Soil water balance components based on real-time multisensor capacitance probes in a sandy. *Soil Science Society of America Journal*, **64**, pp. 311–318, 2000.

[16] Harbeck, GE., A practical field technique for measuring reservoir evaporation utilizing mass-transfer theory. *Geological Survey Professional Paper*, **272–E**, pp. 101–105, 1962.

[17] Jensen, M.E., Burman, R.D. & Allen, R.G., Evapotranspiration and irrigation water requirements, American Society of Civil Engineers, New York, 1990.

[18] Morton, F.I., Evaporation research – a critical review and its lessons for the environmental sciences. *Critical Reviews in Environmental Science and Technology*, **24(3)**, pp. 237–280, 1994.

[19] Monteith, J.L., Evaporation and environment. *The State and Movement of Water in Living Organism*, ed. G.E. Fogg, Academic Press, New York, NY, pp. 205–234, 1965.

[20] Rijtema, P.E., An analysis of actual evapotranspiration. *Agricultural Research Report,* **69**, Center for Agricultural Publication and Documents, Wageningen, The Netherlands, pp. 107, 1965.

[21] Horton, R.E., Analysis of runoff plot experiment with varying infiltration capacity. *Soil Science Society of America Proceedings*, **5**, pp. 399–417, 1939.

[22] Holtan, H.N. & Lopez, N.C., USDAHL-70 Model of watershed hydrology. *Techical Bulletin No. 1435*. USDA-ARS, 1971.

[23] Green, W.H. & Ampt, G., Studies of soil physics, part 1.–the flow of air and water through soils. *Journal of Agricultural Science*, **4**, pp. 1–24, 1911.

[24] Bouwer, H., Infiltration of water into nonuniform soil. *Journal of Irrigation and Drainage Engineering*, **95(IR4)**, pp. 451–462, 1969.

[25] Chu, S.T., Infiltration during unsteady rain. *Water Resources Research*, **14(3)**, pp. 461–466, 1978.

[26] Horton, R.E., An approach toward a physical interpretation of infiltration capacity. *Soil Science Society of America Proceedings*, **5**, pp. 399–417, 1940.

[27] Huggins, L.F., Surface runoff, storage and routing. *Hydrologic Modeling of Small Watersheds,* eds. C.T. Haan, H.P. Johnson & D.L. Brakensiek, ASAE, Michigan, USA, pp. 169–225, 1982.

[28] Lighthill, M.J. & Whitham, G.B., On kinematic waves 1. *Proceedings of Royal Society of London,* Series A, 229, pp. 281–316, 1955.

[29] Woolhiser, D.A. & Liggett, J.A., Unsteady one dimensional flow over a plane– the rising hydrograph. *Water Resources Research*, **3(3)**, pp. 753–771, 1967.

[30] USDA-SCS. *National Engineering Handbook. Part 630 Hydrology*, Section 4, U.S. Government Printing Office, Washington, DC. 1972.

[31] Darcy, H., *Les Fontaines Publiques de la Ville de Dijon:* Paris, Victor Dalmont, pp. 647, 1856.

[32] Vought, L.B.M., Pinay, G., Fuglsang, A. & Ruffinoni, C., Structure and function of buffer strips from a water quality perspective in agricultural landscapes. *Landscape and Urban Planning*, **31**, pp. 323–331, 1995.

[33] Hubbard, R.K. & Sheridan, J.M., Nitrate movement to groundwater in the southeastern Coastal plain. *Journal of Soil and Water Conservation*, **44(1)**, pp. 20–27, 1989.

[34] Foster, G.R., Modeling the erosion process. *Hydrologic Modeling of Small Watershed*, eds. C.T. Haan, H.P. Johnson & D.L. Brakensiek, ASAE, Michigan, USA, pp. 297–370, 1982.

[35] Wischmeier, H. & Smith, D.D., Predicting rainfall erosion losses. *Agriculture Handbook No. 537*, USDA Science and Education Administration, 1978.

[36] Renard, K.G., Foster, G.R., Weesies, G.A. & Porter, J.P., RUSLE: Revised Universal Soil Loss Equation. *Journal of Soil and Water Conservation*, pp. 30–33, 1991.

[37] Chmelova, R. & Sarapatka, B., Soil erosion by water: contemporary research methods and their use. *Geographica*, **37**, 2002.

[38] Merritt, W.S., Letcher, R.A. & Jakeman, A.J., A review of erosion and sediment transport models. *Environmental Modeling & Software*, **18**, pp. 761–799, 2003.

[39] Zhang, L., O'Neill, A. & Lacy, S., Spatial analysis of soil erosion in catchments: a review of modeling approaches. International Congress on Modeling and Simulation (MODSIM95). *Water Resources and Ecology*, **3**, pp. 58–64, 1995.

[40] Lane, L.J. & Nearing, M.A. eds. USDA-Water Erosion Prediction Project: Hillslope Profile Model Documentation, *USDA-ARS-NSERL Report No.* 2, West Lafayette, IN. 1989.

[41] Shirley, E.D. & Lane, L.J., A Sediment yield equation from an erosion simulation model. *Hydrology and Water Resources in Arizona and the Southwest*, **8**, pp. 90–96.10, 1978.

[42] Lane, L.J., Nichols, M.H. & Simanton, J.R., Spatial variability of cover affecting erosion and sediment yield in overland flow. *Effects of Scale on Interpretation and Management of Sediment and Water Quality* (Proceeding of a Boulder Symposium, July 1995). IAHS Pubiication. No. 226, pp. 147–152, 1995.

[43] Lane, L.J., Nichols, M.H. & Paige, G.B., Modeling erosion on Hillslopes: concepts, theory and data. *Proceedings of the International Congress on Modeling and Simulation (MODSIM'95)*. eds. P. Binning, H. Bridgman & B. Williams University of Newcastle, Newcastle, NSW, Australia, Uniprint, Perth, Australia, pp. 1–7, 1995.

[44] Mercer, J.W. & Faust, C.R., Ground-water modeling, *National Water Well Association (NWWA)*, Worthington, Ohio, 1986.

[45] Anderson, M.P. & Woessner, W.W., Applied groundwater modeling, Academic Press, Inc., San Diego, CA, 1992.

[46] Zheng, C. & Bennett, G.D., Applied contaminant transport modeling. *Theory and Practice*, Van Nostrand Reinhold, New York, NY: 1995.

[47] Freeze, R.A. & Cherry, J.A., Groundwater, Englewood Cliffs, NJ, Prentic-Hall, 1979.

[48] Bear, J., *Hydraulics of Groundwater*, McGraw-Hill: New York, 1979.

[49] Javendel, I., Doughty, C. & Tsang, C.F., *Groundwater Transport: Handbook of Mathematical Models*, American Geophysical Union, Washington, DC. *Water Resources Monograph*, pp. 10–228, 1984.

[50] Domenico, P.A. & Robbins, G.A., A new method of contaminant plume analysis. *Ground Water*, **23(4)**, pp. 51–59, 1985.

[51] Bedient, P.B., Rifai, H.S. & Newell, C.J., *Groundwater Contamination: Transport and Remediation*, 2nd edn, PTR publications, Upper Saddle River, NJ, pp. 605, 1999.

[52] Anderson, M.P., Using models to simulate the movement of contaminants through groundwater flow systems. *Critical Reviews In Environmental Control*, **9(2)**, pp. 97–156, 1979.

[53] Wang, H.F. & Anderson, M.P., *Introduction to Groundwater Modeling Finite Difference and Finite-Element Methods*, W.H. Freeman and Company, San Francisco, CA. pp. 237, 1982.

[54] Zhang, W. & Montgomery, D.R., Digital elevation model grid size, landscape representation and hydrological simulations. *Water Resources Research*, **30(4)**: 1019–1028. 1994.

[55] Moore, R.D. & Thompson J.C., Are water table variations in a shallow forest soil consistent with the TOPMODEL concept?, *Water Resources Research*, **32(3)**, pp. 663–669, 1996.

[56] McCuen, R.H., The role of sensitivity analysis in hydrologic modeling. *Journal of Hydrology*, **18**, pp. 37–53, 1973.

[57] Beven, K.J., *Rainfall–Runoff Modeling*, Wiley & Sons, Chichester, pp. 217–225, 2001.

[58] Hamby, D.M., A review of techniques for parameter sensitivity analysis of environmental models. *Environmental Monitoring and Assessment*, **32**, pp. 135–154, 1994.

[59] Campolongo, F. & Braddock, R.D., The use of graph theory in the sensitivity analysis of the model output: a second order screening method. *Reliability Engineering & System Safety*, **64(1)**, pp. 1–12, 1999.

[60] Bastidas, L.A., Gupta, H.V., Sorooshian, S., Shuttleworth, W.J. & Yang, Z.L., Sensitivity analysis of a land surface scheme using multi-criteria methods, *Journal of Geophysical Research*, **104(16)**, pp. 19481–19490, 1999.

[61] SMHI, http://www.smhi.se/sgn0106/if/rc/modeldeveva.htm#evaluation. Accesed on August 01, 2007.

[62] Nash, J.E. & Sutcliffe, J.V., River flow forecasting through conceptual models. Part 1: a discussion of principles. *Journal of Hydrology*, **10**, pp. 282–290, 1970.

[63] Schaake, J.C. Introduction. *Calibration of Watershed Models*, eds. Q. Duan, H.V. Gupta, S. Sorooshian, A.N. Rousseau & R. Turcotte, Water Science and Applications Series, Vol. 6. 2003.

[64] Beven, K. & Binley, A., The future of distributed models: model calibration and uncertainty prediction. *Hydrological Processes*, **6**, 1992.

[65] Refsgaard, J.C., Parameterization, calibration and validation of distributed hydrological models. *Journal of Hydrology*, **198**, pp. 69–97, 1997.

[66] Hassanizadeh, S.M. & Carrera, J., Editorial. Validation of geo-hydrological models (special issue). *Advances in Water Resources*, **15**, pp. 1–3, 1992.

[67] Refsgaard, J.C. & Storm, B., Construction, calibration and validation of hydrological models. *Distributed Hydrological Modeling*, eds. M.B. Abbott & J.C. Refsgaard, Kluwer Academic, Dordrecht, the Netherlands. pp. 41–54, 1996.

[68] Arnold, J.G., Srinivasan, R., Muttiah, R.S. & Williams, J.R., Large area hydrologic modeling and assessment. Part I: Model development. *Journal of America Water Resources Association*, **34(1)**, pp. 73–89, 1998.

[69] Wigmosta, M.S., Vail, L.W. & Lettenmaier, D.P., A distributed hydrology-vegetation model for complex terrain. *Water Resources Research*, **30(6)**, pp. 1665–1679, 1994.

[70] Danish Hydraulic Institute (DHI). MIKE SHE. *Water Movement – User Guide and Technical Reference Manual*, Edition 1.1. 1998.

[71] Kristensen, K.J. & Jensen, S.E., A model for estimating actual evapotranspiration from potential transpiration. *Nordic Hydrology*, **6**, pp. 70–88, 1975.

[72] Young, R.A., Onstad, C.A., Bosch, D.D. & Anderson, W.P., AGNPS, Agricultural nonpoint source pollution model: A large watershed analysis tool. *USDA-ARS Conservation Research Report*, **35**, pp. 77, 1987.

[73] Mockus, V., Estimation of direct runoff from storm rainfall. *National Engineering Handbook*, NEH Notice 4–102, pp. 10.1–10.22, 1972.

[74] Geter, W.F. & Theurer, F.D., AnnAGNPS - RUSLE sheet and rill erosion. *Proceedings of the First Federal Interagency Hydrologic Modeling Conference*. Las Vegas, Nevada. pp. 17–24, 1998.

[75] Panuska, J.C., Moore, I.D. & Kramer, L.A., Terrain analysis: integration into the Agricultural Non-Point Source (AGNPS) pollution model. *Journal of Soil Water Conservation*, pp. 59–64, 1991.

[76] Beven, K.J. & Kirkby, M.J., A physically based variable contributing area model of basin hydrology. *Hydrological Sciences Bulletin*, **24(1)** pp. 43–69, 1979.

[77] Beven, K.J., Infiltration into a class of vertically non-uniform soils, *Hydrological Sciences Journal*, **29**, pp. 425–434, 1984.

[78] Beven, K.J. Chapter 18: TOPMODEL. *Computer Models of Watershed Hydrology*, ed. V.P. Singh, Water Resources Publications, Littleton, CO, 1995.

[79] TOPMODEL:http://www3.bae.ncsu.edu/Regional-bulletins/Modeling-bulletin/TOPMODEL.html. Accessed on August 01, 2007.

[80] Franchini, M., Wendling, J., Obled, C. & Todini, E., Physical interpretation and sensitivity analysis of the TOPMODEL. *Journal of Hydrology*, **175**, pp. 293–338, 1996.

[81] Saulnier, G.-M., Beven, K.J. & Obled, Ch., Digital elevation analysis for distributed hydrological modeling: reducing scale dependence in effective hydraulic conductivity values. *Water Resources Research*, **33(9)**, pp. 2097–2101, 1997a.

[82] Saulnier, G.-M., Obled, Ch. & Beven, K.J., Analytical compensation between DTM grid resolution and effective values of saturated hydraulic conductivity within the TOPMODEL framework. *Hydrology*, **11**, pp. 1331–1346, 1997b.

[83] Bruneau, P., Gascuel-Odoux, C., Robin, P., Merot, Ph., & Beven, K.J., The sensitivity to space and time resolution of a hydrological model using digital elevation data. *Hydrological Processes*, **9**, pp. 69–81, 1995.

[84] Johanson, R.C., Imhoff, J.C., Kittle, Jr. J.L. & Donigian, J.A.S., Hydrological Simulation Program – FORTRAN (HSPF): *User's Manual for Release 8.0,* U.S. EPA, EPA-600/3-84-066, Environmental Research Laboratory, Athens, GA, 1984.

[85] Walton, R. & Hunter, H., Modeling water quality and nutrient fluxes in Johnstone River Catchment, North Queensland. *23rd Hydrology and Resources Symposium*, Sydney, 1996.

[86] Laften, J.M., Elliot, W.J., Flanagan, D.C., Meyers, C.R. & Nearing, M.A., WEPP- predicting water erosion using a process-based model. *Journal of Soil Water Conservation*, **52**, pp. 96–102, 1997.

[87] Williams, J.R. & Nicks, A.D., SWRRB, a Simulator for Water Resources in Rural Basins: an overview. *Proceedings of the Natural Resources Modeling*

Symposium, ed. D.G. DeCoursey, Pingree Park, CO, October 16–21, 1983, USDA-ARS, ARS-30, pp. 17–22, 1985.

[88] Mein, R.G. & Larson, C.L., Modeling infiltration during a steady rain. *Water Resources Research*, **9(2)**, pp. 384–394, 1973.

[89] Ascough, J.C., Nearing, M.A., Flanagan, D.C. & Livingston, S.J., Hydrologic and erosion calculations in the water Erosion Prediction Project (WEPP) watershed model. ASAE Paper. 94–2037, ASAE Summer meeting, Kansas City, MO June 19–22, 1994.

[90] Flanagan, D.C. & Livingston, S.J., USDA-Water Erosion Prediction Project, WEPP user summary conference, August 9–11, Des Moines, Iowa, 1995.

[91] Knisel, W.G., ed. CREAMS: A field-scale model for chemicals, runoff and erosion from agricultural management systems. *USDA Conservation Research Report No. 26*, USDA-ARS, Washington, DC, pp. 640, 1980.

[92] Leonard, R.A., Knisel, W.G. & Still, D.A., GLEAMS: Groundwater Loading Effects of Agricultural Management Systems. *Transaction of ASAE*, **30**, pp. 1403–1418, 1987.

[93] Chow, V.T., Maidment, D.R. & Mays, L.W., *Applied Hydrology*, New York: McGraw-Hill, 1988.

[94] Guntenspergena, G.R., Vairin, B.A. & Burkett, V., Overview of coastal wetland global climate change research. U.S. Geological Survey National Wetlands Research Center, *Biological Science Report Usgsbrdibsr-1998–0002.*

[95] Tsunami impact, http://www.iari.res.in/tsunami/link2.html. Accessed on August 01, 2006.

[96] Volker, R.E. & Rushton, K.R., An assessment of the importance of some parameters for seawater intrusion in aquifers and a comparison of dispersive and sharp interphase modeling approaches. *Journal of Hydrology*, **56**, pp. 239–250, 1982.

[97] Hillier, J.R. Management of saltwater intrusion in coastal aquifers in Queensland, Australia. *AGSO Journal of Australian Geology & Geophysics*, **14(2/3)**, pp. 213–217, 1993.

[98] Murphy, S.F. & Sorensen, R.C., Saltwater intrusion in the Mackay coastal plains aquifer. *Proceeding of Australian Sugar Cane Technology*, **23**, 70–76, 2002.

[99] Narayana, K.A., Schleebergerb, C., Charleswortha, P.B., & Bristowa, K.L., Effects of groundwater pumping on saltwater Intrusion in the lower Burdekin Delta, North Queensland. *MODSIM International Congress on Modelling and Simulation.* Vol. 2, ed. D.A. Post, Modelling and Simulation Society of Australia and New Zealand, pp. 212–217, 2003.

[100] Corbett, C.W., Wahl, M., Porter, D.E., Edwards, D. & Moise, C., Nornpoint source runoff modeling – A comparison of a forested watershed and an urban watershed on the South Carolina coast. *Journal of Experimental Marine Biology and Ecology*, **213**, pp. 133–149, 1997.

[101] Vieux, B.E. & Needham, S., Nonpoint-pollution model sensitivity to grid-cell size. *Journal of Water Resources Planning and Management*, **119(2)**, pp. 141–157, 1993.

[102] Texas Natural Resources Conservation Commission. Regional assessment of water quality in the Nueces Coastal Basins, TNRCC, Austin, TX,1994.

[103] Baird, C., Jennings, M., Ockerman, D. & Dybala, T., Characterization of non-point sources and loadings to the Corpus Christi Bay National Estuary program study area. *Corpus Christi Bay National Estuary Program CCBNEP-05*, Texas Natural Resources Conservation Commission (TNRCC), Austin, Texas, pp. 226, 1996.

[104] Cuo, L., Giambelluca, T.W., Ziegler, A.D., Sutherland, R.A., Nullet, M.A., Larkin, E.D. & Vana, T.T., *Application of Distributed Hydrology-Soil-Vegetation Model (DHVSM) in Pang Khum Experimental Watershed, Chiang Mai Province, Thailand, American Geophysical Union*, Fall Meeting, abstract H21C-0320, 2001.

[105] Wigmosta, M.S. & Perkins, W.A., Simulating the effects of forest roads on watershed hydrology. *Land Use and Watersheds: Human Influence on Hydrology and Geomorphology in Urban and Forest Areas*, Water Science. Application. Series, vol. 2, eds. M.S. Wigmosta & S.J. Burgess, AGU, Washington, DC, pp. 127–143, 2001.

[106] Doten, C.O., Bowling, L.C., Lanini, J.S., Maurer, E.P. & Lettenmaier, D.P., A spatially distributed model for the dynamic prediction of sediment erosion and transport in mountainous forested watershed. *Water Resources Research*, **42(4)**, pp. 4417, 2006.

[107] Ziegler, A.D., Giambelluca, T. W., Sutherland, R. A., Vana, T.T. & Nullet, M. A., Horton overland flow contribution to runoff on unpaved mountain roads: A case study in northern Thailand. *Hydrologic Processes*, **15(6)**, pp. 3203–3208, 2001.

[108] Izuka, S.K. & Delwyn, S.O., Numerical simulation of ground-water withdrawals in the southern Lihue Basin, Kauai, Hawaii: *U.S. Geological Survey Water-Resources Investigations Report,* 01–4200, 2002.

[109] Izuka, S.K. & Gingerich, S.B., Groundwater in the southern Lihue Basin, Kauai, Hawaii: *U.S. Geological Survey Water-Resources Investigations Report* 98–4031, pp. 71, 1998a.

[110] Mair, A., Fares, A. & El-Kadi, A., Effects of rainfall and ground-water pumping on streamflow in Makaha, Oahu, Hawaii. *Journal of the American Water Resources Association*, **43(1)**: 148–159.

[111] Fares, A., Evaluation of AnnAGNPS, a Non-Point Source Pollution Model, in Hanalei Watershed, Kauai. Project Report Submitted to Hawaii-USDA Natural Resources and Conservation Service, pp. 102, 2007.

[112] Ambagis, S. & Jacobi, J., Detailed high resolution vegetation mapping of the Hanalei watershed using object based classification. *The 2006 Hawai'i Conservation Conference*, Honolulu, HI, July 26–28, 2006.

CHAPTER 2

Nutrient bioavailability of soils and sediments in an Australian estuary influenced by agriculture: linking land to sea

K.A.V. Chaston[1], P.W. Moody[2] & W.C. Dennison[1]
[1]*Department of Natural Resources and Environmental Management, University of Hawaii at Manoa, USA.*
[2]*Natural Resource Sciences, Queensland State Department of Natural Resources and Mines, Australia.*

Abstract

Nutrient bioavailability of runoff from agricultural soils was investigated in the Maroochy River watershed, Australia, a coastal watershed influenced by agriculture. Suspended sediments, river and estuarine sediments and deposited sediment in the near-shore coastal ocean were collected and analyzed for nutrient bioavailability using chemical analyses and phytoplankton (*Skeletonema costatum*) bioassays. Suspended sediments in the Maroochy River, which consisted of silt and clay-sized particles, had elevated Fe-oxide-extractable P and total P concentrations comparable to fertilized soil. Similarly, the deposited sediment sampled offshore of the river mouth had elevated total P, Fe-oxide-extractable P and total N concentrations that were much greater than the underlying marine sediment. The deposited offshore sediment contained mainly clay-size particles and appeared to be terrigenous in origin due to its similar composition (total P, Fe oxide P, total N, total carbon, total aluminum, and total silicon) to estuarine suspended sediments and terrestrial soils. This study demonstrated that nutrient-rich clay-sized particles, of terrigenous origin, are being transported and deposited offshore during erosion events. This highlights the need for multifaceted watershed management that encompasses a) erosion control measures that reduce suspended sediment loads of nutrient-rich clay- and silt-sized fractions to coastal waters, and b) nutrient-reduction strategies. Effective management must consider both agricultural productivity and potential environmental impacts, as what is economically viable may not be environmentally sustainable.

1 Introduction

The potential impact of increased nutrient and sediment loads to coral-reef ecosystems, especially inshore reefs of the Great Barrier Reef [1–3] and an increase in algal blooms (toxic and nontoxic) in several Australian estuaries, bays and coastal lakes [4, 5, 6, 7] has necessitated research on the downstream effects of land use on Australian waters. Globally, coral-reef ecosystems are declining due to the impacts of sediment, nutrients and other pollutants attributed to poor land-management practices [8]. Models estimate that 22% of coral reefs world-wide are threatened by soil erosion and land-based pollution [9]. Detrimental effects of phosphorus loss from agricultural land to freshwater rivers and lakes have become increasingly apparent, particularly in North America and Europe [10, 11]. In these regions much research has focused on increasing P retention on land and assessing P bioavailability in agricultural soils and runoff [11, 12].

Agricultural runoff is a major nonpoint source of phosphorus (P) and nitrogen (N) into rivers and estuaries [13]. During rainfall events excess nutrients can be transported from streams to rivers, estuaries and eventually to the coastal ocean. The majority of P (>90%) which is transported from rivers to the ocean is in particulate form [e.g. 14, 15], some fraction of which is desorbable and thus potentially bioavailable [14]. Conversely, most nitrate is lost via leaching from agricultural soils and is readily bioavailable, whilst ammonium is strongly sorbed to soil particles and transported in particulate form [16].

Although several studies have measured sediment nutrients, and/or water quality in estuaries impacted by agriculture [17–19], few have examined nutrients associated with suspended sediments [20], or examined these collectively [21, 22]. Most research has been confined to the watershed, receiving estuary, or near-shore coastal zone, with few studies examining their connectivity [18, 19, 21]. Thus comparisons between nutrient bioavailability of soil, suspended sediments, sediments in the receiving estuary as well as offshore sediments are rare.

Accurate and comparable assessments of sediment nutrient bioavailability have also been troubled by the lack of standard methodology for assessing bioavailable P. An accurate measure of algal available P was identified in Chaston [23], by correlating chemical measures of sediment P with maximum algal biomass of the euryhaline diatom *Skeletonema costatum*. Fe-oxide extractable P [24] and bicarbonate extractable inorganic P [25, 26] were highly correlated with bioavailable P in suspended sediments [23]. The Fe-oxide strip method was recommended for future analyses as it provides a stronger mechanistic basis than chemical extraction for estimating bioavailable P [27]. In addition, the technique has previously been used in both freshwater [27] and marine conditions [28] and is not influenced by sediment or soil type [28, 29], thus making it suitable for assessment of bioavailable P in coastal watersheds that have a broad range of sediment types distributed over a salinity gradient.

The main aim of this study was to assess nutrient bioavailability of runoff from agricultural soils in Australian estuarine and coastal marine ecosystems. Nutrient bioavailability from soil, suspended sediment, estuarine and deposited offshore sediment was determined in a subtropical Australian estuary impacted by agriculture, using chemical analyses and algal bioassays. Results were used to formulate a conceptual model of sediment and nutrient transport in subtropical and tropical Australian estuaries and to demonstrate the link from land to sea.

2 Materials and methods

2.1 Study site

The study was conducted in the Maroochy River Watershed, Sunshine Coast, southeast Queensland, Australia (Fig. 1). The Maroochy River is 27 km long and drains a relatively small (~600 km^2), predominantly rural watershed. Most of the upper watershed is forested (~10%), with the remainder cleared for rural (74%) and urban (~10%) uses, leaving only narrow riparian vegetation in most areas [30]. Major land uses in the watershed include sugar-cane, horticultural tree crop production, tropical fruit, vegetables and pastures for dairy and beef cattle [31]. Sugar-cane is the dominant crop covering 60 km^2, which is approximately 10% of the watershed [30]. Crops are supported on a variety of soils that vary in drainage, P-sorbing ability and bioavailable P content. The major soil types utilized for agriculture in the watershed include Red Kandosols, Yellow Kandosols, Chromosols, Redoxic Hydrosols, Podosols and Yellow Kurosols (classified according to [32]).

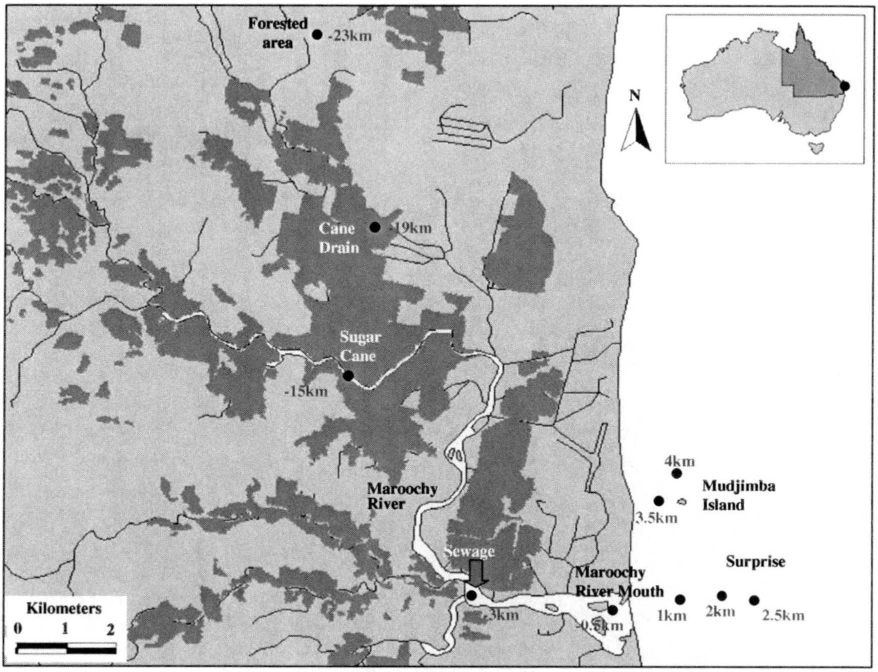

Figure 1: Location of study sites in the Maroochy River and adjacent to the river mouth. Distance is given as kilometers from the river mouth. The shaded area represents cropped land. The patterned area represents the approximate bounds of river plume.

According to the U.S. Taxonomy [33], these soils comprise 5 Alfisols, 1 Ultisol (Redoxic Hydrosol) and 1 Spodosol (Podosol). Fertilisers are applied to these soils in ammonium (N) and orthophosphate (P) forms. The Maroochy River receives nutrients from various diffuse and point sources within the watershed; with sewage outfalls and watershed runoff delivering the most significant loads of N and P at approximately equal loadings [34].

There are several near-shore reefs adjacent to the Maroochy river mouth (Fig. 1), with the reef around Mudjmba Island being a popular SCUBA diving site. The climate of the region is subtropical with typically wet summer and dry winter seasons.

Five study sites that are influenced by various watershed activities were selected in the Maroochy River and offshore of the river mouth (Fig. 1). The five river sites covered the transition from freshwater to marine waters and were located within a forested area (23 km upstream from the river mouth), a cane drain (19 km upstream), a sugar-cane area (15 km upstream), at a sewage-treatment outfall (STP) (3 km upstream) and a well-flushed site close to the river mouth (0.5 km upstream). The freshwater stream located in the forested area flows into Yandina Creek during high flow events only and does not flow in dry conditions.

The sites adjacent to the river mouth were located within the flood plume of the Maroochy River (Fig. 1). The extent of the river plume was assessed by aerial observation during a large flood event in May 1999. The first site was located close to the river mouth (1 km downstream from the mouth), two sites were located at Surprise Reef (2 and 2.5 km downstream) with the remaining two sites at Mudjimba Island (3.5 km and 4 km from the mouth).

The seven major agricultural soil types in the watershed (mentioned above) were also sampled. The surface layer (top 10 cm) of soil was collected from various locations within the watershed (Fig. 2).

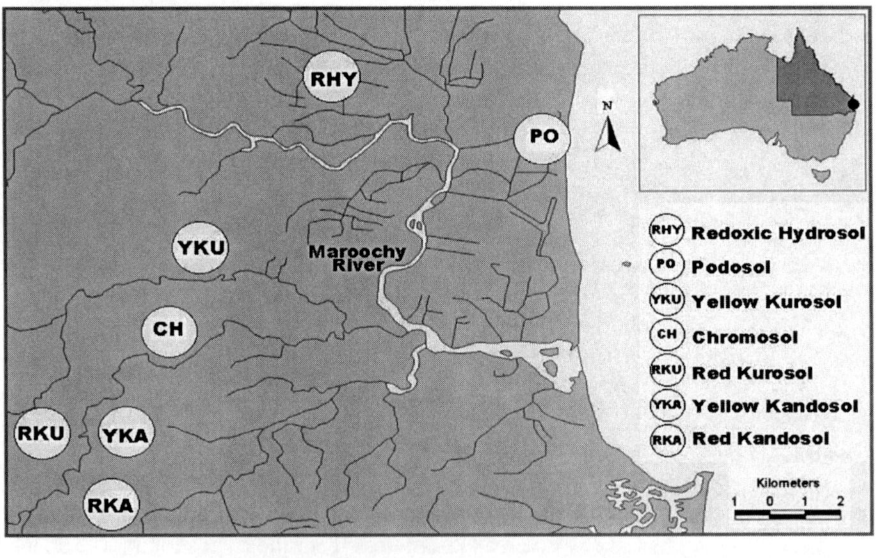

Figure 2: Location of bulk soil samples in the Maroochy River Catchment, Queensland, Australia.

2.2 Sampling strategy

Water and sediment samples were collected during 4 separate trips to the Maroochy River between May 1999 and November 2000 during wet and dry conditions (Fig. 3). The river flooded in May 1999 and a plume stretching past Mudjimba Island was visible from the air (Fig. 1). Due to difficulties in obtaining sufficient suspended sediments for analysis, a transect extending from the Maroochy River mouth to Mudjimba Island was planned to locate and sample sediments that had settled from the plume, or previous run-off events. The trip was delayed for several months to October 1999, due to unsuitable sampling conditions (big seas and strong wind) after the flood. Despite the delay, deposited sediments were located and sampled. Sediment samples from the Maroochy River were taken 1 month later in November 1999 for comparison. Following more than a year of above average rainfall, a drought occurred from August to October 2000. The river was sampled during September 2000 to capture the dry conditions. The final sampling was conducted after the first big rainfall following the drought in November 2000. Bulk soil samples were collected in January 1999.

2.3 Water quality

Physical aspects of water quality measured in this study included total suspended solids (TSS), secchi depth, salinity, chlorophyll *a* (Chl *a*), pH and temperature. Salinity, pH and temperature were measured in the field using a Horiba Water Quality Checker Model U-10. The Horiba was calibrated, as per manufacturers instructions, prior to every sampling. On return from the field, salinity and pH were checked against standard solutions to monitor for instrument drift (which was negligible).

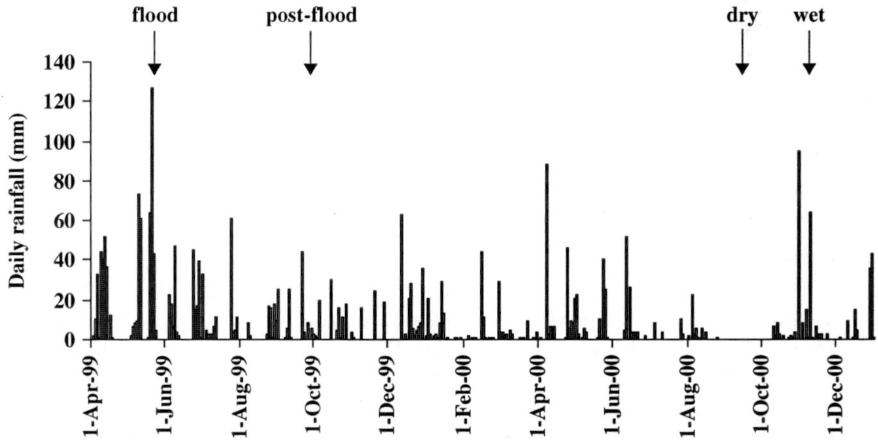

Figure 3: Daily rainfall data (mm) recorded at Maroochydore during the study. Arrows denote time of sampling.

Water was sampled at approximately 0.1 m depth using a bucket on the top of the flooding tide (river sites) and opportunistically at sites that were not tidally influenced. Triplicate water samples for Chl *a* analysis were collected and filtered (Whatman GF/F filters), then stored on dry ice. In the lab, the filter was ground in 90% acetone to extract Chl *a*, spectral extinction coefficients were determined on a spectrophotometer and chlorophyll *a* concentrations calculated according to Parsons *et al.* [35]. Triplicate 2 L samples of water from each site were stored in rinsed plastic containers until return to the laboratory where samples were filtered through a preweighed Whatman GF/F filters for determination of TSS using method SM 2540D [36]. A known volume of water was filtered onto a preweighed and predried (110 °C; 24 h) Whatman GF/C glass-fiber filter. The filter was then oven dried at 60 °C for 24 h and total suspended solids calculated by comparing the initial and final weights [36].

To assess dissolved nutrients (ammonium, nitrate/nitrite, phosphorus) samples were filtered in the field to remove particulate matter using a 60-mL syringe and Sartorius Minisart 0.45 μm disposable membrane filters. Total nutrients (total Kjeldahl nitrogen and total phosphorus) were collected using a 60-mL syringe without a filter in order to obtain a whole water sample. Collected samples were stored in plastic 100-mL bottles. Filtered and unfiltered water samples were frozen immediately using dry ice and transported to the laboratory where they were analyzed using the standard auto-analyzer chemical techniques of Clesceri *et al.* [36].

2.4 Suspended sediments

Bulk water samples (~100-L) were collected at the water surface in 25-L acid-washed opaque plastic drums. Samples were kept out of direct sunlight to minimize heating in the field and stored at 4 °C in the laboratory prior to analysis. Water samples were then centrifuged (in 600-mL aliquots) at 2000 rpm for 15 min. The overlying water was then decanted and the suspended sediment slurry collected and oven-dried. Collected sediments were analyzed for total P (TP) [37], Fe-oxide extractable P (FeO-P) [24], total N (TN) and total organic C (TC) by combustion analyzer and total aluminum (Al) by XRF. Analyses were conducted in triplicate or duplicate (depending on sample availability). Particle size was determined using the laser optical particle-sizing method [38]. Calgon dispersant and ultrasound were used to fully disperse the dry sediment samples prior to analysis with a Malvern laser-diffraction instrument.

2.5 River and oceanic sediment

Sediments were sampled from the Maroochy River and adjacent to the river mouth by divers using SCUBA or snorkeling gear. River samples were taken from subtidal areas of the riverbank, not the scoured floor of the river channel. Near-shore samples were collected from various depths shallower than 15 m. The upper layer (~top 2 cm) of sediment was scraped using a stainless steel scraper and placed directly into a zip-lock plastic bag. The sediment slurry was then transported to the surface and snap frozen using dry ice. Syringes (60 mL) were used to collect the fine clay layer of deposited sediments found at several locations outside the river mouth. This fine clay layer was dark colored and resembled deposited sediment, and was easily distinguished from the underlying marine sediment. Samples were frozen immediately and stored at −20 °C prior to analysis. In the laboratory, the deposited sediment samples were thawed and then shaken to resuspend the sediments. Sand particles were then separated from the finer clay- and silt-sized particles and the remaining mixture centrifuged to collect the deposited sediment.

Sediments were analyzed for total P, Fe-oxide extractable P, total N, total C, total Al and Si as described above. Analyses were conducted in duplicate, where possible (depending on sample volume). Particle size was also determined using the laser optical particle-sizing method described above.

2.6 Soil samples

Bulk soil samples (0–10 cm depth) were collected at each site. Half of each bulk sample was enriched with P (as solution K_2HPO_4) at concentrations comparative to sugar-cane fertilizer applications (solution P concentration of 0.2 mg P L^{-1}). Phosphorus is usually applied as either mono- or di-ammonium phosphate fertilizer. Simulated aquatic sediments (comprising clay- + silt-sized particles <20 μm diameter) were prepared, as follows, from collected soils. Air-dried (<2 mm) soil samples (300 g) and deionized water were added to perspex cylinders (30 cm long × 7 cm internal diameter) to give a final soil and water volume of 1.2 L. Cylinders were then inverted several times over 30 s and allowed to stand without agitation for 4.5 min at 22 °C, after which the top 10 cm of the suspension was removed by suction. Based on Stokes' Law, at 22 °C all particles with diameters >20 μm (i.e. fine and coarse sands) should have vacated the top 10 cm in this time, leaving the suspended clay- + silt-sized (<20 μm diameter) fraction [39]. The suspension was then left for a further 4.5 min without agitation, to enable particles >20 μm to settle out of the next 10 cm, thus enabling further collection of clay + silt-sized particles. Deionized water was then added to the cylinder to bring the volume back to 1.2 L, and the shaking and settling process was repeated four times to collect the remaining particles <20 μm diameter. The collected suspensions were centrifuged (20 min, 4000 g), the supernatant discarded, and the sediment plugs were oven dried at 40 °C for 2 days. Particle size was also determined using the laser optical particle-sizing method of [38].

2.7 Sediment bioassays

Sediment bioavailable P was determined using the sediment bioassay technique developed in Chaston [23]. Nonaxenic unialgal cultures of *Skeletonema costatum* were obtained from the culture collection of CSIRO Marine Research, Tasmania, Australia. Bioassays were conducted in triplicate using 250 mL Erlenmeyer flasks with 100 mL of modified f/2 media (without P) inoculated with approximately 200 000 cells L^{-1} of *Skeletonema costatum.* One hundred mg L^{-1} of air-dry sediments (clay and clay + silt, with and without P enrichment) were used as the sole source of P in the sediment bioassays. Control bioassays that had no sediment additions, only modified f/2 media (f/2–P), were conducted to assess any growth associated with carry over P in the internal nutrient pools of the algal cells or seawater used in the media. Growth under optimum nutrient conditions (complete f/2 media) was also measured as a control response that could be compared between experiments to monitor natural growth fluctuations. Flasks were continually shaken to prevent CO_2 limitation and pH was assumed constant [23].

In-vivo Chl *a* fluorescence was used as a proxy of algal biomass to monitor the bioassay response. Fluorescence was measured at identical circadian times daily with a Turner TD 700 fluorometer fitted with a 13-mm test tube adapter to hold small test tubes, a red photomultiplier, daylight white lamp and broad-band Chlorophyll filter kit (340–500 nm excitation, >665 nm emission). Five-milliliter subsamples were taken aseptically, dark adapted for 30 min prior to fluorescence measurements then discarded. Algal growth was measured daily until a decline in biomass was observed, which usually occurred within 7 days. Maximum growth was defined as the maximum fluorescence standard units (fsu) attained prior to cells reaching stationary growth [when daily growth was not significantly higher ($p < 0.05$)]. Maximum fsu was not used due to the error associated with fsu/cell variability when cells are nutrient limited. The maximum growth (fsu) was then converted into Chl *a* ($\mu g\ L^{-1}$) using the correlation between extracted and *in-vivo* data from different stages of the *S. costatum* growth cycle (Chl $a = 2.5115 \times$ fsu). Chaston [23] determined that direct conversion was possible as long as bioassays were in exponential growth and not nutrient limited. Algal biomass related to internal nutrient pools or seawater used in the media was corrected for by subtracting the maximum control biomass (no added P) from the maximum sediment biomass. This was termed the 'maximum algal biomass'. The maximum algal biomass of each sediment type was then correlated against the measured chemical fractions of P to determine which fractions best reflected algal bioavailable P.

3 Results

3.1 Water column and sediment nutrients

Total water column nutrients (N and P) were higher than dissolved nutrients at all sampling locations along the river and adjacent to the river mouth during low-flow conditions (Fig. 4). Total P was approximately 10 times greater than dissolved reactive P (DRP) in the upstream cane area (–15 km), which was most likely attributable to high suspended sediment (Fig. 4a). Total P and DRP were most elevated at the sewage outfall (–3 km). Phosphorus concentrations dropped in the middle reaches of the river (–15 km) and close to and adjacent to the river mouth. Total P was only 2 times greater than DRP at the oceanic sites.

Sediment Fe-oxide-extractable P followed a similar trend to water column P within the river and was lowest at the upstream cane area (4.2 mg kg^{-1}) and highest at the sewage outfall (7.3 mg kg^{-1}) (Fig. 4c). However, sediment P remained elevated at the river mouth and at several oceanic sites, which varied from 1.5 mg kg^{-1} just outside the river mouth to 7.6 mg kg^{-1} at Surprise Reef.

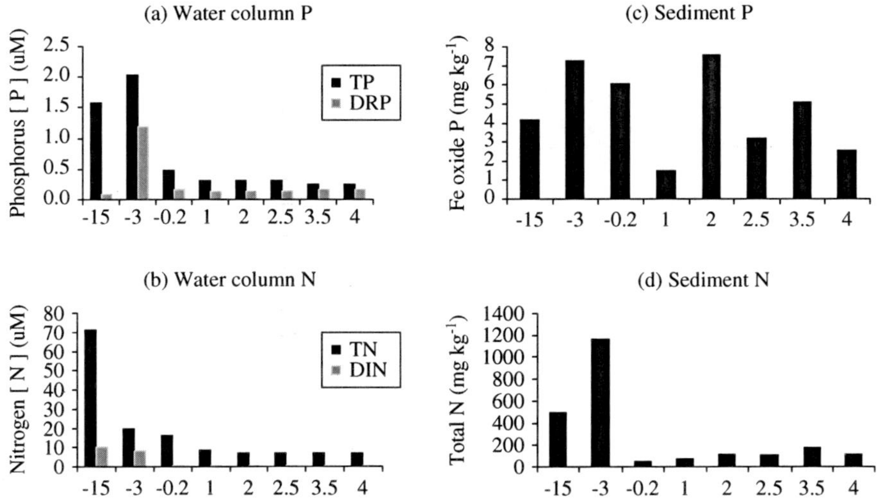

Figure 4: Total and dissolved (a) phosphorus and (b) nitrogen concentrations in the water column and (c) iron-oxide extractable P and (d) total nitrogen in the sediment in the Maroochy River and offshore. Samples were taken in October 1999, during low-flow conditions. DRP = dissolved reactive P; TP = total P; DIN = dissolved inorganic N; TN = total N.

Water column total N decreased from 70 μM upstream (cane area) to 16 μM at the river mouth, averaging 7 μM outside the river (Fig. 4b). In comparison, there was little difference in DIN between the upstream cane area (10 μM) and sewage outfall (8 μM), indicating that elevated TN upstream was most likely attributable to suspended sediments or dissolved organic N, although there was a large drop in DIN from the sewage outfall (8 μM) to the mouth (0.6 μM). Conversely, sediment total N content in the Maroochy River was lowest upstream (500 mg kg^{-1}) and highest at the sewage outfall (1160 mg kg^{-1}) (Fig. 4d), similar to sediment P content. Sediment N content decreased close to the river mouth (50 mg kg^{-1}) and increased with distance outside the mouth to 180 mg kg^{-1} at Mudjimba Island.

3.2 Sediment and soil nutrients

Total P content of the deposited sediment at Surprise Reef and Mudjimba Island (1290 mg kg^{-1}) was much higher than the P content of river (420 mg kg^{-1}) and marine (103 mg kg^{-1}) sediment as well as bulk soil (410 mg kg^{-1}) (Table 1). The total P content was comparable to fertilised soil (888 mg kg^{-1}) and suspended sediment (990 mg kg^{-1}) in the Maroochy River. The Fe-oxide-extractable P content of the deposited sediment (49.7 mg kg^{-1}) was also more elevated than the river (4.2 mg kg^{-1}) and marine sediment (5.3 mg kg^{-1}) as well as soil (7.4 mg kg^{-1}), although less than fertilised soil (68.5 mg kg^{-1}) and suspended sediment (90.2 mg kg^{-1}). It is interesting to note that although the total P content of the river sediment

Table 1: Mean nutrient content of different sediment types in the Maroochy River Catchment.

Sediment type	Total P (mg kg^{-1})	FeO-P (mg kg^{-1})	Total N (mg kg^{-1})	Total C (%)	Total Si (%)	Total Al (%)	Total Fe (%)
Soil	410	7.4	1980	3.5	35.1	4.6	1.7
	(140)	(2.4)	(881)	(1.3)	(2.9)	(1.5)	(0.4)
Fertilized soil	888	68.5	NA	NA	NA	NA	NA
	(346)	(21.7)					
Suspended sediment	990	90.2	NA	NA	26.8	10.0	4.8
	(637)	(46.0)			(1.6)	(0.4)	(0.5)
River sediment	420	4.2	1363	1.9	33.7	6.5	3.3
	(119)	(1.2)	(376)	(0.5)	(2.3)	(1.41)	(1.2)
Marine sediment	103	5.3	110	0.8	42.6	0.5	0.2
	(37)	(1.05)	(18)	(0.4)	(2.8)	(0.2)	(0.1)
Deposited sediment	1290	49.7	3630	4.4	28.3	5.8	3.3
	(159)	(8.8)	(479)	(0.6)	(3.9)	(1.1)	(0.8)

Value in brackets denotes standard error of mean. $N = 3$. NA = not available.

was approximately 4 times higher than the marine sediment, Fe-oxide-extractable P content was similar. This indicates that most bioavailable P in the river sediments is either desorbed/fluxed from the sediment, or associated with organic matter (organic P) that is not measured as Fe-oxide P (as total C is 2–3 times higher in the river than the marine sediments).

Total N content of the deposited sediment (3630 mg kg^{-1}) was approximately two times greater than the soil (1980 mg kg^{-1}), three times higher than the river sediment (1362 mg kg^{-1}), and 30 times higher than the marine sediment (110 mg kg^{-1}). The total carbon content of the deposited sediment (4.4%) was comparable to bulk soil (3.5%) and was greater than the river sediment (1.9%) and marine sediment (0.8%). Total silicon content varied from 26.8% (suspended sediment) to 42.6% (marine sediment). Soil (35.1%) and river sediment (33.7%) had similar silicon contents, as did the suspended sediment (26.8%) and deposited sediment (28.3%).

The total aluminum content of the soil (4.56%), deposited sediment (5.76%) and river sediment (6.46%) were comparable. Suspended sediment had the highest Al content (10.25%) and the marine sediment contained the lowest (0.52%). Similarly, the total iron content of deposited sediment (3.3%) was comparable to the suspended sediment (4.8%) and river sediment (3.3%), and approximately two times greater than the bulk soil (1.7%). The marine sediment was markedly different; with the total Fe content being an order of magnitude lower (0.2%) than the other sediment samples and the bulk soil. Al and Fe are terrestrial elements found commonly in soil [40], and thus are not naturally abundant in marine sediments [41]. The similarities in the composition of deposited sediment with estuarine suspended sediments and bulk soils, particularly Al and Fe, indicates that the deposited sediments are more likely terrigenous in origin than marine.

3.3 Suspended sediment

During wet conditions suspended sediments from the cane area (16 km upstream) had the highest total P content (2260 mg kg^{-1}), which was approximately double the total P in the deposited sediment samples (1290 mg kg^{-1}) and 1.5 times greater than sediments simulated from fertilized soils (silt-sized 1417 mg kg^{-1}, clay 1602 mg kg^{-1}) (Table 2). Total P of the deposited sediment was comparable to the fertilized simulated sediments and greater than the nonfertilized simulated sediments. The forested stream and cane drain had the lowest total P concentrations.

The bioavailable P content of the suspended sediments in the Maroochy River (measured as Fe-oxide-extractable P) was lowest at the forested stream site (3.3 mg kg^{-1}) and cane drain (13.42 mg kg^{-1}), then decreased with proximity to

Table 2: Comparison of suspended sediment nutrient content and grain size between field and simulated suspended sediments and deposited sediment. Total P = total phosphorus, FeO-P = iron-oxide extractable phosphorus, Total N = total nitrogen.

Sample location	Total P (mg/kg)	FeO-P (mg/kg)	Total N (mg/kg)	Mean particle size (μm)
Forested stream (−23 km)	260	3.3 (0.1)	NA	29.9
Cane drain (−19 km)	450	13.4 (0.3)	NA	15.7 (5.8)
Cane area (−16 km)	2260	258 (3.7)	NA	28.3
Sewage outfall (−3 km)	NA	106 (1.5)	NA	27.2
River mouth (−0.5 km)	NA	69.8	NA	32.7
Deposited sediment (+1.5–3 km)	*1290 (159)*	*49.7 (8.8)*	*3630 (480)*	*7.3 (0.0)*
Suspended sediment source				
Simulated silt + clay (silt-sized particles)	677 (142)	13.4 (4.0)	3861 (540)	10.1 (0.9)
Simulated silt + clay (+P) (silt-sized)	1417 (266)	95.5 (25.1)		
Simulated clay (clay-sized particles)	851 (139)	15.1 (4.2)	5413 (805)	4.7 (0.7)
Simulated clay (+P) (clay-sized particles)	1602 (245)	95.6 (22.0)		

Values in brackets indicate standard error of mean when enough sample was available. Some analyses were not replicated or conducted (NA) due to limited sediment quantities.

the river mouth. Suspended sediments collected from the cane area (16 km upstream) had the highest Fe-oxide-extractable P (258.4 mg kg^{-1}) in the Maroochy River (Table 2). Fe-oxide-extractable P was approximately 2.5 times greater than suspended sediments simulated from fertilized soil (silt-sized = 95.5 mg kg^{-1}, clay = 95.6 mg kg^{-1}) and 5 times higher than deposited sediment (49.7 mg kg^{-1}). Suspended sediments collected at the sewage outfall (105.82 mg kg^{-1}) and river mouth (69.8 mg kg^{-1}) had comparable Fe-oxide-extractable P to simulated fertilized sediments. Suspended sediments in the cane drain had comparable Fe-oxide-extractable P to sediments simulated from nonfertilized soils (silt-sized 13.42 mg kg^{-1}; clay 15.1 mg kg^{-1}). The forested stream site had the lowest Fe-oxide-extractable P (3.3 mg kg^{-1}). The deposited sediment samples contained approximately half of the Fe-oxide-extractable P bound to the simulated fertilized sediments and had more than double the P bound to simulated sediments derived from nonfertilized soils.

The deposited sediment and simulated silt-sized sediments had comparable total N contents (3630 and 3861 mg kg^{-1}, respectively). The simulated clay sediments (5412 mg kg^{-1}) had approximately 1.5 times more total N than the deposited sediment and silt-sized sediments, and were approximately 2.5 times more enriched than the soil they were simulated from.

Mean particle size was very fine within the Maroochy River and varied between 16 μm (cane drain) and 33 μm (river mouth). The deposited sediment samples (7.3 μm) were at least half the size of the river samples and were intermediate between the simulated silt-sized (10.1 μm) and clay (4.7 μm) suspended sediments.

Due to the amount of water required to sample suspended sediments during low-flow conditions, insufficient quantities were obtained for sediment analysis except at the sewage outfall during the September 2000 drought. Mean sediment size (9.6 μm) was approximately one third the size of sediments sampled during wet conditions (27.3 μm). There was also variation in particle size between wet sampling [Feb. 1999 (data not presented) and November 2000] at the upstream cane site (28.31 μm Nov. 00, 17.7 μm Feb. 99) and pristine site (29.9 μm Nov. 00, 6.3 μm Feb. 99). The sediment that was sampled was much finer than the sediment captured during the initial 'flush' of sediments in November 2000. This highlights the pulsed nature of sediment transport and the difficulty in accurately capturing the pulse of suspended sediments.

3.4 Deposited sediment bioassays

Sediment deposited at Mudjimba Island and Surprise Reef contained a high content of measured bioavailable P (Fe-oxide-extractable P) (Table 2). When deposited sediment was used as the sole source of P in sediment bioassays, maximum *Skeletonema costatum* biomass reached 450–500 μg Chl *a* L^{-1} (Fig. 5). In comparison the underlying sand at each site only produced a maximum biomass of 150 μg Chl *a* L^{-1}. There was little variation between the sand collected from Surprise Reef (sand 1,2) and Mudjimba Island (sand 3).

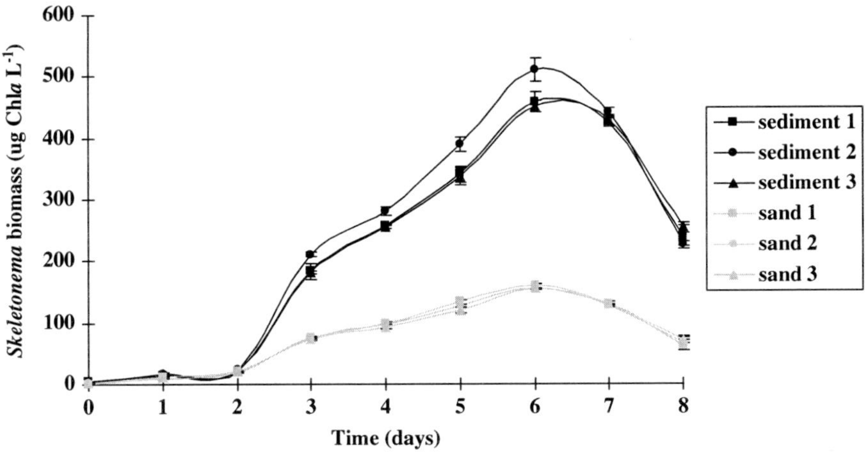

Figure 5: Maximum *Skeletonema costatum* biomass (μg Chl *a* L^{-1}) attained over 8 days with deposited sediment (sediment) and underlying marine sand used as the sole source of phosphorus. Sediment and sand were collected from three different sites offshore. Error bars are standard error of mean.

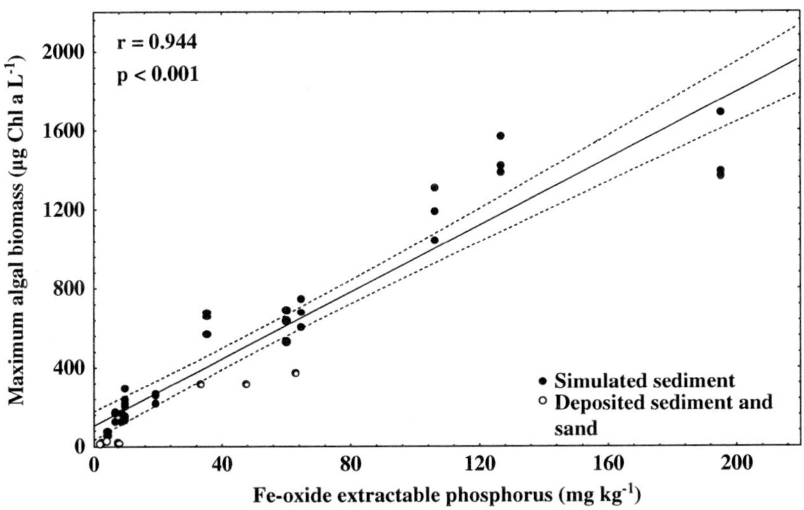

Figure 6: Maximum algal biomass (μg Chl *a* L^{-1}) correlated with bioavailable P (Fe-oxide-extractable P) content (mg kg^{-1}) of deposited sediment, sand and simulated suspended sediment. Dotted line is 95% confidence interval.

Conversely, the maximum biomass differed with deposited sediment type, although there was little variation within replicates. This was attributable to the varying Fe-oxide-extractable P content of the deposited sediment sampled from different sites (Fig. 6). The maximum algal biomass correlated well with Fe-oxide-extractable P content of simulated sediment, sand and deposited sediment

($r = 0.944$ $p < 0.001$) (Fig. 6). This demonstrates the suitability of Fe-oxide P as a measure of bioavailable P in both marine and terrigenous sediments.

3.5 Transport of suspended sediments and deposited sediment

It was apparent from visual observation in the field that suspended sediment load in the Maroochy River was heaviest during heavy rainfall conditions. This was particularly apparent at the well-flushed river mouth, which normally had clear, blue water in dry conditions. The transport of large amounts of suspended sediments outside the river mouth was easily visible.

The influence of freshwater was apparent with a large drop in salinity during the wet season (Fig. 7a). Salinity decreased ~30‰ at the sewage outfall (3 km from the river mouth) to only 5‰. The suspended sediment load varied within the river as well as with sampling time (wet vs. dry) (Fig. 7b). During the wet sampling turbidity was highest in the cane drain (70 mg L^{-1}) and dropped significantly to

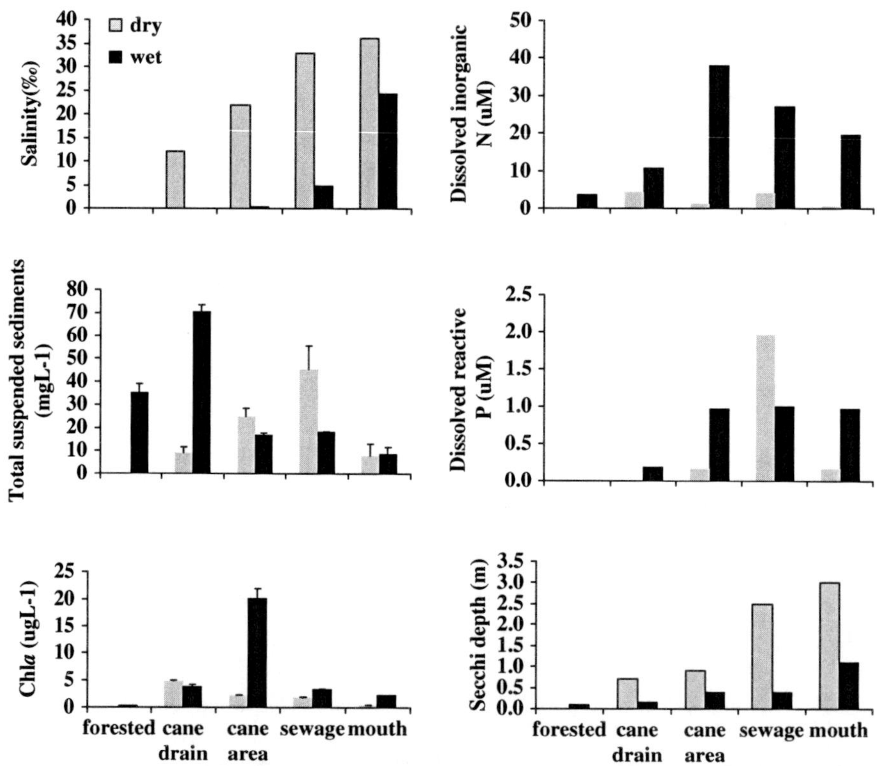

Figure 7: (a) Salinity (‰), (b) total suspended sediments (mg L^{-1}), (c) water column Chl *a* (μg L^{-1}), (d) dissolved inorganic nitrogen (μM), (e) dissolved reactive phosphorus (μM), (f) secchi depth (m) in the Maroochy River during wet and dry conditions.

15 mg L^{-1} in the upstream cane area. Turbidity increased to 20 mg L^{-1} at the sewage outfall and decreased to 10 mg L^{-1} at the mouth. The opposite occurred during the dry sampling, with maximum turbidity at the sewage outfall (~45 mg kg^{-1}) and lowest upstream (10 mg L^{-1}) at the cane drain. The forested stream located 23 km upstream did not flow during dry conditions. The secchi depth was inversely related to water-column turbidity, with the secchi depth increasing in proximity to the river mouth during both dry and wet seasons (Fig. 7f).

The Chl *a* concentration also varied within the river and between sampling times (Fig. 7c). In the wet sampling, the Chl *a* maximum occurred upstream in the cane area (~20 µg Chl *a* L^{-1}). Chl *a* concentration was similar in the cane drain and downstream at the sewage outfall (~5 µg Chl *a* L^{-1}). Chl *a* was not detected at the forested site (–23 km) and was also low at the river mouth (2.5 µg Chl *a* L^{-1}). In the dry sampling the maximum occurred in the cane drain and was only 5 µg Chl *a* L^{-1} (4 times lower than the wet maximum).

Concentration of Chl *a* was inversely related to the suspended sediment load in the water with maximum Chl *a* corresponding to low water-column turbidity. In dry conditions, phytoplankton biomass was highest in the cane drain, which had lower turbidity than sites downstream (except for the mouth). In wet conditions turbidity was highest in the cane drain thus phytoplankton biomass was low as light probably limited growth. During wet conditions the Chl *a* maximum occurred at the upstream cane site, which had lower turbidity than the cane drain and sewage outfall. The chlorophyll maxima also corresponded to DIN, which was greatest in the cane area in the wet season and cane drain in the dry (Fig. 7d). Chl *a* was continually low at the well-flushed river mouth during wet and dry conditions. Although nutrients are flushed through the river mouth, phytoplankton is unlikely to assimilate nutrients within such short water residence times. The forested stream (23 km upstream) also had low Chl *a* biomass. This may be attributable to low dissolved nutrient concentrations and its ephemeral nature as the stream flowed for only short periods. Algal biomass appeared to be influenced by DIN rather than DRP as Chl *a* maximum corresponded to elevated DIN not DRP. However, it should be noted that additional nutrients, particularly P, would be associated with the suspended sediments, which were not quantified. Elevated DIN in the cane area may be caused by fertiliser run-off and leaching of NO_3^- from the land to the water. Elevated P levels are most likely caused by desorption of P from sediment particles eroded from nearby agricultural soil.

4 Discussion

4.1 Delivery of nutrients to the coastal ocean

4.1.1 Phosphorus and nitrogen pathways

Phosphorus (as orthophosphate) and nitrogen (as ammonium) are applied as fertilizers to the seven major agricultural soils in the Maroochy Watershed. It is well known that crops do not assimilate all of the P and N added as fertilizer [42], thus a significant fraction is dissolved in the soil porewater or bound to soil particles.

The major pathways of P and N movement from land to water vary due to different cycling processes in the soil (Fig. 8). Most P will be transported to waterways in particulate form via sorption to soil particles [14, 17, 21, 43], as P sorption readily occurs by covalent bonding of phosphate ions to hydrous oxides of iron (Fe) and aluminum (Al) on particle surfaces [44]. A fraction of the particulate P is desorbable and thus potentially bioavailable (Froelich, [14]). Turnover times for inorganic phosphorus in the water column of the GBR range from hours to weeks [45]. Conversely, the majority of bioavailable N enters the marine environment in dissolved form, either as nitrate or ammonium or adsorbed to particles (ammonium) [13]. Nitrate is the dominant form, transported via surface run-off and through leaching processes into the ground water [13] (Fig. 8). In addition, there is increasing evidence that transport of N by the intertidal and subtidal flux of groundwater is significant for coastal ecosystems [46]. As ammonium is not readily leached, it is either a) dissolved in the soil pore water or b) adsorbed to the surface of clay and organic matter by ion exchange or c) fixed within layers of the clay structure and is not available for biological uptake [16]. However N can be released once sediments are deposited through biological regeneration, remobilization and sediment resuspension [47]. Once transported into marine waters, dissolved ammonium and nitrate are readily assimilated by phytoplankton or other marine plants as N often limits productivity in marine environments [42, 48, 49]. Estimates of N demand by phytoplankton on the GBR indicate dissolved inorganic N pools have turnover times of the order of hours to days [45].

4.1.2 Sediment and nutrient transport

During high-intensity rainfall events P-rich agricultural soils (sand, silt and clay-sized particles) are eroded into nearby streams and rivers (Fig. 8). Coarser soil particles (sand) settle out first and are deposited as river sediments; the finer fractions (silt and clay) eventually deposit close to the coast [50, 51] (Fig. 8). The large difference in particle size between suspended sediments at the Maroochy River mouth (>20 μm) and sediment deposited outside the river mouth (silt-sized <20 μm) (Table 2), indicates that this settling process occurred in the Maroochy River. The suspended sediment and deposited silt-sized particles were enriched with total P comparable to fertilized soil from the watershed. Finer-sized particles have an increased adsorption capacity due to their large surface area [52], thus contain the highest concentration of bioavailable P per unit weight [53]. In addition, the deposited sediment also had elevated TN concentration (3630 mg kg^{-1}) which was almost double the TN measured in the watershed soils. This indicates that nutrient-rich sediments are being eroded from the Maroochy Watershed, flushed into the Maroochy River during heavy rainfall events and deposited offshore. The chemical composition of the deposited offshore sediment was more comparable to the bulk soil, suspended sediment and river sediment than the underlying marine sand (Table 1). The dissimilarities between the deposited sediment and the underlying marine sand, suggest that the deposited sediment was not marine in origin and more likely terrigenous. The large difference in total Al and Total Fe content of the deposited sediment and marine sediment further validates this, as Al and Fe

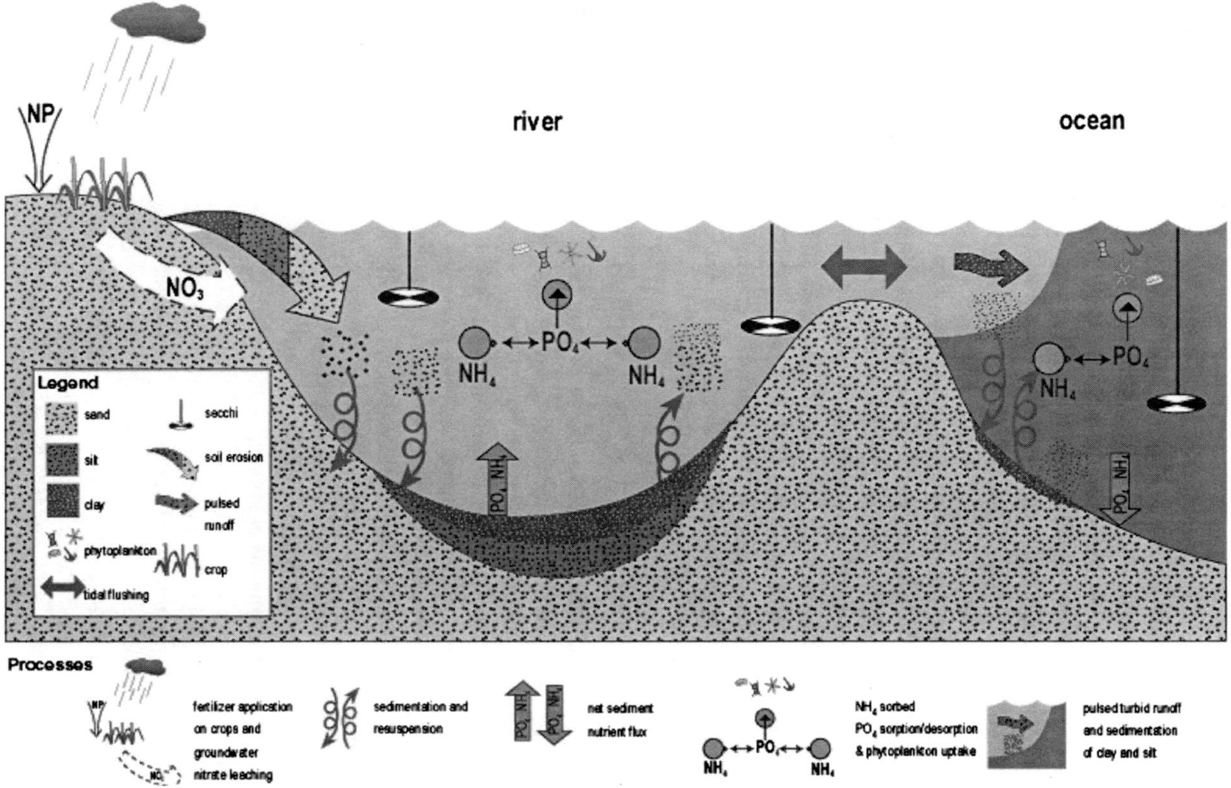

Figure 8: Conceptual model of the movement and processing of sediment and nutrients in the Maroochy River and offshore (see text for explanation).

are terrestrial elements [40], which do not naturally occur in high concentrations in marine sediments [41].

In subtropical and tropical systems, such as the Maroochy Watershed, the transport and deposition of silt and clay and associated nutrients offshore are most often associated with pulsed storm events [20, 22]. During flood conditions, estuarine processes may be bypassed and freshwater, sediments and nutrients discharge directly onto the continental shelf [54, 55]. Additional sediment and nutrients may also be supplied by erosion and scouring of the estuary floor during these conditions [54, 55]. As storm events occur over short time frames, Australian rivers and estuaries only contribute a significant amount of material to the continental shelf for short periods [56]. Although plumes spread in a band up to 50 km from the coast, particulate material is trapped within 10 km of the coast [57]. Fine-sized particles have been shown to deposit and accumulate close to the shore, with silt and clay being found 5–20 km offshore in the Johnstone River watershed (North Queensland, Australia) [19]. In the Maroochy watershed a thin film of clay and silt deposited in grooves and sheltered areas on a rocky near-shore reef located 2.5–4 km from the Maroochy River mouth. Sediment movement may also occur with the continual resuspension and redistribution of sediment at the river mouth due to tidal flushing and wind resuspension [47]. Flushing and exchange of clean oceanic water with estuarine water at the river mouth, improves water clarity (measured by secchi depth) with distance downstream (Fig. 7).

4.1.3 Cycling of nitrogen and phosphorus

Once soils are eroded into waterways, some P may be desorbed from particle surfaces into the water column and be assimilated by marine plants such as phytoplankton [14]. The remaining P may either be deposited with silt and clay particles in the river or transported offshore with suspended clay particles as previously mentioned. Ammonium will also be deposited in the river and transported offshore, due to strong sorption to particle surfaces. This is reflected in the high total N concentration of the deposited sediments outside the Maroochy River (Table 1).

Once sediments (mostly silt and some clay) deposit in the river, microbial mineralization may generate additional release of ammonium and P out of the sediments into the water column. Additional desorption or sorption of P to particles may also occur when sediments are resuspended by wind, waves or tides [14].

The deposition of clay and silt particles offshore provides a source of bioavailable P, total P, total N (Table 1), and ammonium. Although the ammonium content of the sediments was not measured in this study, previous research (Chaston, [23]) found the ammonium concentrations in silt and clay particles collected from the Maroochy watershed were elevated. Although ammonium does not appear to be immediately bioavailable when sediments are in suspension (Chaston, [23]), biological processing in the sediments may facilitate further N release once sediments are deposited [13]. The deposited sediment was a significant source of bioavailable P (Fig. 4), in contrast to the underlying marine sand (Fig. 5). Additional release of sediment bound P and N may occur during sediment resuspension events [58–60].

In the Great Barrier Reef lagoon, algal blooms have been linked to wind resuspension of sedimentary nutrients [61, 62]. The regeneration of P is so efficient in coastal systems that only a small proportion is lost permanently to the sediments by burial [13]. Microbial mineralization of organic N released from resuspended shelf sediments in the GBR has been associated with elevated water-column dissolved inorganic concentrations after a cyclonic disturbance [61]. In the Moresby River estuary (North Queensland), estuarine sediments appeared to be a source of nitrate during the dry season [21].

4.2 Environmental implications

4.2.1 Impact on river and estuary ecosystem health

Although elevated nutrient concentrations may not cause excessive algal growth in turbid waterways where light limits algal growth [63], the movement and deposition of suspended sediments to estuaries and the coastal ocean where light is not limiting, may enhance growth. This is particularly significant in areas where algal productivity is limited by nutrients. Although N usually limits productivity in estuarine and coastal waters [42, 49, 64], increased bioavailable P could further strengthen N limitation, favoring N-fixing cyanobacteria [65]. Nutrient enrichment can also alter the phytoplankton community composition [66], which can have significant impacts on grazers and predators, although this is not well studied [42].

4.2.2 Impact of offshore sediment and nutrient deposition

The impact of sediment and nutrient movement offshore of the Maroochy River is likely to have a localized, small impact due to the dilution of nutrients and dissipation of sediments by the ocean. The Maroochy River drains directly onto the continental shelf of the Pacific Ocean, thus nutrient-rich plume waters are quickly diluted. However, flood plumes extending to Mudjimba Island did persist for several days during the study, and deposited clay and silt was found 5 months after the flood event. Nutrient enrichment is likely to be restricted to localized areas where nutrient-rich clay and silt particles are deposited. Benthic microalgae and macroalgae growing in these areas can potentially take up nutrients from these enriched sediments [67–69]. In addition, marine phytoplankton will quickly assimilate any nutrients that are released into the water column [46]. Although these impacts are localized, they may have a detrimental effect on the recreational value of the near-shore reefs in the area, due to increased turbidity. Continual nutrient enrichment can also lead to the detrimental impacts mentioned above.

In comparison, the movement and deposition of sediments and nutrients offshore to areas with restricted tidal exchange and long residence times, such as enclosed bays and lagoons, may have more significant and detrimental impacts. In other areas of southern Queensland, there have been several reports of very significant impacts of sediment loads on downstream coastal seagrass areas, including extensive loss of seagrass areas in Hervey Bay, following flood events [70], and the loss of seagrass in southern Moreton Bay associated with general increases in turbidity [71]. Nutrient- and sediment-rich river flow, often associated with increased agriculture, has also had significant impacts worldwide. Increased nutrient load in river flow, has seen an increase in hypoxia in the Gulf of Mexico, causing the formation of a huge 'dead zone' [72]. Increased hypoxia has also been reported in the Adriatic Sea due to nutrient-rich water from the Po River [73]. Deterioration of water quality from nutrient enrichment and sediment inputs has also caused hypoxic or anoxic conditions in Chesapeake Bay (USA), causing declines in living resources [74].

Locally, movement and deposition of suspended sediments offshore may also have significant ramifications in the Great Barrier Reef Lagoon, Australia. Terrestrial runoff of sediment and nutrients (mostly N and P) to the GBR has increased 2- to 4-fold over the last century [75, 76]. Many tropical rivers and estuaries discharge into the lagoon, which has been described as a semi-enclosed sea that annually receives an increasing load of nutrients [2]. Gradients in reef and community structure, biodiversity and ecological function in near-shore coral reef systems have been associated with environmental gradients influenced by terrestrial runoff [77].

4.2.3 Nutrient limitation

It is generally accepted that N limits productivity in most coastal marine ecosystems and P limits freshwater productivity [42]. Although N limitation has been shown to occur in southern Queensland estuaries and coastal waters [28, 78, 79], there is increasing evidence that N also limits freshwater productivity in southern Queensland streams [80]. In addition, Australian freshwater systems with long residence times show stoichiometric evidence of N limitation [81].

Low N:P ratios in rivers, estuaries and coastal waters may be reinforced by the episodic transport and deposition of P-rich sediment to Australian coastal ecosystems. The release of bioavailable P from sediments and the release of additional micronutrients such as Fe, may account for the prevalence of N-fixing cyanobacteria such as *Trichodesmium* in coastal seas [2]. Biological cycling of N in the sediment may also result in loss of N via denitrification to N_2 gas, as well as N release through nitrification [13]. Thus, N limitation may be exacerbated if denitrification rates are high.

4.2.4 Watershed management

This study highlights the importance of effective land management and erosion control measures that reduce the suspended sediment load. Although agricultural practices are continually improving to reduce soil erosion and minimize nutrient loss both locally [82] and overseas [83, 84], practices need to encompass a broader approach that emphasizes the hydrological and nutrient link between land and sea. Effective management must consider both agricultural productivity as well as potential environmental impact as practices that are economically viable may not be environmentally sustainable [11]. Mitigation measures should encompass both sediment- and nutrient-reduction strategies in order to reduce particulate and dissolved nutrient transport. Nutrient management should focus on both N and P, as measures that control only one nutrient may enhance loss of the other [85], or cause nutrient limitation to switch between N and P [42].

Soils with high Fe-oxide-extractable P or Colwell P, which correlates with bioavailable P (Chaston, [23]) must be targeted and managed effectively, particularly in areas prone to erosion. This will be specific for each watershed, depending on the dominant erosion processes occurring and watershed hydrology (e.g. stream gradient, watershed size, topography). Hill-slope erosion (sheet and rill erosion) is the dominant erosion process in most of northern Australia, including the Maroochy River watershed [86]. Cultivated land, which covers a large area of the watershed, is an important potential source of hillslope sediments. The majority of P export may be derived from only a small portion of the watershed over a short time period [11, 17]. Thus risk assessments within the watershed should be based on soil type, land use and position in the watershed.

The 7 major agricultural soil types in the Maroochy Watershed varied in chemical composition with the total P content of unfertilized soils ranging from 20-1120 mg P kg^{-1}. Only two of the soils, the Chromosol and Redoxic Hydrosol, had TP concentrations greater than 400 mg kg^{-1}. The average total P content of most Australian surface soils except those derived from basalt is less than 400 mg P kg^{-1} soil [87]. However, the silt- and clay-sized fractions of the soil were more elevated (220–1390 TP mg kg^{-1}), although this was not proportional to TP of the bulk soil. Although total P is commonly used to assess the sediment nutrient status, it does not accurately reflect bioavailable P at low total P concentration and should only be used to identify soils with potentially high bioavailable P, as high total P is often well correlated with high bioavailable P [23]. Identification of soils with high bioavailable P (e.g. Red Kandosol), and thus potential to cause algal blooms when eroded, could be used to set priorities for managing watershed areas and identifying potential problem areas in terms of offsite nutrient transport.

Colwell-extractable P is widely used in Australian soils to assess the availability of P to crops [44]. Colwell P values of less than 10 mg P kg^{-1} are considered very low, while values >100 mg P kg^{-1} are considered very high. Within the intensive land use zone of Australia, 1.6% of surface soils have Colwell P< 10 mg P kg^{-1}, 3.6% have Colwell P >80 mg P kg^{-1}, and the majority (60.9%) have Colwell P ranging between 10 and 30 mg P kg^{-1} [88]. The 7 major agricultural soils in the Maroochy Watershed all fall below 100 mg P kg^{-1} prior to fertilizer application, and three have very low P values (<10 mg P kg^{-1}). However, after application of P fertilizers, most values would be considered very high (up to 430 mg P kg^{-1}). The silt-sized fraction had more elevated Colwell P values, with all values (except 3 nonfertilized soils) considered very high with up to 760 mg P kg^{-1}. This indicates that the fertilized soils in the Maroochy Watershed are all sources of potentially high bioavailable P and need to be carefully managed.

Although suitable chemical analyses of bioavailable P in sediments and silt and clay-sized fractions of the soil have been identified [23] the next step is to relate this to bulk surface soil. Knowledge of the enrichment ratio of P between the bulk soil and resultant sediment will allow an assessment of the risk of sediment being a P source to algae, using analyses of the bulk surface soil, rather than the silt- and clay-sized fraction. Thus, the challenge facing soil scientists will be to develop a realistic nutrient-enrichment ratio that represents sediment moving offsite. Currently, bioavailable P is usually predicted from bulk soils assuming a nutrient-enrichment ratio based on clays only, which is estimated at 10%.

The prioritization of watershed areas could extend to classification of rivers and estuaries, based on the potential for soil erosion, general ecosystem health (background nutrient levels, primary productivity) and residence time of water. Increased residence time caused by low river flow has stimulated algal blooms in Australian rivers and estuaries [4, 89]. Flushing times were also shown to be a dominant factor in controlling the degree of internal processing of nutrients in the Richmond River Estuary, Australia [54]. The amount and duration of rainfall will also influence the amount of nutrients and sediment transported [17].

5 Conclusions

This study indicates that fine, clay-sized particles, rich in total P, total N and bioavailable P, are being transported and deposited offshore during erosion events. These nutrient-rich particles are likely to be terrigenous in origin as their chemical composition is more comparable to the estuarine suspended sediments and surface soils than the underlying marine sand. This demonstrates the connectivity between the watershed, land and sea and highlights the need for multifaceted watershed management that encompasses; a) erosion control measures that reduce the suspended sediment load of nutrient-rich clay- and silt-sized fractions, and b) nutrient-reduction strategies. Effective management must consider agricultural productivity as well as the potential environmental impact, as what is economically viable may not be environmentally sustainable.

Acknowledgements

The authors would like to acknowledge the Australian Research Council (ARC) for supporting this research through an ARC SPIRT Grant to The University of Queensland. We would like to thank Rory Whitehead for conducting the sediment analyses, Dr Cindy Heil for her guidance throughout this study, and the Marine Botany Laboratory at UQ for assisting with field studies.

References

[1] Hutchings, P., Haynes, D., Goudkamp, K. & McCook, L., Catchment to Reef: Water quality issues in the Great Barrier Reef Region – An overview of papers. *Marine Pollution Bulletin,* **51**, pp. 3–8, 2005.

[2] Bell, P.R.F., Eutrophication and Coral Reefs Some Examples in the Great-Barrier-Reef Lagoon. *Water Research,* **26**, pp. 553–568, 1992.

[3] Brodie, J.E., Enhancement of Larval and Juvenile Survival and Recruitment in Acanthaster-Planci from the Effects of Terrestrial Runoff – a Review. *Australian Journal of Marine and Freshwater Research,* **43**, pp. 539–554, 1992.

[4] Hillman, K., Lukatelich, R.J. & McComb, A.J., The impact of nutrient enrichment on nearshore and estuarine ecosytems in Western Australia. *Proceedings of the Ecological Society of Australia,* **16**, pp. 39–53, 1990.

[5] Brodie, J., The problems of nutrients and eturophication in the Australian marine environment. In *The State of the Marine Environment Report for Australia Technical Annex: 1.* (ed. Environment Australia). Environment Australia: Canberra, Australia, 1995.

[6] Hallegraeff, G.M., Marine Phytoplankton communities in the Australian region: current status and future threats. In *The State of the Marine Environment Report for Australia Technical Annex: 1.* (ed. Environment Australia) Environment Australia: Canberra, Australia, 1995.

[7] Brodie, J.E. & Mitchell, A.W., Nutrients in tropical Australian Rivers: Changes with agricultural developments and implications for receiving environments. *Marine and Freshwater Research,* **52**, pp. 279–302, 2005.

[8] Wilkinson, C., *Status of Coral Reefs of the World, Volume 1.* Australian Institute of Marine Science, Townsville, Australia, 2004.

[9] Bryant, D.G., Burke, L., McManus, J. & Spalding, M., *Reefs at Risk: a Map-Based Indicator of Threats to World's Coral Reefs.* World Resources Institute: Washington DC, 1998.

[10] Tunney, H., Carton, O.T., Brookes, P.C. & Johnston, A.E., *Phosphorus Loss from Soil to Water*, CAB International: UK, 1997.

[11] Sharpley, A., Foy, B. & Withers, P., Practical and innovative measures for the control of agricultural phosphorus losses to water: An overview. *Journal of Environmental Quality,* **29**, pp. 1–9, 2000.

[12] Withers, P.J.A., Davidson, I.A. & Foy, R.H., Prospects for controlling non-point phosphorus loss to water: A UK perspective. *Journal of Environmental Quality,* **29**, pp. 167–175, 2000.

[13] Valiela, I., *Marine Ecological Processes.* Springer-Verlag: New York, 1995.

[14] Froelich, P.N., Kinetic control of dissolved phosphate in natural waters and estuaries: a primer on the phosphate buffer mechanism. *Limnology and Oceanography,* **33**, pp. 649–668, 1988.

[15] Sharpley, A.N., Dependence of runoff phosphorus on extractable soil phosphorus. *Journal of Environmental Quality*, **24**, pp. 920–926, 1995.

[16] McLaren, R.G. & Cameron, K.C., *Soil Science – Sustainable Production and Environmental Protection*, Oxford University Press: Auckland, New Zealand, 1996.

[17] Cosser, P.R., Nutrient concentration-flow relationships and loads in the South Pine River, Southeastern Queensland. I. Phosphorus loads. *Australian Journal of Marine and Freshwater Research,* **40**, pp. 613–630, 1989.

[18] Eyre, B., Nutrients in the sediments of a tropical North-Eastern Australian Estuary, Catchment and Nearshore Coastal Zone. *Australian Journal of Marine and Freshwater Research,* **44**, pp. 845–866, 1993.

[19] Pailles, C., McConchie, D.M., Arakel, A.V. & Saenger, P., The distribution of phosphate in sediments of the Johnstone Rivers Catchment Estuary System, North Queensland, Australia. *Sedimentary Geology,* **85**, pp. 253–269, 1993.

[20] Hossain, S., Eyre, B. & McConchie, D., Suspended sediment transport dynamics in the sub-tropical micro-tidal Richmond River Estuary, Australia. *Estuarine, Coastal and Shelf Science*, **52**, pp. 529–541, 2001.

[21] Eyre, B., Nutrient Biogeochemistry in the Tropical Moresby River Estuary System North Queensland, Australia. *Estuarine Coastal and Shelf Science,* **39**, pp. 15–31, 1994.

[22] McKee, L., Eyre, B. & Hossain, S., Intra- and inter-annual export of nitrogen and phosphorus in the subtropical Richmond River catchment, Australia. *Hydrological Processes,* **14**, pp. 1787–809, 2000.

[23] Chaston, K.A.V., *Sediment Nutrient Bioavailability in a Sub-tropical Catchment Dominated by Agriculture: The Transition from Land to Sea*. Doctorate Dissertation Thesis. (The University of Queensland, Australia.) (http://adt.library.uq.edu.au/public/adt-QU20040818.160423/) pp. 164 2001.

[24] Guo, F., Yost, R.S. & Jones, R.C., Evaluating iron-impregnated paper strips for assessing available soil phosphorus. *Communications in Soil Science and Plant Analysis*, **27**, pp. 2561–2590, 1996.

[25] Colwell, J.D., The estimation of phosphorus fertiliser requirements of wheat in southern New South Wales by soil analysis. *Australian Journal of Experimental Agriculture and Animal Husbandry,* **3**, pp. 190–8, 1963.

[26] Rayment, G.E. & Higginson, F.R., *Australian Laboratory Handbook of Soil and Water Chemical Methods*. Inkata Press: Melbourne, 1992.

[27] Sharpley, A.N., An innovative approach to estimate bioavailable phosphorus in agricultural runoff using iron-oxide impregnated paper. *Journal of Environmental Quality,* **22**, pp. 597–601, 1993.

[28] Udy, J.W. & Dennison, W.C., Estimating nutrient availability in seagrass sediments. In *Seagrass Biology: Proceedings of an International Workshop, Rottnest Island, Western Australia*, (ed. J. Kuo, R.C. Philips, D.I. Walker, & H. Kirkman,) UWA: Perth, Australia, pp. 163–72, 1996.

[29] Van der Zee, S.E.A.T.M., Fikkink, L.G.J. & van Riemsdijk, W.H., A new technique for assessment of reversibly adsorbed phosphate. *Soil Science Society of America Journal*, **51**, pp. 599–604, 1987.

[30] Schlacher, T. & Carruthers, T., Maroochy River: Chapter 3. In *State of SEQ Waterways Report 2001.* (ed.) E. Abal, G. Moore, W.C. Dennison South East Queensland Regional Water Quality Managment Strategy: Brisbane, Australia, 2001.

[31] Moss, A., Cox, M., Choy, S. & Marshall, J., *Maroochy and Mooloolah Catchments: Water Quality Study 1997*. Queensland Environmental Protection Agency, Environment Technical Report 29, Brisbane: Australia, 1999.

[32] Isbell, R.F., The Australian Soil Classification System. CSIRO Division of Soils, *Australian Soil and Land Survey Handbooks, Vol. 4*, CSIRO Publishing: Melbourne, Australia, 1996.

[33] Soil Survey Staff, *Soil taxonomy: a basic system of soil classification for making and interpreting soil surveys*, 2nd edn. Agriculture Handbook 436. US Department of Agriculture, Soil Conservation Service, Washington, DC, 1999.

[34] Pocock, J. & Thomson, D., Maroochy Mooloolah Loads and Impact Study Phase 1 Report. WBM Oceanics report No. 12622.R1.0. WBM Oceanics Australia: Brisbane, Australia. 2000.

[35] Parsons, T.R., Maita Y. & Lalli. C.M., *A Manual of Biological and Chemical Methods for Seawater Analysis,* Pergamon Press: Oxford, pp. 173, 1984.

[36] Clesceri, L.S., Greenberg, A.E. & Eaton, A.D., Standard Methods for the Examination of Water and Waste Water. 20th edn. APHA-AWWA-WPCF, 1998.

[37] Blakemore, L.C., Searle, P.L. & Daly, B.K., Methods for the chemical analysis of soils. New Zealand Soil Bureau Science Report, 80, New Zealand. 1987.

[38] Bale, A.J., In *Situ* Laser Optical Particle Sizing. *Journal of Sea Research,* **36**, pp. 31–6, 1996.

[39] Tanner, C.B. & Jackson, M.L., Nomographs of sedimentation times for soil particles under gravity of artificial acceleration. *Soil Science American Proceedings,* **34**, pp. 608–610, 1947.

[40] Wild, A., *Soils and the environment: An Introduction*, Cambridge University Press, Cambridge, Great Britain, 1993.

[41] Riley, J.P. & Chester, R., *Introduction to Marine Chemistry*, Academic Press: NY, 1971.

[42] National Research Council, *Clean Coastal Waters: Understanding and Reducing the Efects of Nutrient Pollution,* National Academy Press: Washington, 2000.

[43] Sharpley, A.N. & Rekolainen, S., *Phosphorus in Agriculture and its Environmental Implications*, CAB International Press: Cambridge, UK, 1997.

[44] Moody, P.W. & Bolland, M.D.A., Phosphorus. In *Soil Analysis and Interpretation Manual.* K.I. Peverill, L.A. Sparrow, & Reuter D.J. eds pp. 187–220. CSIRO Publishing: Australia, 1999.

[45] Furnas, M., Mitchell, A., Skuza, M. & Brodie, J., In the other 90%: phytoplankton responses to enhanced nutrient availability in the Great Barrier Reef Lagoon. *Marine Pollution Bulletin,* **51**, pp. 253–265, 2005.

[46] Valiela, I., Foreman, M., LaMontagne, D., Hersh, D., Costa, J., Peckol, B. & Lajtha, K., Couplings of watersheds and coastal waters: Sources and consequences of nutrient enrichments of Waquoit Bay, Massachussetts. *Estuaries,* **15**, 1992.

[47] Thornton, J.A., McComb, A.J. & Ryding, S.O., Chapter 13. The role of the sediments. *Eutrophic Shallow Estuaries and Lagoons.* ed. A.J. McComb, CRC Press Inc.: Florida, USA, 1995.

[48] Vitousek, P.M. & Howarth, R.W., Nitrogen limitation on land and in the sea – how can it occur. *Biogeochemistry,* **13**, pp. 87–115, 1991.

[49] Downing, J.A., Marine nitrogen: Phosphorus stoichiometry and the global N:P cycle. *Biogeochemistry,* **37**, 1997.

[50] Pailles, C. & Moody, P.W., Phosphorus dynamics of sediments from the Johnstone Rivers. In *Downstream Effects of Land Use.* eds H.M. Hunter, A.G. Eyles & G.E. Rayment, Department of Natural Resources Queensland: Queensland, Australia, pp. 359–65, 1996.

[51] Johnson, C.E. & Carter, R.M., Sedimentary evidence on the seaward limits of suspended materials from rivers. *Nutrients in the Barrier Reef Region*. ed. C.L. Baldwin, Townsville: Australia, 1988.

[52] McComb, A.J., Atkins, R.P., Birch, P.B., Gordon, D.M. & Lukatelich, R.J., Eutrophication in the Peel–Harvey Estuarine System, Western Australia. In *Estuaries and Nutrients*. eds B.J. Neilson & L.E. Cronin, Humana Press: New Jersey, USA, pp. 323–42, 1981.

[53] Sharpley, A.N., The contribution of phosphorus leached from the crop canopy to losses in surface runoff. *Journal of Environmental Quality,* **10**, pp. 160–165, 1981.

[54] Eyre, B. & Twigg, C., Nutrient Behaviour During Post-flood Recovery of the Richmond River Estuary Northern NSW, Australia. *Estuarine, Coastal and Shelf Science,* **44**, pp. 311–26, 1997.

[55] McKee, L.J., Eyre, B.D. & Hossain, S., Transport and retention of nitrogen and phosphorus in the sub-tropical Richmond River estuary, Australia: A budget approach. *Biogeochemistry Dordrecht,* **50**, pp. 241–78, 2000.

[56] Eyre, B., Transport, Retention and Transformation of Material in Australian Estuaries. *Estuaries,* **21**, pp. 540–51, 1998.

[57] Delvin, M.J. & Brodie, D., Terrestrial discharge into the Great Barrier Reef Lagoon: nutrient behavior in coastal waters. *Marine Pollution Bulletin,* **51**, pp. 1–22, 2005.

[58] Ogilvie, B.G. & Mitchell, S.F., Does sediment resuspension have persistent effects on phytoplankton? Experimental studies in three shallow lakes. *Freshwater Biology,* **40**, pp. 51–63, 1998.

[59] Cotner, J.B., Intense winter heterotrophic production stimulated by benthic resuspension. *Limnology and Oceanography,* **45**, pp. 1672–6, 2000.

[60] Alongi, D.M. & McKinon, A.D., The cycling and fate of terrestrially-derived sediments and nutrients in the coastal zone of the Great Barrier Reef shelf. *Marine Pollution Bulletin,* **51**, pp. 239–252, 2005.

[61] Furnas, M.J., Cyclonic disturbance and a phytoplankton bloom in a tropical shelf environment. In *Red Tides: Biology, Environmental Science and Toxicology*. ed. T. Okaichi, D.M. Anderson & Nemoto. T. Elsevier: Amsterdam, pp. 271–7, 1989.

[62] Gabric, A.J., Hoffenberg, P. & Boughton, W., Spatio-temporal variability in surface chlorophyll distribution in the central Great Barrier Reef as derived from CZCS satellite imagery. *Australian Journal of Marine and Freshwater Research,* **41**, pp. 313–24, 1990.

[63] Donnelly, T.H., Olley, J.M., Murray, A.S., Caitcheion, G.G., Olive, L.J. & Wallbrink, P.J., Phosphate Sources and Algal Blooms in Australian Catchments. *Downstream Effects of Land Use*. eds H.M. Hunter, A.G. Eyles, & G.E. Rayment. Department of Natural Resources, Queensland: Queensland, Australia, 1996.

[64] Ryther, J.H. & Dunstan, Nitrogen, W.M. Phosphorus, and Eutrophication in the Coastal Marine Environment. *Science,* **171**, pp. 1008–13, 1971.

[65] Mur, L.R., Skulberg, O.M. & Utliken, H., Cyanobacteria in the Environment. *Toxic Cyanobacteria in Water: A Guide to their Public Health Consequences, Monitoring and Management.* eds I. Chorus & J. Bartram. E, FN Spon, London: UK, pp. 15–40, 1999.

[66] Jorgensen, B.B. & Sorensen, J., Seasonal cycles of O_2, NO^{-3} and SO_2^{-4} reduction in estuarine sediments: the significance of an NO^{-3} reduction maximum in spring. *Marine Ecology Progress Series,* **24**, pp. 65–74, 1985.

[67] Russ, G.R. & McCook, L.J., Potential effects of a cyclone on benthic algal production and yield to grazers on coral reefs across the central Great Barrier Reef. *Journal of Experimental Marine Biology and Ecology,* **235**, pp. 237–54, 1999.

[68] Trimmer, M., Nedwell, D.B., Sivyer, D.B. & Malcolm, S.J., Seasonal organic mineralisation and denitrification in intertidal sediments and their relationship to the abundance of Enteromorpha sp and Ulva sp. *Marine Ecology-Progress Series,* **203**, pp. 67–80, 2000.

[69] Miyajima, T., Suzumura, M., Umezawa, Y. & Koike, I., Microbiological nitrogen transformation in carbonate sediments of a coral-reef lagoon and associated seagrass beds. *Marine Ecology-Progress Series,* **217**, pp. 273–86, 2001.

[70] McKenzie, L.J., Roder, C.A., Roelofs, A.J. & J.L.L.W., Post-flood monitoring of seagrass in Hervey Bay and the Great Sandy Strait, 1999: Implications for dugong, turtle and fisheries management. Department of Primary Industries: Cairns, Australia, 2000.

[71] Abal, E. & Dennison, W.C., Seagrass depth range and water quality in southern Moreton Bay, Queensland, Australia. *Marine and Freshwater Research,* **47**, pp. 763–71, 1996.

[72] Turner, R.E. & Rabalais, N.N., Coastal Eutrophication near the Mississippi River Delta. *Nature,* **368**, pp. 619–621, 1994.

[73] Justic, D., Rabalais, N.N., Turner, R.E. and Wiseman, W.J., Seasonal coupling between riverborne nutrients, net productivity and hypoxia. *Marine Pollution Bulletin,* **26**, 184–9. 1993.

[74] Dennison, W.C., Orth, R.J., Moore, A., Court-Stevenson, J., Carter, V., Kollar, S., Bergstrom, P.W. & Batiuk, R.A., Assessing water quality with submerged aquatic vegetation; habitat requirements as barometers of Chesapeake Bay health. *BioScience,* **43**, pp. 86–94, 1993.

[75] Furnas, M.J., Catchments and Corals: Terrestrial Runoff to the Great Barrier Reef. Australian Institute of Marine Science, Townsville. 2003.

[76] Brodie, J., McKergow, L.A., Prosser, I.P., Furnas, M., Hughes, A.O. & Hunter, H., Sources of sediment and nutrient exports to the Great Barrier Reef World Heritage Area. Report No. 03/11, Australian Centre for Tropical Freshwater Research, James Cook University, Townsville, 2004.

[77] Fabricius, K.E. & Death, G., Identifying ecological change and its causes: a case study on coral reefs. *Ecological Applications,* **14**, pp. 1448–1465. 2004.

[78] Dennison, W.C. & Abal, E.G., *Moreton Bay Study: A Scientific Basis for the Healthy Waterways Campaign*. South East Queensland Regional Water Quality Management Strategy Team: Brisbane, Australia, 1999.

[79] O'Donohue, M.J., Glibert, P.M. & W.C., D., Utilization of nitrogen and carbon by phytoplankton in Moreton Bay, Australia. *Marine and Freshwater Research,* **51**, pp. 703–12, 2000.

[80] Udy, J., personal communication, 2001.

[81] Harris, G.P., Biogeochemistry of nitrogen and phosphorus in Australian catchments, rivers and estuaries: effects of land use and flow regulation and comparisons with global patterns. *Marine and Freshwater Research,* **52(1),** pp. 139–149, 2006.

[82] Prove, B.G. & Hicks, W.S., Soil and sediment movements from rural lands of North Queensland. *Proc of a workshop on land use patterns and nutrient loading of the Great Barrier Reef Region.* ed. D. Yellowlees. James Cook University: Townsville, Australia, pp. 67–76. 1991.

[83] Sims, J.T., Edwards, A.C., Schoumans, O.F. & Simard, R.R., Integrating soil phosphorus testing into environmentally based agricultural Management practices. *Journal of Environmental Quality,* **29**, pp. 60–71, 2000.

[84] Withers, P.J.A., Davidson, I.A. & Foy, R.H., Prospects for controlling non-point phosphorus loss to water: A UK perspective. *Journal of Environmental Quality,* **291**, pp. 67–175, 2000.

[85] Heathwaite, L., Sharpley, A. & Gburek, W., A conceptual approach for integrating phosphorus and nitrogen management at watershed scales. *Journal of Environmental Quality,* **29**, pp. 158–66, 2000.

[86] Lu, H., Gallant, J., Prosser, I., Moran, C. & Priestley, G., Prediction of sheet and rill erosion over the Australian continent, incorporating monthly soil loss distribution. CSIRO Land and Water, Technical Report 13/01. Australia, 2001.

[87] Norrish, K. & Rosser, H., Mineral Phosphate. *Soils: an Australian viewpoint.* Division of Soils, CSIRO. ed. CSIRO Academic Press: Melbourne, Australia, pp. 335–361, 1983.

[88] National Land and Water Resources Audit, Australian Agriculture Assessment 2001, Vol. 1. Canberra, Australia, 1991.

[89] McComb, A.J., Qiu, S., Lukatelich, R.J. & McAuliffe, T.F., Spatial and temporal heterogeneity of sediment phosphorus in the Peel Harvey Estuarine System. *Estuarine Coastal and Shelf Science,* **47,** pp. 561–577, 1998.

CHAPTER 3

Sediment tracing techniques and their application to coastal watersheds

A. Kimoto, A. Fares & V. Polyakov
Department of Natural Resources and Environmental Management, CTAHR, University of Hawai'i at Mānoa, USA.

Abstract

In recent years there has been an increasing interest in identifying sediment sources and tracking sediment movement in watersheds. Sediment tracing allows for monitoring of sediment movement and obtaining spatially distributed information on erosion and deposition rates based on tracer inventories. The sediment tracing techniques dicussed in this chapter includes artificial and cosmogenic radionuclides, exotic particles, fingerprinting, and rare earth elements. Definitions, principles, case studies, advantages and limitations of particular tracers are presented. The chaptzer also gives an overview of application of sediment tracing approaches in coastal areas. Sediment tracing techniques have great advantages over traditional methods of erosion inventory. Further work is needed to identify and refine sediment tracing techniques in order to address sediment-related problems and to develop strategies for watershed management in coastal areas.

1 Introduction

Non-point inputs are the major source of water pollution in the USA. Half of the approximately 5 billion tons of soil eroded every year in the United States reaches streams [1]. This sediment is an important vehicle for transport of chemicals, such as nitrogen, phosphorus and pesticides, into waterways and estuaries [2]. Sediment and associated nutrients and pollutants cause various environmental problems in coastal areas among them are decline of water quality, increase of turbidity, and damage to marine and freshwater ecosystems including fish, sea grasses, and coral reefs [3]. Conservation efforts require large investments, thus there is a need for technology that is capable of correctly estimating erosion rates and precisely identifying problem areas within. Solutions to the problem of non-point source pollution must invariably involve effective modeling and techniques to monitor sediment movement.

There are increasing efforts to develop innovative ways to identify sediment sources and monitor sediment movement in watersheds. One of these is the tagging of sediment with tracers. Assuming that subsequent redistribution of the tracer is due to sediment translocation, this technique allows spatial identification of sediment sources and sinks based on tracer inventories. Sediment sources are locations on the watershed undergoing various types of erosion (e.g. soil erosion, bank collapse etc.). Sediment sinks are locations within or outside the watershed where the sediment is deposited. Tracking of transport process and erosion measurement is also possible by collecting and measurement of suspended sediment samples in streams.

The ideal use of tracers for monitoring soil translocation should be based on the following assumptions [4]: 1) the local distribution of the tracer is uniform (or not uniform, but known), 2) the tracer strongly binds with soil, 3) subsequent redistribution is due to sediment movement, and 4) estimated erosion rates can be derived from tracer inventories.

Tracers have been successfully utilized in the natural sciences to study transport processes and the fate of various pollutants. The set of properties needed for an ideal tracer were defined by Zhang *et al.* [5] as the following: strong binding with soil, sensitivity to analysis, ease of measurement, low background concentration in soil, lack of interference with soil movement, chemical stability and low plant uptake, environmental safety, and availability of a suit of tracers with similar properties. There are a number of substances that meet these requirements. They can be classified as naturally occurring and artificial. Artificial tracers are either introduced into the soil as a result of fallout from atmospheric nuclear weapons rests (e.g. ^{137}Cs) [6] or deliberately, either by tagging soil particles with trace elements (e.g. ^{59}Fe) [7] or by incorporating trace particles (e.g. magnetic or glass beads, lanthanide oxides) into the soil body [8, 9]. Alternatively, various soil chemical and physical properties, such as particle shape and color [10], overall grain-distribution of sediments [11], mineral magnetism [12–14], and the combination of several physical and chemical properties of sediments [15–19] have been used for sediment fingerprinting.

This chapter introduces various sediment tracing techniques and gives an analysis of their advantages and drawbacks. It includes information on artificial and cosmogenic radionuclide methods, exotic particle methods, fingerprinting technique, and rare earth element method. The chapter also summarizes previous studies that used sediment tracing techniques in coastal areas to address sediment and sediment-related problems.

2 Sediment tracing techniques

2.1 Radionuclides

2.1.1 Artificial radionuclide

Artificial radionuclides are the most widely used type of sediment tracers today. For nearly half a century, scientists utilized them to measure soil erosion and deposition on landscapes quickly and accurately. The popularity of radioactive tracers is due to their relative ease of detection and rapid and strong association with soil particles. Although biological and chemical processes can move small amounts of these tracers through various biogeochemical cycles, erosion and sedimentation are the dominant forces that transfer them through the landscape.

Of all sediment tracers, cesium-137 (^{137}Cs) has been studied most extensively [4, 6, 20]. It is an artificial radionuclide produced by atmospheric tests of atomic weapons during the mid 1950s to early 1970s and distributed globally, primarily in the northern hemisphere. It has a half-life of 30.1 years, which makes it suitable for medium term estimates of erosion, and it is easily detectable through gamma-ray emission. Deposition of ^{137}Cs on the ground is predominantly dependent on the weather conditions and the pattern of rainfall [21]. The tracing technique is based on two assumptions: firstly, that ^{137}Cs is strongly and rapidly adsorbed onto clay, first of all illite and bentonite [22]; and secondly, that it is essentially non-exchangeable [23]. The concentration of ^{137}Cs exponentially declines with depth [24]. Because vegetation retention and uptake of ^{137}Cs is low [25], its redistribution is mostly attributed to soil movement.

Erosion and deposition are calculated by conversion models through comparison of: 1) loss or gain of ^{137}Cs inventories at a sampling location relative to a reference inventory at an undisturbed site or 2) comparison of ^{137}Cs inventories at the same location at different points in time. The most commonly used models are the empirical relationship, the simple linear relationship, and the proportional method [26].

The empirical approach uses a third order polynomial equation [27] or logarithmic relationship [28] between soil loss and ^{137}Cs loss of the form:

$$SL = Kx^b \tag{1}$$

where SL is the soil loss, x is the percentage reduction of total soil ^{137}Cs, K and b are coefficients. These relationships are site specific.

The simple linear relationship [29] is in the form:

$$SL = \rho h(C_0 - C_i) / C_0 \tag{2}$$

where ρ is the bulk density of the soil, h is the thickness of the layer of soil in which the tracer is present, C_0 and C_i are the concentrations of the tracer on an undisturbed site and eroded site, respectively. This linear relationship is based on the assumption of uniform distribution of tracer in a tillage layer.

More complex relationships [30] predict the amount of ^{137}Cs in the soil as a function of time and erosion rate. They also account for deposition, decay, tillage dilution, erosional transport, and seasonal fluctuation in ^{137}Cs deposition and erosion rates.

Kachanoski [26] used a proportional method to estimate erosion rate.

$$E = M_s(C_i - C_f) - (t_i - t_f)C_i \tag{3}$$

where E is predicted erosion rate, M_s is specific mass of the plough layer, t_i and t_f is time of initial and final soil ^{137}Cs sampling, respectively, C_i and C_f is total soil ^{137}Cs corrected for radioactive decay at time t_i and t_f, respectively.

Govers *et al.* [31] evaluated a two-component soil erosion model using ^{137}Cs measurements from two, complex-shaped fields in the UK. The model consisted of a component for water erosion and a component for tillage erosion. Tillage erosion was modeled as a diffusive process, and water erosion was modeled as an overland flow process without a diffusion component. They found that the model results best aligned with the measured spatial data when both water erosion and tillage erosion were included. In their discussions of the results, the authors speculated that the diffusive processes associated with water erosion, which they considered to be limited to splash, were negligible. Other researchers have also used the ^{137}Cs technique for evaluating different erosion models [32–37]. For example Ritchie *et al.* [32] compared ^{137}Cs data and model prediction from the USLE (Universal Soil Loss Equation). Quine [37] used ^{137}Cs-derived erosion rate in model validation for getter understanding of tillage erosion.

The ^{137}Cs technique has been extensively used to study soil erosion and sediment redistribution in agricultural lands [6, 38–41], forested lands [42, 43], and rangelands [44, 45].

The advantage of the ^{137}Cs technique is the ability to provide spatially-distributed information on soil redistribution, without long-term monitoring of soil movement [46]. Thus it has been used to validate physically-based distributed erosion models [31, 37, 47]. Also, because the rates of erosion and deposition obtained by the ^{137}Cs technique reflect the integration of past soil movements within a watershed, this information is useful for evaluating the stability of soil resources.

However, there are limitations of the ^{137}Cs technique. Cesium-137 fallout occurred in a complex temporal pattern and its non-uniform distribution is related to rainfall, topography, soil density, and infiltration [4]. This technique is unsuitable in acid organic soils [48, 49] where ^{137}Cs ions could become mobile [22]. Preferential adsorption of ^{137}Cs on fine clay particles makes it prone to sorting during transport by water. Bernard *et al.* [50] reported the enrichment ratio (enrichment of ^{137}Cs levels in sediments compared with the soil in the original location) for this tracer varying between 1.8 and 3.0. Enrichment occurs due to preferential transport of fine particles such as clay, which have greater affinity for ^{137}Cs, during sediment transport processes [51]. This technique is also unable to provide short-term estimates of erosion and deposition rates, because the redistribution of ^{137}Cs gives time-integrated decadal scale average erosion and deposition rates. Furthermore high coefficient of variations (up to 40%) for cesium inventories [52] often call for intensive sampling in reference locations in order to avoid a significant bias in estimation of sediment redistribution [53, 54].

Several sediment labeling techniques with deliberate introduction of radionuclides that are able to preserve the properties of the material under study have been developed. Soil particles were labeled with various radioisotopes (^{46}Sc, ^{110}Ag, ^{32}P), mixed with natural soil, and then placed on hillsides [55, 56] to study soil and sediment movement. Those tracers are retained in silt and clay fractions. The radioactive ion ^{59}Fe in solution can be applied directly to soil [7], while ^{60}Co requires to be tagged with soil aggregates before applying to natural soil [57].

While there are a lot of advantages of using radionuclide as a sediment tracer, radiological risk is a major concern of using the technique. To eliminate this drawack, a neutron activation method can be used that employs various stable isotopes. In order to be detected, these isotopes are activated by neutron irradiation [58]. The technique is based on the fact that stable isotopes can incorporate addiional neutrons into their nuclei, resulting in radioactive nuclides that emit beta- or gamma-radiation. Irradiation is conducted in the laboratory after sample collection, posing less threat to the environment. Noble metals (e.g. Au, Ag, In, Ir) are comonly used because they are only present in trace amounts in the natural envionment. Wheatcroft *et al.* [59] and Olmes and Pink [60] employed thermal diffusion to introduce Au and Ag into the crystalline lattice of clay minerals with subsequent leaching of the excessive tracer. Aggregate size of sediments was not affected by these procedures, thus minimizing the error usually associated with tracer enrichent. While excessive heating can dramatically alter soil chemical composiion, especially clay and organic matter fractions. This method is also relatively complex and costly if a large quantity of tracers is needed.

2.1.2 Cosmogenic radionuclides

More commonly used to study bioturbation, stratigraphy, and sedimentation dating, naturally occurring radionuclides have been incorporated into soil erosion studies. Cosmogenic radionuclides have a less complex input function than their bomb-produced counterparts; hence, the assumption that there exists a constant and uninterrupted supply of these tracers to the soil surface is more valid. Cosmogenic radionuclides are especially useful and often used in combination with artificial ones.

Beryllium-7 and 10 (^{7}Be and ^{10}Be) are cosmogenic radionuclides with a half-life of 53 days and 1.5 10^{6} years, respectively [61]. They are derived from the spallation of oxygen and nitrogen atoms in the upper atmosphere [62] and reach the Earth's surface via precipitation or dry deposition. The fallout has pronounced seasonal pattern due to different degrees of mixing with the troposphere. Beryllium has high affinity to clays and large spatial variability because of interception by vegetation. In general, beryllium does not penetrate far into the soil profile [63]. Wallbrink (unpublished manuscript, 1989) reported mean soil penetration depth at a Canberra site ranging between 0.7 and 10 mm. Beryllium has been used as a stand alone tracer [63] or in conjunction with ^{137}Cs [64] whose typical penetration depth is 10–20 mm from soil surface. The different penetration depths of two nuclides is used to determine the origins of sediments and to identify possible dominant erosion processes. Wallbrink and Murray [63] provided a model to determine dominant erosion processes using the concentration of ^{137}Cs and ^{7}Be. High concentrations of both nuclides in sediments indicate that sheet erosion is the dominant process, while low concentrations of both nuclides suggest that eroded materials have not been subjected to fallout, such as a gully collapse. High ^{137}Cs concentration with low ^{7}Be concentration indicates rill or deep sheet erosion has occurred. Conversely, low ^{137}Cs concentration with high ^{7}Be suggests that eroded material has originated from the site where ^{137}Cs has been lost or never accumulated, but is exposed to ^{7}Be, such as a gully floor [62, 63]. Bonniwell *et al.* [65] used atmospherically-delivered Be-7, Pb-210, and Cs-137 accumulated in the snowpack to trace suspended sediment in a mountain stream. The authors were able to determine the sediment total flux, residence times, and erosion rates. ^{10}Be, a radioactive nuclide (half-life 1.5 million years) is especially suitable for long-term estimates due to its longer half-life [66] compared to ^{7}Be. Thus, Heimsath *et al.* [67] used in situ produced cosmogenic nuclide concentrations of ^{10}Be and ^{26}Al to determine Late Quaternary rates of apparent soil production in highlands of Australia.

Lead-210 (^{210}Pb, half-life 22 years) is also used in conjunction with ^{137}Cs and Be^{7} to monitor erosion [65]. ^{210}Pb originates from the decay of ^{222}Rn in-situ and in the atmosphere. The latter is continually precipitated to the soil surface by rainfall. ^{210}Pb that is derived in-situ must be corrected for in erosion calculations. The typical penetration depth of ^{210}Pb is from 100 to 400 mm [68, 69]. It has been used to determine sediment accumulation rates in marine and freshwater sediments [70]. Results showed a good fit to a steady-state accumulation model.

While radionuclide tracers have been used successfully worldwide, they have several drawbacks. Both artificial and some cosmogenic redionuclides have complex input patterns that reduce the accuracy of tracing methods. Widely distributed in the atmosphere and soil, they cannot be used (with the exception of deliberately introduced non-fallout isotopes) to study point-source sedimentation. Further, cosmogenic radionuclides cannot be manipulated to tag a specific component (particle fraction, etc.) of the studied sediment [60].

3 Exotic particles

To eliminate some shortcomings of radionuclide tracers, various exotic particles have been developed to study sediment movement. Unlike radionuclide tracers, which tag existing soil aggregates, exotic particles are foreign objects introduced into a soil mass to mimic soil particles. However, they must have distinctive properties that make them easily identifiable in displaced sediment by optical, fluorometric, magnetic, or other physical methods. Exotic particles are often used for small-scale studies and quick identification of localized sediment sources.

Fluorescent dyes, incorporated into 44 to 2,000 μm diameter glass beads that can be ground to resemble soil particles of different sizes [71], have been used under simulated rainfall on 4×10 m plots. This technique can detect soil movement, including splash movement and runoff. Fluorescent particles can be detected visually either on-site or in the laboratory under ultraviolet light, or by using fluorometry, although the process is tedious and time consuming. Wheatcroft [72] used plastic beads to study horizontal mixing due to bioturbation. Plante *et al.* [73] successfully used ceramic pills with dysprosium for tracing sediment. The ceramic pills simulate soil aggregates better than glass or plastic beads because of their density. Parsons *et al.* [13] used magnetite for tracing sediment movement in interrill overland flow. Movement of the magnetite was monitored on an experimental plot during three artificial rainfalls. They showed the feasibility of magnitite for studying sediment movement in overland flow; however, they pointed out that the higher density of magnetite to natural soil could affect its detachment and transport. To overcome this problem, Ventura *et al.* [8, 74] used a magnetic tracer consisting of polystyrene plastic beads embedded with a powder magnetite. These authors summarized major advantages of this tracer technique as follows: 1) the technique requires no destructive sampling for measurements; 2) magnetometer readings can be taken from soil surface in a completely non-destructive manner; 3) the technique is extremely simple, quick, and inexpensive. Similarly, Borselli and Torri [75] used steel nuts to allow quick detection of soil movement with a magnetometer in the field without lengthy laboratory analysis. However, the latter method is only applicable for translocation of soil by tillage operations and on limited areas. Fly ash, a product of fossil fuel burning, primarily coal, consisting of siliceous glass with less than 20 μm particle size, was found to be useful in soil loss determination [76, 77]. When comparing its amount in the soil profile at different landscape positions on a cultivated field and reforested site, the authors found deposition rich in fly ash on the footslope, indicating net sediment gain at that point.

However, according to Zhang *et al.* [78], the major problems with using exotic particles as tracers are: i) exotic particles may not bind with soil particles or soil aggregates and therefore are transported separately, ii) they may differ in size distribution, particle density, shape, surface morphology, and surface chemical properties from soil particles, and iii) a large quantity of tracers are needed to study soil and sediment movement on hillslopes and in watersheds.

4 Fingerprinting

Fingerprinting techniques do not rely on foreign particles or chemical elements, but instead use unique soil properties to trace sediment. This approach has been widely used in geology, sedimentology, and stratigraphy and is now applied to identify and quantify the sources of sediment in watersheds.

The techniques include two basic steps: 1) select physical or chemical properties which clearly differentiate potential sediment sources; 2) compare the measurements of the same property obtained from suspended sediment with the corresponding values for potential sources to identify the sources of suspended sediment [79, 80]. Color [81], clay mineralogy [82, 83] mineral magnetic properties [84–86], and radionuclide activity [87] have been used for fingerprint properties.

Many early fingerprinting studies utilized single component signatures as a marker of sediment sources [81–83]. The major limitation of these approaches is their inability to quantify the relative contribution of multiple sediment sources [85].

Quantitative composite fingerprinting approaches use several chemical and physical properties in combination. Shankar *et al.* [88] used chemical, physical, and isotopic properties and an unmixing algorithm to differentiate sediment sources in a watershed and to calculate the relative source contributions to stream bedload. A composite fingerprinting approach, which incorporates statistically verified multi-component signatures and a multivariate sediment-mixing model, has been developed and utilized, primarily in the UK [79, 89–91].

Several physical and chemical properties, such as trace metals, heavy metals, base cations, other organic and inorganic materials, and absolute particle size are used for composite fingerprinting. Selection of fingerprint properties may be based on available analytical equipment, available data previously collected in watersheds, or financial restriction, but the optimum composite fingerprint comprises properties selected from different groups, such as physical, chemical or mineralogical [92, 79]. The approach is capable of determining the spatial origin of suspended sediment in a large river basin and providing quantitative estimates of the relative contribution of each origin to sediment load [93]. Composite fingerprinting also increases the reliability of results and allows discrimination of a greater range of potential sources that are more representative of the source material mixtures [94].

Erosion or deposition of soil can be identified from its magnetic susceptibility, which is the magnetizing ratio of material to the magnetic field inducing it [76]. Magnetic susceptibility is a function of topography, climate, parent material and time [95]. For example, magnetic susceptibility of organic matter and quartz is low; clays with transitional elements in their structure are weakly positive, magnetite and maghemite are strongly positive. Eroded soil and deposited sediment having different composition possess different magnetic susceptibility. For example, eroded areas on summits and slopes tend to have higher magnetic susceptibility than depositional areas in landscape depressions.

While fingerprinting techniques have largely contributed to information on suspended sediment sources and bedload, they have certain limitations. As pointed out by Collins *et al.* [93], fingerprinting techniques are best suited to heterogeneous basins where contrasting geological types contribute sediment characterized by distinctive composite signatures. In other words, the techniques require heterogeneous soil properties within a watershed. This constraint usually limits the application of the method to relatively large basins.

5 Rare earth elements

The lanthanides or rare earth element (REE) group consists of 15 elements with periodic number 57 through 71, which have similar chemical properties. REEs' trivalent state and ionic radiuses ranging between 0.861Å (Lu^{3+}) and 1.03Å (La^{3+}), similar to that of Ca^{2+}, allow them to be easily adsorbed to clay [96]. They occur in a variety of minerals, such as monazite, apatite, and titanite. REEs are found in many soils in concentrations of up to tens of parts per million [97] but usually in several parts per billion, with organic soils generally having higher concentrations than mineral soils. REEs are practically insoluble in water, however the solubility tends to increase with decreasing pH, a key factor in their mobility in soil [98]. REE compounds have low toxicity and often are accumulated by plants [99], although the uptake is too low to noticeably change concentration of REE in soil.

REEs are often used in different areas of biological, environmental and earth sciences that deal with mass transport such as sediment transport. For example, Krezoski [100] used REE oxides to trace lateral transport of sediments in Lake Superior. Similarly, Mahler *et al.* [96] successfully used lanthanide-labeled clay to characterize mass transport in karst. Application of REE on agricultural soils recently received increased attention. They have been used as multiple tracers (i.e. a group of tracers used simultaneously) both in the laboratory under simulated rainfall [78, 101] and in the field [9, 102–107].

Zhang *et al.* [5] tested the feasibility of using REE oxides as a sediment tracer by examining their binding ability with soil materials. They reported that the REEs were nearly uniformly incorporated into different sizes of soil aggregates for loess-derived silt loam soils, and leaching of the tracers was not observed. Direct mixing of these tracers with soil did not substantially affect physical properties of soil aggregates. Kimoto *et al.* [107] also reported little leaching of the REEs and their strong binding capability with the gravelly, sandy loam soils. Most importantly, since the REEs constitute a suite of 14 different elements, they allow individual labeling of various areas of a study site or a watershed. This permits identifying the sources of sediment, evaluating the relative contribution of various watershed parts to total soil loss, and tracing the fate of eroded material. Low detection limits and relatively precise measurement of REEs with inductively coupled plasma mass spectrometry (ICP-MS) adds to the accuracy of the estimation of sediment movement. These properties make REEs suitable as sediment tracers.

Polyakov and Nearing [101] used five REE oxides applied in bands across the slope in a 4×4 m laboratory plot with a silt loam soil. Depending on the slope location, the authors reported 4 to 40% relative error of soil loss estimation by REE method compared with the direct measurement of surface by laser scanner. Polyakov *et al.* [9] in a field study in Ohio, USA, divided a small watershed into six morphological units, each tagged with one of six rare earth element oxides. Using runoff and spatial surface samples it was possible for the first time to itemize the sediment budget for landscape elements into three components: 1) the soil from the element that left the watershed with runoff, 2) soil from the element that was re-deposited on lower positions, with the spatial distribution of that deposition, and 3) soil originating from the upper positions and deposited on the element, with quantification of relative source areas.

A possible limitation of REE technique is the relatively high cost of materials and sample analysis using ICP-MS, which may preclude the application of this technique on a large scale. Another limitation is non-uniform binding of REEs for coarse-textured soils. Kimoto *et al.* [107] examined the applicability of REEs as a sediment tracer for gravelly, sandy loam soils, and showed that the REEs preferentially bound with small size classes of soil particles, which resulted in overestimation of the soil loss. However, the non-uniform binding of the tracers does not preclude their use on coarse-textured soils, as sediment sorting can be quantified and taken into consideration. Kimoto *et al.* [106], analyzing the data of a nearly four-year field experiment on a small agricultural watershed near Coshocton, Ohio, USA, reported that the REE technique had a reasonable potential for studying sediment sources for extended period of time. However, there are two potential sources of error associated with it. One is the selective depletion of a tracer from areas where concentrated flow causes soil loss below the depth of tracer incorporation. This results in underestimation of the proportion of sediment from that area. The other potential source of error is the contamination of downslope areas with tagged sediments from upslope areas. The REE technique is not capable of differentiating between sediment directly transported from its original position and sediment that come from places of re-deposition. The adverse effect of contamination, however, should be offset over time by the continuous process of re-deposition and re-entrainment of soil particles, which maintain a quasi-equilibrium state [9]. While the cumulative amount of soil reaching the outlet increases with every storm event, the amount of sediment in transition (temporarily re-deposited) should remain relatively constant.

The REE tagging method offers a powerful and practical way to trace the movement of soil under erosive forces. It allows one to identify sources and sinks of sediment and itemize the sediment budget of any given location by its major components: sediment loss, deposition from upslope locations, and re-deposition on downslope locations. Additional investigation is needed to determine how soil properties, rainfall characteristics and tracer placement relative to the outlet influence the tracer enrichment ratio.

6 Application of sediment tracing techniques to coastal areas

Limited studies have been conducted to identify sediment sources or to trace sediment movements in coastal watersheds. Hill *et al.* [108] measured concentrations of aerosolic quartz and ^{137}Cs to estimate hillslope erosion during two years of a highway construction project on the island of Oahu, Hawaii. It was reported that the erosion rates of the hillslope during the highway construction ranged from 0.1 to 0.3 mm yr^{-1}, which was within the range of denudation rates estimated previously for drainage basins of Oahu. The authors concluded that the quartz aerosol was useful for identification of fluvial sediment loads in similar semi-tropical drainage basins. In addition, Hill *et al.* [109] used a sediment fingerprinting technique to qualitatively evaluate compensating errors in an annual fine sediment budget and to determine the sources of fluvial sediment transported from the valley during highway construction. They measured concentrations of two aerosols, aeolian quartz and ^{137}Cs in sediment sources and fluvial sediments. The reported differences between sediment-budget and aerosol-budget imbalances were relatively small (25%). They also showed that most fine sediment was produced by erosion of channel margins and attrition of coarse particles, and the contribution of hillslope erosion was relatively small.

Many sediment tracing studies conducted in coastal areas have focused on sediment accumulation rate in seas. For example, Callaway *et al.* [110] used the ^{137}Cs method to assess vertical sediment accretion rate in the North and Baltic Seas. They reported that vertical accretion rates ranged from 0.26 to 0.85 cm year^{-1} during the period of 1963 to 1986, while they ranged from 0.30 to 1.90 cm yr^{-1} during the period of 1986 to 1991. The authors concluded that the ^{137}Cs from the 1986 Chernobyl nuclear accident could be successfully used as a sediment marker to study marine sedimentation processes. Cundy and Croudace [111] measured the vertical distributions of a series of radionuclides (^{210}Pb, ^{137}Cs, ^{238}Pu, $^{239,240}Pu$, ^{241}Am, and ^{60}Co) to estimate the rate of sediment accumulation in Southern England estuaries. They indicated that the rates calculated using the radionuclides showed reasonably good agreement and were corroborated by dating using the natural radionuclide ^{210}Pb. They pointed out that a number of chemically different or inert radionuclides should be used to eliminate possible bias in calculated sediment accumulation rates. Huh and Su [112] examined the profiles and inventories of ^{210}Pb, ^{137}Cs and $^{239,240}Pu$ in the East China Sea and assessed rates of sediment accumulation and mixing, the pathways of sediment transport, and budgets of these nuclides and sediments in this marginal sea. They showed that the sedimentation rates calculated by ^{137}Cs method varied from 0.1 to 2 cm yr^{-1} and generally decreased southward from the Yangtze River estuary and eastward toward the open ocean. They noted that this pattern was consistent with the point of input and expected pathways of sediment transport by the current and tidal systems. Hong *et al.* [113] estimated sediment accumulation rates in the southwestern part of the Sea of Japan by measuring the excess ^{210}Pb activity profiles. They reported that the sediment accumulation rate ranged from 0.02 to 0.2 cm yr^{-1} and pointed out that the magnitude and geographical trends of ^{210}Pb-derived sediment accumulation rates were consistent with long-term sedimentation rates determined by dated volcanic ashes and seismic stratigraphy.

An integrated approach that includes monitoring and modeling is vital for successful management of coastal ecosystems. One of the main components of this approach is sediment tracing, which is becoming an increasingly popular tool for environmental monitoring. The unique feature of coastal areas is the presence of dry land to water body interface. Various tracing techniques need to be adapted to find broader application in coastal watersheds to provide effective and accurate estimates of erosion and sedimentation rates.

7 Conclusion

Sediment and sediment-related pollution is a serious problem in coastal areas. Sediment causes inundation of estuaries, decline of water quality, and damage to coral reefs and marine and freshwater ecosystems. Sediment tracing techniques offer a powerful and practical way to identify sources and sinks of sediment, track sediment movements, and evaluate spatial distribution of erosion and deposition rates. Soil tracers have been used successfully on multiple scales from uniform slope in the laboratory to large complex watersheds.

Radionuclides (^{137}Cs, ^{10}Be, ^{7}Be, ^{46}Sc, ^{110}Ag, ^{32}P, etc.) are the most widely used type of sediment tracers today. The popularity of radioactive tracers is due to their relative ease of detection and good association with soil particles. However, most (except Be^{7}) of the radionuclide methods cannot provide short-term estimates of erosion and deposition rates.

Exotic particles are foreign objects introduced into the soil mass to mimic soil particles. They have distinctive properties, which make them easily identifiable in displaced sediment by optical, fluorometric, magnetic, or other physical methods. Exotic particles became popular for small-scale studies and quick identification of localized sediment sources. However, their difference from natural soil particles in grain-size distribution, density, and surface morphology, could result in different transport properties that limit their application.

Fingerprinting techniques use a native soil's own unique properties to trace sediment. This approach has been widely used in geology, sedimentology, and stratigraphy. It is now applied to identify and quantify the sources of sediment in watersheds. Fingerprinting techniques are applicable only in watersheds with heterogeneous physical and chemical soil properties.

Sediment tracing by REEs is a relatively new method that offers a powerful and practical way to monitor the movement of soil under erosive forces. A set of REEs incorporated into soil in different watershed locations allows the identification of sources and sinks of sediment. An itemized sediment budget is possible for any given location based on its major components: sediment loss, deposition from upslope locations, and re-deposition on downslope locations. The major limitations of the REE technique are the high cost of materials and sample analysis, and possible tracer enrichment.

Properties needed for a successful tracer are: low background concentrations, uniform distribution in soil, sensitivity to analysis, ease of measurement, strong binding with soil, and absence of interference with soil movement. Sediment tracer techniques have a number of advantages over traditional methods of erosion measurement. They allow for tracking the path of non-point source pollutants beyond dry land and into water bodies, a unique quality that no other soil erosion measurement method can offer. Although a tracer is recognized as a powerful tool to study sediment movements, the application of sediment tracers to coastal watersheds has been limited. More research is needed to identify and refine sediment tracing techniques that are the most suitable for addressing sediment-related problems and for developing strategies for watershed management in coastal areas.

References

[1] USDA. The second RCA appraisal: analysis of conditions and trends. U.S. Govt. Printing Office: Washington, DC, 1989.

[2] Barling, R.D. & Moore, I.D., 1994. Role of buffer strips in management of waterway pollution – a review. *Environmental Management*, **18**, pp. 543–558.

[3] Neil, D.T., Orpin, A.R., Ridd, E.V. & Yu, B.F., Sediment yield and impacts from river catchments to the Great Barrier Reef lagoon. *Marine and Freshwater Research*, **53**, pp. 733–752, 2002.

[4] Di Stefano, C., Ferro, V. & Porto, P., Linking sediment yield and Cesium-137 spatial distribution at basin scale. *J. Agric. Eng. Res.*, **74**, pp. 41–62, 1999.

[5] Zhang, X.C., Fredrich, J.M., Nearing, M.A. & Norton, L.D., Potential use of Rare Earth oxides as tracers for soil erosion and aggregation studies. *Soil Sci. Soc. Am. J.*, **65**, pp. 1508–1515, 2001.

[6] Ritchie, J.C. & McHenry, J.R., Application of radioactive fallout Cesium-137 for measuring soil erosion and sediment accumulation rates and patterns: a review. *J. Environ. Qual.*, **19**, pp. 215–233, 1990.

[7] Wooldridge, D.D., Tracing soil particle movement with Fe-59. *Soil Sci. Soc. Am. Proc.*, **29**, pp. 469–472, 1965.

[8] Ventura, E., Nearing, M.A. & Norton, L.D., Developing a magnetic tracer to study soil erosion. *Catena*, **43**, pp. 277–291, 2001.

[9] Polyakov, V.O., Nearing, M.A. & Shipitalo, M., Tracking sediment redistribution in a small watershed: Implications for agro-landscape evolution. *Earth Surface Processes and Landforms*, **29**, pp. 1275–1291, 2004.

[10] Krein, A., Petticrew, E. & Udelhoven, T., The use of fine sediment fractal dimensions and colour to determine sediment sources in a small watershed. *Catena*, **53**, pp. 165–179, 2003.

[11] Kurashige, Y. & Fusejima, Y., Source identification of suspended sediment from grain-size distributions: I. Application of nonparametric statistical tests. *Catena*, **31**, pp. 39–52, 1997.

[12] Caitcheon, G.G., The significance of various sediment magnetic mineral fractions for tracing sediment sources in Killimicat Creek. *Catena*, **32**, pp. 131–142, 1998.

[13] Parsons, A.J., Wainwright, J. & Abrahams, A.D., Tracing sediment movement in interrill overland flow on a semi-arid grassland hillslope using magnetic susceptibility. *Earth Surface Processes and Landforms*, **18**, pp. 721–732, 1993.

[14] Motha, J.A., Wallbrink, P.J., Hairsine, P.B. & Grayson, R.B., Unsealed roads as suspended sediment sources in an agricultural catchment in south-eastern Australia. *Journal of Hydrology*, **286**, pp. 1–18, 2004.

[15] Carter, J., Owens, P.N., Walling, D.E. & Leeks, G.J.L., Fingerprinting suspended sediment sources in a large urban river system. *The Science of the Total Environment*, **314–316**, pp. 513–534, 2003.

[16] Collins, A.L., Walling, D.E., Sichingabula, H.M., & Leeks, G.J.L., Suspended sediment source fingerprinting in a small tropical catchment and some management implications. *Applied Geography*, **21**, pp. 387–412, 2001.

[17] Collins, A.L. & Walling, D.E., Selecting fingerprint properties for discriminating potential suspended sediment sources in river basins. *Journal of Hydrology*, **261**, pp. 218–244, 2002.

[18] Krause, A.K., Franks, S.W., Kalma, J.D., Loughran, R.J. & Rowan, J.S., Multi-parameter fingerprinting of sediment deposition in a small gullied catchment in SE Australia. *Catena*, **53**, pp. 327–348, 2003.

[19] Walling, D.E., Owens, P.N. & Leeks, G.J.L., Fingerprinting suspended sediment sources in the catchment of the River Ouse, Yorkshire, UK. *Hydrological Processes*, **13**, pp. 955–975, 1999.

[20] Ritchie, J.C. & Ritchie, C.A., Bibliography of publications of ^{137}Cs studies related to erosion and sediment deposition. USDA-ARS, Hydrology and Remote Sensing Laboratory, Paper HRSL-2005-01.

[21] Sukhorukov, F.V., Gavshin, V.M., Malikova, I.N., Kovalev, S.I., Malikov, Y.I. & Romashkin, P.A., Cesium-137 in the environment of the Altay Region (Russia). *Water, Air, Soil Pollut.*, **118**, pp. 395–406, 2000.

[22] Maguire, S., Pulford, I.D., Cook, G.T. & Mackenzie, A.B., Caesium sorption-desorption in clay-humic acid system. *J. Soil Sci.*, **43**, pp. 689–696, 1992.

[23] Tamura, T. Consequences of activity release: selective sorption reactions of cesium with soil minerals. *Nuclear Safety*, **5**, pp. 262–268, 1964.

[24] Walling, D.E. & He, Q., Use of fallout ^{137}Cs in investigations of overbank sediment deposition on river floodplain. *Catena*, **29**, pp. 263–282, 1997.

[25] Garten, C.T., Jr., Gentry, J.B., Pinder, V., Sharitz, R.R. & Smith, M.H., Radioacecium dynamics in a contaminated flood plain ecosystem in the southeastern United States. *Impact of Nuclear Releases into the Aquatic Environment*, Int. Atomic Energy Agency, Vienna, pp. 331–347, 1975.

[26] Kachanoski, R.G., Comparison of measured soil 137-Cesium losses and erosion rate. *Canadian Journal of Soil Science*, **67**, pp. 199–203, 1987.

[27] Bajracharya, R.M., Lal, R. & Kimble, J.M., Use of radioactive fallout ^{137}Cs to estimate soil erosion on three farms in west central Ohio. *Soil Sci.*, **163**, pp. 133–142, 1998.

[28] Walling, D.E. & Quine, T.A., Calibration of Ceasium-137 measurements to provide quantitative erosion rate data. *Land Degrad. Rehab.*, **2**, pp. 161–175, 1990.

[29] De Jong, E., Begg, C.B.M. & Kachanoski, R.G., Estimates of soil erosion and deposition for some Saskatchewan soils. *Can. J. Soil Sci.*, **63**, pp. 607–617, 1983.

[30] Kachanoski, R.G. & DeJong, E., Predicting the temporal relationship between soil cesium-137 and erosion rate. *Journal of Environmental Quality*, **13**(2), pp. 301–304, 1984.

[31] Govers, G., Quine, T.A., Desmet, P.J. & Walling, D.E., The relative contribution of soil tillage and overland flow erosion to soil redistribution on agricultural land, *Earth Surface Processes and Landform*, **21**, pp. 929–946, 1996.

[32] Ritchie J.C., Spraeberry, J.A. & McHenry, J.R., Estimating soil erosion from the redistribution of fallout ^{137}Cs. *Soil Sci. Soc. Am. Proc.*, **38**, pp. 137–139, 1974.

[33] Rose, C.W., McCallan, M.E., O'Leary, B.M. & Sander, G., Testing a model of field-scale soil erosion and deposition on a plane land element using the fallout isotope ^{137}Cs. Reprint of Papers no. 80/7. *Proc. Conf. on Agricultural Engineering*, Geelong, The Institute of Engineers, Australia, 1980.

[34] De Roo, A.P.J. & Walling, D.E., Validating the ANSWERS soil erosion model using ^{137}Cs. ed. Rickson, R.J., *Conserving Soil Resources*, European Perspectives. CAB International, Cambridge, pp. 246–263, 1994.

[35] Sidorchuk, A.J. & Golosov, V.N., Calibration of soil-erosion models based on study of radioactive fallout. *Eurasian Soil Sci.*, **28**, pp. 383–395, 1996.

[36] Ferro, V., Di Stefano, C., Giordano, G. & Rizzo, S., Sediment delivery process and ^{137}Cs spatial distributionin a small Sicilian basin. *Hydrol. Process.*, **12**, pp. 701–711, 1998.

[37] Quine, T.A., Use of Ceasium-137 data for validation of spatially distributed erosion models: the implication of tillage erosion. *Catena*, **37**, pp. 415–430, 1999.

[38] Lu, X.X. & Higgitt, D.L., Estimating erosion rates on sloping agricultural land in the Yangtze Three Gorges, China, from caesium-137 measurements, *Catena*, **39**, pp. 33–51, 2000.

[39] Schuller, P., Ellies, A., Castillo, A. & Salazar, I., Use of ^{137}Cs to estimate tillage- and water-induced soil redistribution rates on agricultural land under different use and management in central-south Chile. *Soil and Tillage Research*, **69**, pp. 69–83, 2003.

[40] Quine, T.A., Walling, D.E., Chakela, Q.K., Mandiringana, O.T. & Zhang, X., Rates and patterns of tillage and water erosion on terraces and contour strips: evidence from Caesium-137 measurements. *Catena*, **36**, pp. 115–142, 1999.

[41] Walling, D.E. & Quine, T.A., The use of ^{137}Cs measurements to investigate soil erosion on arable fields in the UK: potential applications and limitations, *Journal of Soil Science*, **42**, pp. 147–165, 1991.

[42] Nyléna, T. & Grip, H., The origin and dynamics of the ^{137}Cs discharge from a coniferous forest catchment, *Journal of Hydrology*, **192**, pp. 338–354, 1997.

[43] Wallbrink, P.J., Roddy, B.P. & Olley, J.M., A tracer budget quantifying soil redistribution on hillslopes after forest harvesting. *Catena*, **47**, pp. 179–210, 2002.

[44] Porto, P., Walling, D.E. & Ferro, V., Validating the use of caesium-137 measurements to estimate soil erosion rates in a small drainage basin in Calabia, Southern Italy. *Journal of Hydrology*, **248**, pp. 93–108, 2001.

[45] Ritchie, J.C., Herrick, J.E. & Ritchie, C.A., Variability in soil redistribution in the northern Chihuahuan Desert based on Cesium measurements. *Journal of Arid Environments*, **55**, pp. 737–746, 2003.

[46] Walling, D.E., Use of ^{137}Cs and other fallout radionuclides in soil erosion investigations: progress, problems and prospects, IAEA-TECDOC-1028 pp. 39–62, 1998.

[47] Walling, D.E., He, Q. & Whelan, P.A., Using ^{137}Cs measurements to validate the application of the AGNPS and ANSWERS erosion and sediment yield models in two small Devon catchments. *Soil and Tillage Research*, **69**, pp. 27–43, 2003.

[48] Livens, F.R. & Loveland, P.J., The influence of soil properties on the environmental mobility of caesium in Cumbria. *Soil Use Manage*, **4**, pp. 69–75, 1988.

[49] Mitchell, P.I., Schell, W.R., McGarry, A., Ryan, T.P. & Sanchezcabeza, J.A., Vidal-Quadras, A., Studies of the vertical distribution of ^{134}Cs, ^{137}Cs ^{23}Spu, $^{239'24°}pu$, ^{241}Pu, ^{241}Am and $^{21°}Pb$ in ombrogenous mires at mid-latitudes. *J. Radioanal. Nucl. Chem., Articles*, **156**, pp. 361–387, 1992.

[50] Bernard, C., Mabit, L., Wicherek, S. & Laverdiere, M.R., Long-term soil redistribution in a small French watershed as estimated from ^{137}Cs data. *J. Environ. Qual*, **27**, pp. 1178–1183, 1998.

[51] Watters, R.L., Hakonson, T.E. & Lane, L.J., The behavior of actinides in the environment. *Radiochim. Acta*, **32**, pp. 89–103, 1983.

[52] Wallbrink, P.J., Olley, J.M. & Murray, A.S., Measuring soil movement using Cs-137: implications of reference site variability. pp. 95–103. In Variability in Stream Erosion and Sediment Transport. IAHS Publ. 224, 1994.

[53] Sutherland, R.A., Examination of caesium-137 areal activities in control (uneroded) locations. *Soil Technol*, **4**, pp. 33–50, 1991.

[54] Sutherland, R.A., Caesium-137 soil sampling and inventory variability in reference locations: A literature survey. *Hydrol. Process*, **10**, pp. 43–53, 1994.

[55] McHenry, J.R. & McDowell, L.L., The use of radioactive tracers in sedimentation research. *J. Geophys. Res*, **67**, pp. 1465–1471, 1962.

[56] Meek, B.D., MacKenzie, A.J. & Stockinger, K.R., Evaluation of a radioactive tracer method for measuring water intake in soils. *Soil Sci. Soc. Amer. Proc*, **28**, pp. 153–155, 1964.

[57] Toth, S.J. & Alderfer, R.B., A procedure for tagging water-stable soil aggregates with Co-60. *Soil Sci*, **89**, pp. 36–37, 1960.

[58] Riebe, B., Monitoring the translocation of soil particles using a neutron activated tracer. ed. K.H. Hartge & B.A. Stewart, *Soil structure: its development and function*, CRC Press: Boca Raton, FL, pp. 277–297, 1995.

[59] Wheatcroft, R.A., Olmez, I. & Pink, F.X., Particle bioturbation in Massachusetts Bay: preliminary results using a new deliberate tracer technique. *J. Mar. Res*, **52**, pp. 1129–1150, 1994.

[60] Olmes, I. & Pink, F.X., New particle-labeling technique for use in biological and physical sediment transport studies. *Environ. Sci. Technol*, **28,** pp. 1487–1490, 1994.

[61] Brown, L., Stensland, G. & Klein, J., Atmospheric deposition of ^{7}Be and ^{10}Be, *Geochimica Et Cosmochimica Acta*, **53**, pp. 153–142, 1989.

[62] Olley, J.M., Murray, A.S., Mackenzie, D.H. & Edwards, K., Identifying sediment sources in a gullied catchment using natural and anthropogenic radioactivity. *Water Resources Research*, **29**, pp. 1037–1043, 1993.

[63] Wallbrink, P.J. & Murray, A.S., Use of fallout radionuclides as indicators of erosion processes. *Hydrological Processes*, **7**, pp. 297–304, 1993.

[64] Matisoff, G., Bonniwell, E.C. & Whiting, P.J., Soil erosion and sediment sources in an Ohio watershed using beryllium-7, cesium-137, and lead-210. *J. Environ. Qual*, **31**, pp. 54–61, 2002.

[65] Bonniwell, E.C., Matisoff, G. & Whiting, P.J., Determining the times and distances of particle transit in a mountain stream using fallout radionuclides. *Geomorphology*, **27**, pp. 75–92, 1999.

[66] Brown, E.T., Pavich, M.J., Hickman, R.E., Klein, J. & Middleton, R., Erosion of the Eastern United States observed with Be-10. *Earth Surf. Process. Landforms*, **13**, pp. 441–457, 1988.

[67] Heimsath, AM., Chappell, J., Dietrich, WE., Nishiizumi, K. & Finkel, RC., Late Quaternary erosion in southeastern Australia: a field example using cosmogenic nuclides. *Quaternary International*, **83–5**, pp. 169–185, 2001.

[68] Nozaki, Y., DeMaster, D.J., Lewis, D.M. & Turekian, K.K., Atmospheric ^{210}Pb fluxes determined from soil profiles. *Journal of Geophysical Research*, **83**, pp. 4047–4052, 1997.

[69] Olsen, C.R., Larsen, I.L., Lowry, P.D., Cutshall, N.H., Todd, J.F., Wong, G.T.F. & Casey, W.H., Atmospheric fluxes and marsh-soil inventories of ^{7}Be and ^{210}Pb. *Journal of Geophysical Research*, **90**, pp. 10487–10496, 1985.

[70] Anderson, R.F. & Schiff, S.L., Determining sediment Accumulation and mixing rates using Pb-210, ^{137}Cs and other tracers: problems due to post-depositional mobility or coring artifacts. *Can. J. Fish. Aquat. Sci*, **44,** pp. 231–249, 1987.

[71] Young, R.A. & Holt, R.F., Tracing soil movement with fluorescent glass particles. *Soil Sci. Soc. Am. J.*, **32**, pp. 600–602, 1968.

[72] Wheatcroft, R.A., Conservative tracer study of horizontal sediment mixing rates in a bathyal basin, California borderland. *J. Mar. Res*, **49**, pp. 565–588, 1991.

[73] Plante, A.F., Duke, M.J.M. & McGill, W.B., A tracer sphere detectable by neutron activation for soil aggregation and translocation studies. *Soil. Sci. Soc. Am. J.*, **63**, pp. 1284–1290, 1999.

[74] Ventura, E., Nearing, M.A., Amore, E. & Norton, L.D., The study of detachment and deposition on a hillslope using a magnetic tracer. *Catena*, **48,** pp. 149–161, 2002.

[75] Borselli, L. & Torri, D., Measurements of soil translocation by tillage using a non invasive electromagnetic method. *J. Soil Water Cons*, **56**, pp. 106–111, 2001.

[76] Hussain, I., Olson, K.R. & Jones, R.L., Erosion patterns on cultivated and uncultivated hillslopes determined by soil fly ash contents. *Soil Sci*, **163**, pp. 726–738, 1998.

[77] Olson, K.R., Gennadiev, A.N., Jones, R.L. & Chernyanskii, S., Erosion patterns on cultivated and reforested hillslopes in Moscow Region, Russia. *Soil Sci. Soc. Am. J.*, **66**, pp. 193–201, 2002.

[78] Zhang , X.C., Nearing, M.A., Polyakov, V.O. & Friedrich, J.M., Using rare earth oxide tracers for studying soil erosion dynamics. *Soil Science Society Journal*, **67**, pp. 279–288, 2003.

[79] Collins, A.L., Walling, D.E. & Leeks, G.J.L., Fingerprinting the Origin of Fluvial Suspended Sediment in Larger River Basins: Combining Assessment of Spatial Provenance and Source Type. Geografiska Annaler: Series A, *Physical Geography*, **79**, pp. 239–254, 1997.

[80] Walling, D.E., Tracing suspended sediment sources in catchments and river systems. *Science of The Total Environment*, **344,** pp. 159–184, 2005.

[81] Grimshaw, D.L. & Lewin, J., Source identification for suspended sediments. *Journal of Hydrology*, **47**, pp. 51–62, 1980.

[82] Klages, M.G. & Hsieh, Y.P., Suspended solids carried by the Galatin River of Southwestern Montana: II. Using mineralogy for inferring sources. *Journal of Environmental Quality*, **4**, pp. 68–73, 1975.

[83] Wall, G.J. & Wilding, L.P., Mineralogy and related parameters of fluvial suspended sediments in northwestern Ohio. *Journal of Environmental Quality*, **5**, pp. 168–73, 1976.

[84] Slattery, M.C., Burt, T.P. & Walden, J., The application of mineral magnetic measurements to quantify within-storm variations in suspended sediment sources. Tracer Technologies for Hydrological Systems. (Proceedings Boulder Symposium, July 1995). IAHS Publication, vol. 229. Wallingford7 IAHS Press: pp. 143–151, 1995.

[85] Oldfield, F, Maher, B.A., Donoghue, J. & Pierce, J., Particle-size related, mineral magnetic source sediment linkages in the Rhode River catchment, Maryland, *USA. J Geol Soc (Lond)*, **142**, pp. 1035–1046, 1985.

[86] Walden, J, Slattery, M.C. & Burt, T.P., Use of mineral magnetic measurements to fingerprint suspended sediment sources: approaches and techniques for data analysis. *Journal of Hydrology*, **202**, pp. 353–372, 1997.

[87] Walling, D.E. & Woodward, J.C., Use of radiometric fingerprints to derive information on suspended sediment sources. *IAHS Publication*, **210,** pp. 153–164, 1992.

[88] Shankar, R., Thompson, R. & Galloway, R.B., Sediment source modelling: Unmixing of artificial magnetization and natural radioactivity measurements. *Earth and Planetary Science Letters*, **126**, pp. 411–420, 1994.

[89] Yu, L. & Oldfield, F., A multivariate mixing model for identifying sediment source from magnetic measurements. *Quaternary Research*, **32,** pp. 168–181, 1989.

[90] Walling, D.E. & Woodward, J.C., Tracing sources of suspended sediment in river basins: A case study of the River Culm, Devon, UK. *Marine and Freshwater Research*, **46**, pp. 327–336, 1995.

[91] Russell, M.A. Walling, D.E., & Hodgkinson, R.A., Suspended sediment sources in two small lowland agricultural catchments in the UK. *Journal of Hydrology*, **252**, pp. 1–24, 2001.

[92] Walling, D.E., Woodward, J.C. & Nicholas, A.P., A multi-parameter approach to fingerprinting suspended sediment sources. Tracers in hydrology (Proceedings Yokohama symposium, July, 1993). IAHS Publication, 215. Wallingford7 IAHS Press: pp. 329–338, 1993.

[93] Collins, A.L., Walling, D.E. & Leeks, G.J.L., Use of composite fingerprints to determine the provenance of the contemporary suspended sediment load transported by rivers. *Earth Surface Processes and Landforms*, **23**, pp. 31–52, 1998.

[94] Gruszowski, K.E., Foster, I.D.L., Lees, J.A. & Charlesworth, S.M., Sediment sources and transport pathways in a rural catchment, Herefordshire, UK. *Hydrological Process*, **17**, pp. 2665–81, 2003.

[95] Singer, M.J. & Fine, P., Pedogenic factors affecting magnetic-susceptibility of northern California soils. *Soil Sci. Soc. Am. J.*, **53**, pp. 1119–1127, 1989.

[96] Mahler, B.J., Bennett, P.C. & Zimmerman, M., Lanthanide-labled clay: a new method for tracing sediment transport in karst. *Ground Water*, **36,** pp. 835–843, 1998.

[97] Markert, B., The pattern of distribution of lanthanide elements in soils and plants. *Phytochemistry*, **26**, pp. 3167–3170, 1987.

[98] Land, M., Ohlander, B., Ingri, J. & Thunberg, J., Solid speciation and fractionation of rare earth elements in a spodosol profile from northern Sweden as revealed by sequential extraction. *Chem. Geo.*, **160**, pp. 121–138, 1999.

[99] Wyttenbach, A., Furrer, V., Schileppi, P. & Tobler, L., Rare earth elements in soil and in soil-grown plants. *Plant Soil* 199, pp. 267–273, 1998. http://hydrolab.arsusda.gov/cesium/Cesium137bib.htm#ftnote, 2005.

[100] Krezoski, J.R., Sediment reworking and transport in eastern Lake Superior: in situ rare earth element tracer studies. *J. Great Lakes Res.*, **15**, pp. 26–33, 1989.

[101] Polyakov, V.O. & Nearing, M.A., Rare earth element oxides for tracing sediment movement. *Catena*, **55**, pp. 255–276, 2004.

[102] Tian, J.L., Zhou, P.H. Liu & P.L., REE tracer method for studies on soil erosion. *Int. J. Sediment Res.*, **9**, pp. 39–46, 1994.

[103] Matisoff, G., Ketterer, M.E., Wilson, C.G., Layman, R. & Whiting, P.J., Transport of rare earth element-tagged soil particles in response to thunderstorm runoff. *Environmental Science Technology*, **35**, pp. 3356–3362, 2001.

[104] Wei, S., Puling, L., Mingyi, Y. & Yazhou, X., Using REE tracers to measure sheet erosion changing to rill erosion. *Journal of Rare Earths*, **21,** pp. 587–590, 2003.

[105] Liu, P.L., Tian, J.L., Zhou, P.H., Yang, M.Y. & Shi, H., Sable rare earth element tracers to evaluate soil erosion. *Soil and Tillage Research*, **76,** pp. 147–155, 2004.

[106] Kimoto, A., Nearing, M.A., Shipitalo, M.J. & Polyakov, V., Multi-year tracking of sediment sources in a small agricultural watershed using rare earth elements. *Earth Surface Processes and Landforms* **31,** pp. 1763–1774, 2006a.

[107] Kimoto, A., Nearing, M.A., Zhang, X.C. & Powell, D.M., Applicability of rare earth element oxides as sediment tracers for coarse-grained, Alluvial Soils. *Catena*, **65**, pp. 214–221, 2006b.

[108] Hill, B.R., Fuller, C.C. & DeCarlo, E.H., Hillslope soil erosion estimated from aerosol concentrations, North Halawa Valley, Oahu, Hawaii. *Geomorphology*, **20**, pp. 67–79, 1997.

[109] Hill, B.R., DeCarlo, E.H., Fuller, C.C. & Wong, M.F., Using sediment fingerprints to assess sediment-budget errors, North Halawa Valley, Oahu, Hawaii, 1991–92. *Earth Surface Processes and Landforms*, **23**, pp. 493–508, 1998.

[110] Callaway, J.C., DeLaune, R.D. & Patrick, J. W.H., Chernobyl ^{137}Cs Used to determine sediment accretion rates at selected Northern European Coastal Wetlands. *Limnology and Oceanography*, **41**, pp. 444–450, 1996.

[111] Cundy, A.B. & Croudace, I.W., Sediment accretion and recent sea-level rise in the Solent, Southern England: inferences from radiometric and geochemical studies. *Estuarine, Coastal and Shelf Science*, **43**, pp. 449–467, 1996.

[112] Huh, C.A. & Su, C.C., Sedimentation dynamics in the East China Sea elucidated from ^{210}Pb, ^{137}Cs and ^{239,240}Pu. *Marine Geology*, **160**, pp. 183–196, 1999.

[113] Hong, G.H., Kim, S.H., Chung, C.S., Kang, D.J., Shin, D.H., Lee, H.J. & Han, S.J., ^{210}Pb-derived sediment accumulation rates in the southwestern East Sea (Sea of Japan). *Geo-Marine Letters*, **17**, pp. 126–132, 1997.

CHAPTER 4

Coastal wetlands: function and role in reducing impact of land-based management

G.L. Bruland
Natural Resources and Environmental Management Department, University of Hawaii Mānoa, USA.

Abstract

Coastal wetlands are among the most productive, valuable, and yet most threatened ecosystems in the world. They provide a variety of functions that reduce the impact of land-based management on the coastal zone such as slowing the flow of water from the mountains to the sea, trapping of sediments, and retaining or transforming nutrients. Numerous studies have reported that increased soil erosion and nutrient export from land-based management are threatening estuaries and coastal zones. Coastal wetlands are located at a critical interface between the terrestrial and marine environments and are ideally positioned to reduce impacts from land-based sources. There are various types of coastal wetlands including riparian wetlands, tidal freshwater marshes, tidal salt marshes, and mangroves. Some classification systems also consider seagrass beds and coral reefs to be wetlands. Coastal wetland ecosystems vary in their ability to reduce impacts from land-based management in both space and time. These wetlands can retain, and transform, or sometimes even act as sources of nutrients and sediments. Some wetland types are more effective at sediment retention and others at nutrient retention. Watershed size, climate, and position of wetlands in a watershed are other important factors that determine the effectiveness of coastal wetlands in reducing the effects of land-based activities. Wetlands do not appear to be infinite sinks for sediment or nutrients. Once critical sediment and nutrient loading thresholds have been crossed, coastal wetlands are subject to degradation and even loss. While many coastal nations have developed coastal-zone management policies and legislation, degradation and losses of coastal wetlands continue to occur due to altered hydrology, increased sediment and nutrient loading, urban development, agriculture, and aquaculture. While we have made significant progress in our ability to restore and create tidal marshes and mangroves, other coastal systems such as seagrass beds and coral reefs appear to be much harder to restore. Thus, it is important that existing natural coastal wetlands be prioritized for conservation and that best management plans be developed to reduce sediment and nutrient losses from terrestrial watersheds.

1 Introduction and current status of coastal wetlands

Changes in terrestrial land-use patterns such as agricultural intensification and urban expansion have resulted in increased sediment and nutrient loadings that are transported into coastal areas [1–4]. For example, the change in nitrogen (N) loading to coastal areas since preindustrial times has increased fourfold in the Mississippi River, eightfold in rivers of the northeastern United States, and tenfold in rivers draining to the North Sea [2]. Human activities have increased sediment loads in rivers by 20% compared to preindustrial times [5]. However, reservoirs and diversions trap approximately 30% of the total sediment load in rivers from reaching the ocean. While the overall sediment load to the coastal zone has decreased by 10% [5], there are hotspots in places like the Philippines, Indonesia, and Madagascar where sediment loads to coastal zones have dramatically increased in recent years. Other pollutants such as herbicides and heavy metals, that are generated from land-based activities, have been shown to suppress photosynthesis in seagrasses and corals and suppress coral fertilization at concentrations of a few tens of parts per billions [5, 6, 7].

Coastal wetlands provide a critical interface between terrestrial and marine environments, and their importance to global sediment and nutrient budgets is much greater than their proportional surface area on earth would suggest [8]. It has been estimated that wetlands provide \$4.88 trillion (US) yr^{-1} in ecosystem services [9]. These ecosystem services include disturbance regulation, water supply, water quality maintenance, pollination, biological control, food production, and others. According to Costanza *et al.* [9], wetlands are 75% more valuable in terms of ecosystem services than lakes and rivers, 15 times more valuable than forests, and 64 times more valuable than grassland or rangelands. Despite the many ecosystem services that wetlands provide, they have been subject to conversion to other land uses for millennia and especially during the last 200 years.

Some of the best data on these conversion rates come from the United States. For example, it has been estimated that the state of California has lost over 90% of its original wetland area due to conversion to urban and agricultural land uses [10]. Iowa, Illinois, Indiana, and Ohio each have lost greater than 85% of their wetlands mainly due to conversion to agriculture [10]. In Florida, the Greater Everglades Ecosystem comprises less than half its original extent due to drainage and conversion to agricultural and urban land uses [11]. The state of Louisiana is expected to lose another 181,300 ha of wetlands in next 50 years or an area equal to the size of Washington DC./Baltimore metropolitan region [12]. Wetland losses in Louisiana continue to occur due to channelization of the Mississippi River, sea-level rise, herbivory from nutria (*Myocastor coypu*), storm surge from hurricanes, and impacts due to oil and gas production. Louisiana accounted for approximately 90% of the coastal marsh loss in the continental states during 1990s (not including Alaska).

On a more positive note, wetland loss rates in the United States have decreased over the last 30 years [13]. In the period from 1986 to 1997, 98% of wetland losses in the United States were from forested and freshwater wetlands, while only 2% of the losses were from estuarine wetlands [13]. While conversion rates of coastal wetlands have declined in the United States, globally conversion of coastal wetlands continues and in some places has even accelerated [5].

The coastal land at the continental margin accounts for less than 5% of the Earth's land area, yet 17% of the earth's human population lives within this zone [14]. Furthermore, approximately 4 billion people live within 60 km of the world's coastlines [15]. Table 1 lists the share of the total and coastal population that live within 50 km of different coastal wetland types.

Specifically, 27% of the earth's human population lives within 50 km of an estuary (Table 1). Coastal population densities have been estimated to be 100 people per square kilometer compared to only 38 people per square kilometer in inland areas [5]. This not only causes damage to coastal wetlands but also to adjacent seagrass beds and coral reefs (classified by some as wetlands and others as deepwater habitats; see the next section). Seagrass beds are currently threatened by physical disturbance, ship activities, dredging, landfill, erosion from terrestrial sources, growth of aquaculture, and eutrophication [16]. Large-scale declines in seagrass beds have been observed at over 40 locations, 70% of which were due to human-induced degradation [16]. Coral reefs are also in serious decline. Thirty per cent of all reefs are severely damaged and close to 60% may be lost by 2030; furthermore, it has been stated that there are no pristine reefs remaining [17].

The rest of this chapter will include information on wetland classification, compare and contrast different types of coastal wetlands, examine the coverage and position of wetlands in the watershed, explore the role of coastal wetlands in trapping sediment and retaining nutrients, compare soils of natural wetlands to created and restored wetlands in a case-study example, and identify future research needs and directions. Specifically, the major wetland classification systems will be identified and their classification of coastal wetlands will be discussed. This will be followed by comparing and contrasting the dominant types of coastal and deepwater

Table 1: Share of world and coastal populations living within 50 km of estuaries, mangroves, seagrasses, and coral reefs in 1995 [14]†.

Types	Human population [millions]	Share of world population [%]	Share of coastal population [%]
Estuaries	1,599	27	71
Mangroves	1,033	18	45
Seagrasses	1,146	19	49
Coral Reefs	711	12	31
Total ‡	5,596	–	–

† Based on spatially referenced population data.

‡ Due to overlap of some habitat types the figures do not add up to 100%.

wetlands including riparian wetlands, tidal freshwater marshes, salt marshes, mangroves, seagrass beds, coral reefs, and kelp forests. The coverage and position of wetlands in the watershed will then be examined in light of reducing impacts from land-based management. Next the methods for assessing sediment retention in coastal wetlands will be discussed followed by a summary of studies that investigated sediment retention. A similar section on methods for assessing nutrient retention and summary of nutrient retention studies will follow. This chapter will conclude with a case study that compares soils of natural wetlands to created and restored wetlands as well as a discussion of future research needs and directions.

2 Wetland classification

Before moving into a discussion of the function and role of wetlands in reducing the impacts of land-based management, it is important to understand that there are different definitions of wetlands and that various systems are used for their classification. Currently, the three most prominent hierarchical wetland classification systems include the U.S. Fish and Wildlife Service (USFWS) Classification System, the Canadian Classification System, and the Ramsar Convention System. The USFWS classification system, titled "Classification of Wetland and Deepwater Habitats of the United States" was published in 1979 [18]. This system includes both wetlands and deepwater habitats. The USFWS System defines wetlands as "Lands transitional between terrestrial and aquatic systems where the water table is usually at or near the surface." This definition does not include areas with permanent standing water greater than two m deep. Such areas would be defined as aquatic or marine habitats [18]. The levels of the USFWS System include *Systems*, *Subsystems*, *Classes*, and *Subclasses*. *Systems* are the broadest level of the classification scheme and include Marine, Estuarine, Riverine, Lacustrine, and Palustrine. Coastal wetlands are usually classified in the Estuarine *System*.

The Canadian Wetland Classification System defines a wetland as a "land that is saturated with water long enough to promote wetland or aquatic processes as indicated by poorly drained soils, hydrophytic vegetation and various kinds of biological activity which are adapted to a wet environment" [19]. Classification in the Canadian System is hierarchical and has three main levels: *Classes*, *Forms*, and *Types* [19]. The five *Classes* of wetlands in this system include Bogs, Fens, Swamps, Marshes, and Shallow Water Marshes. Coastal wetlands are found in all of the *Classes* except for Bog.

The Ramsar Convention defines wetlands as "Areas of marsh, fen, peatland, or water, whether natural or artificial, permanent or temporary, with water that is static, flowing, fresh, brackish, or salt, including areas of marine water where the depth at low tide does not exceed six m" [20]. This is a more inclusive definition of wetlands that incorporates coral reefs and seagrass beds that are not defined as wetlands in the USFWS or Canadian systems. The Ramsar System groups wetlands into *Classes* based on their location in the landscape and vegetation [20]. It has 32 *Classes* that are divided into marine/coastal and inland groups.

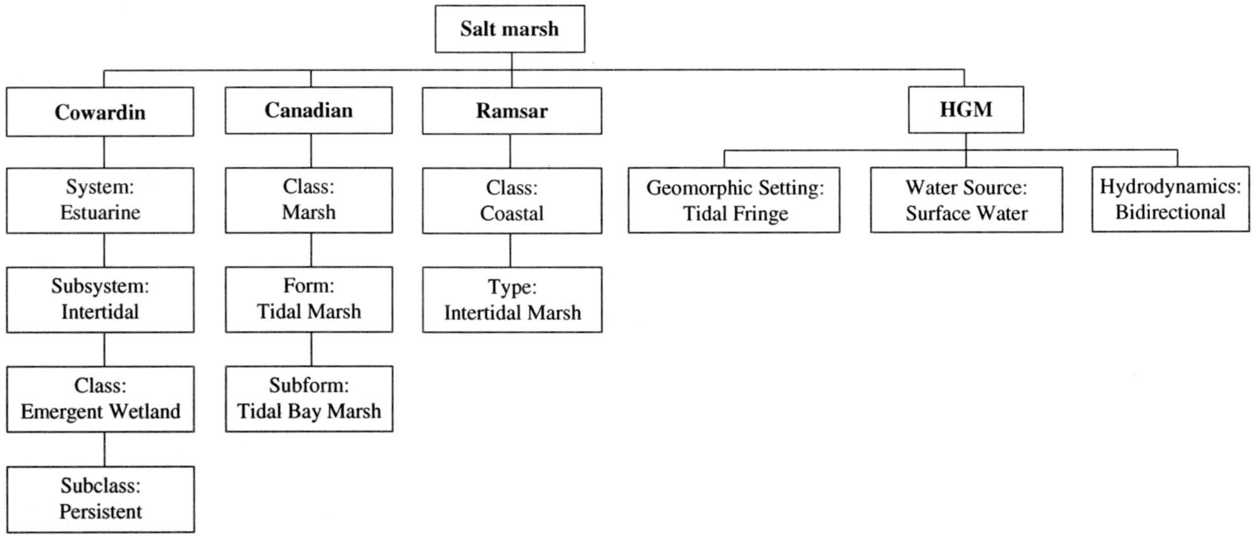

Figure 1: Classification of a common coastal wetland, the salt marsh, according to the Cowardin, Canadian, Ramsar, and Hydrogeomorphic systems.

The Hydrogeomorphic (HGM) Classification System [21] is functional classification system that is also worth mentioning. The HGM System is focused on evaluating physical, chemical, and biological functions of wetlands in the field simply, rapidly, and inexpensively [21]. The HGM system emphasizes two abiotic controls in maintaining wetland functions, hydrology and geomorphology [21]. Hydrology controls the amount, source, and season of water entering the wetland whereas geomorphology controls where the water comes from and whether or not it leaves. The HGM system includes 7 different *Geomorphic Settings*: depressional, riverine, lacustrine fringe, tidal fringe, slope, mineral soil flats, and organic soil flats [21]. It recognizes three *Water Sources*: precipitation, surface water, and groundwater as well as three types of *Hydrodynamics*: vertical, unidirectional, and bidirectional [21]. Coastal wetlands can have either riverine or tidal fringe geomorphic settings, all three types of water sources, and unidirectional (riverine) or bidirectional (tidal fringe) hydrodynamics. Figure 1 illustrates how the four systems mentioned above would classify a common coastal wetland, the tidal salt marsh.

3 Types of coastal wetlands

There are a variety of types of coastal wetlands that occur in different landscape positions, have different vegetative communities, and provide different functions in terms of reducing the effects of land-based pollution in coastal zones. Spanning a continuum from salt to fresh water, these include mangroves, tidal salt marshes, tidal freshwater marshes, and riparian wetlands (see Fig. 2a). Deepwater habitats include seagrass beds, coral reefs, and kelp forests. These ecosystems are often nested across the hierarchy of the larger estuarine-coastal system (Fig. 2b).

3.1 Riparian wetlands

Riparian wetlands occur as ecotones or interfaces between aquatic and upland ecosystems, have distinct vegetation and soil characteristics [22, 23], and perform important functions at the watershed scale [24]. Riparian wetlands are not easily classified or delineated but instead are comprised of mosaics of landforms and communities within the larger landscape [23]. Brinson *et al.* [24] stated that riparian wetlands are characterized by an abundance of water and fertile alluvial soils. They also list three major features that separate riparian wetlands from other types of wetlands and upland ecosystems that include:

1. Their linear form as a consequence of their proximity to rivers and streams.
2. Energy and material from the surrounding landscape [upstream watershed] converge and pass through riparian wetlands in much greater amounts than those of any other freshwater wetland systems.
3. Riparian wetlands are functionally connected to upstream and downstream wetlands and are laterally connected to upslope [upland] and downslope (aquatic) ecosystems [24].

Thus, riparian wetlands are dynamic, open systems that are subject to large inputs of surface water, sediments, and nutrients from forested, agricultural, and urban

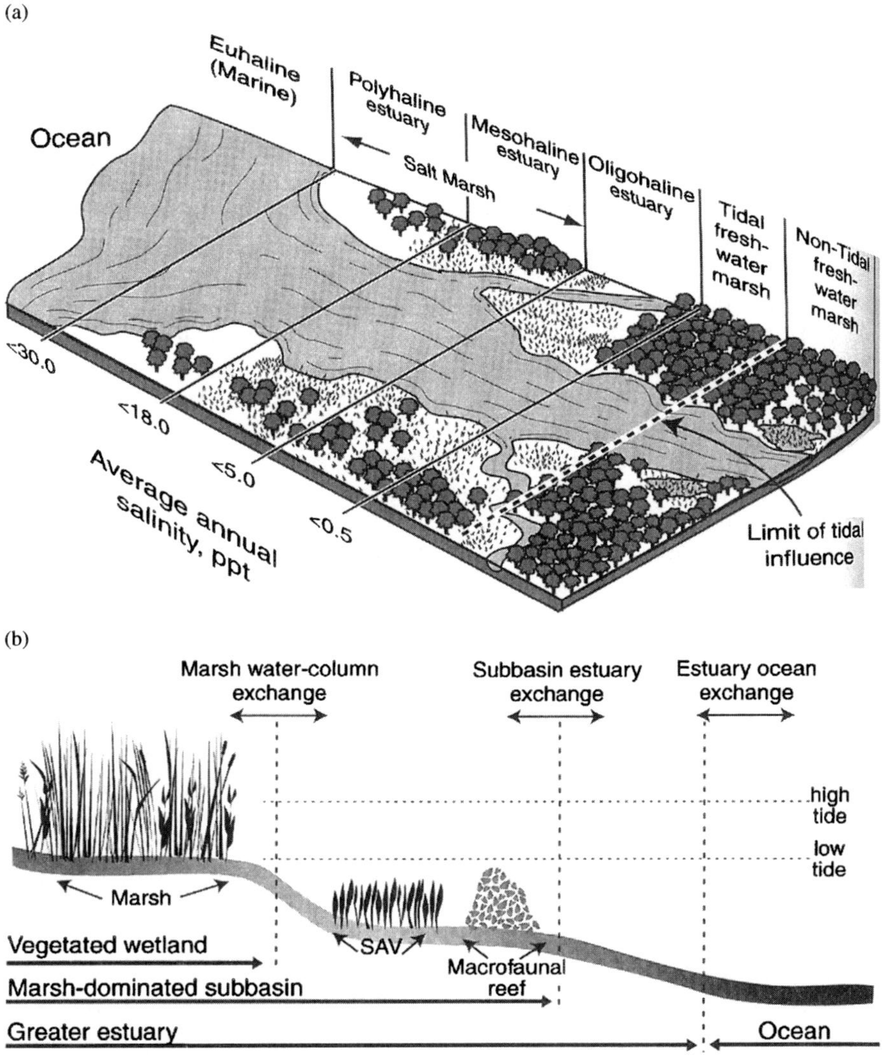

Figure 2: (a) The estuarine salinity gradient. (b) Hierarchy of the estuarine-coastal landscape. Copywrite (2000) from *Wetlands*, 3rd Edition by W.J. Mitsch and J.G. Gosselink [27]. Reproduced by permission of John Wiley and Sons, Inc.

areas upstream in the watershed. These riparian systems have the capacity to retain and transform large quantities of these inputs and keep them from being transported to coastal zones.

In the United States, the most extensive riparian wetland ecosystems are the bottomland hardwood forests of the Southeast. These bottomlands stretch across the Gulf and Atlantic coastal plains from Texas to Maryland and are associated with rivers such as the Mississippi, Appalachicola-Chattahoochee, Ogeeche,

Altamaha-Ocmulgee, Pee Dee-Yadkin, Neuse, Tar-Pamlico, and Roanoke. Bottomland hardwood forests in the southeastern U.S. have been, and continue to be, converted to other land uses such as agriculture and urban developments. The Nature Conservancy [25] estimated that in 1991 about 2.0 million ha of bottomland hardwood forested remained in the Mississippi River Alluvial Plain; this area supported about 8.5 million ha of bottomland hardwood forest prior to European settlement. The Atlantic coastal plain that stretches from Florida to Maryland also has extensive areas of riparian hardwood forest lining many of the rivers that flow from the Piedmont to the ocean. Some of these riparian forests remain intact, others have been logged, others have regenerated, and many are in the process of being restored [26]. At the global scale there are various other forested wetland systems that are found in floodplains or riparian zones. The Amazon River floodplain in Brazil is an example of one of these types of systems as is the Zaire Swamps in Africa. These systems perform similar functions to the bottomland hardwood systems described above but are located in a tropical rather than a temperate setting.

Riparian wetlands located in temperate regions of the Western United States, or in tropical climates such as those in the Hawaiian and other Pacific Islands, are generally quite different from the bottomland hardwood forests described above. These riparian wetlands are located in steep, narrow, and dynamic riparian zones. While floodplain forests often have subtle changes in elevation and vegetation, the gradients in steep-sloped riparian wetlands are usually sharp and the visual distinctions between community types are clear [27]. These riparian wetlands have also been extensively modified by human activity [27]. Logging, grazing, conversion to agriculture and urban development have been widespread. These riparian areas are often the only flat lands available for cultivation and home building. They also tend to concentrate grazers due to the flat terrain and presence of water. Logging also continues to have a major impact on these wetlands. Often the streams in these riparian zones are used to move logs from the forest to estuarine holding pens. As many streams were too small to move logs efficiently, they were dammed and their banks were cleared to enhance logging and transport of timber [27]. This has lead to extensive erosion from upland areas and deposition in riparian ecosystems with subsequent transport of sediment into coastal zones. In some cases, destruction of riparian zones has resulted in floods and burial of natural estuarine habitats under tons of silt and enriched sediment [28]. The value of the ecosystem services provided by riparian wetlands and floodplains has been estimated to be \$19,580 ha^{-1} yr^{-1} [9], which is one of the highest values for any type of ecosystem.

3.2 Tidal freshwater marshes

Tidal freshwater marshes are close enough to the oceans to experience tides, but are above the reach of oceanic saltwater [27]. These coastal wetlands combine many of the features of salt marshes and freshwater marshes. They are similar in structure and function to salt marshes, with the major difference being a greater diversity in biota due to the reduction in salt stress. Plant diversity is high and more birds use these marshes than any other marsh type [27]. Often, the boundary between tidal freshwater marshes and salt marshes is difficult to determine.

Tidal freshwater marshes have been studied much less than salt marshes or inland freshwater marshes. Three major types of tidal freshwater marshes have been recognized: 1) mature marshes, 2) floating marshes, and 3) new marshes in prograding deltas [27].

In a cross-section, elevation in tidal freshwater marshes usually increases slowly from the stream edge to the adjacent upland areas. They typically have a slightly elevated levee along the stream bank where the overflowing water deposits much of its sediment load. The sediments in these marshes are fairly organic, especially in the floating marshes [27]. The European Union Habitats Directive has declared the conservation or coastal freshwater wetlands a priority [29].

3.3 Tidal salt marshes

Tidal salt marshes are found in coastal areas in the middle and high latitudes. They are common wherever accumulation of sediment is equal to or greater than the rate of land subsidence and where there is adequate protection from destructive waves and storms [27]. From afar, tidal salt marshes appear to be vast fields of a single species, often salt marsh cordgrass (*Spartina alterniflora*). While their physiognomy is much simpler than a bottomland hardwood forest, the vegetation of tidal salt marshes does vary across salinity and flooding gradients and provides habitat for a variety of plants, animals, microbes that are adapted to deal with the stresses of this environment. Numerous studies have shown salt marshes to be highly productive and to support the spawning and feeding of various marine organisms [27]. Thus, salt marshes represent a critical interface between terrestrial and marine ecosystems [27].

Chapman [30] divided the world's salt marshes into the following major geographical groups: artic, northern Europe, Mediterranean, Eastern North America, Western North America, Australasia, eastern Asia, Australia, South America, and the tropics. Although different plant associations are dominant in the different geographic groups, the ecological structure and function of salt marshes is similar around the world [27].

Salt marshes are predominantly intertidal and found in areas that are at least occasionally inundated at high tide but not flooded during low tide [27]. The upper and lower boundaries of these marshes are usually set by the tidal range. The lower boundary is determined by physical stresses such as depth and duration of flooding and the mechanical effects of waves, sediment availability and erosional forces [30, 31]. The upper boundary was thought to be set by the limit of flooding on extreme tides [32], but more recent research has indicated that the upper boundary is set by plant competition [31]. Based on elevation and flooding patterns, salt marshes can be divided into two zones: the high marsh and the low marsh [27]. The high marsh is flooded irregularly and can experience at least 10 days of continuous exposure to the atmosphere, while the low marsh is flooded almost daily and there are never more than 9 continuous days of exposure [27]. Competitively superior plants tend to dominate the high marsh habitats while stress-tolerant plants dominate the low marsh habitats [31].

Another prominent feature of salt marshes is the presence of tidal creeks. These creeks are often found in the low marsh. As the flow in tidal creeks is bidirectional, the channels tend to remain stable [27]. Generally, the banks of these tidal creeks are characterized by greater vegetative production than the interior areas of the marsh due to better flushing of salts and toxins from the tides, more oxygen in the soils due to higher elevation, and higher nutrient concentrations.

Tidal salt marshes are among the most productive ecosystems in the world, producing up to 80 metric tons ha^{-1} of plant material ($8000\,g\,m^{-2}\,yr^{-1}$) in the southern Coastal Plain of North America [27]. They have also been estimated to be one of the most economically valuable ecosystems due to the many services that they provide. One study estimated the value of tidal salt marshes and mangroves to be $9990 $ha^{-1}\,yr^{-1}$ [9]. Unfortunately, due to increased construction of dams and reservoirs, sediment delivery to estuarine zones and salt marshes has decreased considerably, and in some cases large areas of salt marsh cannot keep pace with rising sea levels and are experiencing subsidence [5]. In the Southeastern United States, recent studies have shown that overharvesting of the blue crab (*Callenectes sapidus*) has led to explosive increases in the populations of the periwinkle snail (*Littoraria irrorata*) [33]. Freed of predatory control, the snails have engaged in destructive grazing of the salt marsh cordgrass and caused large-scale die-offs in salt marshes [33]. Thus, human alterations of salt-marsh trophic dynamics may decrease the ability of these systems to trap sediments and retain nutrients from land-based sources.

3.4 Mangroves

Mangroves swamps replace salt marshes along coastlines in subtropical and tropical regions [27]. In a few transitional situations (e.g. Florida) mangroves and salt marshes coexist. Global mangrove forest cover is estimated to be between 16 and 18 million hectares [34, 35]. In the region between 25° N and 25° S latitude, mangroves dominate approximately 75% of the world's coastline [36]. Mangroves are defined as areas of trees, shrubs, and other plants found in the intertidal zones and estuarine margins that have adapted to living in saline waters, continually or at high tides [37]. Mangroves are known for their seemingly impenetrable maze of woody vegetation and their unique adaptations to the double stresses of flooding and salinity [27]. Mangrove swamps provide many critical ecosystem services such as exporting organic matter to adjacent coastal zones, serving as nursery habitat for marine organisms, trapping sediments and nutrients, stabilizing shorelines, buffering land from storms, and providing safe havens for humans in the 118 coastal countries in which they occur [27, 34].

The frequency and severity of frosts are the main factors that limit the extension of mangroves beyond tropical and subtropical climates [38]. Mangroves are particularly dominant in the Indo-West Pacific region where they contain the greatest diversity of species [27]. Three main types of mangroves have been identified based on dominant physical processes and geomorphological characteristics: tide-dominated riverine mangroves, river-dominated riverine fringe mangroves, and interior basin mangroves [39]. Up to 95% of the detritus generated by the vegetation may be exported from riverine mangroves to adjacent estuaries and coastal zones, while only 21% of the detritus in basin mangroves may be exported [36].

Some of the most intact mangrove forests in the world are found in Malaysia and Micronesia. Recent studies have shown how important these ecosystems are to the local economies in these countries [39]. The importance of these ecosystems is related to the goods and services they provide, such as trapping of sediment, processing of nutrients and organic matter, providing food and habitat for animals, protecting shorelines, and providing plant products such as fuelwood, building materials, woodchips, tannins, honey, and medicinal products [39]. Mangroves are rapidly being converted to other land uses (i.e. urban development, rice fields, oil palm plantations, and aquaculture sites) in this region at an average rate of 1% per year [40]. In some countries, greater than 80% of the original mangrove cover has been lost due to deforestation [34]. Currently, the conversion to aquaculture accounts for 52% of mangroves losses, forest use accounts for 26% of losses, and freshwater diversions account for 11% of losses [35]. The value of ecosystem services provided by mangroves has been estimated to be \$9990 ha^{-1} yr^{-1} [9].

The Hawaiian Archipelago provides an interesting example of mangrove ecology and management. Hawai'i has no native mangrove species despite having both suitable climate and geomorphic settings [41]. However, since their introduction in 1902, mangroves have flourished to such a degree that many people have been concerned about their impacts especially their dramatic effects on native plant community structure [42]. Expensive projects have been undertaken for their removal [41] with varying degrees of success.

3.5 Seagrass beds

Seagrass beds are defined as areas with aquatic flowering plants that live fully submerged in saline waters with a sediment substrate [36]. They occur over soft sediments worldwide, from the tropics to the boreal margins of every ocean [16]. In higher latitudes, eelgrass (*Zostera* spp.) forms dense meadows, while in the tropics, manatee grass (*Thalassia testudinum*) and turtle grass (*Syringodium filiforme*) are dominant [5]. Seagrasses are estimated to cover about 0.1–0.2% of the global ocean [43]. They are adapted to continuous and complete immersion in saline water, and as such, they grow in tidal zones although some can also grow in intertidal waters [16]. Seagrasses provide important ecosystem functions such as biological productivity, habitat for various marine organisms, trapping sediment, carbon sequestration, and buffering of wave action [36, 44]. The combined productivity of seagrasses and associated algae make seagrass beds among the most productive ecosystems on earth [45]. Seagrasses are also ecosystem engineers that provide physical structure that transforms featureless sediment bottoms into diverse and complex habitats for coastal biota [44].

As with salt marshes, the presence of a few plants has the effect of slowing water movement and permitting the settlement of even more sediment [36]. Since seagrasses grow submerged in seawater, they are more likely to be light-limited than salt marsh or mangrove vegetation [36]. Often, the distance to which seagrasses extend from the shoreline is related to the slope of the sea floor. When adequate sediment is present, seagrasses will extend seaward to the depth at

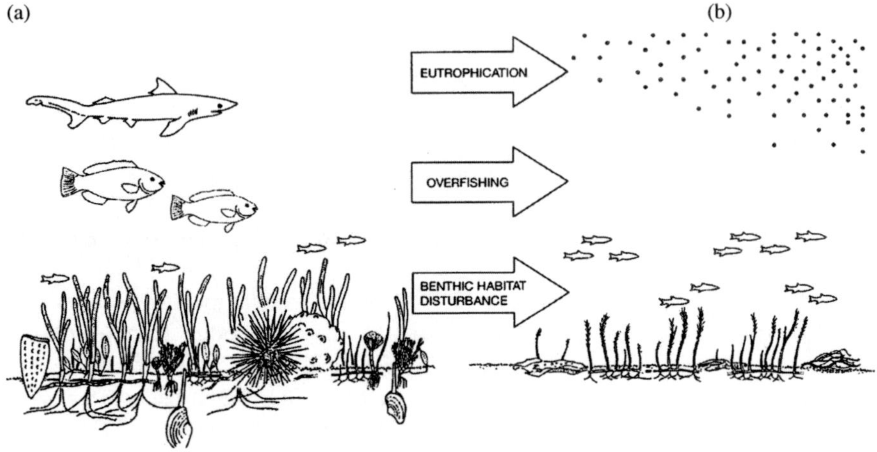

Figure 3: Comparison of a pristine (a) and a degraded (b) seagrass bed. Copywrite (2006) from J.E. Duffy [44]. Reproduced by permission of Inter Research.

which annual light flux is just sufficient to support a balance of photosynthesis over respiration [36]. However, if the water becomes more turbid due to increases in suspended sediment loads, light penetration is reduced, and the depth at which seagrasses can survive will also be reduced [46]. In undisturbed conditions, the maximum depth at which seagrass beds are found is about 30 m [47].

Increased sediment and nutrient loading from land-based sources are major threats to seagrass beds. Increased siltation is a particularly acute problem in coastal zones of Southeast Asia that receive the high amounts of sediment delivery as a result of soil erosion caused by extensive deforestation and changes in land-use in this region [48]. Under high nutrient loading from watersheds dominated by agricultural or urban land uses, growth of epiphytic algae on the seagrasses or blooms of phytoplankton in the water column can lead to decreased light availability to seagrasses and reduction in their extent [49]. Figure 3 shows the difference in structure of a pristine and a degraded seagrass bed.

The majority of remaining seagrass beds are found on the coastline of tropical countries, most of which are experiencing rapid rates of land-use conversion and environmental degradation [43]. The Pacific Islands are the only region where losses of seagrass beds are expected to be lower due to strict enforcement of zero-loss policies and small populations relative to the region's coastline [43]. An economic evaluation of ecosystem services provided by seagrass and algal bed ecosystems was quite high, estimated at \$19,004 ha^{-1} yr^{-1} [9].

3.6 Coral reefs and kelp forests

Coral reefs occur in tropical coastal waters with a minimum temperature of 18 °C, suitable light conditions, and high salinity [5, 36]. They can occur in association with tropical seagrass beds (as they do in coastal areas such as Florida Bay, USA or

Shark Bay and the Gulf of Carpenteria in Australia) and removed from them. Coral reefs are particularly abundant where sediment loading and freshwater inputs are minimal [5]. Thus, coral reefs are quite susceptible to changes in terrestrial land use and losses of riparian wetlands, tidal marshes, mangroves, and seagrass beds that can trap sediments from land-based sources. Major areas of coral reefs occur in the Pacific and Indian Oceans and the Caribbean Sea. Three main types of coral reefs have been defined [36]:

1. fringing reef: a reef found growing as a fringe attached to a land mass.
2. barrier reef: a reef that occurs at some distance out to sea and creates a shallow lagoon between the reef and the land.
3. atoll: an isolated structure surrounded by deep water that tends to form a ring of coral with a central lagoon (Fig. 4).

Coral reefs also provide a number of ecosystem functions such as serving habitat for a tremendous diversity of marine species, supporting coastal fisheries, and protecting coastal areas from storms and marine erosion [36]. Most coral reefs occur along the coasts of developing countries where the most intensive coastal degradation is occurring [50]. Coral reefs are at risk from global change processes such as bleaching and sea temperature increases, as well as various human activities such as coastal development, overfishing, sediment and sewage inputs that lead to eutrophication, dumping of debris and toxic wastes, and oil spills [5]. One study has suggested that all current coral reefs will disappear by 2040 due to warming sea temperatures [17]. Despite the fact that they provide habitat for a tremendous diversity of species, the valuation of ecosystem services provided by coral reefs was estimated to be \$6075 ha^{-1} yr^{-1} [9], which is less than one third of the amounts estimated for swamps/floodplains and seagrass/algae beds.

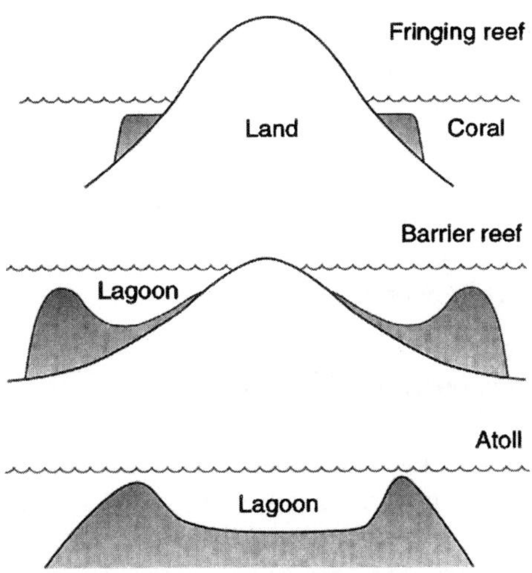

Figure 4: Types of coral reefs. Copywrite (2000) from *Ecology of Coastal Waters* by K.H. Mann. Reproduced by permission of John Wiley and Sons, Inc.

A temperate counterpart to the coral reef is the kelp forest. Kelp forests are temperate marine ecosystems dominated by large brown algae (*Macrocystis* spp.). They are characterized by high productivity and diversity [36]. For example, large brown algae can grow 45 cm d^{-1}, and extend to 60 m in length. Kelp forests are also remarkably resilient to disturbances from wave impacts, storm surges, and other extreme oceanographic events [51].

4 Wetlands in different types of watersheds

Wetlands in temperate versus tropical watersheds can differ in their ability to reduce impacts from land-based activities (i.e. logging, agriculture, urbanization) on sediment and nutrient loading to the coastal zone. Plant uptake and microbial activity in temperate and boreal wetlands has been shown to decrease dramatically, if not totally, during the cold winter months [52]. In contrast, plant uptake and microbial immobilization may occur throughout the year in tropical wetlands. However, the intense precipitation that is often experienced in tropical watersheds has been shown to be much more erosive and generate greater sediment export from terrestrial areas [53]. For example, on the island of Maui (Hawai'i), erosivity can vary from less than 100 to greater than 1800 erosivity units over a span of 20 km [54]. Furthermore, in these small tropical watersheds, the terrestrial and marine environments are intimately connected. Here, human land-use activities are quickly translated to coastal areas because of high amounts of rainfall (sometimes over 5000 mm yr^{-1}) and steep stream gradients [55]. Rainfall on the ridgetop can result in increased surface water inputs to the estuary and coral reefs in a matter of hours. Thus, all lands may be considered coastal in the Pacific Islands [54]. This is quite a contrast to the Mississippi River Watershed that covers about 40% of the continental U.S. and for which it would take many days for precipitation in the headwaters of the upper Midwest to reach the rivermouth at the Gulf of Mexico.

Furthermore, many tropical watersheds are located in areas with steep topography and subject to intense land-use conversion. The combination of erosive rainfall, steep topography, and conversion of forests to agriculture and rangeland land has lead to massive increases in sediment transport through tropical watersheds. The coastal wetlands in these watersheds may be unable to handle these large fluxes of sediments and nutrients, or are themselves subject to conversion to more intensive land-use such as agriculture or aquaculture.

5 Coverage and position of wetlands in a watershed

In a study in the MidAtlantic United States, Novitski [56] found that when the per centage of the watershed in lakes and wetlands dropped below 10%, there were rapid increases in flooding. Thus watersheds appear to have critical thresholds of wetland area for flood control and most likely also for sediment and nutrient retention. The position of a wetland in the watershed also influences its ability to retain sediments and nutrients. Faber *et al.* [57] delineated three main watershed zones,

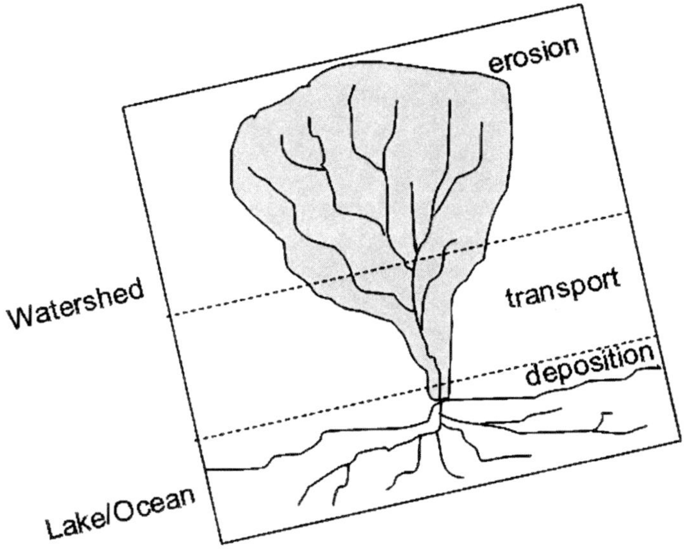

Figure 5: The erosional, storage and transport, and depositional zones of a watershed. After [57].

the zone of erosion, the zone of storage and transport, and the zone of deposition (Fig. 5). Most coastal wetlands are located in the zone of storage and transport (riparian wetlands, tidal freshwater marshes) or the zone of deposition (tidal salt marshes, mangroves, seagrass beds). As the human activity in a watershed increases, wetlands are often converted to urban, agricultural, or other land uses.

These changes often reduce the ability of the wetlands in the watershed to store water, sediment and nutrients.

Another interesting assessment of the position of wetlands in the watershed revealed that overbank flooding increases in importance with increasing stream order (Fig 6a), while overland flow is more important in headwater streams (Fig. 6b) [58].

Assuming that riparian transport is the most critical step in water-quality improvement of nonpoint-source runoff, Fig. 6c suggests that more emphasis should be placed on avoiding impacts to wetlands associated with lower-order streams than those associated with higher-order streams [58]. A given area of disturbance to these headwater wetlands will affect a greater proportion of the watershed than the same disturbance to a floodplain wetland. In other words, headwater riparian wetlands are converted to agriculture or other more intensive land uses, losses of sediments and nutrients are felt throughout the entire downstream watershed. Thus, it has been argued that it is not only the surface area of wetlands in a watershed, but also the length of wetlands along the streams that matters [58]. The large differences in average length with increasing stream order (Table 2) indicate that the riparian wetland length may be a better index of water-quality maintenance than wetland area.

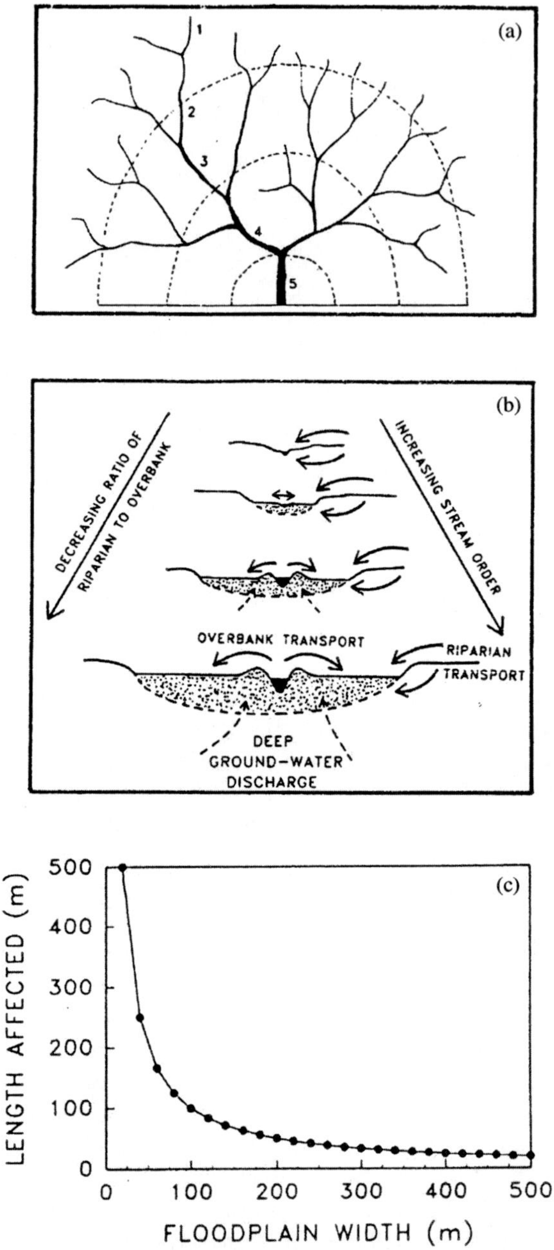

Figure 6: (a) A stream drainage network following the nomenclature of Strahler [174]. (b) Cross sections of riparian wetland floodplains showing how riparian transport and overbank flooding vary with Strahler stream order. (c) Change in length of floodplain affected by 1 hectare of disturbance as a function of floodplain width. Modified from [58] with permission of *Wetlands*.

Table 2: Relationships between stream order and other dimensions of stream configuration. First four columns are from Leapold *et al.* [173] and last two columns are from Brinson [58].

Stream order	Number	Average length [km]	Total length [km]	Estimated floodplain width [m]	Floodplain Surface Area [km^2]
1	1,570,000	1.6	1,526,130	3	7578
2	350,000	3.7	1,295,245	6	7771
3	80,000	8.5	682,216	12	8187
4	18,000	19.3	347,544	24	8341
5	4200	45.1	189,218	48	9082
6	950	103.0	97,827	96	9391
7	200	236.5	47,305	192	9082
8	41	543.8	22,298	384	8562
9	8	1250.2	10,002	768	7681
10	1	2896.2	2896	1536	4449

Ultimately, wetlands are needed in both the headwater and the depositional zones in terms of habitat and biodiversity but also in terms of flood control, water quality, sediment retention, and biogeochemical cycling. However, to meet the goal of improved water quality from a reduction in nonpoint-source pollution from land-based sources, restoration of wetlands along lower order streams may be the best strategy [58].

6 Methods for quantifying sediment accumulation in coastal wetlands

Short-term (monthly to annual rates) sediment accumulation in wetlands has been quantified by a combination of techniques that include horizon markers and sediment traps. Horizon markers have been used to quantify sediment accumulation in various wetland types [59–61]. This method involves laying down marker horizons at various points in the wetland. Feldspar, a white material composed of silt- and clay-sized particles [60], is commonly used as a marker horizon, although other studies have employed glitter or sand [62]. Sampling consists of taking a small core of sediments from the surface, down through the marker horizon, and measuring the thickness of sediments deposited above these highly visible marker horizons. A specialized coring device called a cryocorer can be used to collect the cores [63]. The amount of sediment that has been deposited on top of the marker corresponds to the sediment accumulation. Unfortunately, this method only estimates sediment depth and not the mass of sediment per unit area.

Mass of sediment per unit area can be measured with sediment traps. Sediment traps consist of tiles, petri dishes, or plastic containers that are deployed on the soil or sediment surface [64–67]. The material that collects on the trap can be quantitatively removed, weighed, and even analyzed for chemical composition.

Long-term sedimentation rates in wetlands have been quantified using radioisotopic dating techniques. Both ^{137}Cs and ^{210}Pb profiles have been shown to be effective for these purposes [68–70]. This process involves collecting deep soil cores (to bedrock when possible) and sectioning them into fine (i.e. 2 cm) intervals. The ^{137}Cs activity in each sample can be measured with a germanium detector. Peak concentrations of ^{137}Cs levels within each core correspond to peak fallout levels from atmospheric nuclear weapons testing in 1964 [69, 71]. The sediment located above the ^{137}Cs peak is equal to the amount of sediment that has been trapped in these wetlands since 1964. ^{210}Pb levels can be used to estimate sediment accretion rates over an even longer time intervals, up to 100 years before present [70–72]. Excess ^{210}Pb has been shown to accumulate in depositional environments from both atmospheric deposition and sedimentation [70, 73]. ^{210}Pb is considered more reliable than ^{137}Cs because it is polyvalent, and thus bound more tightly to mineral and organic soil particles [74].

7 Role of coastal wetlands in trapping sediment

In an analysis of eight temperate watersheds containing wetlands, Phillips [75] found that less than 65% of the sediment eroded from upland areas was transported out of the watersheds. Of the sediment that reached streams within the watersheds, 23–93% was retained by wetlands through which these streams flowed [75]. Thus, in watersheds that have sufficient wetland surface area and stream lengths associated with wetlands, sediment transport to coastal zones may be low. However, as watersheds develop, construction activities and the increase in the amount of impervious surface in the watershed have been shown to cause accelerated silt loading to neighboring estuaries [15, 76]. A number of studies in the tropics have reported that increased soil erosion in terrestrial watersheds is threatening estuaries and coastal coral reefs [77–81]. The steep topography, intense precipitation, and extensive land-use changes in these watersheds leads to high sediment loads that appear to have exceeded the sediment retention capacity of the coastal wetlands. When this occurs, sediment deposition within estuarine and reef zones can smother adult corals, kill juvenile corals, and prevent larval recruitment [81].

Mangroves have been shown to initially play a passive role in sediment accumulation [27]. Once mangrove vegetation has been established, it then acts to prevent erosion and trap sediments. The stems and leaves of mangrove and salt marsh vegetation slow water velocity and promote sediment deposition, roots and rhizomes increase the stability of the sediment, algae help trap fine sediments, oyster colonies modify the flow of water and sediments, and macroinvertebrates trap suspended detritus [27].

A study of a tropical watershed in Palau, Micronesia reported the mangrove fringe zone trapped about 44% of the riverine fine sediment flux, which was still not enough to prevent degradation of the associated coral reefs [79]. Another study on the island of Moloka'i (Hawai'i), reported that turbidity was lower on coral reefs adjacent to mangroves than on reefs with no adjacent mangroves [82]. A study of the Heeia Swamp in Kaneohe Bay, on the island of Oahu (Hawaii), reported that 10 cm of sediment was deposited in 16 months in the mangrove areas [83].

8 Methods for quantifying nutrient retention and transformation in coastal wetlands

A number of processes must be considered when quantifying nutrient transformation and retention in coastal wetlands. Plant uptake can be determined by harvesting plant tissue throughout the growing season and analyzing plant tissues for N and P concentration [36]. Microbial immobilization is difficult to quantify but can be measured in laboratory dosing studies and with radioisotopic techniques for P [84].

One of the dominant N transformation processes in coastal wetlands is denitrification. This process can be measured with *in-situ* core techniques but there are many difficulties associated with these methods due to the high background concentration of nitrogen gas in the atmosphere. The denitrification enzyme activity (DEA) [85, 86] is commonly used as an index of denitrification potential. The DEA is useful for site comparisons because it offers a method by which the denitrification potential can be compared across different soil types [87]. The DEA involves amending sieved, field-moist soils with solutions of glucose and potassium nitrate to ensure nonlimiting substrate conditions, and chloramphenicol to inhibit protein synthesis. The resulting slurries are made anaerobic by repeated flushing with N_2 gas. The anaerobic slurries are then injected with acetylene to inhibit N_2 production [85] and then shaken for a specific time (i.e. 90 min). At multiple time intervals (i.e. 30, 60, and 90 min), gas samples are collected from sample jars. Nitrous oxide (N_2O) concentrations of these gas samples can be determined with a gas chromatograph. Nitrous oxide fluxes are then calculated as the time-linear rate of concentration increase in the headspace of the sample jars. The DEA is then calculated as the short-term rate of N_2O production in the jars and is indicative of the size of the denitrifying enzyme pool present in the soil [85].

Another important nutrient retention function that occurs in wetlands is P sorption. The P sorption index (PSI) has been widely used to estimate the P retention capacity of wetland soils. The index was developed by Bache and Williams [88] and used by Richardson [89] in a seminal study of P sorption across various wetland types. Numerous studies have established that the PSI: 1) serves as a reliable gauge of a wetland soil's P sorption potential, 2) is less time consuming to measure than multiple-point P sorption isotherms, and 3) facilitates comparison with related soil properties [89–93]. The PSI can be determined by shaking a sterilized soil sample with a known P concentration for 24 h. The difference in concentration of inorganic P between the initial and final concentration represents the amount of P sorbed. The index is then calculated as $X \cdot (\log C)^{-1}$ where X = amount of P sorbed (mg P100 g soil^{-1}) and C = the final inorganic P concentration in solution (mg PL^{-1}).

Phosphorus fractionation is another common method that is used to quantify P storage in wetland soils and involves the quantification of the distribution of

P across various soil pools. These pools are operationally defined, but have generally been equated with bioavailable-P, Ca and Mg bound P, Al and Fe bound P, and residual pools [93–96].

9 Role of coastal wetlands in retaining and transforming nutrients

As P is generally considered to be the limiting nutrient in freshwater ecosystems, N is considered to be the limiting nutrient in marine ecosystems, and coastal wetlands represent transitional zones of both N and P limitation [8, 97], the results described in this section involve processes dealing with transformation and retention of N and P. Retention of N and P in wetlands results from cumulative fluxes into storage compartments of wetland ecosystems such as microbes, vegetation, plant litter, and soils [98]. Uptake of N and P by emergent wetland plants may be high during the growing season, but much of this N is released upon senescence in the late fall and winter [99, 100]. Rooted wetland plants tend to preferentially take up reduced ammonium nitrogen rather than oxidized nitrate nitrogen [52, 101]. In terms of climate, plant uptake is lowest in cold northern-hemisphere zones and wetlands with stagnant hydrology, whereas plant uptake is greatest in tropical zones and wetlands with active hydrology [52]. Unlike emergent vegetation, trees in forested wetlands provide long-term nutrient storage [99, 102]. Microbial uptake of N and P has also been thought to be a short-term rather than a long-term nutrient sink [84, 89, 98]. A comprehensive study of the effect of wetlands on water quality in Minnesota found that wetlands were more effective in removing suspended solids, total phosphorus and ammonia during high-flow periods, but were more effective at removing nitrate in low-flow periods [103]. This suggests that the ability of wetlands to retain and store nutrients may not only vary on a seasonal scale but also during storm flow compared to the baseflow.

Organic matter accumulation and denitrification are two of the more dominant and long-term N transformation and retention mechanisms. Organic matter accumulation involves the storage of N and P in soil organic matter (SOM) pools. These pools are generated from litter inputs, and often escape decomposition due to the anaerobic soil conditions. Soils with high organic carbon generally also have high organic N and P [104]. This processes mentioned above are common to all coastal wetland types. In addition, different coastal wetlands vary in their ability to retain and transform nutrients. These differences will be explored in the next sections of the chapter.

9.1 Retention and transformation of N and P in riparian wetlands

As a result of their location in the landscape, riparian wetlands interact with both upstream and upslope sources of nonpoint-source runoff and have the ability to reduce inputs of N and P to coastal waters [58, 105, 106]. Thus, riparian wetlands may be sinks for N and P at the watershed scale and if so, play a central role in maintaining regional water quality [107]. Retention and transformation of N and P by wetlands involves a combination of biogeochemical processes such as denitrification, P sorption, sedimentation, and organic matter accumulation, plant uptake, microbial immobilization [27, 52, 108].

Riparian wetlands have been shown to support high denitrification rates and, in certain cases, to transform the majority of nitrate inputs to nitrogen gases [109, 110]. Denitrification in these riparian wetlands has also shown high spatial variability [111–115] due to the presence of patches of organic matter and anaerobic microsites in the soil profile [116–119]. The term "hot spots" was coined to describe these areas of high denitrification [116, 120]. Recent studies have shown significantly lower denitrification potential in wetlands with nutrient-poor substrates (sand, light till) than in wetlands with nutrient-rich substrates (alluvium, dark till) [121], significantly lower DEA levels in restored/created wetlands compared to natural wetlands [86, 115], and a positive relationship between plant species richness and denitrification potential [122].

Soils of riparian wetlands have been shown to have higher P sorption capacities than adjacent uplands or streambanks [90, 107]. Long-term P storage in wetlands is believed to be controlled by three main processes: (1) deposition of sediment-bound P; (2) sorption of dissolved phosphate; or (3) the storage of organic P by peat accretion [96, 108]. While significant amounts of P can be stored by sedimentation [98], these sediments may be resuspended in future hydrologic events. Consequently, sorption and peat accretion are believed to represent the most important long-term P-retention pathways [89, 108].

In alkaline wetland soils, P sorption has been shown to be significantly correlated with calcium and magnesium content, and to a lesser degree with aluminum, iron and SOM content [108]. In acid wetland soils, P sorption has been shown to be significantly correlated with amorphous Al and Fe content [89, 93, 106], and to a lesser degree SOM and particle size [90, 93]. While iron phosphates are solubilized when Fe(III) is reduced to Fe(II) under anaerobic conditions, aluminum phosphates are unaffected by changes in redox potential [123]. Thus, soluble iron and phosphate may be lost from wetland soils in reducing conditions, while aluminum phosphates persist in acid wetland soils [107].

Another layer of complexity is added when we consider that amorphous Al and Fe, SOM, and texture exhibit significant spatial and temporal variability in riparian wetlands [93, 107, 124, 125]. Beyond a locally random aspect, this spatial variability may be related to the combined action of physical, chemical, or biological processes that operate at different spatial scales [126]. In natural riparian wetlands, these processes might include overbank flooding, sediment deposition, surface runoff, erosion, groundwater inputs, fire, tree-throw, root activity, litter production, and activity of macro and micro soil fauna. Each of these processes may influence particular locations of the riparian zone with varying degrees of intensity. For example, Fig. 7 shows the spatial variability of per cent clay, oxalate extractable aluminum, and the PSI of two riparian wetlands in the North Carolina Coastal Plain [93].

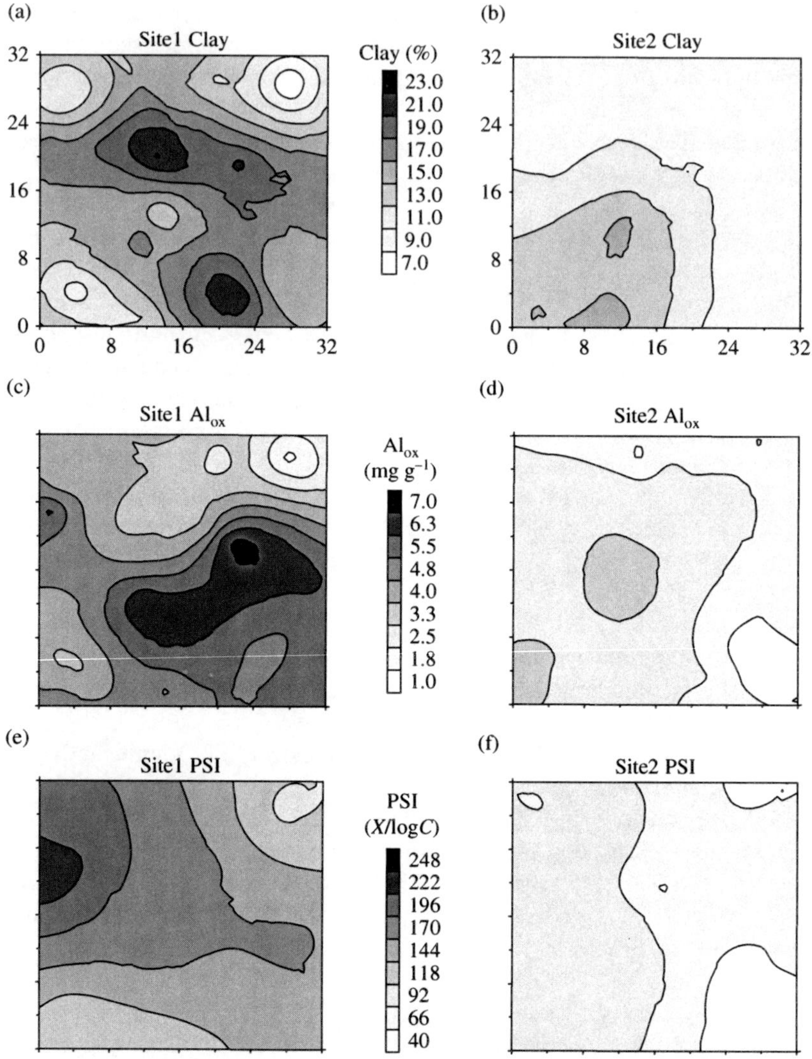

Figure 7: Spatial distribution of% clay (a), oxalate extractable aluminum (Al_{ox}) (c), and the phosphorus sorption index (PSI) (e) at Site 1, and% clay (b), Al_{ox} (d), and PSI (f) at Site 2. Plot size is 32 m by 32 m. Modified from [92].

9.2 Retention and transformation of N and P in tidal marshes

Salt marshes are thought to act as N transformers, importing dissolved oxidized inorganic forms of N and exporting dissolved and particulate reduced forms of N [127]. Salt marshes appear to be sinks for total P, but remobilization of phosphate in the sediments can lead to small net exports of phosphate from salt marshes [127].

Many studies have documented the fact that salt marshes are net exporters of organic material [27]. Salt marshes are generally thought to be N limited [3], but a recent study indicated that while the vegetation in the salt marshes was N limited, the microbial community was P limited [128].

In terms of P retention, a study along an estuarine salinity gradient along the Cooper River in South Carolina, demonstrated that there was a trend of decreasing P sorption capacity of intertidal marsh sediments with increasing salinity [129]. Specifically, the freshwater marsh site had the highest P sorption capacity followed by the two brackish marshes with intermediate P sorption. The salt marsh sedimentshad the lowest P sorption capacity. The results were attributed to the decrease in soil surface area across the salinity gradient as well as the changes in mineralogy, ionic strength, and redox chemistry of Al and Fe [129]. Interestingly, under freshwater conditions, Al and Fe hydroxides carry a net positive charge that facilitates P sorption, whereas under saltwater conditions, Al and Fe hydroxides carry a net negative charge that inhibits P sorption [130, 131].

9.3 Retention and transformation of N and P in mangroves

While the tidally dominated fringe mangroves provide protection of shorelines from marine wave action and storms, riverine and basin mangroves are more involved in trapping of sediment and retention of nutrients from terrestrial sources [39]. Mangroves are considered to be in a steady-state balance between N loss and N fixation, but they are capable of being stimulated to higher levels of production from local additions of fertilizers [36]. In addition to retaining sediments, Walsh [83] reported that the high nitrate and phosphate levels in Heeia Stream were reduced significantly in the upper reaches of the swamp, indicating that the mangroves may serve as sinks for these nutrients ask well. Mangroves have also been shown to have a high capacity to absorb and adsorb heavy metals and other toxic substances in effluents [132].

9.4 Retention and transformation of N and P in seagrass beds and coral reefs

While seagrass beds have been shown to trap sediments from terrestrial sources, their ability to retain and transform nutrients is largely unquantified [133]. Sediments underlying seagrass beds have been shown to have higher carbon content and lower redox potentials [134] indicating that they may be more effective at denitrification than adjacent areas without seagrasses or with rocky bottoms. Research on the deposition of seagrass castings and macroalgae remnants on beaches has shown these sources to be important for nutrient provisioning to coastal invertebrates and shorebirds [5]. Likewise, it was found that over 6 million kilograms dry weight of seagrass and algal detritus (20% of the annual production) is deposited each year on the 9.5 km beach of Mombasa Marine Park in Kenya [134]. These studies provide evidence that seagrasses may act as nutrient sources in coastal zones.

Nutrient cycling in coral reefs is quite complex. Primary producers have been shown to take up ammonium and nitrate from the waters surrounding the reef [36]. Consumers in the reef environment commonly excrete ammonia [36]. Reefs support a diverse community of bacteria that are involved in processes that produce, transform, or consume N such as N fixation, ammonification, nitrification, denitrification, and processing of organic N compounds [135]. Thus, it is difficult to generalize about the role the coral reefs plat in nutrient transformation and retention. However, it is clear that when reefs are subject to high sediment and nutrient loads, they do not respond favorably [17]. Thus, there is a great need to effectively manage the agriculture, pasture land, rangeland, and forestland in terrestrial watersheds, as well as to conserve and restore coastal wetlands that are so important for sediment and nutrient retention at the watershed scale.

10 Case study: comparison of soils from created, restored and natural wetlands

The United States Clean Water Act (CWA) of 1972 mandates mitigation whenever natural wetlands are impacted by development. The Army Corps of Engineers (ACoE) has jurisdiction over this process and requires created and restored wetlands to meet specific vegetative and hydrologic criteria during a five-year monitoring period to be considered successful [136]. Vegetative criteria require survival of a certain per centage of planted species per acre. Hydrologic criteria require that the water table be within 30 cm of the soil surface for a consecutive period of at least 12.5% of the growing season. The current process does not require any monitoring of soil properties or processes [136, 137]. It is interesting that soil has been omitted from the mitigation process, as soil plays an integral part in the definition of a wetland as stated in the USFWS Wetland Classification System [18]. Hydric soil, along with hydrophytic vegetation, and wetland hydrology are also the three criteria used to delineate jurisdictional wetlands as defined by the ACoE [138]. The lack of consideration of edaphic characteristics in the wetland mitigation process is a cause for concern for a number of reasons: 1) soil forms the foundation of these developing ecosystems; 2) inadequate soil properties can be detrimental to vegetative survival and the establishment of wetland hydrology; and 3) soil is the medium for biogeochemical processes that transform and retain nutrients [137]. Without suitable soil properties, created wetlands (CWs) and restored wetlands (RWs) may never replace the nutrient transformation and retention functions of the natural wetlands (NWs) that were destroyed.

As few CWs or RWs are assessed beyond what is needed to meet hydrologic and vegetative success criteria [139], the ability of these wetlands to replace natural wetland functions is a topic of considerable debate [86, 140–142]. It has been stated that the definitive test of success for CWs and RWs is how closely they function like NWs [143]. Unfortunately, only a few studies have attempted to determine whether wetland functions in CWs and NWs are equivalent to those of NWs [86]. The assumption that wetland function follows wetland structure, which underlies the ACoE monitoring process, is also largely untested [144].

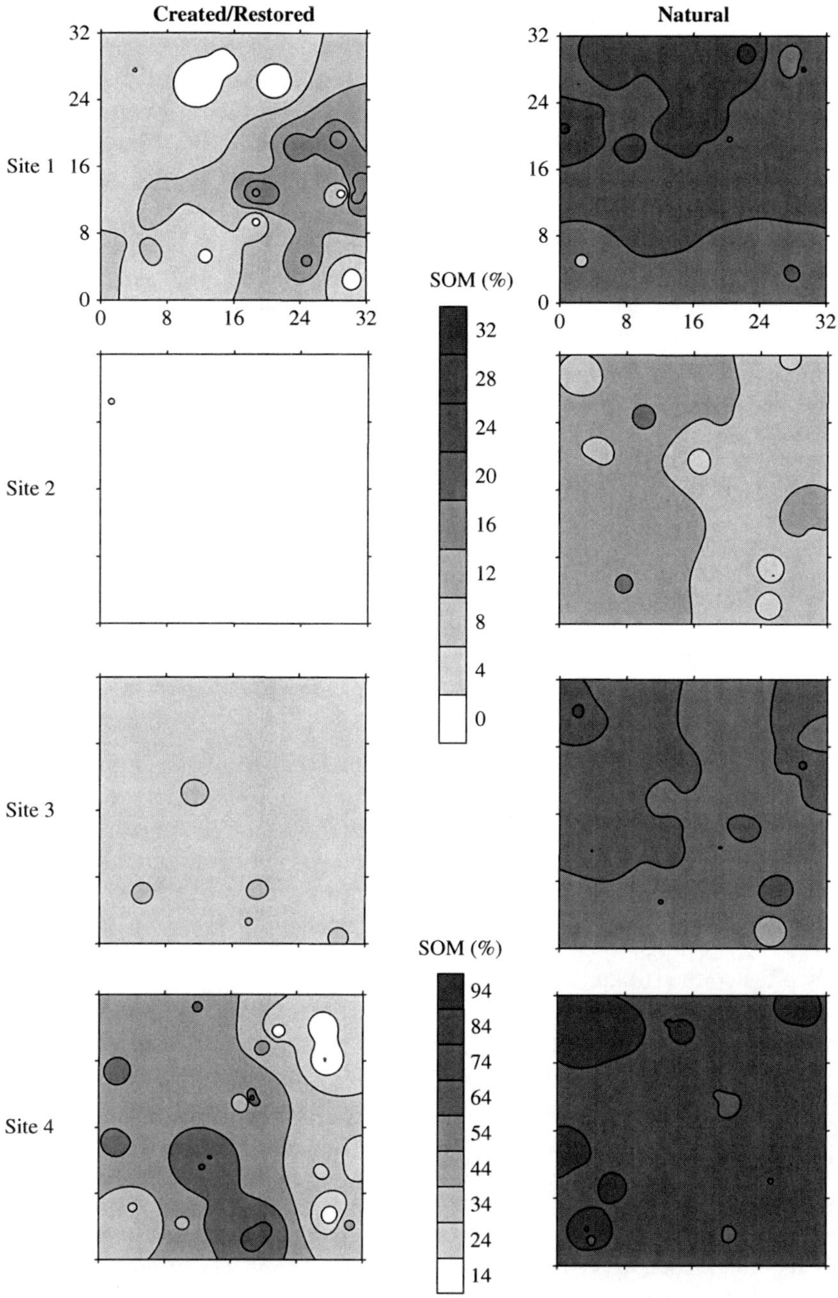

Figure 8: Spatial distribution of soil organic matter (SOM) at Site 1 restored (a) and natural wetland (b), at Site 2 created (c) and natural wetland (d), at Site 3 (e) and natural wetland (f), and at Site 4 restored (g) and natural wetland (h) plots. The upper scale applies to Sites 1–3, while the lower scale applies to Site 4. Plot size is 32 m by 32 m. Modified from [153].

For example, soil properties of CWs and RWs have almost always been shown to differ from NWs [145]. Created wetlands typically have higher sand and lower clay content than NWs [141, 146, 147]. This has important implications for wetland function, as coarse-textured soils typically have lower water-holding and nutrient-retention capacities than fine-textured soils [148, 149]. Created wetlands and RWs also usually have lower levels of SOM and higher bulk densities than NWs [143, 146, 147, 150, 151]. Such soil conditions can lead to low growth and survival of planted and colonizing species. Litter layers in CW/RWs are often poorly developed or absent in comparison to that of NWs [86, 144]. As a result of low organic matter and sparse litter, it has been speculated that the microbial communities in the soil of CWs are much less viable than those of NWs [152]. Additionally, soil temperatures were reported to be significantly higher in CWs than in NWs as a result of a lack of shading by mature trees [141]. Furthermore, microtopography has been reported to be considerably lower in CW/RWs than in NWs [141, 151].

Bruland and Richardson [153] conducted a study in which they compared patterns of spatial variability of soil properties in CWs/RWs and paired NWs in the North Carolina coastal plain. They found that the spatial variability of SOM was lower at some but not all of the CWs/RWs (Fig. 8), and concluded that prior land use and mitigation activities could decrease, increase, or cause no change in the spatial variability of soil properties in CWs/RWs compared to NWs. In another study, Bruland *et al.* [115] found that the spatial variability of predicted DEA values in was much lower in the CWs/RWs than in the NWs for the riverine sites but not for the nonriverine sites

11 Future research needs and directions

Existing landscape patterns contain information about the processes that generated these patterns [154]. In various coastal wetlands around the world, the processes of hydrologic modification, sedimentation, and nutrient loading have not only affected the structure and function of these ecosystems but also their spatial extent and distribution [153, 155–158]. The use of remote sensing and geographic information systems (GIS) have enhanced scientists' capacity to describe patterns in nature over larger spatial scales and at finer levels of detail than ever before [159]. These methods can be used to quantify anthropogenic impacts to wetlands at multiple scales, providing valuable information to aid in the design of wetland and watershed restoration projects. These types of approaches may be especially useful for the monitoring of seagrasses and coral reefs. Traditional assessment of cover and density of seagrass and coral cover along transects and quadrats generally have an associated error >30% about the mean [43]. This makes it difficult to detect reliable changes in seagrass or coral cover. Typically, changes can only be detected when they are greater than 50–80%. Thus there is a great need to develop remote-sensing techniques to more accurately monitor changes in seagrass beds over large spatial scales [43].

Significant strides have been made in coastal-zone management in the last few decades and many of the world's 123 coastal countries now have some form of coastal management plans and legislation [5]. However, countries with well-developed coastal-zone management plans are still facing loss of coastal wetlands, overexploitation of coastal resources, user conflicts, and indirect degradation from activities occurring sometimes hundreds of km from the coastal zone itself [5]. Thus, management and policy have not been able to keep pace with increasing degradation of coastal ecosystems.

One bright spot in this picture is that coastal wetlands such as tidal marshes and mangroves can often be restored once they have been degraded. The science of salt-marsh and mangrove restoration is relatively advanced compared to restoration of other ecosystems. Research has shown that the establishment or re-establishment of proper elevation, topography, tidal flushing, and the planting of a few key species can restore salt marshes and mangroves on relatively short time scales [144, 160]. Mangroves appear to be the most amenable to restoration [15] of the coastal wetlands considered in this chapter. Restoration of mangroves has been accomplish by transplanting vegetative propagules, young trees, or mature trees with considerable success in India and Southeast Asia [161]. Coastal wetland restoration in the United States has been most successfully achieved in estuarine salt-marsh systems [162, 163]. However, studies have shown that, while the vegetative structure of created and restored marshes appears to approximate that of natural marshes, the soil properties and macroinverbrate communities in the created and restored marshes may take much longer time scales to develop [164, 165]. Restoration of seagrass meadows although increasingly common, has been fraught with difficulties; a worldwide success rate for restoration of these systems was estimated to be only 45% [161, 166]. Restoration of coral reefs may be the most difficult challenge of any ecosystem, and it has only been practiced at small scales and with limited success [133].

Future research, management, restoration, and policy needs to occur at the watershed or regional scale [167] and involve an interdisciplinary approach to assessing the role and functions of coastal wetlands in reducing the impacts of land-based management. According to Vivian-Smith [168], wetland restoration should consider ecological processes and structure at multiple spatial scales (Fig. 9). Hydrologic, vegetative, and edaphic heterogeneity increases the probability that an optimal habitat will exist at a restoration site for the intended species and better approximate the structure and function of natural wetlands [15, 169]. On a positive note, there are currently a number of landscape-scale coastal wetland restoration projects in various phases of planning and implementation that are designed to ameliorate the effects of hydrologic alteration and nutrient loading including the Chesapeake Bay [170], Mississippi River basin [171], Florida Everglades [172], and the wetlands of Iraq [157]. However, it has been estimated that a tripling of the area of riparian forests and buffer systems would be needed in the Upper Mississippi River and Ohio River Basins in order to cause significant reduction in the N load of the Mississippi River to the Gulf of Mexico [171].

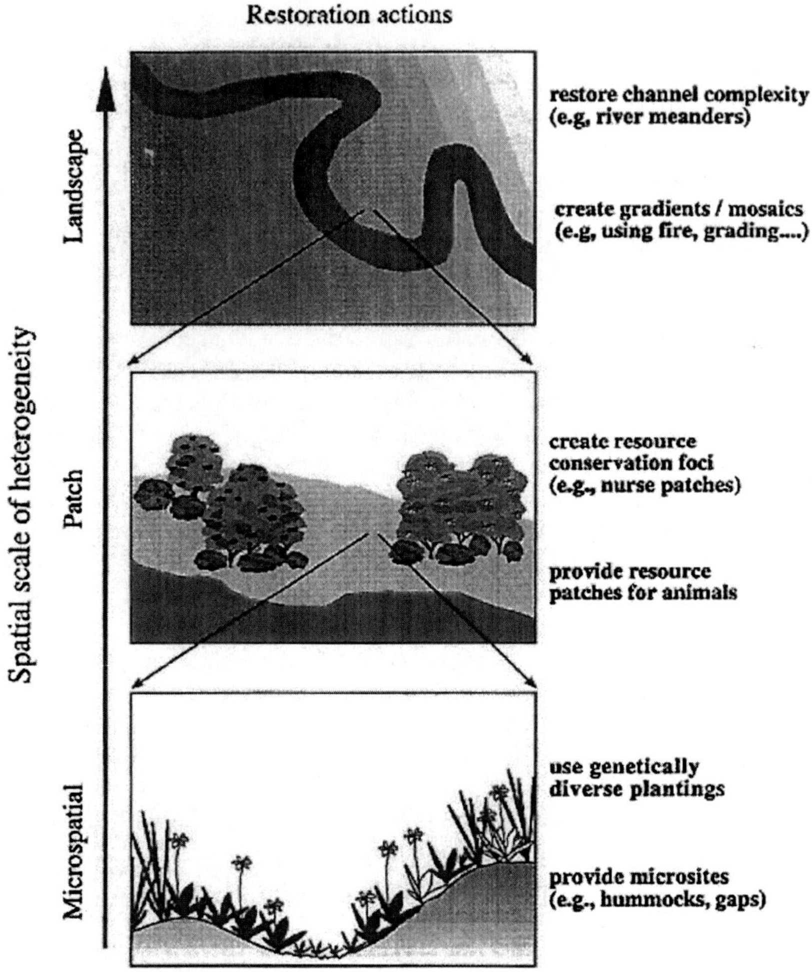

Figure 9: Restoration actions in the coastal wetland landscape focusing on habitat heterogeneity at different spatial scales (microscale, patch, and landscape) that influence key processes and accelerate ecosystem development. Copyright (2001) from, Developing a Framework for Restoration, by G. Vivian-Smith [168]. Reproduced by permission of Taylor & Francis, a division of Informa plc.

Another move in the right direction appears to be the coupling of coastal-zone management with watershed management as has occurred with the European Water Framework Directive and projects under the LOICZ (Land-Sea Interactions in the Coastal Zone) initiative [5]. Also, the U.S. Coral Reefs Task Force has established local action strategies to assess land-based threats to coral reefs specifically looking at solutions to prevent erosion and nutrient loading to coastal zones from terrestrial sources. With future population trends suggesting that the world's coastal population is expected to approach 6 billion people by 2025 [15], our ability to preserve, protect, manage, and restore watersheds, coastal wetlands, and near-shore habitats is becoming increasingly important.

References

[1] Nixon, S.W., Coastal marine eutrophication: a definition social causes, and future concerns. *Ophelia*, **41**, pp. 199–219, 1995.

[2] Howarth, R.W., Billen, G., Swaney, D., Townsend, A., Jaworski, N., Lajtha, K., Downing, J.A., Elmgren, R., Caraco, N., Jordan, T., Berendse, F., Freney, J., Kudeyarov, V., Murdoch, P. & Zhu, Z.L., Regional nitrogen budgets and riverine N and P fluxes for the drainages to the North Atlantic Ocean: Natural and human influences. *Biogeochemistry*, **35**, pp. 75–79, 1996.

[3] Vitousek, P.M., Aber, J.D., Howarth, R.W., Likens, G.E., Matson, P.A., Schindler, D.W., Schlesinger, W.H. & Tilman, D.G., Human alteration of the global nitrogen cycle: sources and consequences. *Ecological Applications*, **7**, pp. 737–50, 1997.

[4] Rablais, N.N., Turner, R.E. & Scavia, D., Beyond science into policy: Gulf of Mexico hypoxia and the Mississippi River. *BioScience*, **52**, pp. 129–142, 2002.

[5] Agardy, T. & Alder, J., Coastal Systems. *Millennium Ecosystem Assessment, Ecosystems and Human Well-Being: Current State and Trends* [Vol. 1], Island Press: Washington, DC, pp. 513–549, 2005.

[6] Reichelt-Brushett, A.J. & Harrison, P.L., The effect of copper, zinc, and cadmium on fertilization success of gametes from scleractinian reef corals. *Marine Pollution Bulletin*, **38**, pp. 182–187, 1999.

[7] Jones, R.J., Muller, J., Haynes, D. & Schreiber, U., Effects of herbicides diuron and atrazine on corals of the Great Barrier Reef, Australia. *Marine Ecology Progress Series*, **251**, pp. 153–167, 2003.

[8] Schlesinger, W.H., *Biogeochemistry*, 2nd edn, Academic Press: San Diego, 1997.

[9] Costanza, R., d'Arge, R., de Groot, R., Farber, S., Grasso, M., Hannon, B., Limburg, K., Naeem, S., O'Neill, R.V., Paruelo, J., Raskin, R.G., Sutton, P. & van den Belt, M., The value of the world's ecosystem services and natural capital. *Nature*, **387**, pp. 253–260, 1997.

[10] Dahl, T.E. & Johnson, C.E., Wetland status and trends in the conterminous United States mid-1970s to mid-1980s. U.S. Department of the Interior, Fish & Wildlife Service: Washington, DC, 1991.

[11] Noe, G.B., Childers, D.L. & Jones, R.D., Phosphorus biogeochemistry and the impact of phosphorus enrichment: Why is the Everglades so unique? *Ecosystems*, **4**, pp. 603–624, 2001.

[12] United States Geological Survey (USGS). 100 + Years of Land Change for Coastal Louisiana. National Wetlands Research Center, Lafayette. Map ID: USGS-NWRC 20030-03-085, 2003.

[13] Millennium Ecosystem Assessment (MEA), *Ecosystems and human well-being: wetlands and water synthesis*. World Resources Institute: Washington, DC, 2005.
[14] Center for International Earth Science Information Network (CIESIN). Gridded population of the world [GPW] version 3 beta, http://quin.unep-wcmc.org/MA/index.cfm.
[15] Kennish, M.J., Environmental threats and environmental future of estuaries. *Environmental Conservation*, **29**, pp. 78–107, 2002.
[16] Hemminga, M.A. & Duarte, C.M., *Seagrass ecology*, Cambridge University Press, Cambridge, 2000.
[17] Hughes, T.P., Baird, A.H., Bellwood, D.R., Card, M., Connolly, S.R., Folke, C., Grosberg, R., Hoegh-Guldberg, O., Jackson, J.B.C., Kleypas, J., Lough, J.M., Marshall, P., Nysrom, M., Palumbi, S.R., Pandolfi, J.M., Rosen, B. & Roughgarden, J., Climate change, human impacts, and the resilience of coral reefs. *Science*, **301**, pp. 929–933, 2003.
[18] Cowardin, L.M., Carter, V., Golet, F.C. & LaRoe, E.T., *Classification of Wetlands and Deepwater Habitats of the United States. U.S. Department of the Interior, Fish & Wildlife Service FWS/OBS-79/31*: Washington, DC, 1979.
[19] Warner, B.G. & Rubec, C.D.A., *The Canadian Wetland Classification System*, 2nd edn. Wetlands Research Center, University of Waterloo, Waterloo, 1997.
[20] Navid, D., The international law of migratory species: The Ramsar Convention. Natural *Resources Journal*, **29**, pp. 1001–1016, 1989.
[21] Brinson, M.M., A hydrogeomorphic classification for wetlands. Wetlands Research Program Technical Report WRP-DE-4. U.S. Army Corps of Engineers Waterways Experiment Station, Vicksburg, 1993.
[22] Johnson, R.R. & McCormick, J.F., Tech. Coords., Strategies for the Protection and Management of Floodplain Wetlands and Other Riparian Ecosystems. Proceedings of the Symposium, Callaway Gardens, Georgia, Dec. 11–13, 1978. General Technical Report WO-12, U.S. Forest Service: Washington, DC, 1979.
[23] Gregory, S.V., Swanson, F.J., McKee, W.A. & Cummins, K.W., An ecosystem perspective of riparian zones. *BioScience*, **41**, pp. 540–551, 1991.
[24] Brinson, M.M., Swift, B.L., Plantico, R.C. & Barclay, J.S., Primary productivity, decomposition, and consumer activity in freshwater wetlands. *Annual Review of Ecology and Systematics*, **12**, pp. 123–161, 1981.
[25] The Nature Conservancy, The forested wetlands of the Mississippi River: an ecosystem in crisis. *Nature Conservancy*: Baton Rogue, 1992.
[26] Bernhardt, E.S., Palmer, M.A., Allan, J.D., Alexander, G., Barnas, K., Brooks, S., Carr, J., Clayton, S., Dahm, C., Follstad-Shah, J., Galat, D., Gloss, S., Goodwin, P., Hart, D., Hassett, B., Jenkinson, R., Katz, S., Kondolf, G.M., Lake, P.S., Lave, R., Meyer, J.L., O'Donnell, T.K.O., Pagano, L., Powell, B. & Sudduth, E., Synthesizing U.S. river restoration efforts. *Science*, **305**, pp. 636–637, 2005.
[27] Mitsch, W.J. & Gosselink, J.G., *Wetlands*, 3rd edn, John Wiley & Sons, Inc.: New York, 2000.

[28] Teal, J.M. & Teal, M., *Life and Death of a Salt Marsh*, Audobon/Ballantine Books: New York, 1969.

[29] Ledoux, L., Vermaatb, J.E., Bouwerb, L., Salomons, W. & Turnera, R.K., ELOISE research and implementation of the EU policy in the coastal zones, http://130.37.129.100/english/o_o/instituten/IVM/research/eloise/pdf/btbch1.pdf.

[30] Chapman, V.J., *Salt Marshes and Salt Deserts of the World*. Interscience: New York, 1960.

[31] Bertness, M.D., Ewanchuk, P.J. & Silliman, B.R., Anthropogenic modification of New England salt marsh landscapes. *Proceedings of the National Academy of Sciences of the United States of America*, **99**, pp. 1395–1398, 2002.

[32] Beeftick, W.G., Salt marshes. *The Coastline,* (ed.) R.S.K. Barnes, John Wiley & Sons: New York, p. 93–121, 1977.

[33] Silliman, B.R. & Bertness, M.D., A trophic cascade regulates salt marsh primary production. *Proceedings of the National Academy of Sciences of the United States of America*, **99**, pp. 10500–10505, 2002.

[34] Spalding, M.D., Blasco, F. & Field, C.D., *World Mangrove Atlas*, The International Society for Mangrove Ecosystems: Okinawa, 1997.

[35] Valiela, I., Bowen, J.L. & York, J.K., Mangrove forests: One of the world's threatened major tropical environments. *Bioscience*, **51**, pp. 807–815, 2001.

[36] Mann, K.H., *Ecology of Coastal Waters*, 2nd edn. Blackwell Science: Malden, 2000.

[37] Duke, N.C., Mangrove floristics and biogeography. *Tropical Mangrove Ecosystems*, (ed.) A.I. Robertson & D.M. Alongi, American Geophysical Union: Washington, DC, pp. 63–100, 1992.

[38] Twilley, R.R., Mangrove Wetland. *Southern Forested Wetlands: Ecology and Management*, (ed.) M.G. Messina & W.H. Conner, Lewis Publishers: Boca Raton, pp. 445–473, 1998.

[39] Ewel, K.C., Twilley, R.R. & Ong, J.E., Different kinds of mangrove forests provide different goods and services. *Global Ecology & Biogeography*, **7**, pp. 83–94, 1998.

[40] Ong, J.E., The ecology of mangrove conservation and management. *Hydrobiologia*, **295**, pp. 342–351, 1995.

[41] Allen, J.A. Mangroves as alien species: the case of Hawaii. *Global Ecology and Biogeography Letters*, **7**, pp. 61–71, 1998.

[42] Ewel, J.J. & Bigelo, S.W., Plant life forms and tropical ecosystem functioning. *Biodiversity and Ecosystem Processes in Tropical Forests*, (ed.) G. Orians, R. Dirzo & J.H. Cushman, Springler Verlag: Berlin, 1996.

[43] Duarte, C.M., The future of seagrass meadows. *Environmental Conservation*, **29**, pp. 192–206, 2002.

[44] Duffy, J.E., Biodiversity and the functioning of seagrass ecosystems. *Marine Ecology Progress Series*, **311**, pp. 233–250, 2006.

[45] Duarte, C.M. & Cebrian, J., The fate of marine autotrohic production. *Limnology & Oceanography*, **41**, pp. 1758–1766, 1996.

[46] Anderson, E.E., Economic benefits of habitat restoration: Seagrass and the Virginia hardshell blue crab fishery. *North American Journal of Fisheries Management*, **9**, pp. 140–149, 1989.

[47] den Hartog, C., *The Seagrasses of the World*, North Holland: Amsterdam, 1970.

[48] Miliman, J.D. & Meade, R.H., World-wide delivery of river sediment to the oceans. *Journal of Geology*, **91**, pp. 1–21, 1993.

[49] Shepherd, S.A., McComb, A.J., Bulthuis, D.A., Neverauskas, V., Stefensen, D.A. & West, R., Decline of seagrasses. *Biology of Seagrasses*, (ed.) A.W.D. Larkum, A.J. McComb & S.A. Shepherd. Elsevier: Amsterdam, the Netherlands, pp. 346–393, 1989.

[50] Burke, L., Kura, Y., Kassem, K., Ravenga, C., Spalding, M. & McAllister, D., *Pilot Assessment of Global Ecosystems: Coastal Ecosystems*, World Resources Institute: Washington, DC, 2001.

[51] Dayton, P.K., The importance of the natural sciences to conservation. *American Naturalist*, **162**, pp. 1–10, 2003.

[52] Bowden, W.B., The biogeochemistry of nitrogen in freshwater wetlands. *Biogeochemistry*, **4**, pp. 313–348, 1984.

[53] Hudson, N., *Soil Conservation*, 2nd edn. Cornell University Press: Ithaca, 1981.

[54] El-Swaify, S.A., Operative Processes for sediment-based watershed degradation in small, tropical volcanic island ecosystems. *Integrated Watershed Management in the Global Ecosystem*, (ed.) R. Lal. CRC Press: Boca Raton, pp. 35–49, 2000.

[55] DeCarlo, E.H., Beltran, V.L. & Tomlinson, M.S., Composition of water and suspended sediment in stream of urbanized subtropical watersheds in Hawaii. *Applied Geochemistry*, **19**, pp. 1011–1037, 2004.

[56] Novitski, P.R. Wetland hydrology. *Wetlands Ecology and Conservation: Emphasis on Pennsylvania*, (ed.) S.K. Majumdar, R.P. Brooks, F.J. Benner & R.W. Tiner, Pennsylvania Academy of Science: Easton, pp. 47–64, 1989.

[57] Faber, P.A., Keller, E., Sands, A. & Masser, B.M., The Ecology of Riparian Habitats of the Southern California Coastal Region: A Community Profile. Biological Report 85 [7.27], U.S. Fish & Wildlife Service: Washington, DC, 1989.

[58] Brinson, M.M., Changes in the functioning of wetlands along environmental gradients. *Wetlands*, **13**, pp. 65–74, 1993.

[59] Hupp, C.R. & Bazemore, D.E., Spatial and temporal aspects of sediment deposition in West Tennessee forested wetlands. *Journal of Hydrology*, **14**, pp. 179–196, 1993.

[60] Kleiss, B.A., Sedimentation in a bottomland hardwood wetland in eastern Arkansas. *Wetlands*, **16**, pp. 321–333, 1996.

[61] Noe, G.B. & Hupp, C.R. Carbon, nitrogen, and phosphorus accumulation in floodplains of Atlantic Coastal Plain Rivers, USA. *Ecological Applications*, **15**, pp. 1178–1190, 2005.

[62] Harter, S.K. & Mitsch, W.J., Patterns of short-term sedimentation in a freshwater created marsh. *Journal of Environmental Quality*, **32**, pp. 325–334, 2003.

[63] Cahoon, D.R., Lynch, J.C. & Knaus, R.M., Improved cryogenic coring device for sampling wetland soils. *Journal of Sedimentary Research*, **66**, pp. 1025–1027, 1996.

[64] DeLaune, R.D., Reddy, C.N. & Patrick, W.H., Jr., Accumulation of plant nutrients and heavy metals through sedimentation processes and accretion in a Louisiana salt marsh. *Estuaries*, **4**, pp. 328–334, 1981.

[65] Boumans, R.M. & Day, J.W., Jr., The effects of two Louisiana marsh management plans on water and materials flux and short-term sedimentation. *Wetlands*, **14**, pp. 247–261, 1994.

[66] Fennessy, M.S., Brueske, C.C. & Mitsch, W.J., Sediment deposition patterns in restored freshwater wetlands using sediment traps. *Ecological Engineering*, **3**, pp. 409–428, 1994.

[67] Hensel, P.F. & Day, J.W., Jr., Short-term sedimentation dynamics in the Rhone River Delta France: The importance of riverine pulsing. *Estuaries*, **21**, pp. 52–65, 1998.

[68] Robbins, J. A. & Edgington, D.N., Determination of recent sedimentation-rates in Lake Michigan using Pb-210 and Cs-137. *Geochimicha et Cosmochimica Acta*, **39**, pp. 285–295, 1975.

[69] Craft, C.B. & Richardson, C.J., Peat accretion and N, P, and organic C accumulation in nutrient-enriched and unenriched Everglades peatlands. *Ecological Applications*, **3**, pp. 446–458, 1993.

[70] Craft, C.B. & Richardson, C.J., Recent [^{137}Cs] and long-term [^{210}Pb, ^{14}C] organic soil accretion and nutrient accumulation in the Everglades. *Soil Science Society of America Journal*, **62**, pp. 834–843, 1998.

[71] Richie, J.C. & McHenry, J.R., Determination of fallout Cs-137 and naturally occurring gamma ray emitters in sediments. *International Journal of Applied Radioactive Isotopes*, **24**, pp. 575–578, 1973.

[72] Armentano, T.V. & Woodwell, G.M., Sedimentation rates in a Long Island marsh determined by Pb-210 dating. *Limnology and Oceanography*, **20**, pp. 452–456, 1975.

[73] Brenner, M., Peplow, A.J. & Scheleske, C.L., Disequilibrium between ^{226}Ra supported ^{210}PB in a sediment core from a shallow Florida Lake. *Limnology & Oceanography*, **39**, pp. 1222–1227, 1994.

[74] Binford, M.W., Calculation and uncertainty analysis of ^{210}Pb dates for PIRLA project lake sediment cores. *Journal of Paleolimnology*, **3**, pp. 253–267, 1990.

[75] Phillips, J.D. 1989. Fluvial sediment storage in wetlands. *Water Resources Bulletin*, **25**, pp. 867–877.

[76] Edgar, G.J. & Barrett, N.S., Effects of catchment activities on macrofaunal assemblages in Tasmanian estuaries. *Estuarine, Coastal & Shelf Science*, **50**, pp. 639–654, 2000.

[77] Dubinsky, Z. & Stambler, N., Marine pollution and coral reefs. *Global Change Biology*, **2**, pp. 511–526, 1996.

[78] Wolanski, E. & Spagnol, S., Environmental degradation by mud in tropical estuaries. *Regional Environmental Change*, **1**, pp. 152–162, 2000.

[79] Golbuu, V., Victor, S., Wolanski, E. & Richmond, R.H., Trapping of fine sediment in an enclosed bay, Palau, Micronesia. *Estuarine, Coastal, & Shelf Science*, **57**, pp. 941–949, 2003.

[80] Wolanski, E., Richmond, R.H., Davis, G. & Bonito, V., Water and fine sediment dynamics in transient river plumes in a small, reef-fringed bay, Guam. *Estuarine, Coastal & Shelf Science*, **56**, pp. 1029–1040, 2003.

[81] Victor, S., Neth, L., Golbuu, Y., Wolanski, E. & Richmond, R.H., Sedimentation in magroves and coral reefs in a wet tropical islands, Pohnpei, Micronesia. *Estuarine, Coastal & Shelf Science*, **66**, pp. 409–416, 2006.

[82] Bigelow, K.A., Alspach, K., Lohle, R., McDonough, T., Ravetto, P., Rosenfeld, C., Stender, G. & Wong, C., Assessment of mangrove ecosystem of west Moloka'i, Hawaii, with additional surveys of Moauni Beach Park and Ualapu'e Fishpond. Marine Option Program, University of Hawaii: Honolulu, 1989.

[83] Walsh, G.E., An ecological study of a Hawaiian mangrove swamp. ed. G.H. Lauff, *Estuaries*, American Association for the Advancement of Science Publication No. 83: Washington, DC, pp. 420–431, 1967.

[84] Richardson, C.J. & Marshall, P.R., Processes controlling movement, storage, and export of phosphorus in a fen peatland. *Ecological Monographs*, **56**, pp. 279–302, 1986.

[85] Tiedje, J.M., Denitrification. *Methods of Soil Analysis, Part 2. Chemical and Microbiological Properties,* ASA-SSSA, Madison, 1982.

[86] Hunter, R.G. & Faulkner, S.P., Denitrification potential in restored and natural bottomland hardwood wetlands. *Soil Science Society of America Journal*, **65**, pp. 1865–1872, 2001.

[87] Groffman, P.M., Gold, A.J. & Simmons, R.C., Nitrate dynamics in riparian forests: microbial studies. *Journal of Environmental Quality*, **21**, pp. 666–671, 1992.

[88] Bache, B.W. & Williams, E.G., A phosphate sorption index for soils. *Journal of Soil Science*, **22**, pp. 289–301, 1971.

[89] Richardson, C.J., Mechanisms controlling phosphorus retention capacity in freshwater wetlands. *Science*, **228**, pp. 1424–1427, 1985.

[90] Axt, J.R. & Walbridge, M.R., Phosphate removal capacity of palustrine forested wetlands and adjacent uplands in Virginia. *Soil Science Society of America Journal*, **63**, pp. 1019–1031, 1999.

[91] Bridgham, S.D., Johnston, C.A. & Schubaurer-Berigan, J.P., P sorption dynamics in soils and coupling with surface and pore water in riverine wetlands. *Soil Science Society of America Journal*, **65**, pp. 577–588, 2001.

[92] Bruland, G.L. & Richardson, C.J., A spatially explicit investigation of P sorption and related soil properties in two riparian wetlands. *Journal of Environmental Quality*, **33**, pp. 785–794, 2004.

[93] Bruland, G.L. & Richardson, C.J., An Assessment of the phosphorus sorption capacity of wetlands in the Painter Creek Watershed, Minnesota, USA. *Water, Air & Soil Pollution*, **171**, pp. 169–184, 2006.

[94] Hedley, M.J., Stewart, J.W.B. & Chauhan, B.S., Changes in inorganic and organic soil phosphorus fractions by cultivation practices and by laboratory incubations. *Soil Science Society of America Journal*, **46**, pp. 970–976, 1982.

[95] Tiessen, H. & Moir, J.O., Characterization of available P by sequential extraction, *Soil Sampling and Methods of Analysis*, (ed.) M.R. Carter, Lew is Publishers: Boca Raton, pp. 75–86, 1993.

[96] Qualls, R.G. & Richardson, C.J., Forms of soil phosphorus along a nutrient enrichment gradient in the northern Everglades. *Soil Science Society of America Journal*, **160**, pp. 183–198, 1995.

[97] Correll, D.L., Phosphorus: a rate limiting nutrient in surface waters. *Poultry Science*, **78**, pp. 674–682, 1999.

[98] Johnston, C.A., Sediment and nutrient retention by freshwater wetlands: Effects on surface water quality. *Critical Reviews in Environmental Control*, **21**, pp. 495–565, 1991.

[99] Reddy, K.R. & DeBusk, W.F., Nutrient storage capacities of aquatic and wetland plants, (ed.) K.R. Reddy & W.H. Smith, Magnolia Publishers, pp. 337–357, 1987.

[100] Mitsch, W.J., Cronk, J.K., Wu, X. & Nairn, R.W., Phosphorus retention in constructed freshwater riparian marshes. *Ecological Applications*, **5**, pp. 830–845, 1995.

[101] Howard-Williams, C. & Downes, M.T., Nitrogen cycling in wetlands. *Nitrate: Processes Patterns and Management*, (ed.) T.P. Burt, A.L. Heathwaite, S.T. Trudgill, John Wiley & Sons, Ltd.: New York, pp. 141–167, 1993.

[102] Mitsch, W.J., Dorje, C.L. & Wiemhoff, J.R., Ecosystem dynamics and a phosphorus budget of an alluvial cypress swamp in Southern Illinois. *Ecology*, **60**, pp. 1116–1124, 1979.

[103] Johnston, CA., Deitenbeck, N.E. & Nieme, G.J., The cumulative effect of wetlands on stream water quality and quantity. A landscape approach. *Biogeochemistry*, **10**, pp. 105–141, 1990.

[104] Reddy, K.R., Kadlec, R.H., Flaig, E. & Gale, P.M., Phosphorus retention in Stream and Wetlands: a review. *Critical Reviews in Environmental Science & Technology*, **29**, pp. 83–146, 1999.

[105] Walbridge, M.R., Functions and values of forested wetlands in the Southern United States. *Journal of Forestry*, **91**, pp. 15–19, 1993.

[106] Lockaby, B.G. & Walbridge, M.R., Biogeochemistry. *Southern Forested Wetlands: Ecology and Management*, (ed.) M.G. Messina & W.H. Conner, Lewis Publishers: Boca Raton, pp. 149–172, 1998.

[107] Darke, A.K. & Walbridge, M.R., Al and Fe biogeochemistry in a floodplain forest: Implications for P retention. *Biogeochemistry*, **51**, pp. 1–32, 2000.

[108] Richardson, C.J., The role of wetlands in storage, release, and cycling of phosphorus on the landscape: A 25–year retrospective. *Phosphorus Biogeochemistry of Sub-Tropical Ecosystems*, (ed.) K.R. Reddy, CRC Press: Boca Raton, pp. 47–68, 1999.

[109] Jacobs, T.C. & Gilliam, J.W., Riparian losses of nitrate from agricultural drainage waters. *Journal of Environmental Quality*, **14**, pp. 472–478, 1985.

[110] Lowrance, R., Groundwater nitrate and denitrification in a Coastal Plain riparian forest. *Journal of Environmental Quality*, **41**, pp. 401–405, 1992.

[111] Schipper, L.A., Cooper, A.B., Harfoot, C. G. & Dyck, W. J., Regulators of denitrification in an organic riparian soil. *Soil Biology & Biochemistry*, **25**, pp. 925–933, 1993.

[112] Pinay, G., Black, V.J., Planty-Tabacchi, A.M., Gumiero, B. & Decamps, H., Geomorphic control of denitrification in large river floodplain soils. *Biogeochemistry*, **50**, pp. 63–182, 2000.

[113] Casey, R.E., Taylor, M.D. & Klaine, S.J., Mechanisms of nutrient attenuation in a subsurface flow riparian wetland. *Journal of Environmental Quality*, **30**, pp. 1732–1737, 2001.

[114] Dhont, K., Boeckx, P., Hofman, G. & Van Cleemput, O., Temporal and spatial patterns of denitirifcation and nitrous oxide fluxes in three adjacent vegetated riparian buffer zones. *Biology & Fertility of Soils*, **40**, pp. 243–251, 2004.

[115] Bruland, G.L., Richardson C.J. & Whalen, S., Spatial variability of denitrification potential and related soil properties in created, restored and paired natural wetlands. *Wetlands*, **26**, pp. 1042–1056, 2006.

[116] Parkin, T.B., Soil microsites as a source of denitrification variability. *Soil Science Society of America Journal*, **51**, pp. 1194–1199. 1987.

[117] Gold, A.J., Jacinthe, P.A., Groffman, P.M., Wright, W.R. & Puffer, R.H., Patchiness in groundwater nitrate removal in a riparian forest. *Journal of Environmental Quality*, **27**, pp. 146–155, 1998.

[118] Jacinthe, P.A., Groffman, P.M., Gold, A.J. & Mosier, A., Patchiness in groundwater nitrate removal in a riparian forest. *Journal of Environmental Quality*, **27**, pp. 156–164, 1998.

[119] Casey, R.E., Taylor, M.D. & Klaine, S.J., Localization of denitrification activity in macropores of a riparian wetland. *Soil Biology & Biochemistry*, **36**, pp. 563–569. 2004.

[120] Christensen, S., Simkins, S. & Tiedje, J.M., Spatial variation in denitrification: dependency of activity centers in the soil environment. *Soil Science Society of America Journal*, **54**, pp. 1608–1613, 1990.

[121] Groffman, P.M. & Hanson G.C., Wetland denitrification: Influence of site quality and relationships with wetland delineation protocols. *Soil Science Society of America Journal*, **61**, pp. 323–329, 1997.

[122] Chabrerie, O., Poudevigne, I., Burdeau, F., Vinceslas-Akpa, M., Nebbache, S., Aubert, M., Bourcier, A. & Alard, D., Biodiversity and ecosystem functions in wetlands: A Case study in the estuary of the Seine River, France. *Estuaries*, **24**, pp. 1088–1096, 2001.

[123] Ponnamperuma, F.N., The chemistry of submerged soils. *Advances in Agronomy*, **24**, pp. 29–97, 1972.

[124] Lyons, J.B., Gorres, J.H. & Amador, J.A., 1998. Spatial and temporal variability of phosphorus retention in a riparian forest soil. *Journal of Environmental Quality*, **27**, pp. 895–903.

[125] Johnston, C.A., Bridgham, S.D. & Schubaurer-Berigan, J.P., Nutrient dynamics in relation to geomorphology of riverine wetlands. *Soil Science Society of America Journal*, **65**, pp. 557–577, 2001.

[126] Goovaerts, P., Geostatistical tools for characterizing the spatial variability of microbiological and physico-chemical soil properties. *Biology and Fertility of Soils*, **27**, pp. 315–334, 1998.

[127] Nixon, S.W., Between coastal marshes and coastal waters – A review of twenty years of speculation and research on the role of salt marshes in estuarine productivity and water chemistry. *Estuarine & Wetland Processes*, (ed.) P.K. Hamilton & K.B. MacDonald, Plenum Press: New York, pp. 437–525, 1980.

[128] Sundareshwar, P.V., Morris, J.T., Koepfler, E.K. & Fornwalt, B., Phosphorus limitation of coastal ecosystem processes. *Science*, **299**, pp. 563–565, 2003.

[129] Sundareshwar, P.V. & Morris, J.T., Phosphorus sorption characteristics of intertidal marsh sediments along an estuarine salinity gradient. *Limnology & Oceanography*, **44**, pp. 1693–1701, 1999.

[130] Stumm, W. & Morgan, J.J., *Aquatic Chemistry*, 2nd edn, Wiley Interscience: New York, pp. 625–640, 1981.

[131] Smith, K.S., Ranville, J.F., Plumlee, G.S. & Macalady, D.L., Predictive double-layer modeling of metal sorption in mine-drainage systems, *Adsorption of Metals by Geomedia,* (ed.) E.A. Jenne, Academic Press, pp. 521–547, 1998.

[132] Lacerda, L.D. & Abrao, J.J., Heavy metal accumulation by mangrove and salt marsh intertidal sediments. *Revista Basileira de Botanica*, **7**, pp. 49–52, 1984.

[133] Moberg, F. & Rönnbäck, P., Ecosystem services of the tropical seascape; interactions, substitutions and restoration. *Ocean & Coastal Management*, **46**, pp. 27–46, 2003.

[134] Ochieng, C.A. & Erftemeijer, P.L.A., The seagrasses of Kenya and Tanzania. *World Atlas of Seagrasses*, (ed.) E.P. Green & F.T. Short, University of California Press: Berkeley, 2003.

[135] D'Elia, C.F & Wiebe, W.J., Biogeochemical nutrient cycles in coral-reef ecosystems. *Coral Reefs*, (ed.) Z. Dubinsky, Elsevier: Amsterdam, pp. 49–74, 1990.

[136] Rheinhardt, R.D. & Brinson, M.M., An evaluation of the effectiveness of existing NCDOT wetland mitigation sites. CTE/NCDOT Joint Environmental Research Program, 2000.

[137] Bruland, G.L., An observational, geostatistical, and experimental assessment of edaphic properties and processes in created, restored, and natural wetlands of the Southeastern Coastal Plain. Ph.D. dissertation. Duke University; Durham, 2004.

[138] Environmental Laboratory. *Corps of Engineers Wetlands Delineation Manual*. U.S. Army Corps of Engineers, Wetland Research Program Technical Report Y-87-1: Washington, DC, 1987.

[139] Zedler, J.B., Progress in wetland restoration ecology. *Trends in Ecology & Evolution*, **15**, pp. 402–407, 2000.

[140] Zedler, J.B & Calloway, J.C., Tracking wetland restoration: do mitigation sites follow desired trajectories. *Restoration Ecology*, **7**, pp. 69–73, 1999

[141] Stolt, M.H., Genthner, M.H., Daniels, W.L., Groover, V.A., Nagle, S. & Haering, K.C., Comparison of soil and other environmental conditions in constructed and adjacent palustrine reference wetlands. *Wetlands*, **20**, pp. 671–683, 2000.

[142] Spencer, D.R., Perry, J.E. & Silberhorn, G.M., Early secondary succession in bottomland hardwood forests of southeastern Virginia. *Environmental Management*, **27**, pp. 559–570, 2001.

[143] Galatowitsch, S.M. & Van der Valk, A.G., Vegetation and environmental conditions in recently restored wetlands in the prairie pothole region of the USA. *Vegetatio*, **126**, pp. 89–99, 1996.

[144] Clewell, A.F. & Lea, R., Creation and restoration of forested wetland vegetation in the Southeastern United States. *Wetland Creation and Restoration: The Status of the Science*, (ed.) J.A. Kusler & M.A. Kentula, Island Press: Washington, DC, pp. 195–232, 1990.

[145] Verhoeven, J.T.A., Whigham, D.F., van Logtestijn, R. & O'Neill, J., A comparative study of nitrogen and phosphorus cycling in tidal and non-tidal riverine wetlands. *Wetlands*, **21**, pp. 210–222, 2001.

[146] Bishel-Machung, L., Brooks, R.P., Yates, S.S. & Hoover, K.L., Soil properties of reference wetlands and wetland creation projects in Pennsylvania. *Wetlands*, **16**, pp. 532–541, 1996.

[147] Campbell, D.A., Cole, C.A. & Brooks, R.P., A comparison of created and natural wetlands in Pennsylvania, USA. *Wetlands Ecology & Management*, **10**, pp. 41–49, 2002.

[148] Sopher, C.D. & Baird, J.D., *Soils and Soil Management,* Reston Publishing Company, Inc.: Reston, 1978.

[149] Poach, M.E. & Faulkner, S.P., Soil phosphorus characteristics of created and natural wetlands in the Atchafalaya Delta, LA. *Estuarine, Coastal & Shelf Science*, **46**, pp. 195–203, 1998.

[150] Shaffer, P.W. & Ernst, T.L., Distribution of SOM in freshwater wetlands in the Portland, Oregon area. *Wetlands*, **19**, pp. 505–516, 1999.

[151] Whittecar, G.R. & Daniels, W.L., Use of hydrogeomorphic concepts to design created wetlands in southeastern Virginia. *Geomorphology*, **31**, pp. 355–371, 1999.

[152] Duncan, C.P. & Groffman, P.M., Comparing microbial parameters in natural and constructed wetlands. *Journal of Environmental Quality*, **23**, pp. 298–305, 1994.

[153] Bruland, G.L. & Richardson, C.J., Spatial variability of soil properties in created, restored and paired natural wetlands. *Soil Science Society of America Journal*, **69**, pp. 273–284, 2005.

[154] Holling, C.S. & Gunderson, L.H., Resilience and adaptive cycles. *Understanding Transformations in Human and Natural Systems*, (ed.) L.H. Gunderson & C.S. Holling, Panarchy: Island Press: Washington, DC, pp. 25–62, 2002.

[155] DeBusk, W.F. Newman S & Reddy K.R. Spatial patterns of soil phosphorus in Everglades Water Conservation Area 2A. *Soil Science Society of America Journal*, **58**, pp. 543–552. 1994.

[156] Newman, S., Reddy, K.R., DeBusk, W.F., Wang, Y., Shih, G. & Fisher, M.M., Spatial Distribution of soil nutrients in a northern Everglades marsh: Water Conservation Area 1. Soil *Science Society of America Journal*, **61**, pp. 1275–1283, 1997.

[157] Richardson, C.J., Reiss, P., Hussain, N.A., Alwash, A.J. & Pool, D.J., The restoration potential of the Mesopotamian marshes of Iraq. *Science*, **307**, pp. 1307–1311, 2005.

[158] Bruland, G.L., Grunwald, S., Osborne, T.Z., Reddy, K.R. & Newman, S., Spatial distribution of soil properties in Water Conservation Area 3 of the Everglades. *Soil Science Society of America Journal*, **70**, pp. 1662–1676, 2006.

[159] Miller, J.R., Turner, M.G., Smithwick, E.A.H., Dent, C.L. & Stanley, E.H., Spatial extrapolation: the science of predicting ecological patterns and processes. *BioScience*, **54**, pp. 310–320, 2004.

[160] Kaly, U.L. & Jones, G.P., Mangrove restoration: A potential tool for coastal management in tropical developing countries. *Ambio*, **27**, pp. 656–661, 1998.

[161] Alongi, D.M., *Coastal Ecosystem Processes*, CRC Press: Boca Raton, 1998.

[162] Kusler, J.A. & Kentula, M.E., (ed.), *Wetland Creation and Restoration: The Status of the Science*. Island Press: Washington, DC, 1990.

[163] Kennish, M.J., (ed.), *Estuary Restoration and Maintenance: The National Estuary Programme*. CRC Press: Boca Raton, 2000.

[164] Sacco, J., Seneca, E.D. & Wentworth, T.R., Infaunal community development of artificially established salt marshes in North Carolina. *Estuaries*, **17**, pp. 489–500, 1994.

[165] Craft, C.B., Reader, J., Sacco, J.N. & Broome, S.W., Twenty-five years of ecosystem development of constructed *Spartina alterniflora* [Loisel] marshes. *Ecological Applications*, **9**, pp. 1405–1419, 1999.

[166] Berger, J.J., (ed.), *Environmental Restoration: Science and Strategies for Restoring the Earth*. Island Press: Washington, DC, 1990.

[167] Zedler, J.B., Coastal mitigation in southern California: the need for a regional restoration strategy. *Ecological Applications*, **6**, pp. 84–93, 1996.

[168] Vivian-Smith, G., Developing a framework for restoration. *Handbook for Restoring Tidal Wetlands*, (ed.) J.B. Zedler, CRC Press: Boca Raton, pp. 39–88, 2001.

[169] Bruland, G.L. & Richardson, C.J., Hydrologic, edaphic, and vegetative responses to microtopographic reestablishment in a restored wetland. *Restoration Ecology*, **13**, pp. 515–523, 2005.

[170] Teal, J.M. & Peterson, S., Restoration benefits in a watershed context. *Journal of Coastal Research*, **40**, pp. 132–140, 2005.

[171] Mitsch, W.J., Day J.W., Jr., Gilliam, J.W., Groffman, P.M., Hey, D.L., Randall, G.W. & Wang, N., Reducing nitrogen loading to the Gulf of Mexico from the Mississippi River basin: Strategies to counter a persistent ecological problem. *Bioscience*, **51**, pp. 373–388, 2001.

[172] Sklar, F.H., Chimney, M.J., Newman, S., McCormick, P.V., Gawlik, D., Miao, S., McVoy, C., Said, W., Newman, J., Coronado, C., Crozier, G., Korvela, M. & Rutchey, K., The ecological-societal underpinnings of Everglades restoration. *Frontiers in Ecology*, **3**, pp. 161–169, 2005.

[173] Leapold, L.B., Wolman, M.G. & Miller, J.P., *Fluvial Processes in Geomorphology*. W.H. Freeman & Co.: San Francisco, 1964.

[174] Strahler, A.N., Hypsometric [area-altitude] analysis of erosional topography. *Bulletin of the Geological Society of America*, **63**, pp. 1117–1142, 1952.

CHAPTER 5

Fine particles in small steepland streams: physical, ecological, and human connections

Nira L. Salant & Marwan A. Hassan
Department of Geography, University of British Columbia, Canada.

Abstract

Fine particle dynamics in small steepland streams can be severely impacted by human activities, with ensuing effects on ecological and physical processes. Four components of fine particle dynamics are reviewed: sources and supply mechanisms; instream transport and deposition; biological impacts; and spatial and temporal scales of study and variability. Included within each topic is information on inorganic and organic particles, measurement and modelling techniques, and the impacts of human activities. Lastly, several remaining research needs are identified.

1 Introduction

Fine particulate matter can be an important component of many physical and biological processes in streams. In particular, deposition of fine sediment (<2 mm) has been repeatedly shown to degrade benthic habitat for fish and other organisms [1–3] and reduce water quality [4]. Fine-particulate organic matter (FPOM; 50–100 μm) can be an important flux linking up- and downstream reaches [5, 6] and can supply a significant amount of carbon to benthic invertebrates [7]. Furthermore, FPOM may play a large role in the movement and deposition of sorbed contaminants and nutrients [8]. However, both fine sediment and FPOM dynamics are impacted by human activities, such as logging and flow regulation. For example, logging can alter the dynamics of organic matter delivery and channel storage of fine particles [9–11].

Fine sediment and FPOM are often studied independently, despite the fact that suspended particles in many streams are composites of mineral and organic matter [12–14]. Interest in these composite particles (also known as aggregates or flocs) has increased over the past decade, but most of the existing literature separates suspended load into inorganic and organic components, or considers all fine particles generally as "sediment", usually assumed to be inorganic. In part, this one-sided focus may be due to disciplinary biases, but may also be logical for a given study, or necessary due to a limited knowledge of particle composition. In this review, we summarize information on fine sediment and organic particles as they are referred to in the literature – either as generic sediment or as independent components of the total suspended load. Although our review is somewhat limited in this regard, future study will benefit from attempts to link the dynamics of inorganic, organic, and aggregate particles. Progress in this field depends upon improved understanding of the interactions, differences, and similarities of these particles and the factors that control them. To this end, we attempt throughout our review to identify how the behavior of particles differ and how they interact.

A comprehensive review by Wood and Armitage [15] provides information on sedimentation and human activity in riverine systems, including the nature and origin of fine sediments, processes of sediment transport and deposition, and biological problems associated with increased sediment loads. Anthropogenic effects on fine sediment dynamics and related management issues are also reviewed in detail by Owens *et al.* [16]. Gomi *et al.* [17] also give a review of suspended sediment dynamics and the biological effects of forest harvesting in small streams of the Pacific Northwest. Reviews of general geomorphic processes in small, steepland streams in relation to forest harvesting can be found elsewhere [18–20]. We complement these previous reviews by considering new material and focusing on small, steep, forested streams in several regions. We present recent contributions to this topic, including advancements in the fields of source identification, particle storage and residence time, streambed infiltration, and biological response. We also consider the importance of scale and variability to the study of fine particle dynamics. Thus, this chapter reviews fine particle dynamics in small, forested streams in relation to human activities, focusing on four general areas: 1) fine particle supply; 2) fine particle transport and deposition; 3) biological significance of fine particles; and 4) spatial and temporal variability. General characteristics of fine inorganic and organic particles are summarized and compared in Table 1.

Before we begin, it is important to consider how small streams are defined, an issue of surprising complexity and incongruity in the literature. Small streams are often defined as 1st- or 2nd-order streams in the Horton–Strahler channel-ordering system [21]. However, classification by order usually requires the analysis of topographic maps and many small channels may be excluded if map resolution is low or canopy cover is high [22, 23]. Many alternative classification schemes have been developed, based on the dominant geomorphic or hydrologic processes [24–26] and a range of quantitative and qualitative criteria. For example, Church [27] defines a fundamental distinction between 'small channels', scaled by the size of individual grains, and 'large channels', scaled by the size of grain aggregates or structures. Montgomery and Foufoula-Georgiou [28] identify a transition from debris-flow dominated, colluvial and low gradient alluvial channels at a drainage basin area of ~1 km^2, suggesting an alternative small/large distinction. Problems

Table 1: Characteristics of fine inorganic and organic particles in streams.

	Inorganic	Organic
Sources		
Internal	Bank/bar erosion	*Same as inorganic particles plus:*
	Bed interstices	Zoo- and phytoplankton
	Surficial deposits	Feces
	Backwater areas	Biotic decay
	Pools	DOM flocculation
	Log jams/large woody debris	Feeding byproducts
	Vegetation surfaces	
External	Runoff erosion (gullies, hillslopes, soils)	*Same as inorganic particles plus:*
	Mass movements (landslides, debris flows, earthflows, debris avalanches)	Allocthonous inputs (leaf litter, woody debris) and mechanical breakdown
	Atmospheric/Aeolian deposition	
Transport/deposition/storage		
Controlling factors	Suspended particle concentration	*Same as inorganic particles plus:*
	Channel morphology (stream size, storage zone size)	Water temperature
	Bed composition (roughness, physical barriers)	Invertebrates
	Channel or water surface gradient, flow discharge, stream power	Particle size and composition (density, geometry, surface charge)
	Particle size and composition (density, degree of flocculation)	Season and source
	Large woody debris	
	Frequency and timing of flow events	

(Continued)

Table 1: Continued

	Inorganic	Organic
Characteristics	Hysteresis	*Same as inorganic particles*
	Highly variable relationship with flow	
	Dominated by mass movements	
	Little low flow transport	
	Long residence times	
External influences		
Forest harvesting		
Changes to availability	Expose soil	*Same as inorganic particles plus:*
	Reduce slope stability	Remove canopy cover
	Damage streambanks	Reduce wood delivery
	Accelerate mass movements	Change species composition (leaf-litter type, decomposition rate, timing of input)
		Increase productivity (increased sunlight, temperature, nutritional quality)
Changes to hydrology	Reduce transpiration and interception; raise water tables and soil moisture; increase hydrologic connectivity, streamflow and delivery	*Same as inorganic particles*
Flow regulation	Reduce magnitude of high-flow events	*Same as inorganic particles plus:*
	Trap particles; reduce sediment discharge	Alter temperature and nutrient regime in turn changing instream community composition and OM processing
		Increase seston concentrations
Spawning salmon	Excavate buried particles	*Same as inorganic particles plus:*
	Accelerate downstream transport	Carcass decay supplies POM/DOM

with this definition arise, however, when considering the strong influence of local climate, geology, and history on stream size, shape, and the processes that dominate. For a complete discussion of the problems associated with defining small streams, see Benda *et al.* [18]. For simplicity, small streams in this review are considered those with a bankfull width less than 2–3 m or catchment area <1 ha. For the most part, we restrict our review to studies of small, forested streams in mountainous regions, although limited data availability may require us to occasionally draw from studies in larger systems or geographic locations.

2 Sources, supply mechanisms, and source identification

2.1 Fine particle sources

Fine sediment and FPOM sources may be divided into two main categories: internal, or in-channel sources, and external, or nonchannel sources (Table 1). Supply mechanism, particle composition, and sediment load will vary according to source type. Internal sources and supply mechanisms include bank or bar erosion; piping from sub-surface flow; high flow mobilization from bed interstices, surficial deposits, backwater areas, log jams, or pools; and release of particles upon senescence of aquatic vegetation. Organic particles also include small biota such as zoo- and phytoplankton, biotic waste products, or biotic decay. External sources are derived from the catchment, including runoff erosion from gullies, hillslopes, and exposed soils, mass movements such as landslides and debris flows, and atmospheric deposition. Allochthonous inputs of leaves and woody debris provide an external supply of organic matter that is further broken into fine particles by in-channel biotic activity.

Fine sediment dynamics in small, steep, forested streams can differ from large rivers in several ways. Because of their close proximity to the terrestrial environment, sediment sources reflect a mix of hillslope and channel processes, including episodic release due to mass movements, bank erosion during and between floods, and release of internal sediment stored in pools formed by large bed materials or woody debris. In mountainous regions, gullies and hillslopes are the main source of sediment, supplied to the stream via mass movements or rill erosion of exposed surfaces [18, 29, 30]. Little sediment is contributed from surface erosion by overland flow [31, 32] because of the protective effect of dense vegetation cover and well-developed organic soil horizons [33]. Sediment supply occurs by two main mechanisms: debris flows and fluvial transport, but unlike higher-order streams mass movements dominate delivery to the channel.

Entirely covered by canopy, forested streams receive litter and wood inputs comparable to that of the forest floor. Because of a greater edge-to-area ratio, bankside inputs are also high, such that small streams have higher organic matter inputs per unit area compared to larger streams in the same forest type. In-channel productivity, however, is low because little sunlight penetrates the canopy and groundwater inputs reduce stream temperatures [34]. Most organic matter in the stream is in the form of low-nutrition wood [35], though nutritional quality increases when organic matter is broken down and colonized by microbes. Fine-particulate organic matter constitutes a large portion of the particulate organic matter pool and is comprised of leaf fragments, invertebrate faeces [36], small wood fragments [37], or flocs of DOM [38]. There is a large seasonal variation in the quality of FPOM due to the type and timing of inputs [11, 39] as well as forest age and type [11]. In regions dominated by deciduous trees, organic matter input is higher during autumn leaf-drop, such that the particle composition of the sediment load may vary over the course of the year. In contrast, coniferous inputs are evenly dispersed throughout the year. Differences in vegetation may influence organic matter dynamics [40]; for example, coniferous needles decompose more slowly than deciduous leaves [41, 42], due in part to less microbe colonization, protective chemicals, and low stream water nutrients and temperature [43]. Slow decomposition may affect the rate at which FPOM is produced, thus limiting export to downstream systems. In some areas, such as the west coast of North America, particulate and dissolved organic matter are also supplied by the decomposing carcasses of spawning salmon [44]. Excavation by redd-building salmon can also release fine particles from the streambed and accelerate transport for short distances downstream [45–47].

2.2 Source identification

Identifying the nature and relative contribution of suspended sediment sources is key to constructing watershed sediment budgets [48, 49], estimating sediment yields [50, 51] and designing effective management strategies for reducing sediment pollution [52, 53]. A wide range of source-identification techniques have been developed, but due to the complexity of factors governing sediment mobilization and supply, results are often conflicting and problematic. Because suspended sediment sources are highly spatially and temporally variable, sampling schemes, limited by logistics and costs, are often insufficient to provide representative and reliable data [54]. A comprehensive review of common approaches to source identification and the problems associated with each was provided most recently by Collins and Walling [55].

The source tracing, or fingerprinting, approach has been proposed as a less problematic, more direct technique for sediment sourcing that uses a variety of diagnostic properties to characterize and link suspended sediment samples to potential source areas [56–58]. A range of physical and chemical properties may be selected, depending on watershed and potential source characteristics, such as particle mineralogy and size [59, 60], sediment chemistry [61] mineral magnetism [56], or environmental radionuclides [58, 62, 63]. Advancements in fingerprinting techniques, including the use of multiple diagnostic properties [13, 58, 64], quantitative mixing models, and discriminant statistical tests [64–66], have enabled researchers to determine the relative contribution of source areas and supply mechanisms in many lowland and some upland rivers. A detailed review of the development and application of source fingerprinting techniques is given by Walling [67]; here we provide background on the use of environmental radionuclides as source tracers to facilitate discussion of their use as tracers of in-channel sediment transport in the next section.

Several studies have used both lithogenic and fallout radionuclides to identify sediment sources and temporal changes in supply and to construct sediment budgets [62, 68, 69]. Since the 1980s, a large number of *in-situ* produced long-lived cosmogenic radionuclides, such as beryllium-10 (^{10}Be) and aluminum-26 (^{26}Al) have been used to measure surface-exposure ages, erosion rates, and regolith production [70–72]. Long-lived lithogenic radionuclides, produced *in-situ* from the uranium-238 (^{238}U) and thorium-232 (^{232}Th) decay series, have also been used in fluvial systems as source tracers, linking hillslope and channel processes [73–75]. Shorter-lived fallout radionuclides, including caesium-137 (^{137}Cs), excess lead-210 (^{210}Pb) and, more recently, beryllium-7 (^{7}Be) have been used to quantify sediment erosion, mobilization, transport and storage over shorter time scales and even individual events [63, 76–78].

Radionuclide activity of sediment is an advantageous diagnostic property because it is independent of geology and can be used to differentiate between surface and subsurface soil, as well as cultivated and uncultivated soil. Longer-lived ^{137}Cs and ^{210}Pb have been used to calculate decadal sedimentation rates [79], biological mixing [80], and soil erosion [81], while the short-lived ^{7}Be has been used to quantify sediment transport [63, 77, 82], resuspension [83, 84], and deposition [85, 86] at event- to month-long time scales.

Additional background and details of the source tracing technique for fallout radionuclides can be found in several papers [81, 87] and in the chapter of this book entitled "Effects of land-use changes and groundwater pumping on salt water intrusion in coastal watersheds" [258]. Essentially, the approach begins by assuming that radionuclide fallout is uniform across the landscape; upon deposition, radionuclides strongly sorb to fine particles [88–90], thus movement of these radionuclides through the watershed reflects the mobilization of soil and sediment. By comparing radionuclide activity of a sample with an undisturbed reference, rates of erosion and deposition can be estimated [68]. Reference activity can be determined from direct measurement of radionuclide delivery, analysis of precipitation samples, or collection of soil or snowpack cores [14, 63, 68, 77, 91]. Atmospheric ^{210}Pb and ^{7}Be are derived naturally, but ^{137}Cs fallout results from nuclear weapons testing in the 1950s and 1960s; all enter the ecosystem primarily through wet deposition [92–94]. Some of the variability in precipitation-delivered radionuclides can be corrected through the use of the $^{7}Be/^{210}Pb$ ratio [77, 95, 96]. Because fallout radionuclides are atmospherically derived, activity typically declines exponentially with soil depth [13, 58, 68], allowing differentiation between surface and subsoil-derived sediment. In addition, because plowing and tilling of soil mixes high-activity surface soil into lower layers, sediment derived from the surface of cultivated soils will have lower activity levels than uncultivated soils. Each radionuclide distributes differently in the soil, thus sediment source areas can be distinguished by the relative amounts of different radionuclides corresponding to land use and depth [14, 97, 98]. Furthermore, upon entering the river, sediment is no longer exposed to the atmosphere and the radionuclide activity begins to decay. Thus the activity of fluvial sediment reflects mixing between landscape-, bank- and streambed-derived sediment, storage times within the channel, and transport rates through the river system. Application of fallout radionuclides to channel processes such as transport and deposition will be discussed further in the next section.

2.3 Impact of human activities on sources

Due to the spatiotemporal variability of sediment sources, human activities that increase fine sediment input to streams are often regulated as non-point sources of pollution [99]. Thus, techniques that improve our ability to accurately identify the location of dominant sediment sources, and thus the activities associated with them, will greatly aid management efforts. Small, forested watersheds are particularly impacted by forest harvesting practices, such as logging, road building and slash burning, which can both indirectly and directly affect the sources and supply of sediment to streams (Table 1). The main impact of forest harvesting activities is an increase in sediment availability; all activities disturb and expose soil, alter slope stability, and damage streambanks, increasing sediment mobility from surface sources and destabilizing stored sediment. In particular, mass movements are accelerated post-harvest due to reduced hillslope stability when roots are removed from streamside areas [100, 101]. Both mass movements and surface runoff from logging roads can increase sediment load to stream channels [102, 103], due in part to construction on unstable terrain and poor road drainage [104]. Although some studies have shown that roads can also act as depositional and storage sites for sediment [105], roads generally increase sediment production and input to fluvial systems, depending on hillslope position, timing of logging activity and large storms, and road-management practices [104–107].

Tree removal also reduces transpiration and interception, increasing soil moisture and water table levels [108, 109], which can increase connectivity between perennial and ephemeral streams and subsequent sediment delivery. Despite extensive study, uncertainty still remains over the relative influence of hydrologic changes versus increased sediment supply following harvesting. In particular, sediment yield to headwater streams appears to increase due to changes in flow rather than sediment supply [110, 111], but this is largely unexplored.

Tree removal from streamside areas also alters the type, amount, and timing of organic matter delivered to streams by altering tree-species composition, removing canopy coverage, and reducing wood delivery. The transition from old growth conifers to young, deciduous regrowth changes the type of organic matter inputs – from needles to leaves – and the timing of inputs – from year-round to seasonal [10]. Riparian buffers are commonly proposed as a strategy for intercepting sediment input via overland flow and minimizing physical disturbance adjacent to the stream. In most cases, riparian buffers reduce increases in sediment yield following harvesting (see Gomi *et al.* [17] for a review) but are ineffective at intercepting sediment generated outside the riparian zone or by mass movements and road erosion [17, 112, 113].

Table 2: Characteristics of three theoretical models representing fine particle distribution and movement that differ in application, approach and intended focus.

	Rouse equation	Advection-dispersion model	Local Exchange Model
General form	Vertical solution (1–D) $C_z = C_a \left[\frac{z}{z_a} \cdot \frac{h - z_a}{h - z} \right]^{-\hat{S}}$	Continuity equation (3–D) $\frac{\partial c}{\partial t} + \nabla \cdot (\mathrm{cu} - \mathrm{K}\nabla c) = 0$ Longitudinal solution (1–D) $\frac{\partial c}{\partial t} = \frac{-Q}{A}\frac{\partial c}{\partial x} + \frac{1}{A}\frac{\partial}{\partial x}\left[\frac{AK(x)C}{\partial x} \right] + \frac{Q_L}{A}(C_L - C) + a(C_S - C)$ and $\frac{\partial c_s}{\partial t} = -a\frac{A}{A_s}(C_s - C)$	Stochastic-diffusion equation $\frac{\partial f}{\partial t} + \frac{\partial}{\partial z}\left[(\mu - \frac{1}{2}v')f - \frac{1}{2}v\frac{\partial f}{\partial z} \right] = 0$ Normalized stationary vertical profile (1–D) $\frac{d}{dz}\left[u_z c^* - K\frac{dc^*}{dz} \right] = 0$
Variables	Rouse number $\left(S = \frac{V_{\mathrm{fall}}}{\beta \kappa u_*} \right)$	Concentration gradient vector (∇c) Advective velocity vector (u) Matrix of longitudinal, transverse, and vertical dispersion coefficients (K)	Infinitesimal variance and mean in vertical direction $v(z) = 2K(z)$ $\mu(z) = u_z(z) + K'(z)$

(Continued)

Table 2: Continued

	Rouse equation	Advection-dispersion model	Local Exchange Model
	Particle fall velocity *(V_{fall})* Particle diffusivity *(β)* von Karman's constant *(κ)* Bed shear velocity $u* = \sqrt{\tau_0 / \rho}$ Bed shear stress (τ_0) Near-bed concentration *(C_a)* Reference elevation *(z_a)* Water depth *(h)* Elevation *(z)* Concentration at $z(C_z)$	Discharge *(Q)* Cross-sectional area of the stream *(A)* Diffusion coefficient in the downstream direction *$K(x)$* Groundwater or tributary inflow *(Q_L)* Solute concentration of inflow *(C_L)* Area of the transient storage zones *(A_s)* Concentration of solute in the transient storage zone (C_s) Coefficient of exchange with the transient storage zones *(α)*	Vertical dispersion rate $K(z) = \psi \left\{ \frac{M}{2} + \sqrt{\left(\frac{M}{2}\right)^2 + l(z)^2 u_*^2 (1 - z/H)} \right\}$ Constant > 0 *(ψ)* Kinematic molecular viscosity (M) Mixing length (measure of kinematic eddy viscosity) *($l(z)$)* First derivative of *$K(z)$ ($K'(z)$)* Vertical component of the advective velocity ($u_z(z)$) (= –s(z), vertical component of V_{fall}) Normalized steady-state concentration ($c*$) $c*(z) = \frac{c(z)}{\int_0^u c(\zeta) d\zeta}$
Assumptions	2D steady, uniform flow Particles of uniform size, shape, density Flat, hydrodynamically smooth bed	Particles of uniform size, shape, density No variation in parameters with location or concentration	Particles of uniform size, shape, density No particle–particle interactions Flat, hydrodynamically smooth bed

	Straight channel Mean particle size used in computation of v_{fall} Depth-independent variables (β and κ) No particle–water or particle–particle interactions *Downward velocity of particles equals particle fall velocity* *Particle diffusivity equals turbulent momentum diffusivity of water*	No particle–water or particle–particle interactions Longitudinal particle velocity equals local water velocity Downward velocity of particles equals particle fall velocity Particle diffusivity equals turbulent momentum diffusivity of water	Downward velocity of particles (uz(z)) equals particle fall velocity *Principle of local exchange: upward and downward movements of water always balance (no net transport)*
Application	Inorganic particles Theoretical studies	FPOM Field studies	FPOM; living, motile organisms Theoretical, field, or flume studies
Dimensions	1D (vertical) Cross-sectional scale	3D (usually only longitudinal) Reach scale	3D (initially only vertical) Particle scale
Advantages	Simple; few parameters	Incorporate stream morphology, groundwater and tributary inputs, transient storage, and immobilization of nonconservative solutes Estimate extent of transient storage and particle exchange	Considers individual particle movement as stochastic-diffusion process Incorporates effect of flow on particles with depth-dependent vertical dispersion coefficient ($K(z)$) Includes motile particles Provides equations for concentration profile Provides equations for probability distributions of particle-hitting time and hitting distance and dependence on initial elevation and fall velocity

(Continued)

Table 2: Continued

	Rouse equation	Advection-dispersion model	Local Exchange Model
Limitations	Flows will distribute particles nonuniformly due to differences in fall velocity Natural beds are not perfectly smooth C_a difficult to measure Particle–water interactions influence turbulent flows, streamwise velocities, and particle diffusivity Particle diffusivity varies with depth, density of suspended particles and bed conditions Purely physical model; no biological influences *Downward particle velocities do not always equal fall velocities* *Particle diffusivity varies with depth, density of suspended particles and bed conditions*	Particle–water interactions influence turbulent flows, streamwise velocities, and particle diffusivity Biological influences only represented indirectly *Streamwise velocities vary with depth* *Downward particle velocities do not always equal fall velocities* *Particle diffusivity varies with depth, density of suspended particles and bed conditions*	Particle–water interactions influence turbulent flows and streamwise velocities Natural beds are not perfectly smooth No terms for morphology, flow inputs, transient storage, or non-conservative solutes Purely physical model; no biological influences (except motile particles) *Downward particle velocities do not always equal fall velocities*
References	Bagnold [236], Einstein and Chien [122], Halbronn [237], Hunt [238], Rouse [239], Tanaka and Fugimoto [240]	Bencala and Walters [242], Valett *et al.* [243], Webster and Ehrman [244], Cushing *et al.* [152], Minshall *et al.* [113], Newbold *et al.* [159], Paul and Hall [148]	McNair *et al.* [175], McNair [158], McNair and Newbold [159]

3 Particle transport, deposition, and streambed infiltration

Upon entering the fluvial system, fine particles may be transported downstream or deposited on the surface of the streambed. Once deposited, these particles may be retained, accumulating or infiltrating into the bed, consumed, or entrained back into the water column. A number of physical and biological factors determine the fate of fine particles (Table 1), including suspended particle concentrations [114], bed composition [115], channel morphology [116, 117], benthic ecology [118], particle size and composition [119, 120], and flow discharge [121, 122]. For decades, both physical and ecological researchers have quantified and modeled particle transport and deposition, greatly enhancing our understanding of particle movement and storage. A high degree of variability and uncertainty among results, however, arises from the inherent complexity of factors that govern particle dynamics.

Most simply, the mode and rate of particle transport is a function of particle size, density, and stream discharge. Generally, fine particles such as silts, clays, or FPOM are carried in suspension; therefore, under normal flow conditions they are unlikely to frequently interact with the bed or have long residence times within the stream. In contrast, larger particles roll and saltate along the bed as bed load [123] and have longer residence times. The boundary between these two transport modes is transitional; depending on flow magnitude, medium and coarse sand (0.25–2 mm) may move either short distances in suspension or roll along the bed as bed load. In practice, however, these physically based distinctions are blurred. Because of instrumental and practical limitations, measurements of sediment transport are divided into bed or suspended load; transitional material is included in one or the other depending on flow and channel conditions. Under typical flow conditions, more than 90% of suspended load is composed of very fine particles such as silt and clay, whereas particles 0.25–2 mm (bed material load) are retained in the channel bed. Numerous physical models and measurement techniques have been developed to quantify suspended and bed-load transport [124–127]. Further information on the long history of this field can be found in several papers, including Hassan *et al.* [19].

Focusing primarily on small, steep streams, we first review recent studies of fine particle transport and deposition that attempt to address the complexities inherent in these processes, including ecological studies concerned with the movement and storage of FPOM. Secondly, we discuss recent insights into the mechanisms of streambed infiltration and retention, a topic of increasing interest due to its potentially serious biological consequences. Thirdly, we present methods of tracing particle transport used in the physical and ecological sciences, highlighting recent advancements into the use of fallout radionuclides to track particle movement. Fourthly, we introduce three theoretical models of particle movement and vertical distribution that differ in their application, approach and intended focus (Table 2). Lastly, we close this section by evaluating how human activities in forested watersheds change in-channel particle dynamics.

3.1 Fine particle transport and vertical movement in the water column

Most fine particle transport from small streams occurs during snowmelt periods and single floods, usually of short duration and high magnitude, displaying greater inter-event and intra-event variability than lowland streams [122] and a highly variable relationship between suspended load and water discharge. Transport is highly related to discharge and often exhibits hydrograph hysteresis, with concentrations at a given flow on the rising limb much greater than the corresponding flow on the falling limb [30, 128, 129]. Patterns of hysteresis in the relation between suspended load and water discharge are related to types and locations of active sources [30, 122]. High-flow events also change the transport, storage, and characteristics of organic particles. Lateral flooding can deposit significant amounts of organic matter onto floodplains and banks, reducing overall export [130]. Extensive mass transfer and exchange of organic matter can occur during flood events, increasing or decreasing surficial organic particles [131–135], depending on organic matter storage and load, as well as the retention capacity of a given section of streambed. Organic content, bioavailability, and metal affinity also differ between particles generated during a flood event and those accumulated during low-flow periods [136], with possible nutritional effects on the growth and metabolism of benthic organisms.

Particle deposition and storage in small streams occurs primarily in pools associated with physical barriers in the stream, so that total storage capacity is less than downstream alluvial reaches with floodplains, bars, and side channels. Large structural elements such as boulders and large woody debris control the amount of particle storage and channel stability [137, 138]. Wood and inorganic substrates form a high degree of roughness relative to stream depth that forms steps, pools and depositional areas, influencing the movement of inorganic and organic particles [130, 139, 140]. Episodic mass movements such as landslides and debris flows dominate transport in these systems [141], with little to no transport during low flows. As a result, these streams may act as inorganic and organic particle reservoirs for long periods of time, but the timing and frequency of storage release is highly unpredictable. Thus, periods of particle storage may be much longer and transport more temporally variable in small streams than in larger channels.

Ecologists have long been interested in the factors controlling FPOM transport and storage because of its implications for ecosystem productivity and diversity. Most studies of FPOM dynamics in streams focus primarily on longitudinal (up- to downstream) linkages and the quantification of particle transport distance (S), depositional velocity (V_{dep}), and residence time. These studies have elucidated many of the factors that govern FPOM dynamics, including bed and channel roughness [7, 115]; channel gradient, flow discharge, and stream power [115, 121, 142–144]; water temperature [145]; debris dams [146]; invertebrates [118, 147, 148]; stream and storage zone size [115]; the frequency and timing of flow events [7, 128, 148]; and particle size, geometry, surface charge and density [149, 150], in turn a function of organic matter season and source (i.e. allochthonous vs. autochthonous) [151, 152].

Downward movement through the water column and the rate of streambed deposition can be quantified by particle depositional velocity (V_{dep}), the rate at which a released particle settles on the bed. Depositional velocity is primarily controlled by the particle's still-water settling velocity (V_{fall}), which is in turn a function of particle size, shape, and density [153]. In theory, the measured V_{dep} should be approximately equal to V_{fall}, but most field studies show no consistent relationship between these two parameters [115, 154–156]. The ratio of V_{dep} to V_{fall} in these studies ranges from 0.04 to 1690, depending on particle properties and stream characteristics, but generally increases with decreasing particle size and density. Fluorescently labelled bacteria has the highest ratio (1690) [155] while natural FPOM and corn pollen are much lower (0.04–0.56).

In the cases where V_{dep} is $\ll V_{fall}$ [115, 144, 150], it is proposed that turbulent mixing and resuspension factors overwhelm gravitational settling in controlling particle movement. This is represented in theoretical models that predict $V_{dep} < V_{fall}$ when local shear stresses exceed a critical threshold for resuspension [157]. The ratio of gravitational velocity to turbulent mixing velocity (due to bed shear) is expressed by the Rouse number ($\hat{s}$). There is little effect of gravity at values of $\hat{s} < 0.1$ but gravitational factors increasingly dominate as $\hat{s}$ approaches 1 [158, 159]. Georgian *et al.* [160] calculate $\hat{s} \ll 0.01$ and low V_{dep}: V_{fall} for both FPOM and pollen in the field and $\hat{s}$ approaching 0.1 for FPOM in a flume, demonstrating the importance of shear forces over gravity in the field and increased importance of gravitational settling in a flume. A large V_{dep} may occur *in situ*ations where energy at the streambed is dissipated by plunge pools or stagnant zones, therefore decreasing bed shear stress and turbulent resuspension. Minshall *et al.* [115] report $V_{dep} \sim V_{fall}$ in small streams with high bed complexity, while Georgian *et al.* [160] report that V_{dep} in a smooth-bedded flume is less than V_{dep} of the same particles in the field, both suggesting that bed complexity and its effect on turbulence may strongly influence the deposition of particles. Particles with very low settling velocities and larger depositional velocities (i.e. bacteria) may also be deposited via advective transport into interstitial spaces, hyporheic entrainment, or adhesion to the substrate.

Traditional models explain the discrepancy between V_{dep} and V_{fall} by the hydrodynamic/gravitational mechanisms described above, but so far, measurements of bed roughness and shear stress have not revealed a consistent or significant relationship with V_{dep} [115]. Furthermore, conflicting and limited results have not fully elucidated the mechanistic role of transient storage zones in FPOM deposition. For example, Minshall *et al.* [115] report that V_{dep} was positively correlated with the relative size of transient storage zones and a coefficient of transient storage exchange, but not related to the advective exchange of water into these zones. Paul and Hall [150] report no relationship between V_{dep} and any measure of transient storage or exchange. Newbold *et al.* [161] measure brief retention of FPOM at a rate similar to that of water retention in transient storage zones, indicating simple advective transport to and from these zones without deposition. Based on earlier findings that hyporheic exchange increases particle removal from the water column [162], Newbold *et al.* [161] propose that most transient storage may not occur in hyporheic zones, but rather in deep lateral areas where turbulence is high enough to keep particles suspended. Differences between studies may be due to stream-to-stream variation, the size and composition of particles, the relative amount of in-channel versus hyporheic zone storage, or biological properties of the streambed. Newbold *et al.* [161] also calculate that average residence times of deposited particles are longer than turbulent fluctuations, suggesting the influence of biological retention processes. Evidence for the role of biological mechanisms in particle retention is limited, but suggestive. For example, biofilm adhesion is proposed as one explanation for the discrepancy between V_{dep} and V_{fall} [163, 164], while invertebrate manipulation [148] or removal by filter feeders may account for a measurable proportion of total deposition [131, 165, 166]. Particle composition may also explain much of the observed discrepancy, an explanation that has received increasing attention in the past decade. As noted above, most suspended particles are composites of organic and inorganic components. Composite particles may enter the stream from the watershed as aggregates, retaining their structure during transport, or may form in-channel via physical, biological, or chemical flocculation processes [176–178].

Cohesive properties of organic matter tend to enhance flocculation [119, 120]; however some evidence exists for electrochemical flocculation in glacial meltwaters that lack organic matter [178]. Particle size and hydrodynamic properties of composite particles differ considerably from their mineral components. Thus, predictions of deposition rates assuming single-grain settling are likely to be inaccurate when composite particles predominate [176–179].

3.2 Fine particle deposition, retention and infiltration in the streambed

Rates and mechanisms of particle deposition, retention, and streambed infiltration determine the type and degree of impact fine particles will have on benthic ecology, including biotic metabolism, nutrient cycling [180], and contaminant delivery [8, 181–185]. Collectively termed streambed clogging, particle deposition and infiltration into the hyporheic zone reduces interstitial habitat and the exchange of water and solutes with subsequent effects on benthic community structure. Biological consequences of clogging are discussed further in the next section and are given a full review by Gayraud *et al.* [186].

Retention of inorganic and organic particles can occur by several physical mechanisms, including filtration, sedimentation, burial, and vertical hydraulic entrainment. On an armored bed, small sand grains may infiltrate the pore spaces of the bed surface, moving only when the entire bed is entrained [116]. Filtration into hyporheic zone sediments varies spatially and temporally, depending on geomorphic and hydraulic conditions of the streambed, in turn forming patches of differing exchange capacities and transient storage characteristics [187]. Sedimentation of particles occurs in regions of low flow, such as pools, eddies or backwaters, created by structural elements within the channel [154, 188–190]. High sedimentation and low resuspension rates produce zones of accumulation. Burial by shifting sediments occurs in lowland, sand-bedded rivers when migrating bedforms bury organic particles deposited in the lee of the structure [191–193]. Vertical hydraulic entrainment occurs in well-sorted, open-frame sediments, where local differences in water pressure produce vertical hydraulic gradients of up- and downwelling water, mixing materials between the surface and interstitial water. Several studies have documented vertical exchange in riffle-pool sequences [194–197], where even single boulders can increase vertical interactions [198] and FPOM concentrations in the hyporheic zones of riffles [199, 200]. Bed-roughness elements, whether inorganic or organic, can produce vertical fluxes of water and materials between the stream and bed sediments [192, 193, 201], with ensuing affects on nutrient and carbon dynamics. Downward fluxes of oxygenated water can maintain aerobic conditions deep within the sediment. Combined with the flux of organic particles, this phenomenon contributes to the entrainment, storage, and microbial transformation of these particles within bed sediments.

3.3 Measuring fine particle transport and infiltration

Because of their small size and stochastic behavior, tracking the movement and storage of individual fine particles has until recently, remained elusive. Previous studies primarily focused on the mode of fine sediment transport; whether fine particles are transported in a series of steps and jumps [202–205] or travel the length of a river in a single hydrograph [206, 207]. However, several recent studies have used radionuclides to calculate downstream transport distances and channel-bed mixing of suspended sediment [66, 77, 82, 208] and transitional bed material [209]. Using ^{7}Be, Bonniwell *et al.* [77] found that suspended fine particles have short residence times and travel long distances during high flow periods in high gradient streams. Using the activity of ^{7}Be, ^{137}Cs, and excess ^{210}Pb in source soils and suspended sediment, Whiting *et al.* [63] found that the radionuclide signature of suspended sediment reflects the relative contributions of upland soil and bank erosion and that transport distances increased with basin size. Both studies calculated transport distances from the exponential decrease in activity from source areas to suspended sediment samples downstream. Alternatively, the radionuclide signature of particles can be used to track movement through the channel or watershed. For example, Salant *et al.* [209] tracked the movement of a pulse of sediment released from storage behind a dam, where it had become depleted in ^{7}Be, and thus determined sediment transport velocities. Results from these studies demonstrate that radionuclides can be effectively used as fine particle tracers for determining transport and deposition rates.

Meanwhile, in the field of ecology, more than two decades of research has led to the development of a range of field techniques for tracking the source and movement of FPOM. In most cases, a known concentration of particles, either ^{14}C-labeled natural FPOM [115, 156, 160, 210, 211] or a FPOM analog or surrogate, is released into the stream at a fixed point and the concentrations of water samples are measured at subsequent downstream stations. These concentrations are used in an exponential decay model to determine the loss of particles from the water column with distance [144, 150, 154, 211–213]:

$$N_x = N_0 e^{-kx}, \tag{1}$$

where N_x and N_0 are the quantities in suspension at a distance x and at the point of introduction, respectively, and k is the longitudinal loss rate of particles. Average transport distance (S) is the inverse of k. The paired release of a conservative tracer allows for the calculation of flow discharge. In order to correct for differences in water depth and velocity that may exist between streams, transport distances are converted to a related parameter, deposition velocity (V_{dep}) [152, 148, 142], by the equation

$$V_{dep} = hV_{water}k = \frac{hV_{water}}{S}, \tag{2}$$

where h and V_{water} are the depth and velocity of water, respectively. The mean time in suspension (T) is estimated as S/V_{water}.

Fine sediment infiltration can be measured using coring and pumping techniques or sediment collectors. Sediment collectors, also known as infiltration traps, only measure fine sediment accumulation, whereas coring and pumping techniques are also used to collect chemical and invertebrate samples. Although quantitatively the most robust and reliable, coring techniques are typically complex, time-consuming, labor-intensive, and expensive. In small streams dominated by coarse particles, freeze cores are most commonly used. Freeze-coring techniques are ideal for quantifying vertical variability in subsurface material because they allow for the removal of a vertical section of streambed that can be further divided by depth. Core installation generally follows a standard protocol [214], in which the evaporation of liquid carbon is used to freeze sediments around a probe inserted into the bed. Upon removal, cores are typically subdivided and analysed for particle-size distribution and organic matter content. Problems with this technique do exist, however, including bed disruption [215], bias due to an irregular sample boundary [216], and small sample sizes that result in high variability among individual cores, bedforms, and reaches. Previous studies have shown that a minimum sample size with total weight of 20 kg, including 5 cores from each sample site, provides reproducible results and a subdivision depth of 15 cm reduces the error in subdividing cores [214, 217–219]. Furthermore, biases of the freeze-core technique are negligible when only a sample of the fine matrix fraction is sought [214, 217, 219].

Less intensive and less expensive than coring, the Bou–Rouch pumping method is commonly used for sampling hyporheic water chemistry, sediment, and invertebrate fauna. In this method, water is pumped from the streambed at several depths using temporary or permanent standpipe wells [220]. Although biases may occur towards invertebrate types due to sample volume, sediment filtration, well permanence, attraction of predators, recovery time of fauna, and pumping rate [169, 221–224], the sampling of fine sediment content by this method produces similar results to coring [225]. Other concerns exist, however: well installation, shape, and size can modify substrate conditions [226], pumping cannot collect particles >1 mm [225], and streambed location can be uncertain in cases of high sediment heterogeneity (Palmer and Strayer, [220]).

Sediment collectors or infiltration traps offer the simplest, least expensive method for assessing streambed infiltration. Several varieties of traps exist, but most are composed of a solid-base, mesh container filled with clean gravel. Traps are embedded for a period of time, usually over the course of one or more flow events, then removed and analysed for fine sediment content [216, 227–234]. In addition to assessing the amount and characteristics of sediment collected, traps have also been used to determine controls on infiltration, such as suspended and bed-load transport rates or the relative size of transported and streambed sediment [227, 229, 235–239]. Detailed descriptions of commonly used instruments can be found in several papers [216, 230, 233]; two recently developed methods are briefly described here. Lachance and Dube [240] designed a collector from two solid-based, 1–L cylindrical buckets set inside each other, drilled with multiple 1–3 cm matching holes to allow the flux of water and sediment, and sealed before removal by rotating the outer bucket. Although these traps appear adequate for assessing and comparing infiltration rates and sediment types, discrepancies between sediment accumulation into collectors and natural salmon redds indicates that these traps may overestimate actual streambed infiltration, possibly due to lower sand content and large pore volume [234]. Newly developed by Levasseur *et al.* [241], the infiltration cube is proposed as an efficient and reliable instrument for sampling large (~65 kg) volumes of bed sediment, with minimal bias to infiltration rates. Modified from the previously developed infiltration bag [216], the cube is formed of a rectangular metal frame with a folded plastic bag at its base. Installation and burial of the cube is designed to mimic redd construction of a spawning salmon, producing a structure with grain size and morphological characteristics typical of natural redds. Four wires at the corners of the bag protrude from the streambed to raise the bag around the frame upon removal, ensuring retention of all collected material. Although intended for assessments of salmonid spawning habitat and the effect of fine sediment on embryo survival, this technique may be applied to studies of sediment dynamics by modifying installation procedures and filling the cube with gravel of a known composition and porosity. As yet, only a single published study has used this instrument, demonstrating successful retention of even the finest particles upon removal of the sample, permitting the assessment of very small-scale variations in the amount and effect of this size fraction [242]. Further studies for a range of environments and applications are needed to fully test this technique.

3.4 Models of vertical particle distribution and exchange

Although early studies of FPOM dynamics focused primarily on longitudinal transport, ecologists are increasingly recognizing the importance of vertical particle exchange, particularly rates of particle deposition, retention, and resuspension, to stream functioning. Three models are introduced and briefly described here: the Rouse equation, the advection-dispersion model, and the Local Exchange Model; these are summarized and compared in Table 2. Limitations of these three models are also discussed.

3.4.1 The Rouse equation

In the physical sciences, decades of research has led to the development of robust theoretical equations for the vertical distribution of suspended particles in water, based primarily on mathematical representations of turbulent flow, molecular diffusion, and particle movement. Although confirmed by early studies in uniformly turbulent tanks known as "turbulence jars" [243–245], these equations rely on empirically derived constants and several assumptions that may not be valid under natural conditions. Several such equations exist [124, 167–171, 246], each attempting to improve those developed before, but all containing the same fundamental model, basic assumptions, and empirical constants. Rouse [170] developed the original form from a differential equation of suspended sediment (Table 2) that assumes two-dimensional, steady, uniform flow and particles with uniform size, shape, and density. Steady flow implies that the average concentration of particles remains constant, such that the flux of particles in all three dimensions is balanced; in other words, for a given volume, sediment fluxes in and out are equal. Thus, particle distribution in the vertical direction represents a balance between the downward settling of particles due to gravity and the upward movement of particles due to diffusion [247].

However, despite the theoretically robust nature of this model, it is difficult to apply in practice because near-bed concentration (C_a) and the Rouse number ($\hat{s}$) (Table 2) are poorly constrained, particularly over beds with a wide range of particle sizes. In theory, both C_a and $\hat{s}$ are dependent on the flow level and the particles available for transport, which is in turn a function of bed composition. Previous work has used a mean particle size for the computation of particle fall velocity in the equation for $\hat{s}$, despite the fact that given a wide range of particle sizes and densities, flows will distribute particles non-uniformly according to their fall velocities; flows will selectively transport fine particles higher in the water column, particularly as flow intensity decreases. The Rouse number at any point in the profile may differ depending on the grain size and density of suspended particles. Variation in sediment diffusivity (β), normally assumed constant, can also change the value of $\hat{s}$. Several studies have measured suspended sediment profiles and demonstrated wide variation in sediment diffusivity depending on the region of the flow depth [248, 249], the density of suspended particles [114], and the conditions of the bed [250]. These discrepancies indicate that the Rouse equation may be inappropriate for application to most natural systems.

3.4.2 The advection-dispersion model

Similar to the Rouse equation, the advection-dispersion model [172–174] describes solute or particle flux in three spatial dimensions as a balance between the transport driven by particle fall velocity (V_{fall}) and flow velocity (advection) and that driven by molecular diffusion and turbulence (dispersion) (Table 2). In this framework, the concentration of an assemblage of particles is modeled as a deterministic diffusion process. Integration of the advection-dispersion model over the flow depth produces an estimate of the average concentration of suspended particles.

The advection-dispersion model can be expanded to include other stream characteristics such as morphology, groundwater and tributary inputs, and transient storage [174] (Table 2). Transient, or temporary, storage of a solute or particle can occur in slower moving areas of the stream, such as pools or hyporheic zones [172]. In the case of a nonconservative solute, abiotic and biotic exchanges take place between the water column and the stream substrate (i.e. adsorption, plant uptake). Immobilization, the removal of solutes from the water column, can be incorporated into a model of nonconservative solute dynamics with an additional term. Terms within the model that represent transient storage and immobilization describe the interaction between the free-flowing stream channel and hydrologic storage zones. A number of studies have used this model to investigate the relationship between particle dynamics and the extent of transient storage [115, 150, 154, 161].

Prediction of suspended sediment transport by the advection-dispersion model requires three major assumptions: 1) the streamwise particle velocity is equal to the local water velocity, 2) the downward velocity of particles in turbulent flow is equal to the particle fall velocity, and 3) particle diffusivity is equal to the turbulent momentum diffusivity of water [114]. However, these assumptions ignore the small-scale mechanics of suspended particles, implying that particles and water act as a single phase or mixture. As a result, predictions based on these three assumptions may be of limited validity, because particle–water and particle–particle interactions can influence the dynamics of turbulent flows and streamwise velocities. Several studies have demonstrated that because of these interactions the above assumptions are in fact questionable [114, 251–254]. For example, using image-based techniques to separately quantify particle and water movement, Muste *et al.* [114] measured streamwise particle velocities less than the velocity of water in the upper region of the flow and a reverse relationship in the lower region, violating the first assumption. They also found that vertical velocities of sand particles differed from water velocities throughout the flow depth. Downward vertical velocities of sand particles were lower than water velocities in the upper and near-bed regions, but greater in the lower flow region. Although this study did not compare these velocities directly to particle fall velocities, the variation in downward velocities contradicts assumption two. This study also showed that sediment diffusivity was less than that of water in the mid- and upper regions of the flow, contradicting assumption three. These findings demonstrate that the advection-dispersion model is limited because it does not consider particle–water interactions or the depth-dependent nature of model parameters.

3.4.3 The Local Exchange Model

McNair *et al.* [175] propose an alternative approach to modeling suspended particle dynamics that considers the behavior of an individual particle, rather than an assemblage of particles. Like the advection-dispersion model, the focus and intended application of the McNair *et al.* [175] approach is FPOM. The authors argue that theoretical models developed for suspended inorganic particle dynamics are inadequate for ecological applications because of their focus on inorganic particles with high fall velocities, passive gravitational settling, and physical forces, as well as an interest in large-scale processes. In contrast, ecological studies emphasize the biological significance of particles that often vary in shape, composition, and density and may be deposited and mobilized via behavioral means. McNair *et al.* [175] describe the process of fine particle transport as including four key components: the attachment problem, the entrainment problem, the hitting-time problem, and the hitting-distance problem. The attachment and entrainment problems address how a particle at the bed/water interface becomes fixed to the substrate and its residence time before resuspension into the water column. The hitting-time/distance problems consider the temporal and spatial dimensions of longitudinal transport; namely, how long a particle remains in the water column and the distance it travels. Time spent in the water column may be relevant to free-swimming consumers, while travel distance is important in determining the rate of downstream dispersal.

Fine particles in turbulent water move along irregular trajectories, buffeted up and down by fluid eddies, thus vertical particle movement and elevation may be considered a stochastic process [175, 255]. McNair *et al.* [175] provided a discrete representation of this stochastic process by considering the motion of a neutrally buoyant, nonmotile particle as occurring in two ways: particles can be propelled by molecular collisions or may be incidentally carried by the turbulent transport of water. Because of the complex, nonlinear structure of turbulent fluid motion, a simplified approximation, known generally as a stochastic-diffusion process, can be used to model turbulent transport. All stochastic-diffusion processes are defined by forward and backward Kolmogorov equations, which include the infinitesimal mean and variance functions (Table 2). Once specified, these functions convert the abstract stochastic-diffusion process into a meaningful description of the process of interest, i.e. the dynamics of particle motion. For more detailed explanation of the background and equations of the Local Exchange Model, see McNair [156], McNair and Newbold [157], and McNair *et al.* [175].

Although the Local Exchange Model improves upon the advection-dispersion model by addressing the small-scale dynamics of individual particle motion and the inclusion of depth-dependent parameters, it retains assumptions that limit the model's validity. The Local Exchange Model does represent the effect of fluid motion on particles by a vertical dispersion coefficient that varies with depth ($K(z)$) (Table 2), but it does not consider the reverse effect of particles on the flow that can occur throughout the flow depth regardless of particle density [114]. As noted by McNair *et al.* [175], close agreement between vertical profiles predicted by the Local Exchange Model and data from the literature does not fully validate the model because the available empirical evidence is limited to flume data and one field case that only considered particles of high fall velocities [256–258]. Vertical profiles from field measurements and a range of particle fall velocities are needed to test the model. Furthermore, the model contains a number of assumptions and approximations that may not hold under all conditions, most importantly the assumption of a flat, hydrodynamically smooth bed. Extension of the model to a complex bed topography, where the presence of retention structures or transient storage zones may greatly alter particle dynamics and the profile of vertical mixing, would be difficult and highly uncertain. The influence of bed topography may be greatest for small particles whose vertical movement is strongly affected by turbulent flows. It should also be noted that despite being developed within an ecological framework, the Local Exchange Model is a purely physical-based model that does not consider biological influences.

3.5 Impact of human activities on particle transport

Human activities, such as dam operation and forest harvesting, can alter the timing and magnitude of hydrologic events, altering discharge regimes and sediment transport (Table 1). Dams both store water and capture sediment, so that the downstream impact differs whether considering the dam's effect on sediment discharge or its effect on transport capacity. Reduced sediment loads may enhance channel armoring, while reduced flows may lead to sediment deposition if the tributaries and banks contribute more sediment than the mainstem has the capacity to transport [259]. Small streams are typically regulated by flood-control dams that both trap sediment and reduce the magnitude of high-flow events; the effect on sediment transport is therefore variable and unpredictable [260]. Several reviews have shown that logging and road building increase sediment yield [17, 261–263], primarily via increases in sediment availability due to soil disturbance [102], bank destabilization [264] or landslide acceleration [106, 265]. However, in some cases, post-harvest increases in sediment export may be due to increases in streamflow that result from reduced evapotranspiration rates and interception losses; high channel flows in turn activate in-channel sources of sediment [110, 111]. Others have further proposed that higher water tables throughout the watershed also increase hydrologic connectivity by activating zero-order basins and ephemeral reaches, resulting in more frequent sediment delivery to perennial channels [109]. Only a few studies have examined the effects of harvesting practices on streambed infiltration in small streams, and with variable results; one found significant increases in bed sediment following logging [266], but others found only road construction increased infiltration [267] and only at low flows [268].

4 Biological significance

Fine particles play a major role in stream ecology, but their specific impact depends on particle composition and stream characteristics, including discharge level, bed and channel morphology, and invertebrate-community composition, among others. Organic and inorganic particles may have distinct effects, due to differences in density and nutritional quality. Particles stored in the bed will impact benthic communities, while those suspended in the water column will affect free-swimming consumers and water quality. We begin by reviewing the biological impact of organic and inorganic particle infiltration and then discuss the effect of suspended particles.

4.1 Impacts of fine particle infiltration into the streambed and hyporheic zone

Numerous studies have documented the deleterious impact of increased fine particle deposition to benthic habitats, the hyporheic zone, and associated organisms. Deposition and infiltration of fines into the hyporheic zone, referred to collectively as "streambed clogging," modifies substrate conditions, trophic resources, and predator activity with ensuing effects on community structure, including an increase in drift and decline in abundance, followed by higher mortality and lower productivity [186]. Progressively, burrowing and fines-adapted assemblages replace invertebrates that require interstitial spaces for habitat [186, 269, 270]. Several studies have shown that fine sediment deposition will decrease macroinvertebrate diversity and abundance [3, 271–273]. An increase in egg and larval mortality of fish species is also commonly linked to the degradation of spawning habitat by fine particle deposition and infiltration into gravel- and cobble-bed sediments [274–276]. In addition, homogenization of the substrate by fine sediment deposition reduces the productivity of algae and the respiration of benthic biofilm [277]. In contrast, the increase in particulate organic matter from decaying post-spawning salmon in western North America has been shown to enhance the growth of stream bacteria and algae [278, 279]. Zones of FPOM accumulation correspond to high microbial activity and degradation rates. The amount of organic matter produced in-channel and biological activity is directly related to the efficiency of particle retention [117, 280].

In addition to direct effects on benthic organisms, fine particle infiltration also blocks intergravel flow and restricts the exchange of oxygen, water, and nutrients vital to benthic organisms [180, 186, 281, 282]. Oxygen supply may be restricted by a layer of sand at the surface of the streambed reducing surface–subsurface exchange [239] or by fines within the gravel pore space that reduce interstitial flow volume and oxygen delivery. Both inorganic and organic particles can reduce oxygen supply via these two mechanisms, but oxygen levels are further reduced by the metabolism of organic particles by bacterial communities [283]. How much organic matter infiltrates into gravel spaces depends on the type and amount of organic inputs as well as the degree of flocculation; flocculation with mineral substances serves to enhance the settling and storage of low-density organic particles [284, 285]. In turn, flocculated fine particle size and density changes according to season and the type of organic matter source; studies from salmon streams of British Columbia demonstrate that the largest, least dense particles are associated with salmon carcasses and die-off periods [285, 286].

4.2 Impacts of suspended particles

High concentrations of particles suspended in the water column can harm the feeding habits of free-swimming consumers such as filter-feeding invertebrates and fish [287, 288] and degrade water quality. Increases in fine particle loads may lead to higher turbidity, as well as eutrophication and high toxicity of both the stream and receiving water bodies. Water quality is typically assessed according to concentrations of suspended particles and dissolved solutes, including nitrogen (N), phosphorous (P), organic carbon (DOC), and major cations and anions, and occasionally turbidity, conductivity, black disk visibility, pH, alkalinity, or temperature [289–291]. Fine particles, especially those <63 micrometers and those with high organic content, are highly electronegative and therefore strongly associated with the sorption and transport of hydrophobic pollutants and nutrients, including polychlorinated biphenyls (PCBs), dioxins, radionuclides, heavy and trace metals, and nutrients such as N and P [8, 183–185, 292–94].

4.3 Impacts of anthropogenic changes to particle dynamics

Numerous studies have documented a decline in water and bed sediment quality, with ensuing effects on fish and invertebrates, following land-use changes – including agricultural activity, wildfire, dam building, and forest harvesting practices [291, 295–299]. A review of forest harvesting impacts on streams in North America is provided by Binkley and Brown [99]. Typically, any form of vegetation removal and fertilization will degrade water quality by increasing nitrate and suspended-particle concentrations, but consequences vary widely. When intact, floodplain forests act as a filter for pollutants in surface runoff to streams by biological and physical adsorption [300]. High sediment loads and subsequent degradation of spawning habitat following land-use changes has been documented by many researchers [274, 276, 301]. Furthermore, some of these practices (i.e. forest harvesting and road construction) have been linked to an increase in the organic matter content of bed sediments [302, 303], likely due to the erosion of soil organic matter or the delivery of dead plant material from the riparian zone. In some cases, high levels of bed organic matter may not be associated with high inorganic particle yields but may still result in oxygen depletion (see Owens *et al.* [16] for review).

Some ecological effects may be unique to certain regions; for example, in western North America, logged forests are recolonized by alders, which fix nitrogen and thus produce high-quality, low C:N ratio litter. Stream productivity may increase as a result of this high quality input [304] and higher-quality FPOM, also characteristic of most young, regenerating forests [11]. Higher productivity may also be due to increased decay rates, a consequence of easily decomposed deciduous leaves and an increase in stream temperatures due to the removal of canopy cover and increased sunlight penetration [305–307].

Changes to particle dynamics due to flow regulation can also have profound ecological effects, degrading benthic and floodplain habitat, altering the community composition of aquatic organisms, and decreasing productivity [308–310]. Depending on the style and type of management, dams can also alter thermal regimes or resource availability, with subsequent effects on downstream biotic communities [311] and thus organic matter production. Small streams are typically regulated by dams built for powering mills or flood control, many of which maintain a permanent reservoir. In addition to indirect geomorphic effects on ecology, these dams and associated reservoirs may also directly alter the quantity and nutritional quality of organic particles and the organisms that feed on them. Reservoir effects at dam outlets, including altered water temperatures and increased FPOM availability, have been well-documented; these are typically linked to an increase in filter-feeding invertebrates downstream [312]. An increase in organic particles below outlets is due to releases from the surface of the reservoir containing lake-derived plankton, otherwise known as seston. Seston quantity and quality then decline with downstream distance, due to factors such as selective depletion by these filter-feeding consumers [313, 314] or dilution by low-quality particles [313, 315].

5 Variability at different spatial and temporal scales

All elements of fine particle dynamics vary across a range of spatial and temporal scales. As a result, inconsistencies between or even within studies may arise from scale-related differences in process or sampling design. Spatial scales range from individual particles to watershed to geographic region. Temporal variations occur on within-year, seasonal scales or over multi-decadal periods of climate and hydrologic change. Nevertheless, relatively few studies have considered the effect of scale on the variability of fine particle dynamics in streams and even fewer have addressed temporal and spatial variation concurrently.

5.1 Spatial scales and variability

Sampling limitations typically restrict most studies to the reach or channel scale. Few measurement techniques operate at a fine enough scale for the monitoring of individual particle movement, while watershed- or larger-scale sampling may be prohibitively expensive and time-consuming. Most studies are confined to a single spatial scale and stream size, although some have attempted to study fine particle dynamics at multiple scales and over large geographic areas. Water quality classification schemes, for example, have been developed to extrapolate data collected at a few sites to an entire region or to subdivide large areas into smaller zones (see review by Roberston and Saad [316]). Three broad types of extrapolation/classification schemes exist, including those based on similar environmental characteristics, regression equations, and mechanistic models, as well as a combination of these three. However, variability in these classification schemes arises due to uncertainties in data representativeness, within-zone variation, and the selected environmental factor or water-quality metric (e.g. total N, P, or DOC). A recently developed approach, known as SPARTA, attempts to avoid these problems by using a regression-tree analysis and GIS coverage to subdivide an area into homogenous zones specific to a given water-quality metric and based on the environmental factor most statistically important to that metric [316].

Management and restoration of degraded watersheds necessitates an understanding of which factors, and which scales, have the greatest effect on stream conditions, yet debate remains over which spatial scale contains the best indicators of water quality and in-channel ecological conditions. In recent years, advancements in geographic information systems (GIS) techniques have allowed researchers to determine landscape metrics over large areas and at fine scales [317]. Combined with in-channel measurements, spatial land-cover and land-use databases can be used to determine which variables best explain stream conditions. Several studies indicate that watershed or landscape factors, such as soil type, geology or land use, are better predictors than local stream characteristics, such as riparian condition [181, 317–319], but the reverse has also been found [320]. A combination of effects at both the local and watershed scale may also explain in-channel biological conditions [321, 322]. Landscape metrics can be successful predictors of water quality, as shown by Jones *et al.* [323] who find that metrics generated from easily obtained spatial data, such as agricultural land cover, extent of riparian forests, and amount of atmospheric nitrate deposition explain most of the variability in dissolved N, P, and suspended sediment. Most studies addressing this issue, however, have been limited to only one or a few adjacent watersheds with similar climate, land use, and fish species. In contrast, Meador and Goldstein [290] use data from 20 major U.S. river basins and a multimetric approach to assess the relationship between land use, riparian condition, water physicochemistry, and fish community structure condition over a broad regional geographic scale. Across large geographic scales, they suggest that water physicochemistry indices are better indicators of fish community condition than watershed land use. Results also indicate that across all the regions studied, riparian condition and associated practices regulate the delivery of nutrients, suspended particles, and dissolved solids to streams, in turn affecting the fish community condition. Despite their relative success, predictive models used in these types of studies contain a certain degree of spatial variability that is inherent to the processes of particle supply and transport.

5.2 Temporal scales, trends, and variability

Temporal variability adds another dimension of complexity that can be exacerbated by ill-suited sampling schemes and techniques. As a result, researchers have suggested that sampling frequency and technique be selected according to nutrient status [324] or statistic of interest and study duration [325]. Long-term monitoring programs typically take manual samples at fixed intervals (i.e. monthly), while short-term studies may be more intensive, taking continuous, automatic samples during flood events and fixed-interval manual samples during supposedly stable intervening flows. From these samples, various statistics, such as annual load or maximum concentration, and rating relations may be calculated to assess water quality. A review and analysis of the cost-effectiveness, precision, and accuracy of approaches for monitoring water quality in small streams is provided by Robertson and Richards [316].

For a long time, researchers have recognized that water quality and discharge characteristics vary seasonally [326, 327], but recent studies have also documented large fluctuations during rain events [328, 329]. Typically, seasonal variability is controlled by precipitation patterns and factors related to land cover, such as evapotranspiration, interception, and infiltration [330]. During rain events, rising discharge can either increase suspended particles via entrainment from the bed or increased surface runoff [331] or dilute cations and suspended biota if little excess is supplied from the watershed [332–334]. Biological retention and physical sorption processes may also decline with increased discharge levels [335, 336].

Hydrologic variability typically changes with basin size; groundwater inputs to small streams and high water volume in large rivers dampen temporal fluctuations in discharge and water quality, such that mid-sized streams are assumed to have the greatest environmental variability [34], with implications for biological productivity and diversity [34, 337]. Despite the influence of river size on hydrologic variability, however, a recent study by Chatelet and Pick [324] demonstrates that at regional and watershed scales, the seasonal variability of water-quality characteristics does not change with river size. Instead, temporal variability and the requisite sampling frequency depend on the water-quality metric, the time of year, and in some cases, the mean concentration. Other spatial characteristics may also influence temporal variability, such as spatial patterns of vegetative cover, subsurface flows, and soil moisture, which control rates of surface erosion and runoff to the stream [186, 330].

Although intra-annual variability has been relatively well studied, documentation of long-term temporal changes in hydrology and associated sediment and solute dynamics is limited by a paucity of long-term data sets. Nevertheless, several methods, including the analysis of paleoenvironmental archives (i.e. archaeological remains, tree rings, documents and instrumental records), numerical simulations (i.e. runoff and erosion models), and long-term monitoring programs, allow researchers to elucidate temporal trends in discharge, particle, and solute dynamics over sub-decadal [338–341] to millennial [342, 343] time scales, often focusing on the impact of climate change and human activities. Paleoenvironmental studies and long-term monitoring programs provide information on past conditions that can facilitate understanding of present-day systems and future trajectories, as well as provide parameterizations for numerical simulation models [344, 345]. Simulation models are typically used to reconstruct or predict how changes in environmental variables, such as climate or land disturbance, affect particle and solute dynamics [346, 347].

Over the past century, regions throughout the world have experienced severe changes in land use and climate. Although anthropogenically induced climatic warming over the past century has had a significant impact on hydrologic and sediment regimes in systems throughout the world [345, 348, 349], climate change often acts concurrently with human activities such as water diversion and consumption [341, 345] or land disturbance [350–352]. For this reason, determining the relative impacts of climatic versus land-use changes can be difficult, but is often attempted through multiple regression models [341] or spatial and chronological links [340, 350, 351].

Results from small streams suggest that land-use changes and local catchment conditions have a dominant effect on long-term temporal trends of sediment transport and water quality, reflecting the strong link between stream and hillslope processes [338]. For example, despite increases in precipitation, reforestation has been shown to reduce soil erosion rates and overall sediment yield [350, 351]. In contrast, sediment cores from headwater lakes in several UK catchments chronologically link high sedimentation rates to land drainage, land conversion, and afforestation, but not to temporal climate changes [340]. Non-climate-related human impacts are not unique to small streams, however. Walling and Fang [345] assembled long-term records of annual sediment load and runoff for 145 major rivers and found highly variable, mostly insignificant temporal trends. Evidence from this and other studies indicates that reservoir construction and the consumption or diversion of water by humans, may be the most important influence on large river sediment fluxes [341].

Several studies demonstrate that the magnitude, direction, and mechanism of change will depend on catchment factors such as river size [338, 352], local climate [352–354], or underlying geology [348, 349]. Decreasing sediment yields contradict climate models that predict increasing global precipitation and temperatures will subsequently increase erosion and runoff rates [339, 350], perhaps reflecting either unchanging controlling variables or buffering mechanisms within the watershed [345, 354]. Increasing sediment yields is reported in several areas with intense human activity, such as deforestation, agriculture, and road building, which increase erosion rates [352, 355, 356], but climate appears to play only a secondary role. Relatively consistent sediment yields [357, 358], despite significant climatic and environmental changes, may be due to alluvial storage and remobilization, which keeps the fluvial system in a state of dynamic stability. It is proposed that any external changes are buffered by internal reorganizations in sediment deposition and negative feedbacks on weathering rates [359].

Natural climatic fluctuations occurring on interannual to millennial timescales are shown to influence the temporal variability of sediment dynamics and geomorphic processes [360]. Several studies document the impacts of climate changes induced by sub-decadal La Nina and El Nino/Southern Oscillations and Pacific Decadal Oscillations [348, 349, 361–363], but results vary between regions. Strong El Nino events produce significant increases in sediment yield in some areas [349, 362] and decreases in others [348, 363, 364]. Events can also alter the composition of sediment loads, including the concentration of particulate nutrients, metals, and organic matter, with effects on productivity and contaminant concentrations of receiving waters [349, 361, 364, 365]. In several cases, factors such as basin geology [348, 349], vegetative cover [364], or impoundment [363], influence the degree of climatic effect by controlling erosion rates or water fluxes.

As revealed in the discussions above, spatial and temporal effects are closely linked; both must be incorporated into a study design. Selecting the appropriate scale for study of a given system requires understanding which factors and processes dominate the subject of interest and the degree of variability inherent to each. Factors such as human activities, geology, location, vegetation, river size, and climate can influence the response of the system and should be considered when selecting measurement techniques, sampling design, and data analysis.

6 Research needs

One major research need in the field of fine particle dynamics are the factors influencing the vertical movement and exchange of fine particles, particularly streambed deposition and infiltration. As demonstrated in this chapter, a complex combination of biological and physical factors will influence the distribution, deposition, and infiltration of particles. A major remaining question is why measured rates of particle deposition often differ markedly from particle settling velocities; although several mechanisms have been proposed to explain this discrepancy, it is not fully understood. Still largely unexplored are the roles of transient storage zones, biological controls such as particle trapping by biofilm or removal by invertebrates, and particle composition. Also largely unexplored is the relative importance of particle flocculation and organic matter content on particle transport and deposition. Recent studies suggest that flocculation may affect near-bed concentrations [366], bed-sedimentation rates [367, 368], and metal adsorption [369]. Organic matter content will also affect biological consequences; higher nutritional quality may increase benthic productivity, but decomposition may lead to more pronounced oxygen depletion within the hyporheic zone. Thus, both the physical and biological impacts of flocculation are still uncertain. In addition, only a few studies have measured the response of streambed deposition and infiltration rates to changes in sediment yield, organic matter inputs, and water discharge following forest harvesting.

A second major area in which more research should focus is the development of more accurate and representative models of vertical exchange and distribution. As a consequence of our limited understanding and the complexity of fine particle dynamics, current models of vertical exchange and distribution are constrained by assumptions and conditions unrealistic for most natural systems. In particular, no available model provides an explicit representation of bed composition or bed complexity, particle composition, or benthic ecology. Traditional hydrodynamic models of vertical particle distribution (Rouse equation, advection-dispersion model) are limited because they do not consider the small-scale mechanics of particle motion, particularly the effect of particle–water interactions on turbulent flows and particle movement. The Local Exchange Model improves upon these models by describing the movement of an individual particle due to turbulence as a stochastic diffusion process that includes a depth-dependent measure of vertical dispersion. Nevertheless, this model is still limited because it does not consider the effect of particles or the effect of complex bed topography on turbulent flows. An alternative approach to modeling the vertical movement of suspended particles has recently been presented by Muste *et al.* [114] in which particle and water movement are treated as two distinct phases rather than as a single mixed fluid. However, no theoretical model exists that can fully encapsulate the numerous factors influencing vertical particle exchange. A conceptual model of particle exchange that includes physical and biological mechanisms is needed as a framework on which to base our predictions. As a starting point to developing this model, we need more accurate measurements of the vertical distribution of particles under a range of hydrological, geomorphic, and biological conditions.

A third major research need is the development and refinement of instruments and measurement techniques for particle movement and infiltration. Newly introduced devices such as the infiltration cube require further testing for different conditions and study objectives. In addition, the use of radionuclides for tracking particle movement is a promising new field that should be explored. Short-lived radionuclides have been used to quantify the longitudinal transport of fine sediment, but may also serve as tracers in the vertical direction, indicating rates of deposition, depths of infiltration, and degree of streambed mixing. Short- and long-lived radionuclides could be used in combination to identify particle residence times in both long-term and transient storage zones, as well as spatial variability in particle storage across the stream channel.

Clearly, despite extensive study on the dynamics of fine inorganic and organic particles in streams, gaps in our knowledge still remain. Many complex and interrelated factors govern the movement, storage, and impact of fine particles, factors that vary greatly over spatial and temporal scales. Human activities can also profoundly alter and further complicate these processes. Generally, the most significant limitation to our understanding of fine particle dynamics is the predominantly one-sided focus of most studies. Increasingly, researchers are recognizing the important links between physical and biological components, but truly cross-disciplinary studies are rare. One of the most important advancements in coming years will be the successful integration of the approaches, methods, and knowledge of these two fields.

References

[1] Lowe, W.H. & Bolger, D.T., Local and landscape-scale predictors of salamander abundance in New Hampshire headwater streams. *Conservation Biology*, **16(1)**, pp. 183–193, 2000.

[2] Nerbonne, B.A. & Vondracek, B., Effects of local land use on physical habitat, benthic macroinvertebrates, and fish in the Whitewater River, Minnesota, USA. *Environmental Management*, **28(1)**, pp. 87–99, 2001.

[3] Waters, T.F., Sediment in streams: Source, biological effects, and control. *American Fisheries Society Monograph 7*, American Fisheries Society: Bethesda, Maryland, pp. 251, 1995.

[4] Reiser, D.W., Sediment in gravel bed rivers; ecological and biological considerations. *Gravel-bed rivers in the Environment*, **4**, pp. 199–228, 1998.

[5] Wallace, J.B., Ross, D.H., Meyer, J.L., Seston and dissolved organic-carbon dynamics in a southern Appalachian stream. *Ecology*, **63(3)**, pp. 824–838, 1982.

[6] Webster, J.R., Meyer, J.L., Stream organic matter budgets. *Journal of the North American Benthological Society*, 16(1), pp. 3–4, 1997.

[7] Webster, J.R., Benfield, E.F., Golladay, S.W., Hill, B.H., Hornick, L.E., Kazmierczak, R.F., Perry, W.B., Experimental studies of physical factors affecting seston transport in streams. *Limnology and Oceanography*, **32**, pp. 848–863, 1987.

[8] Karichkhoff, R.F., Organic pollutant sorption in aquatic systems. *Journal of Hydraulic Engineering – American Society of Civil Engineers*, **110**, pp. 707–735, 1984.

[9] Benenati, E.P., Shannon, J.P., Hagan, J.S. & Binn, D.W., Drifting fine particulate organic matter below Glen Canyon Dam in the Colorado River, Arizona. *Journal of Freshwater Ecology*, **16(2)**, pp. 235–248, 2001.
[10] Bilby, R.E. & Bisson, P.A., Allochthonous versus autochthonous organic matter contributions to the trophic support of fish populations in clear-cut and old-growth forested streams. *Canadian Journal of Fisheries and Aquatic Sciences*, **49(3)**, pp. 540–551, 1992.
[11] Bonin, H.L., Griffiths, R.P. & Caldwell, B.A., Nutrient and microbiological characteristics of fine benthic organic matter in mountain streams. *Journal of the North American Benthological Society*, **19(2)**, pp. 235–249, 2000.
[12] Droppo, I.G. & Ongley, E.D., The state of suspended sediment in the freshwater fluvial environment – a method of analysis. *Water Research*, **26(1)**, pp. 65–72, 1992.
[13] Walling, D.E. & Woodward, J.C., Use of a field-based water elutriation system for monitoring the in-situ particle-size characteristics of fluvial suspended sediment. *Water Research*, **27(9)**, pp. 1413–1421, 1993.
[14] Woodward, J.C. & Walling, D.E., A field sampling method for obtaining representative samples of composite fluvial suspended sediment particles for SEM analysis. *Journal of Sedimentary Petrology*, **62(4)**, pp. 742–744, 1992.
[15] Wood, P.J. & Armitage, P.D., Biological effects of fine sediment in the lotic environment. *Environmental Management*, **21(2)**, pp. 203–217, 1997.
[16] Owens, P.N., Batalla, R.J., Collins, A.J., Gomez, B., Hicks, D.M., Horowitz, A.J., Kondolf, G.M., Marden, M., Page, M.J., Peacock, D.H., Petticrew, E.L., Salomons, W. & Trustrum, N.A., Fine-grained sediment in river systems: Environmental significance and management issues. *River Research and Applications*, **21(7)**, pp. 693–717, 2005.
[17] Gomi, T., Moore, R.D. & Hassan, M.A., Suspended sediment dynamics in small forest streams of the Pacific Northwest. *Journal of the American Water Resources Association*, **41(4)**, pp. 877–898, 2005.
[18] Benda, L., Hassan, M.A., Church, M. & May, C.L., Geomorphology of steepland headwaters: The transition from hillslopes to channels. *Journal of the American Water Resources Association*, **41(4)**, pp. 835–851, 2005.
[19] Hassan, M.A., Church, M., Lisle, T.E., Brardinoni, F., Benda, L. & Grant, G.E., Sediment transport and channel morphology of small, forested streams. *Journal of the American Water Resources Association*, **41(4)**, pp. 853–876, 2005.
[20] Hassan, M.A., Hogan, D.L., Bird, S.A., May, C.L., Gomi, T. & Campbell, D., Spatial and temporal dynamics of wood in headwater streams of the Pacific Northwest. *Journal of the American Water Resources Association*, **41(4)**, pp. 899–919, 2005.
[21] Strahler, A.N., Quantitative analysis of watershed geomorphology. *Transaction of the American Geophysical Union*, **38**, pp. 913–920, 1957.
[22] Meyer, J.L. & Wallace, J.B., Lost linkages and lotic ecology: rediscovering small streams. *Ecology: Achievement and Challenge*, eds. M. Pages, N. Huntly & S. Levin, Blackwell Science: Oxford, United Kingdom, pp. 295–317, 2001.

[23] Morisawa, M.E., Accuracy of determination of stream lengths from topographic maps. *Transactions of the American Geophysical Union*, **38**, pp. 86–88, 1957.

[24] Church, M., Channel morphology and typology. *The Rivers Handbook: Hydrological and Ecological Principles*, Blackwell Science: London, United Kingdom, pp. 126–143, 1992.

[25] Grant, G.E., Swanson, F.J. & Wolman, M.G., Pattern and origin of stepped-bed morphology in high-gradient streams, western Cascades, Oregon. *Geological Society of America Bulletin*, **102(3)**, pp. 340–352, 1990.

[26] Montgomery, D.R. & Buffington, J.M., Channel-reach morphology in mountain drainage basins. *Geological Society of America Bulletin*, **109(5)**, pp. 596–611, 1997.

[27] Church, M. & Hassan, M.A., Size and distance of travel of unconstrained clasts on a streambed. *Water Resources Research*, **28(1)**, pp. 299–303, 1992.

[28] Montgomery, D.R. & Foufoulageorgiou, E., Channel network source representation using Digital Elevation Models. *Water Resources Research*, **29(12)**, pp. 3925–3934, 1993.

[29] Benda, L. & Dunne, T., Sediment routing by debris flows. *Erosion and Sedimentation in the Pacific Rim,* ed. R.L. Beschta, T. Blinn, G.E. Grant, G.G. Ice & F.J. Swanson, International Association of Hydrological Sciences Press: Wallingford, pp. 213–223, 1987.

[30] Nistor, C.J. & Church, M., Suspended sediment transport regime in a debris-flow gully on Vancouver Island, British Columbia. *Hydrological Processes*, **19(4)**, pp. 861–885, 2005.

[31] Cheng, J.D., Subsurface stormflows in the highly permeable forested watersheds of southwestern British Columbia. *Journal of Contaminant Hydrology*, **3**, pp. 171–191, 1988.

[32] Harr, R.D., Water flux in soil and subsoil on a steep forested slope. *Journal of Hydrology*, **33**, pp. 37–58, 1977.

[33] Roberts, R.J. & Church, M., The sediment budget in severely disturbed watersheds, Queen Charlotte Ranges, British Columbia. *Canadian Journal of Earth Sciences*, **16**, pp. 1092–1106, 1986.

[34] Vannote, R.L., Minshall, G.W., Cummins, K.W., Sedell, J.R. & Cushing, C.E., River continuum concept. *Canadian Journal of Fisheries and Aquatic Sciences*, **37(1)**, pp. 130–137, 1980.

[35] Jones, J.B., Benthic organic matter storage in streams: Influence of detrital import and export, retention mechanisms, and climate. *Journal of the North American Benthological Society*, **16(1)**, pp. 109–119, 1997.

[36] Wotton, R.S. & Malmqvist, B., Feces in aquatic ecosystems. *BioScience*, **51**, pp. 537–544, 2001.

[37] Naiman, R.J., The influence of stream size on the food quality of seston. *Canadian Journal of Zoology-Revue Canadienne de Zoologie*, **61(9)**, pp. 1995–2010, 1983.

[38] Petersen, R.C., In situ particle generation in a southern Swedish stream. *Limnology and Oceanography*, **31**, pp. 432–437, 1986.

[39] Bilby, R.E., Beach, E.W., Fransen, B.R., Walter, J.K. & Bisson, P.A., Transfer of nutrients from spawning salmon to riparian vegetation in western Washington. *Transactions of the American Fisheries Society*, **132(4)**, pp. 733–745, 2003.

[40] Ross, H.H., Stream communities and terrestrial biomes. *Archiv für Hydrobiologie*, **59**, pp. 235–242, 1963.

[41] Richardson, J.S., Shaughnessy, C.R. & Harrison, P.G., Litter breakdown and invertebrate association with three types of leaves in a temperate rainforest stream. *Archiv für Hydrobiologie*, **159**, pp. 309–325, 2004.

[42] Webster, J.R. & Benfield, E.F., Vascular plant breakdown in fresh-water ecosystems. *Annual Review of Ecology and Systematics*, **17**, pp. 567–594, 1986.

[43] Beschta, R.L., Long-term patterns of sediment production following road construction and logging in the Oregon Coast Range. *Water Resources Research*, **14**, pp. 1011–1016, 1978.

[44] Helfield, J.M. & Naiman, R.J., Effects of salmon-derived nitrogen on riparian forest growth and implications for stream productivity. *Ecology*, **82(9)**, pp. 2403–2409, 2001.

[45] Gottesfeld, A.S., Hassan, M.A., Tunnicliffe, J.F. & Poirier, R.W., Sediment dispersion in salmon spawning streams: the influence of floods and salmon redd construction. *Journal of the American Water Resources Association*, **40(4)**, pp. 1071–1086, 2004.

[46] Hames, S.D.E., Peterson, P.N., Conrad, R. & Quinn, T.P., Patterns of gravel scour and fill after spawning by chum salmon in a western Washington stream. *North American Journal of Fisheries Management*, **20**, pp. 610–617, 2000.

[47] Kondolf, G.M., Sale, M.J. & Wolman, M.G., Modification of fluvial gravel size by spawning salmonids. *Water Resources Research*, **29(7)**, pp. 2265–2274, 1993.

[48] Reid, L.M. & Dunne, T., *Rapid evaluation of sediment budgets*, Geo-Ecology paperbacks, Catena Verlag: Germany, 1996.

[49] Walling, D.E., Collins, A.L., Sichingabula, H.M. & Leeks, G.J.L., Integrated assessment of catchment suspended sediment budgets: A Zambian example. *Land Degradation & Development*, **12(5)**, pp. 387–415, 2001.

[50] Dedkov, A.P. & Moszherin, V.T., Erosion and sediment yield in mountain areas of the world. *Erosion, Debris Flows and Environment in Mountainous Regions*, ed. D.E. Walling, T.R. Davies & B. Hasholt, International Association of Hydrological Sciences Press: Wallingford, 1992.

[51] Summer, W., Klaghofer, E. & Hintersteiner, K., Trends in soil erosion and sediment yield in the alpine basin of the Austrian Danube. *Erosion and Sediment Yield: Global and Regional Perspectives*, ed. D.E. Walling & B.W. Webb, International Association of Hydrological Sciences Press: Wallingford, pp. 473–479, 1996.

[52] Collins, A.L., Walling, D.E., Sichingabula, H.M. & Leeks, G.J.L., Suspended sediment source fingerprinting in a small tropical catchment and some management implications. *Applied Geography*, **21(4)**, pp. 387–412, 2001.

[53] EPA, U.S., Protocol for developing sediment TMDLs. Office of Water (4503F) United States Environmental Protection Agency: Washington, DC, **841-B-99-004**, 1999.

[54] Stocking, M.A., Measuring land degradation. *Land degradation and society*, eds. P. Blaikie, H. Brookfield, Methuen: London, pp. 49–63, 1987.

[55] Collins, A.L. & Walling, D.E., Documenting catchment suspended sediment sources: problems, approaches and prospects. *Progress in Physical Geography*, **28(2)**, pp. 159–196, 2004.

[56] Oldfield, F., Rummery, T.A., Thompson, R. & Walling, D.E., Identification of suspended sediment sources by means of magnetic measurements – some preliminary results. *Water Resources Research*, **15(2)**, pp. 211–218, 1979.

[57] Peart, M.R. & Walling, D.E., Techniques for establishing suspended sediment sources in two drainage basins in Devon, UK: a comparative assessment. *Sediment Budgets. International Association of Hydrological Sciences*, ed. M.P. Bordas & D.E. Walling, International Association of Hydrological Sciences Press: Wallingford, pp. 269–279, 1988.

[58] Walling, D.E. & Woodward, J.C., Use of radiometric fingerprints to derive information on suspended sediment sources. *Erosion and Sediment Transport Monitoring Programmes in River Basins (Proceedings Oslo Symposium)*. International Association of Hydrological Sciences Press: Wallingford, **210**, pp. 153–164, 1992.

[59] Klages, M.G. & Hsieh, Y.P., Suspended solids carried by Gallatin River of southwestern Montana. 2. Using mineralogy for inferring sources. *Journal of Environmental Quality*, **4(1)**, pp. 68–73, 1975.

[60] Walling, D.E. & Amos, C.M., Source, storage and mobilization of fine sediment in a chalk stream system. *Hydrological Processes*, **13(3)**, pp. 323–340, 1999.

[61] Wall, G.J. & Wilding, L.P., Mineralogy and related parameters of fluvial suspended sediments in northwestern Ohio. *Journal of Environmental Quality*, **5(2)**, pp. 168–173, 1976.

[62] Olley, J.M., Murray, A.S., Mackenzie, D.H. & Edwards, K., Identifying sediment sources in a gullied catchment using natural and anthropogenic radioactivity. *Water Resources Research*, **29(4)**, pp. 1037–1043, 1993.

[63] Whiting, P.J., Matisoff, G. & Fornes, W., Suspended sediment sources and transport distances in the Yellowstone River basin. *Geological Society of America Bulletin*, **117(3–4)**, pp. 515–529, 2005.

[64] Collins, A.L., Walling, D.E. & Leeks, G.J.L., Source type ascription for fluvial suspended sediment based on a quantitative composite fingerprinting technique. *Catena*, **29(1)**, pp. 1–27, 1997.

[65] Collins, A.L., Walling, D.E. & Leeks, G.J.L., Fingerprinting the origin of fluvial suspended sediment in larger river basins: Combining assessment of spatial provenance and source type. *Geografiska Annaler Series A-Physical Geography*, **79A(4)**, pp. 239–254, 1997.

[66] Walling, D.E. & Woodward, J.C., Tracing sources of suspended sediment in river basins – a case-study of the River Culm, Devon, UK. *Marine and Freshwater Research*, **46(1)**, pp. 327–336, 1995.

[67] Walling, D.E., Tracing suspended sediment sources in catchments and river systems. *Science of the Total Environment*, **344(1–3)**, pp. 159–184, 2005.

[68] Blake, W.H., Walling, D.E. & He, Q., Using cosmogenic beryllium-7 as a tracer in sediment budget investigations. *Geografiska Annaler Series A-Physical Geography*, **84A(2)**, pp. 89–102, 2002.

[69] Walling, D.E., Russell, M.A., Hodgkinson, R.A. & Zhang, Y., Establishing sediment budgets for two small lowland agricultural catchments in the UK. *Catena*, **47(4)**, pp. 323–353, 2002.

[70] Bierman, P.R., Using in-situ produced cosmogenic isotopes to estimate rates of landscape evolution - a review from the geomorphic perspective. *Journal of Geophysical Research-Solid Earth*, **99(B7)**, pp. 13885–13896, 1994.

[71] Bierman, P.R. & Clapp, E.M., Estimating geologic age from cosmogenic nuclides: An update. *Science*, **271(5255)**, pp. 1606–1606, 1996.

[72] Small, E.E., Anderson, R.S. & Hancock, G.S., Estimates of the rate of regolith production using Be-10 and Al-26 from an alpine hillslope. *Geomorphology*, **27(1–2)**, pp. 131–150, 1999.

[73] Yeager, K.M. & Santschi, P.H., Invariance of isotope ratios of lithogenic radionuclides: more evidence for their use as sediment source tracers. *Journal of Environmental Radioactivity*, **69(3)**, pp. 159–176, 2003.

[74] Yeager, K.M., Santschi, P.H., Phillips, J.D. & Herbert, B.E., Sources of alluvium in a coastal plain stream based on radionuclide signatures from the U-238 and Th-232 decay series. *Water Resources Research*, **38(11)**, 2002.

[75] Yeager, K.M., Santschi, P.H., Phillips, J.D. & Herbert, B.E., Suspended sediment sources and tributary effects in the lower reaches of a coastal plain stream as indicated by radionuclides, Loco Bayou, Texas. *Environmental Geology*, **47(3)**, pp. 382–395, 2005.

[76] Blake, W.H., Walling, D.E. & He, Q., Fallout beryllium-7 as a tracer in soil erosion investigations. *Applied Radiation and Isotopes*, **51(5)**, pp. 599–605, 1999.

[77] Bonniwell, E.C., Matisoff, G. & Whiting, P.J., Determining the times and distances of particle transit in a mountain stream using fallout radionuclides. *Geomorphology*, **27(1–2)**, pp. 75–92, 1999.

[78] Walling, D.E. & He, Q., Using fallout lead-210 measurements to estimate soil erosion on cultivated land. *Soil Science Society Of America Journal*, **63(5)**, pp. 1404–1412, 1999.

[79] Krishnaswami, S., Benninger, L.K., Aller, R.C. & Vondamm, K.L., Atmospherically-derived radionuclides as tracers of sediment mixing and accumulation in near-shore marine and lake-sediments – evidence from Be-7, Pb-210, and Pu-239, Pu-240. *Earth and Planetary Science Letters*, **47(3)**, pp. 307–318, 1980.

[80] Matisoff, G., Mathematical models of bioturbation. *Animal Sediment Relations: The Biotic Alterations of Sediments*, eds. P.L. McCall, M.J.S. Tevesz, Plenem Press: New York, pp. 289–330, 1984.

[81] Ritchie, J.C. & McHenry, J.R., Application of radioactive fallout cesium-137 for measuring soil-erosion and sediment accumulation rates and patterns – a review. *Journal of Environmental Quality*, **19(2)**, pp. 215–233, 1990.

[82] Matisoff, G., Bonniwell, E.C. & Whiting, P.J., Radionuclides as indicators of sediment transport in agricultural watersheds that drain to Lake Erie. *Journal of Environmental Quality*, **31(1)**, pp. 62–72, 2002.

[83] Fitzgerald, S.A., Klump, J.V., Swarzenski, P.W., Mackenzie, R.A. & Richards, K.D., Beryllium-7 as a tracer of short-term sediment deposition and resuspension in the Fox River, Wisconsin. *Environmental Science & Technology*, **35(2)**, pp. 300–305, 2001.

[84] Robbins, J.A. & Eadie, B.J., Seasonal cycling of trace-elements Cs-137, Be-7, and Pu-239 and 240 in Lake Michigan. *Journal of Geophysical Research-Oceans*, **96(C9)**, pp. 17081–17104, 1991.

[85] Oktay, S.D., Brabander, D.J., Smith, J.P., Kada, J., Bullen, T. & Olsen, C.R., WTC geochemical fingerprint recorded in New York Harbor sediments. *Eos Transactions*, **84(3)**, pp. 21–28, 2003

[86] Olsen, C.R., Larsen, I.L., Lowry, P.D., Cutshall, N.H. & Nichols, M.M., Geochemistry and deposition of Be-7 in river-estuarine and coastal waters. *Journal of Geophysical Research-Oceans*, **91(C1)**, pp. 896–908, 1986.

[87] Walling, D.E. & He, Q., Use of fallout Cs-137 in investigations of overbank sediment deposition on river floodplains. *Catena*, **29(3–4)**, pp. 263–282, 1997.

[88] Brown, L., Stensland, G.J., Klein, J. & Middleton, R., Atmospheric deposition of Be-7 and Be-10. *Geochimica et Cosmochimica Acta*, **53(1)**, pp. 135–142, 1989.

[89] Hawley, N., Robbins, J.A. & Eadie, B.J., The partitioning of beryllium-7 in fresh-water. *Geochimica et Cosmochimica Acta*, **50(6)**, pp. 1127–1131, 1986.

[90] Owens, P.N., Walling, D.E. & He, Q.P., The behavior of bomb-derived caesium-137 fallout in catchment soils. *Journal of Environmental Radioactivity*, **32(3)**, pp. 169–191, 1996.

[91] Wallbrink, P.J., Roddy, B.P. & Olley, J.M., A tracer budget quantifying soil redistribution on hillslopes after forest harvesting. *Catena*, **47(3)**, pp. 179–201, 2002.

[92] Olsen, C.R., Larsen, I.L., Lowry, P.D., Cutshall, N.H., Todd, J.F., Wong, G.T.F. & Casey, W.H., Atmospheric fluxes and marsh-soil inventories of Be-7 and Pb-210. *Journal of Geophysical Research-Atmospheres*, **90(ND6)**, pp. 10487–10495, 1985.

[93] Olsen, C.R., Larsen, I.L., Lowry, P.D., McLean, R.I. & Domotor, S.L., Radionuclide distributions and sorption behavior in the Susquehanna-Chesapeake Bay system. *Maryland Department of Natural Resources Report: PPER-R-12*, Power Plant and Environmental Review Division, pp. 111,1989.

[94] Wallbrink, P.J. & Murray, A.S., Fallout of Be-7 over South Eastern Australia. *Journal of Environmental Radioactivity*, **25**, 1994.

[95] Matisoff, G., Wilson, C.G. & Whiting, P.J., The Be-7/Pb-210(xs) ratio as an indicator of suspended sediment age or fraction new sediment in suspension. *Earth Surface Processes and Landforms*, **30(9)**, pp. 1191–1201, 2005.
[96] Wilson, C.G., Matisoff, G., Whiting, P.J. & Klarer, D.M., Transport of fine sediment through a wetland using radionuclide tracers: Old Woman Creek, OH. *Journal of Great Lakes Research*, **31(1)**, pp. 56–67, 2005.
[97] Wallbrink, P.J., Murray, A.S., Olley, J.M. & Olive, L.J., Determining sources and transit times of suspended sediment in the Murrumbidgee River, New South Wales, Australia, using fallout Cs-137 and Pb-210. *Water Resources Research*, **34(4)**, pp. 879–887, 1998.
[98] Whiting, P.J., Bonniwell, E.C. & Matisoff, G., Depth and areal extent of sheet and rill erosion based on radionuclides in soils and suspended sediment. *Geology*, **29(12)**, pp. 1131–1134, 2001.
[99] Binkley, D. & Brown, T.C., Forest practices as nonpoint sources of pollution in North America. *Water Resources Bulletin*, **29(5)**, pp. 729–740, 1993.
[100] Bovis, M.J., Millard, T.H. & Oden, M.E., *Gully Processes in Coastal British Columbia: the Role of Woody Debris.* Ministry of Forest Research: Victoria, BC., pp. 49–76. 1998.
[101] Sidle, R.C., Pearce, A.J. & O'Loughin, C.L., Hillslope stability and land use. *Proceedings of the American Geophysical Union*, Washington, DC, 1985.
[102] Reid, L.M. & Dunne, T., Sediment production from forest road surfaces. *Water Resources Research*, **20(11)**, pp. 1753–1761, 1984.
[103] Schwab, J.W., Mass wasting: October-November 1978 storm, Rennell Sound, Queen Charlotte Islands, British Columbia. Land Management Report No. 91, British Columbia Ministry of Forests and Lands, Victoria, British Columbia, **23**, 1983.
[104] Slaymaker, O., Assessment of the geomorphic impacts of forestry in British Columbia. *Ambio*, **29**, pp. 381–387, 2000.
[105] Wemple, B.C., Swanson, F.J. & Jones, J.A., Forest roads and geomorphic process interactions, Cascade Range, Oregon. *Earth Surface Processes and Landforms*, **26**, pp. 191–204, 2001.
[106] Grant, G.E. & Wolff, A.L., Long-term patterns of sediment transport after timber harvest, western Cascade Mountains, Oregon, USA. *Sediment and Stream Water Quality in a Changing Environment: Trends and Explanation*, International Association of Hydrological Sciences Press: Wallingford, pp. 31–40, 1991.
[107] Madej, M.A., Erosion and sediment delivery following removal of forest roads. *Earth Surface Processes and Landforms*, **26**, pp. 175–190, 2001.
[108] Adams, P.W., Flint, A.L. & Frederickson, R.L., Long-term patterns in soil moisture and revegetation after a clearcut of a Douglas-fir forest in Oregon. *Forest Ecology and Management*, **41**, pp. 249–263, 1991.
[109] Dhakal, A.S. & Sidle, R.C., Pore water pressure assessment in a forest watershed: Simulations and distributed field measurements related to forest practices. *Water Resources Research*, **40(2)**, 2004.

[110] Lewis, J., Mori, S.R., Keppeler, E.T. & Ziemer, R.R., Impacts of logging on storm peak flows, flow volumes and suspended sediment loads in Casper Creek, California. *Land Use and Watersheds: Human Influence on Hydrology and Geomorphology in Urban and Forest Areas*, ed. M.S. Wigmosta & S.J. Burges, American Geophysical Union: Washington, DC, pp. 85–126, 2001.

[111] Troendle, C.A. & Olsen, W.K., Potential effects of timber harvest and water management on streamflow dynamics and sediment transport. *Proc. of the Sustainable Ecological Systems Proceedings*, USDA Forest Service, Rocky Mountain Forest and Range Experiment Station GTR RM-247: Fort Collins, Colorado, pp. 34–41, 1993.

[112] Keim, R.E. & Schoenholtz, S.H., Functions and effectiveness of silvicultural streamside management zones in loessial bluff forests. *Forest Ecology and Management*, **118(1–3)**, pp. 197–209, 1999.

[113] Wemple, B.C., Jones, J.A. & Grant, G.E., Channel network extension by logging roads in two basins, Western Cascades, Oregon. *Water Resources Bulletin*, **32**, pp. 1195–1207, 1996.

[114] Muste, M., Yu, K., Fujita, I. & Ettema, R., Two-phase versus mixed-flow perspective on suspended sediment transport in turbulent channel flows. *Water Resources Research*, **41**, pp. 1–22, 2005.

[115] Minshall, G.W., Thomas, S.A., Newbold, J.D., Monaghan, M.T. & Cushing, C.E., Physical factors influencing fine organic particle transport and deposition in streams. *Journal of the North American Benthological Society*, **19(1)**, pp. 1–16, 2000.

[116] Diplas, P. & Parker, G., Pollution of gravel spawning grounds due to fine sediment, St. Anthony Falls Laboratory Project Report 240, University of Minnesota: Minneapolis, MN, 1985.

[117] Maridet, L., Wasson, J.G., Philippe, M. & Amoros, C., Benthic organic matter dynamics in three streams – riparian vegetation or bed morphology control. *Archiv fur Hydrobiologie*, **132(4)**, pp. 415–425, 1995.

[118] Georgian, T. & Thorp, J.H., Effects of microhabitat selection on feeding rates of net-spinning caddisfly larvae. *Ecology*, **73**, pp. 229–240, 1992.

[119] Droppo, I.G., Leppard, G.G., Flannigan, D.T. & Liss, S.N., The freshwater floc: A functional relationship of water and organic and inorganic floc constituents affecting suspended sediment properties. *Water, Air and Soil Pollution*, **99(1–4)**, pp. 43–53, 1997.

[120] Kranck, K., Dynamics and distribution of suspended particulate matter in the St. Lawrence Estuary. *Naturaliste Canadien*, **106(1)**, pp. 163–173, 1979.

[121] Broekhuizen, N. & Quinn, J.M., Influences of stream size and catchment land-use on fine particulate organic matter retention in streams. *New Zealand Journal of Marine and Freshwater Research*, **32(4)**, pp. 581–590, 1998.

[122] Lenzi, M.A. & Marchi, L., Suspended sediment load during floods in a small stream of the Dolomites (northeastern Italy). *Catena*, **39(4)**, pp. 267–282, 2000.

[123] Knighton, D., *Fluvial Forms and Processes: A New Perspective*, Arnold: London, 1998.

[124] Einstein, H.A. & Chien, N., Second approximation to the solution of the suspended load theory. *Series No. 47*. Institute of Engineering Research: University of California: Berkeley, CA, 1952.

[125] Emmett, W.W., A field calibration of the sediment-trapping characteristics of the Helley-Smith bedload sampler. *U.S. Geological Survey Professional Paper*, **1139**, 1980.

[126] Hassan, M.A., P., E., Use of tracers in fluvial geomorphology. *Tools in fluvial geomorphology*, eds. G.M. Kondolf, H. Piégay, John Wiley & Sons: Hoboken, New Jersey, pp. 397–423, 2002.

[127] Wilcock, P.R., Toward a practical method for estimating sediment-transport rates in gravel-bed rivers. *Earth Surface Processes and Landforms*, **26**, pp. 1395–1408, 2001.

[128] Bilby, R.E. & Likens, G.E., Effect of hydrologic fluctuations on the transport of fine particulate organic-carbon in a small stream. *Limnology and Oceanography*, **24(1)**, pp. 69–75, 1979.

[129] Langlois, J.L., Johnson, D.W. & Mehuys, G.R., Suspended sediment dynamics associated with snowmelt runoff in a small mountain stream of Lake Tahoe (Nevada). *Hydrological Processes*, **19(18)**, pp. 3569–3580, 2005.

[130] Webster, J.R., Benfield, E.F., Ehrman, T.P., Schaeffer, M.A., Tank, J.L., Hutchens, J.J. & D'Angelo, D.J., What happens to allochthonous material that falls into streams? A synthesis of new and published information from Coweeta. *Freshwater Biology*, **41(4)**, pp. 687–705, 1999.

[131] Cummins, K.W., Sedell, J.R., Swanson, F.J., Minshall, G.W., Fisher, S.G., Cushing, C.E., Petersen, R.C. & Vannote, R.L., Organic matter budget for stream ecosystems: problems in their evaluation. *Stream Ecology*, eds. J.R. Barnes, G.W. Minshall, Plenum Publishing Corporation: New York, pp. 299–353, 1983.

[132] Naegeli, M.W., Hartmann, U., Meyer, E.I. & Uehlinger, U., POM dynamics and community respiration in the sediments of a flood prone prealpine river (Necker, Switzerland). *Archiv fur Hydrobiologie*, **133(3)**, pp. 339–347, 1995.

[133] Short, R.A. & Ward, J.V., Benthic detritus dynamics in a mountain stream. *Holarctic Ecology*, **4(1)**, pp. 32–35, 1981.

[134] Smock, L.A., Spatial and temporal variation in organic matter storage in low-gradient, headwater streams. *Archiv für Hydrobiologie*, **118(2)**, pp. 169–184, 1990.

[135] Snaddon, C.D., Stewart, B.A. & Davies, B.R., The effect of discharge on leaf retention in two headwater streams. *Archiv für Hydrobiologie*, **125(1)**, pp. 109–120, 1992.

[136] Galle, T., Van Lagen, B., Kurtenbach, A. & Bierl, R., An FTIR-DRIFT study on river sediment particle structure: Implications for biofilm dynamics and pollutant binding. *Environmental Science and Technology*, **38(17)**, pp. 4496–4502, 2004.

[137] Bilby, R.E. & Ward, J.W., Changes in characteristics and function of woody debris with increasing size of streams in western Washington. *Transactions of the American Fisheries Society*, **118(4)**, pp. 368–378, 1989.

[138] Swanson, F.J., Janda, R.J. & Dunne, T., Summary: sediment budgets and routing in forested drainage basins, *Technical Report Number PNW-141,* United States Department of Agriculture, General Technical Report, 1982.

[139] Bilby, R.E. & Ward, J.W., Characteristics and function of large woody debris in streams draining old-growth, clear-cut, and second-growth forests in southwestern Washington. *Canadian Journal of Fisheries and Aquatic Sciences*, **48(12)**, pp. 2499–2508, 1991.

[140] Heede, B.H., Deteriorated watersheds can be restored – case-study. *Environmental Management*, **3(3)**, pp. 271–281, 1979.

[141] Keller, E.A. & Swanson, F.J., Effects of large organic material on channel form and fluvial processes. *Earth Surface Processes and Landforms*, **4(4)**, pp. 361–380, 1979.

[142] Kiffney, P.M., Richardson, J.S. & Feller, M.C., Fluvial and epilithic organic matter dynamics in headwater streams of southwestern British Columbia, Canada. *Archiv für Hydrobiologie*, **149(1)**, pp. 109–129, 2000.

[143] Naiman, R.J., Characteristics of sediment and organic-carbon export from pristine boreal forest watersheds. *Canadian Journal of Fisheries and Aquatic Sciences*, **39(12)**, pp. 1699–1718, 1982.

[144] Wanner, S.C. & Pusch, M., Use of fluorescently labeled Lycopodium spores as a tracer for suspended particles in a lowland river. *Journal of the North American Benthological Society*, **19(4)**, pp. 648–658, 2000.

[145] Miller, J. & Georgian, T., Estimation of fine particulate transport in streams using pollen as a seston analog. *Journal of the North American Benthological Society*, **11(2)**, pp. 172–180, 1992.

[146] Bilby, R.E., Role of organic debris dams in regulating the export of dissolved and particulate matter from a forested watershed. *Ecology*, **62**, pp. 1234–1243, 1981.

[147] Newbold, J.D., Elwood, J.W., Oneill, R.V. & Sheldon, A.L., Phosphorus dynamics in a woodland stream ecosystem – a study of nutrient spiraling. *Ecology*, **64(5)**, pp. 1249–1265, 1983.

[148] Wallace, J.B., Cuffney, T.F., Webster, J.R., Lugthart, G.J., Chung, K. & Goldowitz, B.S., Export of fine organic particles from headwater streams – effects of season, extreme discharges, and invertebrate manipulation. *Limnology and Oceanography*, **36(4)**, pp. 670–682, 1991.

[149] Kazmierczak, R.F., Webster, J.R. & Benfield, E.F., Characteristics of seston in a regulated Appalachian mountain river, U.S.A. *Regulated Rivers: Research and Management*, **1**, pp. 287–300, 1987.

[150] Paul, M.J. & Hall, R.O., Particle transport and transient storage along a stream-size gradient in the Hubbard Brook Experimental Forest. *Journal of the North American Benthological Society*, **21(2)**, pp. 195–205, 2002.

[151] Atkinson, C.F., Aumen, N.G., Miller, G.L. & Ward, G.M., Influence of particle shapes and size distributions on fine particulate organic matter surface-area in streams. *Journal of the North American Benthological Society*, **11(3)**, pp. 261–268, 1992.

[152] Atkinson, C.F., Miller, G.L. & Aumen, N.G., Influence of season, size, and collection site on surface-area of fine particulate matter in streams. *Journal of the North American Benthological Society*, **12(4)**, pp. 410–417, 1993.

[153] Ferguson, R.I. & Church, M., A simple universal equation for grain settling velocity. *Journal of Sedimentary Research*, **74(6)**, pp. 933–937, 2004.

[154] Cushing, C.E., Minshall, G.W. & Newbold, J.D., Transport dynamics of fine particulate organic matter in two Idaho streams. *Limnology and Oceanography*, **38(6)**, pp. 1101–1115, 1993.

[155] Hall, R.O., Peredney, C.L. & Meyer, J.L., The effect of invertebrate consumption on bacterial transport in a mountain stream. *Limnology and Oceanography*, **41**, pp. 1180–1187, 1996.

[156] Thomas, S.A., Newbold, J.D., Monaghan, M.T., Minshall, G.W., Georgian, T. & Cushing, C.E., The influence of particle size on seston deposition in streams. *Limnology and Oceanography*, **46(6)**, pp. 1415–1424, 2001.

[157] Einstein, H.A. & Krone, R.B., Experiments to determine modes of cohesive sediment transport in salt water. *Journal of Geophysical Research*, **67**, pp. 1450–1461, 1961.

[158] McNair, J.N., Turbulent transport of suspended particles and dispersing benthic organisms: The hitting-time distribution for the local exchange model. *Journal of Theoretical Biology*, **202(3)**, pp. 231–246, 2000.

[159] McNair, J.N. & Newbold, J.D., Turbulent transport of suspended particles and dispersing benthic organisms: the hitting-distance problem for the local exchange model. *Journal of Theoretical Biology*, **209(3)**, pp. 351–369, 2001.

[160] Georgian, T., Newbold, J.D., Thomas, S.A., Monaghan, M.T., Minshall, G.W. & Cushing, C.E., Comparison of corn pollen and natural fine particulate matter transport in streams: Can pollen be used as a seston surrogate? *Journal of the North American Benthological Society*, **22(1)**, pp. 2–16, 2003.

[161] Newbold, J.D., Thomas, S.A., Minshall, G.W., Cushing, C.E. & Georgian, T., Deposition, benthic residence, and resuspension of fine organic particles in a mountain stream. *Limnology and Oceanography*, **50(5)**, pp. 1571–1580, 2005.

[162] Packman, A.I. & MacKay, J.S., Interplay of stream-subsurface exchange, clay particle deposition, and streambed evolution. *Water Resources Research*, **39(4)**, 2003.

[163] Battin, T.J., Kaplan, L.A., Newbold, J.D. & Hansen, C.M.E., Contributions of microbial biofilms to ecosystem processes in stream mesocosms. *Nature*, **426(6965)**, pp. 439–442, 2003.

[164] Lock, M.A., River epilithon – a light and energy transducer. *Perspectives in Running Water Ecology*, ed. M.A. Lock & D.D. Williams, Plenum Press: New York, pp. 4–40, 1981.

[165] McCullough, D.A., Minshall, G.W. & Cushing, C.E., Bioenergenics of lotic filter-feeding insects *Simulium* spp. (Diptera) and *Hydropsyche occidentalis* (Trichoptera) and their function in controlling organic transport in streams. *Ecology*, **60**, pp. 585–596, 1979.

[166] Monaghan, M.T., Thomas, S.A., Minshall, G.W., Newbold, J.D. & Cushing, C.E., The influence of filter-feeding benthic macroinvertebrates on the transport and deposition of particulate organic matter and diatoms in two streams. *Limnology and Oceanography*, **46(5)**, pp. 1091–1099, 2001.

[167] Bagnold, R.A., Experiments on a gravity free dispersion of large solid spheres in a Newtonian fluid under shear. *Proceedings, Royal Society of London, Series A*, London, England, **225**, pp. 49–70,1954.

[168] Halbronn, G., Remarque sur la theorie de l'Austausch applique au transport des materiux en suspension. *Proc. of the International Association of Hydraulic Research, 3rd Meeting*, Grenoble, France, pp. 1–6, 1949.

[169] Hunt, G.W. & Stanley, E.H., An evaluation of alternative procedures using the Bou-Rouch method for sampling hyporheic invertebrates. *Canadian Journal of Fisheries And Aquatic Sciences*, **57(8)**, pp. 1545–1550, 2000.

[170] Rouse, H., Modern conceptions of the mechanics of fluid turbulence. *Transactions – ASCE*, **24(1965)**, pp. 463–543, 1937.

[171] Tanaka, S. & Fugimoto, S., On the distribution of suspended sediment in experimental flume flow. *Memoirs of the Faculty of Engineering,* Kobe University: Kobe, Japan, **5**, 1958.

[172] Bencala, K.E. & Walters, R.A., Simulation of solute transport in a mountain pool-and-riffle stream: a transient storage model. *Water Resources Research*, **19**, pp. 718–724, 1983.

[173] Valett, H.M., Morrice, J.A. & Dahm, C.N., Parent lithology, groundwater-surface water exchange and NO{NE}3- retention in headwater streams. *Limnology and Oceanography*, **41**, pp. 333–345, 1996.

[174] Webster, J.R. & Ehrman, T.P., Solute dynamics. *Methods in stream ecology*, eds. F.R. Hauer, G.A. Lamberti, Academic Press: San Diego, 1996.

[175] McNair, J.N., Newbold, J.D. & Hart, D.D., Turbulent transport of suspended particles and dispersing benthic organisms: How long to hit bottom? *Journal of Theoretical Biology*, **188(1)**, pp. 29–52, 1997.

[176] Droppo, I.G., Jeffries, D., Jaskot, C. & Backus, S., The prevalence of freshwater flocculation in cold regions: A case study from the Mackenzie River Delta, Northwest Territories, Canada. *Arctic*, **51(2)**, pp. 155–164, 1998.

[177] Petticrew, E.L., Sediment aggregation and transport in northern interior British Columbia streams. *Erosion and Sediment Yield: Global and Regional Perspectives*, ed. D.E. Walling & B.W. Webb, International Association of Hydrological Sciences Press: Wallingford, Publication No. 236, pp. 313–319, 1996.

[178] Woodward, J.C., Porter, P.R., Lowe, A.T., Walling, D.E. & Evans, A.J., Composite suspended sediment particles and flocculation in glacial meltwaters: preliminary evidence from Alpine and Himalayan basins. *Hydrological Processes*, **16(9)**, pp. 1735–1744, 2002.

[179] Ongley, E.D., Krishnappan, B.G., Droppo, I.G., Rao, S.S. & Maguire, R.J., Cohesive sediment transport – emerging issues for toxic-chemical management. *Hydrobiologia*, **235**, pp. 177–187, 1992.

[180] Brunke, M., Colmation and depth filtration within streambeds: retention of particles in hyporheic interstices. *International Review of Hydrobiology*, **84**, pp. 99–117, 1999.

[181] Allan, J.D., Erickson, D.L. & Fay, J., The influence of catchment land use on stream integrity across multiple spatial scales. *Freshwater Biology*, **37(1)**, pp. 149–161, 1997.

[182] Bonzongo, J.C., Heim, K.J., Warwick, J.J. & Lyons, W.B., Mercury levels in surface waters of the Carson River Lahontan Reservoir system, Nevada: Influence of historic mining activities. *Environmental Pollution*, **92(2)**, pp. 193–201, 1996.

[183] Foster, I.D.L. & Charlesworth, S.M., Heavy metals in the hydrological cycle: Trends and explanation. *Hydrological Processes*, **10(2)**, pp. 227–261, 1996.

[184] Horowitz, A.J., Elrick, K.A. & Cook, R.B., Effect of mining and related activities on the sediment trace-element geochemistry of Lake Coeur-Dalene, Idaho, USA.1. Surface sediments. *Hydrological Processes*, **7(4)**, pp. 403–423, 1993.

[185] Horowitz, A.J., Elrick, K.A., Robbins, J.A. & Cook, R.B., Effect of mining and related activities on the sediment trace-element geochemistry of Lake Coeur-Dalene, Idaho, USA.2. Subsurface sediments. *Hydrological Processes*, **9(1)**, pp. 35–54, 1995.

[186] Gayraud, S., Herouin, E. & Philippe, M., The clogging of stream beds: A review of mechanisms and consequences on habitats and macroinvertebrate communities. *Bulletin Francais de la Peche et de la Pisciculture*, **(365–66)**, pp. 339–355, 2002.

[187] Pusch, M., Fiebig, D., Brettar, I., Eisenmann, H., Ellis, B.K., Kaplan, L.A., Lock, M.A., Naegeli, M.W. & Traunspurger, W., The role of micro-organisms in the ecological connectivity of running waters. *Freshwater Biology*, **40(3)**, pp. 453–495, 1998.

[188] Hedin, L.O., Factors controlling sediment community respiration in woodland stream ecosystems. *Oikos*, **57(1)**, pp. 94–105, 1990.

[189] Jones, J.B. & Smock, L.A., Transport and retention of particulate organic matter in two low-gradient headwater streams. *Journal of the North American Benthological Society*, **10(2)**, pp. 115–126, 1991.

[190] Mulholland, P.J., Organic-carbon flow in a swamp-stream ecosystem. *Ecological Monographs*, **51(3)**, pp. 307–322, 1981.

[191] Meyer, J.L., Benthic bacterial biomass and production in a blackwater river. *Verhandlungen der Internationalen Vereinigung fur theoretische und angewandte Limnologie*, **23**, pp. 643–651, 1988.

[192] Rutherford, J.C., *River Mixing*, John Wiley and Sons: Chichester, United Kingdom, 1994.

[193] Savant, S.A., Reible, D.D. & Thibodeaux, L.J., Convective-transport within stable river sediments. *Water Resources Research*, **23(9)**, pp. 1763–1768, 1987.

[194] Hendricks, S.P., Microbial ecology of the hyporheic zone – a perspective integrating hydrology and biology. *Journal of the North American Benthological Society*, **12(1)**, pp. 70–78, 1993.

[195] Hendricks, S.P. & White, D.S., Seasonal biogeochemical patterns in surface-water, subsurface hyporheic, and riparian ground-water in a temperate stream ecosystem. *Archiv fur Hydrobiologie*, **134(4)**, pp. 459–490, 1995.

[196] Jones, J.B., Fisher, S.G. & Grimm, N.B., Vertical hydrologic exchange and ecosystem metabolism in a Sonoran desert stream. *Ecology*, **76(3)**, pp. 942–952, 1995.

[197] Valett, H.M., Fisher, S.G. & Stanley, E.H., Physical and chemical characteristics of the hyporheic zone of a Sonoran Desert stream. *Journal of the North American Benthological Society*, **9(3)**, pp. 201–215, 1990.

[198] Williams, D.D., Nutrient and flow vector dynamics at the hyporheic groundwater interface and their effects on the interstitial fauna. *Hydrobiologia*, **251(1–3)**, pp. 185–198, 1993.

[199] Pusch, M., Community respiration in the hyporheic zone of a riffle-pool sequence. *Groundwater/Surface Water Ecotones: Biological and Hydrological Interactions and Management Options*, ed. J. Gibert, J. Mathieu & F. Fournier, Cambridge University Press: Cambridge, pp. 51–56, 1997.

[200] Pusch, M., The metabolism of organic matter in the hyporheic zone of a mountain stream, and its spatial distribution. *Hydrobiologia*, **323(2)**, pp. 107–118, 1996.

[201] Huettel, M., Ziebis, W. & Forster, S., Flow-induced uptake of particulate matter in permeable sediments. *Limnology and Oceanography*, **41(2)**, pp. 309–322, 1996.

[202] Jobson, H.E. & Carey, W.P., Interaction of fine sediment with alluvial streambeds. *Water Resources Research*, **25(1)**, pp. 135–140, 1989.

[203] Verhoff, F.H. & Melfi, D.A., Total phosphorus transport during storm events. *Journal of the Environmental Engineering Division-ASCE*, **104(5)**, pp. 1021–1026, 1978.

[204] Verhoff, F.H., Melfi, D.A. & Yaksich, S.M., Storm travel distance calculations for total phosphorus and suspended materials in rivers. *Water Resources Research*, **15(6)**, pp. 1354–1360, 1979.

[205] Verhoff, F.H., Melfi, D.A. & Yaksich, S.M., An analysis of total phosphorus transport in river systems. *Hydrobiologia*, **91-2(JUL)**, pp. 241–252, 1982.

[206] Cahill, T.H., Imperato, P. & Verhoff, F.H., Evaluation of phosphorus dynamics in a watershed. *Journal of the Environmental Engineering Division – American Society of Civil Engineers*, **100(NEE2)**, pp. 439–458, 1974.

[207] Partheniades, E., Results of recent investigations on erosion and deposition of cohesive soils. *Sedimentation*, ed. H.W. Shen, Colorado State University: Fort Collins, Colorado, 1972.

[208] Matisoff, G., Bonniwell, E.C. & Whiting, P.J., Soil erosion and sediment sources in an Ohio watershed using Beryllium-7, Cesium-137, and Lead-210. *Journal of Environmental Quality*, **31**, pp. 54–61, 2002.

[209] Salant, N.L., Renshaw, C.E., Magilligan, F.J., Kaste, J.M., Nislow, K.H. & Heimsath, A.M., The use of short-lived radionuclides to quantify transitional bed material transport in a regulated river. *Earth Surface Processes and Landforms*, **32(4)**, pp. 509–524, 2007.

[210] Cushing, C.E., Minshall, G.W. & Newbold, J.D., Transport dynamics of fine particulate organic matter in 2 Idaho streams. *Limnology and Oceanography*, **38(6)**, pp. 1101–1115, 1993.

[211] Newbold, J.D., Cushing, C.E. & Minshall, G.W., Use of ^{14}C-label to study fine-particulate organic matter dynamics in flowing water. *Organic Substances and Sediments in Water. Volume 2. Processes and Analysis*, ed. R.A. Baker, Lewis Publishers: Chelsea, Michigan, 1991.

[212] Newbold, J.D., Elwood, J.W., O'Neill, R.V. & Van Winkle, W., Measuring nutrient spiraling in streams. *Canadian Journal of Fisheries and Aquatic Sciences*, **38**, pp. 860–863, 1981.

[213] Reynolds, C.S., White, M.L. & Clarke, R.T., Suspension and settlement of particles in flowing water: comparison of the effects of varying water depth and velocity in circulating channels. *Freshwater Biology*, **24**, pp. 23–34, 1990.

[214] Petts, G.E., Thoms, M.C., Brittan, K. & Atkin, B., A freeze-coring technique applied to pollution by fine sediments in gravel-bed rivers. *Science of the Total Environment*, **84**, pp. 259–272, 1989.

[215] Beschta, R.L. & Jackson, W.L., Intrusion of fine sediments into a stable gravel bed. *Journal of the Fisheries Research Board of Canada*, **36(2)**, pp. 204–210, 1979.

[216] Lisle, T.E. & Eads, R.E., Methods to measure sedimentation of spawning gravels, Research Note US PSW-411, Pacific Southwest Research Station, Forest Service, U.S. Department of Agriculture: Berkeley, California, 1991.

[217] Hughes, N., Coats, J.S. & Petts, G.E., Local variability of gold in active stream sediments. *Journal of Geochemical Exploration*, **54(2)**, pp. 137–148, 1995.

[218] Thoms, M.C., A comparison of grab-sampling and freeze-sampling techniques in the collection of gravel-bed river sediment. *Sedimentary Geology*, **78(3–4)**, pp. 191–200, 1992.

[219] Zimmermann, A., Coulombe-Pontbriand, M. & Lapointe, M., Biases of submerged bulk and freeze-core samples. *Earth Surface Processes and Landforms*, **30(11)**, pp. 1405–1417, 2005.

[220] Palmer, M.A. & Strayer, D.S., Chapter 15. Meiofauna. *Methods in stream ecology*, eds. F.R. Hauer & G.A. Lamberti, Academic Press: New York, pp. 315–337, 1996.

[221] Bretschko, G. & Klemens, W.E., Quantitative methods and aspects in the study of the interstitial fauna of running waters. *Stygologia*, **2**, pp. 297–316, 1986.

[222] Danielopol, D.L., The distribution of the fauna in the interstitial habitats of riverine sediments of the Danube and the Pisting (Austria). *International Journal of Speleology*, **8**, pp. 23–51, 1976.

[223] Hakenkamp, C.C. & Palmer, M.A., Problems associated with quantitative sampling of shallow groundwater invertebrates. *Proceedings of the First International Conference on Groundwater Ecology*, eds. J.A. Stanford & J.J. Simons, American Water Resources Association: Bethesda, MD, pp. 101–110, 1992.

[224] Williams, D.D., The hyporheic zone as a habitat for aquatic insects and associated arthropods. *The Ecology of Aquatic Insects*, ed. V.H. Resh & D.M. Rosenberg, Praeger Publishers: New York, pp. 430–455, 1984.

[225] Mauclaire, L., Marmonier, P. & Gibert, J., Sampling water and sediment in interstitial habitats: a comparison of coring and pumping techniques. *Archiv für Hydrobiologie*, **142(1)**, pp. 111–123, 1998.

[226] Tillman, D.C., Barsotti, A.M., Silliman, S.E. & Lamberti, G.A., A comparison of hyporheic pore water samplers: sample designs and heterogeneous physical environments. *Bulletin of the North American Benthological Society*, **13(261)**, 1996.

[227] Acornley, R.M. & Sear, D.A., Sediment transport and siltation of brown trout (Salmo trutta L.) spawning gravels in chalk streams. *Hydrological Processes*, **13(3)**, pp. 447–458, 1999.

[228] Bond, N.R., A simple device for estimating rates of fine sediment transport along the bed of shallow streams. *Hydrobiologia*, **468(1–3)**, pp. 155–161, 2002.

[229] Carling, P.A., Deposition of fine and coarse sand in an open-work gravel bed. *Canadian Journal of Fisheries and Aquatic Sciences*, **41(2)**, pp. 263–280, 1984.

[230] Fletcher, W.K., McLean, W.E. & Sweeten, T., An instrument to monitor infiltration of fine sediment into stable gravel stream beds. *Aquacultural Engineering*, **14(4)**, pp. 289–296, 1995.

[231] Lisle, T.E., Sediment transport and resulting deposition in spawning gravels, north coastal California. *Water Resources Research*, **25(6)**, pp. 1303–1319, 1989.

[232] Lisle, T.E. & Lewis, J., Effects of sediment transport on survival of salmonid embryos in a natural stream – a simulation approach. *Canadian Journal of Fisheries and Aquatic Sciences*, **49(11)**, pp. 2337–2344, 1992.

[233] Wesche, T.A., Reiser, D.W., Hasfurther, V.R., Hubert, W.A. & Skinner, Q.D., New technique for measuring fine sediment in streams. *North American Journal of Fisheries Management*, **9**, pp. 234 –238, 1989.

[234] Zimmermann, A.E. & Lapointe, M., Sediments infiltration traps: their use to monitor salmonid spawning habitat in headwater tributaries of the Cascapedia River, Quebec. *Hydrological Processes*, **19(20)**, pp. 4161–4177, 2005.

[235] Alonso, C.V., Theurer, F.D. & Zachmann, D.A., Sediment intrusion and dissolved-oxygen transport model (SIDO). *Technical Report (peer reviewed), No. 5*, USDA-ARS National Sedimentation Laboratory: Oxford, MS,1996.

[236] Carling, P.A. & McCahon, C.P., Natural siltation of brown trout (Salmo trutta L.) spawning gravels during low-flow conditions. *Regulated Streams: Advances in Ecology*, eds. J.F. Craig & J.B. Kemper, Plenum Press: New York, pp. 229–244, 1987.

[237] Frostick, L.E., Lucas, P.M. & Reid, I., The infiltration of fine matrices into coarse-grained alluvial sediments and its implications for stratigraphical interpretation. *Journal of the Geological Society*, **141(NOV)**, pp. 955–965, 1984.

[238] Meehan, W.R. & Swanston, D.N., Effects of gravel morphology on fine sediment accumulation and survival of incubation salmon eggs. *USDA Forest Service Research Paper, No. PNW-220*, Pacific Northwest Forest and Range Experiment Station: Portland, OR,1977.

[239] Sear, D.A., Fine sediment infiltration into gravel spawning beds within a regulated river experiencing floods – ecological implications for salmonids. *Regulated Rivers-Research & Management*, **8(4)**, pp. 373–390, 1993.

[240] Lachance, S. & Dube, M., A new tool for measuring sediment accumulation with minimal loss of fines. *North American Journal of Fisheries Management*, **24(1)**, pp. 303–310, 2004.

[241] Levasseur, M., Berube, F. & Bergeron, N.E., A field method for the concurrent measurement of fine sediment content and embryo survival in artificial salmonid redds. *Earth Surface Processes and Landforms*, **31(4)**, pp. 526–530, 2006.

[242] Levasseur, M., Bergeron, N.E., M.F., L. & Bérubé, F., Effects of silt and very fine sand dynamics in Atlantic salmon (Salmo salar) redds on embryo hatching success. *Canadian Journal of Fisheries and Aquatic Sciences (in press)*, 2006.

[243] Dobbins, W.E., Effect of turbulence on sedimentation. *Transactions of the American Society of Civil Engineers*, **109(2218)**, pp. 629–678, 1944.

[244] Hurst, H.E., The suspension of sand in water. *Proceedings, Royal Society of London*, **24**, pp. 196–201, 1929.

[245] Rouse, H., An analysis of sediment transportation in the light of fluid turbulence. *Soil Conservation Service Report No. SCS-TP-25*, United States Department of Agriculture: Washington, DC, 1939.

[246] Hunt, J.N., The turbulent transport of suspended sediment in open channels. *Proceedings, Royal Society of London*, **224**, pp. 322–335 1954.

[247] O'Brien, M.P., Review of the theory of turbulent flow and its relation to sediment transportation. *Transactions-American Society of Civil Engineers*, pp. 487–491, 1933.

[248] Bennett, S.J., Bridge, J.S. & Best, J.L., Fluid and sediment dynamics of upper stage plane beds. *Journal of Geophysical Research-Oceans*, **103(C1)**, pp. 1239–1274, 1998.

[249] Nikora, V.I. & Goring, D.G., Fluctuations of suspended sediment concentration and turbulent sediment fluxes in an open-channel flow. *Journal of Hydraulic Engineering– American Society of Civil Engineers,* **128(2)**, pp. 214–224, 2002.

[250] Graf, W.H. & Cellino, M., Suspension flows in open channels; experimental study. *Journal of Hydraulic Research*, **40(4)**, pp. 435–447, 2002.

[251] Elghobashi, S., On predicting particle-laden turbulent flows. *Applied Scientific Research*, **52(4)**, pp. 309–329, 1994.

[252] Kaftori, D., Hetsroni, G. & Banerjee, S., The effect of particles on wall turbulence. *International Journal of Multiphase Flow*, **24(3)**, pp. 359–386, 1998.

[253] Rouson, D.W.I. & Eaton, J.K., On the preferential concentration of solid particles in turbulent channel flow. *Journal of Fluid Mechanics*, **428**, pp. 149–169, 2001.

[254] Squires, K.D. & Eaton, J.K., Particle response and turbulence modification in isotropic turbulence. *Physics of Fluids A-Fluid Dynamics*, **2(7)**, pp. 1191–1203, 1990.

[255] Denny, M.W. & Shibata, M.F., Consequences of surf-zone turbulence for settlement and external fertilization. *American Naturalist*, **134**, pp. 859–889, 1989.

[256] Colby, B.R., Fluvial sediments – a summary of source, transportation, deposition, and measurement of sediment discharge. *U.S. Geological Survey Bulletin*, **1181-A**, 1963.

[257] Nomicos, G.N., Effects of sediment load on the velocity field and friction factor of turbulent flow in an open channel, Ph.D. Thesis, California Institute of Technology 1956.

[258] Vanoni, V.A. & Brooks, N.H., Laboratory studies of the roughness and suspended load of alluvial streams. *California Institute of Technology Sedimentation Laboratory Report*, **E-68**, 1957.

[259] Andrews, E.D., Downstream effects of Flaming Gorge Reservoir on the Green River, Colorado and Utah. *Geological Society of America Bulletin*, **97(8)**, pp. 1012–1023, 1986.

[260] Magilligan, F.J. & Nislow, K.H., Hydrologic alteration in a changing landscape: effects of impoundment in the Upper Connecticut River Basin, USA. *Journal of the American Water Resources Association*, **36(7)**, pp. 1551–1569, 2001.

[261] Maitland, P.S., Newson, M.D. & Best, G.A., The impact of afforestation and forestry practice on freshwater habitats. *Focus on Nature Conservation*, **(23)**, 1990.

[262] Moffat, A.J., Forestry and soil erosion in Britain – a review. *Soil Use and Management*, **4(2)**, pp. 41–44, 1988.

[263] Soutar, R.G., Afforestation, soil-erosion and sediment yields in British fresh waters. *Soil Use and Management*, **5(4)**, pp. 200–200, 1989.

[264] Stott, T., Leeks, G., Marks, S. & Sawyer, A., Environmentally sensitive plot-scale timber harvesting: impacts on suspended sediment, bedload and bank erosion dynamics. *Journal of Environmental Management*, **63(1)**, pp. 3–25, 2001.

[265] Brown, G.W. & Krygier, J.T., Clear-cut logging and sediment production in the Oregon coast range. *Water Resources Research*, **7**, pp. 1189–1198, 1971.

[266] Davies, P.E. & Nelson, M., The effect of steep slope logging on fine sediment infiltration into the beds of ephemeral and perennial streams of the Dazzler Range, Tasmania, Australia. *Journal of Hydrology*, **150(2–4)**, pp. 481–504, 1993.

[267] Kreutzweiser, D.P. & Capell, S.S., Fine sediment deposition in streams after selective forest harvesting without riparian buffers. *Canadian Journal of Forest Research-Revue Canadienne de Recherche Forestiere*, **31(12)**, pp. 2134–2142, 2001.

[268] Bilby, R.E., Contribution of road surface sediment to a western Washington stream. *Forest Science*, **31**, pp. 827–838, 1985.

[269] Lemly, A.D., Modification of benthic insect communities in polluted streams: Combined effects of sedimentation and nutrient enrichment. *Hydrobiologia*, **87**, pp. 229–245, 1982.

[270] Lenat, D.R., Penrose, D.L. & Eagleson, K.W., Variable effects of sediment addition on stream benthos. *Hydrobiologia*, **79**, pp. 187–194, 1981.

[271] Angradi, T.R., Fine sediment and macroinvertebrate assemblages in Appalachian streams: a field experiment with biomonitoring applications. *Journal of the North American Benthological Society*, **18(1)**, pp. 49–66, 1999.

[272] Richards, C. & Bacon, K.L., Influence of fine sediment on macroinvertebrate colonization of surface and hyporheic stream substrates. *Great Basin Naturalist*, **54(2)**, pp. 106–113, 1994.

[273] Weigelhofer, G. & Waringer, J., Vertical distribution of benthic macroinvertebrates in riffles versus deep runs with differing contents of fine sediments (Weidlingbach, Austria). *International Review of Hydrobiology*, **88(3–4)**, pp. 304–313, 2003.

[274] Adams, J.N. & Beschta, R.L., Gravel bed composition in Oregon coastal streams. *Canadian Journal of Fisheries and Aquatic Sciences*, **37(10)**, pp. 1514–1521, 1980.

[275] Huang, X. & Garcia, M.H., Pollution of gravel spawning grounds by deposition of suspended sediment. *Journal of Environmental Engineering – American Society of Civil Engineers*, **126(10)**, pp. 963–967, 2000.

[276] Scrivener, J.C. & Brownlee, M.J., Effects of forest harvesting on spawning gravel and incubation survival of chum (Oncorhynchus-Keta) and coho salmon (Oncorhyncus-Kisutch) in Carnation Creek, British-Columbia. *Canadian Journal of Fisheries and Aquatic Sciences*, **46(4)**, pp. 681–696, 1989.

[277] Cardinale, B.J., Palmer, M.A., Swan, C.M., Brooks, S. & Poff, N.L., The influence of substrate heterogeneity on biofilm metabolism in a stream ecosystem. *Ecology*, **83(2)**, pp. 412–422, 2002.

[278] Richie, J.E., Perkins, M.A. & Goldman, C.R., Effects of kokannee salmon (Oncorhynchus nerka) decomposition on the ecology of a subalpine stream. *Journal of Fisheries Research Board of Canada*, **32**, pp. 817–820, 1975.

[279] Wold, A.K.F. & Hershey, A.E., Effects of salmon carcass decomposition on biofilm growth and wood decomposition. *Canadian Journal of Fisheries and Aquatic Sciences*, **56(5)**, pp. 767–773, 1999.

[280] Prochazka, K., Stewart, B.A. & Davies, B.R., Leaf litter retention and its implications for shredder distribution in two headwater streams. *Archiv fur Hydrobiologie*, **120(3)**, pp. 315–325, 1991.

[281] Baxter, C.V. & Hauer, F.R., Geomorphology, hyporheic exchange, and selection of spawning habitat by bull trout (Salvelinus confluentus). *Canadian Journal of Fisheries and Aquatic Sciences*, **57(7)**, pp. 1470–1481, 2000.

[282] Chapman, D.W., Critical review of variables used to define effects of fines in redds of large salmonids. *Transactions of the American Fisheries Society*, **117**, pp. 1–21, 1988.

[283] Bjornn, T.C. & Reiser, D.W., Habitat requirements of salmonids in streams. *Influences of Forest and Rangeland Management on Salmonid Fishes and their Habitats*, ed. W.R. Meehan, American Fisheries Society Special Publication: pp. 83–138, 1991.

[284] Droppo, I.G. & Ongley, E.D., Flocculation of suspended sediment in rivers of southeastern Canada. *Water Research*, **28(8)**, pp. 1799–1809, 1994.

[285] Petticrew, E.L. & Biickert, S.L., Characterization of sediment transport and storage in the upstream portion of the Fraser River (British Columbia, Canada). *Modelling Soil Erosion, Sediment Transport and Closely Related Hydrological Processes*, ed. W. Summer, E. Klaghofer & W. Zhang, International Association of Hydrological Sciences Publication 249. International Association of Hydrological Sciences Press: Wallingford, pp. 383–391, 1998.

[286] Petticrew, E.L. & Droppo, I.G., The morphology and settling characteristics of fine-grained sediment from a selection of Canadian rivers. *Contributions to the Hydrological Programme-V by Canadian Experts. Technical Documents in Hydrology No. 33*, United Nations Educational, Scientific, and Cultural Organization: Paris, pp. 111–126, 2000.

[287] Anderson, P.G., Taylor, B.R. & Balch, G.C., Quantifying the effects of sediment release on fish and their habitats, *Canadian Manuscript Report of Fisheries and Aquatic Sciences*, No. 2346, 1996.

[288] Runde, J.M. & Hellenthal, R.A., Effects of suspended particles on net-tending behaviors for Hydropsyche sparna (Trichoptera: Hydropsychidae) and related species. *Annals of the Entomological Society of America*, **93(3)**, pp. 678–683, 2000.

[289] Malmer, A., Streamwater quality as affected by wild fires in natural and manmade vegetation in Malaysian Borneo. *Hydrological Processes*, **18(5)**, pp. 853–864, 2004.

[290] Meador, M.R. & Goldstein, R.M., Assessing water quality at large geographic scales: Relations among land use, water physicochemistry, riparian condition, and fish community structure. *Environmental Management*, **31(4)**, pp. 504–517, 2003.

[291] Quinn, J.M. & Stroud, M.J., Water quality and sediment and nutrient export from New Zealand hill-land catchments of contrasting land use. *New Zealand Journal of Marine and Freshwater Research*, **36(2)**, pp. 409–429, 2002.

[292] Means, J.C., Wood, S.G., Hassett, J.J. & Banwart, W.L., Sorption of polynuclear aromatic-hydrocarbons by sediments and soils. *Environmental Science & Technology*, **14(12)**, pp. 1524–1528, 1980.

[293] Owens, P.N., Walling, D.E., Carton, J., Meharg, A.A., Wright, J. & Leeks, G.J.L., Downstream changes in the transport and storage of sediment-associated contaminants (P, Cr and PCBs) in agricultural and industrialized drainage basins. *Science of the Total Environment*, **266(1–3)**, pp. 177–186, 2001.

[294] Voice, T.C. & Weber, W.J., Sorption of hydrophobic compounds by sediments, soils and suspended-solids.1. theory and background. *Water Research*, **17(10)**, pp. 1433–1441, 1983.

[295] Fossati, O., Wasson, J.G., Hery, C., Salinas, G. & Marin, R., Impact of sediment releases on water chemistry and macroinvertebrate communities in clear water Andean streams (Bolivia). *Archiv fur Hydrobiologie*, **151(1)**, pp. 33–50, 2001.

[296] Grayson, R.B., Haydon, S.R., Jayasuriya, M.D.A. & Finlayson, B.L., Water-quality in mountain ash forests – separating the impacts of roads from those of logging operations. *Journal of Hydrology*, **150(2–4)**, pp. 459–480, 1993.

[297] Minshall, G.F., Brock, J.T., Andrews, D.A. & Robinson, C.T., Water quality, substratum and biotic responses of five central Idaho (USA) streams during the first year following the Mortar Creek fire. *International Journal of Wildland Fire*, **10(2)**, pp. 185–199, 2001.

[298] Sutherland, A.B., Meyer, J.L. & Gardiner, E.P., Effects of land cover on sediment regime and fish assemblage structure in four southern Appalachian streams. *Freshwater Biology*, **47(9)**, pp. 1791–1805, 2002.

[299] Zimmerman, J.K.H., Vondracek, B. & Westra, J., Agricultural land use effects on sediment loading and fish assemblages in two Minnesota (USA) watersheds. *Environmental Management*, **32(1)**, pp. 93–105, 2003.

[300] Golladay, S.W. & Battle, J., Effects of flooding and drought on water quality in gulf coastal plain streams in Georgia. *Journal of Environmental Quality*, **31(4)**, pp. 1266–1272, 2002.

[301] Richardson, J. & Jowett, I.G., Effects of sediment on fish communities in East Cape streams, North Island, New Zealand. *New Zealand Journal of Marine and Freshwater Research*, **36(2)**, pp. 431–442, 2002.

[302] Chamberlain, T.W., Harr, R.D. & Everest, F.H., Timber harvesting, silviculture, and watershed processes. *Influences of Forest and Rangeland Management on Salmonid Fishes and their Habitats*, ed. W.R. Meehan, American Fisheries Society Special Publication: pp. 181–205, 1991.

[303] McNeil, W.J. & Ahnell, W.H., Success of pink salmon spawning relative to size of spawning bed material. *Special Scientific Report, Fisheries 469*, 1964.

[304] Piccolo, J.J. & Wipfli, M.S., Does red alder (Alnus rubra) in upland riparian forests elevate macroinvertebrate and detritus export from headwater streams to downstream habitats in southeastern Alaska? *Canadian Journal of Fisheries and Aquatic Sciences*, **59(3)**, pp. 503–513, 2002.

[305] Anderson, N.H. & Sedell, J.R., Detritus processing by macroinvertebrates in stream ecosystems. *Annual Review of Entomology*, **24**, pp. 351–377, 1979.

[306] Richardson, J.S., Food, microhabitat, or both – macroinvertebrate use of leaf accumulations in a montane stream. *Freshwater Biology*, **27(2)**, pp. 169–176, 1992.

[307] Short, R.A. & Ward, J.V., Leaf litter processing in a regulated Rocky-Mountain stream. *Canadian Journal of Fisheries and Aquatic Sciences*, **37(1)**, pp. 123–127, 1980.

[308] Collier, K.J., Effects of flow regulation and sediment flushing on instream habitat and benthic invertebrates in a New Zealand river influenced by a volcanic eruption. *River Research and Applications*, **18(3)**, pp. 213–226, 2002.

[309] Nislow, K.H., Magilligan, F.J., Fassnacht, H., Bechtel, D. & Ruesink, A., Effects of hydrologic alteration on flood regime of natural floodplain communities in the Upper Connecticut River. *Journal of the American Water Resources Association*, **38**, pp. 1522–1548, 2002.

[310] Poff, N.L. & Hart, D.D., How dams vary and why it matters for the emerging science of dam removal. *Bioscience*, **52(8)**, pp. 659–668, 2002.

[311] Ward, J.V. & Stanford, J.A., *The Ecology of Regulated Streams*, Plenum Press: New York, pp. 398. 1979.

[312] Richardson, J.S. & Mackay, R.J., Lake outlets and the distribution of filter feeders – An assessment of hypotheses. *Oikos*, **62(3)**, pp. 370–380, 1991.

[313] Morin, A., Back, C., Chalifour, A., Boisvert, J. & Peters, R.H., Effect of black fly ingestion and assimilation on seston transport in a Quebec lake outlet. *Canadian Journal of Fisheries and Aquatic Sciences*, **45(4)**, pp. 705–714, 1988.

[314] Sheldon, A.L. & Oswood, M.W., Blackfly (Diptera-Simuliidae) abundance in a lake outlet – test of a predictive model. *Hydrobiologia*, **56(2)**, pp. 113–120, 1977.

[315] Richardson, J.S., Effect of seston quality on the growth of a lake-outlet filter feeder. *Oikos*, **43(3)**, pp. 386–390, 1984.

[316] Robertson, D.M. & Saad, D.A., Environmental water-quality zones for streams: A regional classification scheme. *Environmental Management*, **31(5)**, pp. 581–602, 2003.

[317] Wang, X.H. & Yin, Z.Y., Using GIS to assess the relationship between land use and water quality at a watershed level. *Environment International*, **23(1)**, pp. 103–114, 1997.

[318] Johnson, L.B., Richards, C., Host, G.E. & Arthur, J.W., Landscape influences on water chemistry in Midwestern stream ecosystems. *Freshwater Biology*, **37(1)**, pp. 193–208, 1997.

[319] Roth, N.E., Allan, J.D. & Erickson, D.L., Landscape influences on stream biotic integrity assessed at multiple spatial scales. *Landscape Ecology*, **11(3)**, pp. 141–156, 1996.

[320] Lammert, M. & Allan, J.D., Assessing biotic integrity of streams: effects of scale in measuring the influence of land use/cover and habitat structure on fish and macroinvertebrates. *Environmental Management*, **23(2)**, pp. 257–270, 1999.

[321] Marsh-Matthews, E. & Matthews, W.J., Geographic, terrestrial and aquatic factors: which most influence the structure of stream fish assemblages in the midwestern United States? *Ecology of Freshwater Fish*, **9(1–2)**, pp. 9–21, 2000.

[322] Stauffer, J.C., Goldstein, R.M. & Newman, R.M., Relationship of wooded riparian zones and runoff potential to fish community composition in agricultural streams. *Canadian Journal of Fisheries and Aquatic Sciences*, **57(2)**, pp. 307–316, 2000.

[323] Jones, K.B., Neale, A.C., Nash, M.S., Van Remortel, R.D., Wickham, J.D., Riitters, K.H. & O'Neill, R.V., Predicting nutrient and sediment loadings to streams from landscape metrics: A multiple watershed study from the United States mid-Atlantic Region. *Landscape Ecology*, **16(4)**, pp. 301–312, 2001.

[324] Chetelat, J. & Pick, F.R., Temporal variability of water chemistry in flowing waters of the northeastern United States: does river size matter? *Journal of the North American Benthological Society*, **20(3)**, pp. 331–346, 2001.

[325] Robertson, D.M., Influence of different temporal sampling strategies on estimating total phosphorus and suspended sediment concentration and transport in small streams. *Journal of the American Water Resources Association*, **39(5)**, pp. 1281–1308, 2003.

[326] Kofoid, C.A., The phytoplankton of the lower Illinois River and its basin. Part 1. Quantitative investigations and general results. *Bulletin of the Illinois State Laboratory of Natural History*, **6**, pp. 95–269, 1903.

[327] Reinhard, E.G., The plankston ecology of the upper Mississippi, Minneapolis to Winona. *Ecological Monographs*, **1**, pp. 395–464, 1931.

[328] Munn, N. & Prepas, E., Seasonal dynamics of phosphorus partitioning and export in two streams in Alberta, Canada. *Canadian Journal of Fisheries and Aquatic Sciences*, **43(12)**, pp. 2464–2471, 1986.

[329] Meyer, J.L. & Likens, G.E., Transport and transformation of phosphorus in a forest stream ecosystem. *Ecology*, **60(6)**, pp. 1255–1269, 1979.

[330] Gallart, F., Llorens, P., Latron, J. & Regues, D., Hydrological processes and their seasonal controls in a small Mediterranean mountain catchment in the Pyrenees. *Hydrology and Earth System Sciences*, **6(3)**, pp. 527–537, 2002.

[331] Leopold, L.B., *A View of the River*, Harvard University Press: Cambridge, Massachusetts, 1994.

[332] Webb, B.W. & Walling, D.E., Long-term water temperature behavior and trends in a Devon, UK, River System. *Hydrological Sciences Journal-Journal des Sciences Hydrologiques*, **37(6)**, pp. 567–580, 1992.

[333] Pace, M.L., Findlay, S.E.G. & Lints, D., Zooplankton in advective Environments – the Hudson River community and a comparative-analysis. *Canadian Journal of Fisheries and Aquatic Sciences*, **49(5)**, pp. 1060–1069, 1992.

[334] Jones, F.H., The dynamics of suspended algal populations in the Lower Wye catchment. *Water Research*, **18(1)**, pp. 25–35, 1984.

[335] Klotz, R.L., Temporal relation between soluble reactive phosphorus and factors in stream water and sediments in Hoxie Gorge Creek, New York. *Canadian Journal of Fisheries and Aquatic Sciences*, **48(1)**, pp. 84–90, 1991.

[336] Mulholland, P.J., Newbold, J.D., Elwood, J.W., Ferren, L.A. & Webster, J.R., Phosphorus spiraling in a woodland stream – seasonal variations. *Ecology*, **66(3)**, pp. 1012–1023, 1985.

[337] Minshall, G.W., Petersen, R.C. & Nimz, C.F., Species richness in streams of different size from the same drainage basin. *American Naturalist*, **125(1)**, pp. 16–38, 1985.

[338] Evans, M. & Slaymaker, A., Spatial and temporal variability of sediment delivery from alpine lake basins, Cathedral Provincial Park, southern British Columbia. *Geomorphology*, **61(1–2)**, pp. 209–224, 2004.

[339] Nearing, M.A., Jetten, V., Baffaut, C., Cerdan, O., Couturier, A., Hernandez, M., Le Bissonnais, Y., Nichols, M.H., Nunes, J.P., Renschler, C.S., Souchere, V. & van Oost, K., Modeling response of soil erosion and runoff to changes in precipitation and cover. *Catena*, **61(2–3)**, pp. 131–154, 2005.

[340] Walling, D.E., Owens, P.N., Foster, I.D.L. & Lees, J.A., Changes in the fine sediment dynamics of the Ouse and Tweed basins in the UK over the last 100–150 years. *Hydrological Processes*, **17(16)**, pp. 3245–3269, 2003.

[341] Xu, J.X., The water fluxes of the Yellow River to the sea in the past 50 years, in response to climate change and human activities. *Environmental Management*, **35(5)**, pp. 620–631, 2005.

[342] Eden, D.N. & Page, M.J., Palaeoclimatic implications of a storm erosion record from late Holocene lake sediments, North Island, New Zealand. *Palaeogeography Palaeoclimatology Palaeoecology*, **139(1–2)**, pp. 37–58, 1998.

[343] Thomas, M.F., Late Quaternary sediment fluxes from tropical watersheds. *Sedimentary Geology*, **162(1–2)**, pp. 63–81, 2003.

[344] Dearing, J.A., Battarbee, R.W., Dikau, R., Larocque, I. & Oldfield, F., Human-environment interactions: learning from the past. *Regional Environmental Change*, **6(1–2)**, pp. 1–16, 2006.

[345] Walling, D.E. & Fang, D., Recent trends in the suspended sediment loads of the world's rivers. *Global and Planetary Change*, **39(1–2)**, pp. 111–126, 2003.

[346] Lopes, V.L. & Canfield, H.E., Effects of watershed representation on runoff and sediment yield modeling. *Journal of the American Water Resources Association*, **40(2)**, pp. 311–319, 2004.

[347] West, T.O. & Wali, M.K., Modeling regional carbon dynamics and soil erosion in disturbed and rehabilitated ecosystems as affected by land use and climate. *Water Air and Soil Pollution*, **138(1–4)**, pp. 141–163, 2002.

[348] Hudson, P.F., The influence of the El Nino Southern Oscillation on suspended sediment load variability in a seasonally humid tropical setting: Panuco Basin, Mexico. *Geografiska Annaler Series A-Physical Geography*, **85A(3–4)**, pp. 263–275, 2003.

[349] Inman, D.L. & Jenkins, S.A., Climate change and the episodicity of sediment flux of small California rivers. *Journal of Geology*, **107(3)**, pp. 251–270, 1999.

[350] Korytny, L.M., Bazhenova, O.I., Martianova, G.N. & Ilyicheva, E.A., The influence of climatic change and human activity on erosion processes in sub-arid watersheds in southern East Siberia. *Hydrological Processes*, **17(16)**, pp. 3181–3193, 2003.

[351] Liebault, F., Gomez, B., Page, M., Marden, M., Peacock, D., Richard, D. & Trotter, C.M., Land-use change, sediment production and channel response in upland regions. *River Research and Applications*, **21(7)**, pp. 739–756, 2005.

[352] Lu, X.X., Spatial variability and temporal change of water discharge and sediment flux in the Lower Jinsha tributary: Impact of environmental changes. *River Research and Applications*, **21(2–3)**, pp. 229–243, 2005.

[353] Bogaart, P.W., Tucker, G.E. & de Vries, J.J., Channel network morphology and sediment dynamics under alternating periglacial and temperate regimes: a numerical simulation study. *Geomorphology*, **54(3–4)**, pp. 257–277, 2003.

[354] Woo, M.K. & McCann, S.B., Climatic variability, climatic change, runoff, and suspended sediment regimes in northern Canada. *Physical Geography*, **15(3)**, pp. 201–226, 1994.

[355] Restrepo, J.D. & Syvitski, J.P.M., Assessing the effect of natural controls and land use change on sediment yield in a major Andean river: The Magdalena drainage basin, Colombia. *Ambio*, **35(2)**, pp. 65–74, 2006.

[356] Wallbrink, P.J., Martin, C.E. & Wilson, C.J., Quantifying the contributions of sediment, sediment-P and fertilizer-P from forested, cultivated and pasture areas at the landuse and catchment scale using fallout radionuclides and geochemistry. *Soil & Tillage Research*, **69(1–2)**, pp. 53–68, 2003.

[357] Burbank, D.W., Lave, J., Meigs, A.J., Fielding, E.J. & Blyth, A.E., Consistent long and short term denudation patterns in the San Gabriel and Himalayan Mountains. *AGU Fall Meeting, F338*, 1998.

[358] Schaller, M., von Blanckenburg, F., Veldkamp, A., Tebbens, L.A., Hovius, N. & Kubik, P.W., A 30 000 yr record of erosion rates from cosmogenic Be-10 in Middle European river terraces. *Earth and Planetary Science Letters*, **204(1–2)**, pp. 307–320, 2002.

[359] Phillips, J., Alluvial storage and the long-term stability of sediment yields. *Basin Research*, **15(2)**, pp. 153–163, 2003.

[360] Viles, H.A. & Goudie, A.S., Interannual, decadal and multidecadal scale climatic variability and geomorphology. *Earth-Science Reviews*, **61(1–2)**, pp. 105–131, 2003.

[361] Dean, J.M. & Kemp, A.E.S., A 2100 year BP record of the Pacific Decadal Oscillation, El Nino Southern Oscillation and Quasi-Biennial Oscillation in marine production and fluvial input from Saanich Inlet, British Columbia. *Palaeogeography Palaeoclimatology Palaeoecology*, **213(3–4)**, pp. 207–229, 2004.

[362] Kawahata, H. & Gupta, L.P., El Nino Southern Oscillation (ENSO) related variations in particulate export fluxes in the western and central equatorial Pacific. *Journal of Oceanography*, **59(5)**, pp. 663–670, 2003.

[363] Wang, H.J., Yang, Z.S., Saito, Y., Liu, J.P. & Sun, X.X., Interannual and seasonal variation of the Huanghe (Yellow River) water discharge over the past 50 years: Connections to impacts from ENSO events and dams. *Global and Planetary Change*, **50(3–4)**, pp. 212–225, 2006.

[364] Villar, C.A., Stripeikis, J., D'Huicque, L., Tudino, M. & Bonetto, C., Concentration and transport of particulate nutrients and metals in the Lower Parana River during extreme flooding. *Archiv für Hydrobiologie*, **153(2)**, pp. 273–286, 2002.

[365] Thunell, R.C., Seasonal and annual variability in particle fluxes in the Gulf of California: A response to climate forcing. *Deep-Sea Research Part I-Oceanographic Research Papers*, **45(12)**, pp. 2059–2083, 1998.

[366] Lick, W., Lick, J. & Ziegler, C.K., Flocculation and its effect on the vertical transport of fine-grained sediments. *Hydrobiologia*, **235**, pp. 1–16, 1992.

[367] Mangini, S.P., Prendes, H.H., Amsler, M.L. & Huespe, J., The importance of flocculation for suspended load sedimentation in the Parana river, Argentina. *Ingenieria Hidraulica En Mexico*, **18(3)**, pp. 55–69, 2003.

[368] Phillips, J.M. & Walling, D.E., The particle size characteristics of fine-grained channel deposits in the River Exe Basin, Devon, UK. *Hydrological Processes*, **13(1)**, pp. 1–19, 1999.

[369] Ross, J.M. & Sherrell, R.M., The role of colloids in trace metal transport and adsorption behavior in New Jersey Pinelands streams. *Limnology and Oceanography*, **44(4)**, pp. 1019–1034, 1999.

CHAPTER 6

Effect of nitrogen best management practices on water quality at the watershed scale

D.J. Mulla
Dept. Soil, Water and Climate, University of Minnesota, St. Paul, Minnesota, USA.

Abstract

Coastal marine ecosystems are important for their productivity and because nearly 40% of the world's population live in close proximity to them. Nutrient overenrichment leading to hypoxia of coastal marine ecosystems is accelerating at an unprecedented rate. This chapter summarizes the effectiveness of various agricultural Best Management Practices (BMPs) to reduce nitrate-N losses from tile drained areas of the Midwestern Corn Belt. These BMPs can be classified into three general categories: hydrologic modification, nutrient management and landscape diversification. Hydrologic modification includes changes in tile drain depth and spacing, and a practice known as controlled drainage. These practices have been shown to reduce nitrate-N losses through tile drains by from 15–96%. Nutrient management includes changes in rate and timing of application for either fertilizer or manure. These practices have been shown to reduce nitrate-N losses through tile drains by from 6–58%. Landscape diversification includes changes in cropping systems, planting of cover crops or riparian buffer strips, agroforestry, prairie restoration or wetland restoration. These practices have been shown to reduce nitrate-N losses by from 5–97%.

1 Introduction

Coastal marine ecosystems are important for their productivity and because nearly 40% of the world's population live in close proximity to them [1]. Nutrient over-enrichment of coastal marine ecosystems is accelerating at an unprecedented rate. There are currently 146 coastal zones with documented periods of low dissolved oxygen concentrations (hypoxia) for extended periods of time, more than 16 times the number of hypoxic zones identified during the 1960s [2]. Most of the hypoxic zones off the coast of the United States, Europe and China are caused by anthropogenic sources of nitrogen and phosphorus flowing from rivers and airsheds into coastal regions. Riverine nutrient flows to affected coastal regions have increased since the industrial revolution by 15-fold in North Sea watersheds, 11-fold in northeastern U.S. watersheds, 10-fold in the Yellow River basin, and 6-fold in the Mississippi River basin [1]. Global riverine exports of nitrogen come from a variety of sources, it is estimated that by 2030, 57% will come from natural sources, 34% from agricultural sources and 9% from human sewage [3]. According to Helly and Levin [4], marine hypoxia on continental margins is also due to natural causes. They identified over 1.1 million km^2 of coastal regions that are naturally afflicted by oxygen concentrations less than 0.5 ml/L (hypoxia). The Arabian Sea and Bay of Bengal (Indian Ocean) account for 59% of this area, 31% occurs in the eastern Pacific Ocean and only 10% occurs in the southeastern Atlantic.

Hypoxia in the Gulf of Mexico covers the second largest area of any hypoxic zone in the world, currently the area affected by Gulf of Mexico hypoxia averages about 15,000 km^2 [5, 6]. As with other hypoxic zones around the world, hypoxia in the Gulf of Mexico is commonly attributed to nutrient overenrichment, particularly from nitrates entering the Gulf of Mexico via the Mississippi-Atchafalaya Rivers [7]. According to Goolsby *et al.* [7], the Upper Mississippi River basin, comprising the states of Iowa, Illinois, Minnesota, Missouri and Wisconsin, drains roughly 18% of the entire Mississippi River basin, but contributes a disproportionately high 35% of the nitrate-N loads at the mouth of the Mississippi River. This Upper Midwestern region is characterized by extensive row crop agriculture, deep soils rich in organic matter, high rates of nitrogen applications from fertilizer and manure, extensive networks of artificial drainage, and intense rainstorms from April to July. All of these factors combined explain the disproportionate nitrate-N loads originating in the Upper Mississippi River basin [8].

Hypoxia in the Gulf of Mexico causes a number of detrimental ecologic effects, including accelerated growth of phytoplankton, large-scale depletion of bottom-level oxygen concentrations, degradation of habitat, mass fish migrations, and increased mortality of crabs, worms and shrimp. The Mississippi River/Gulf of Mexico Watershed Nutrient Task Force [9] established a goal for a 30% reduction in the average area of the hypoxic zone (to 5000 km^2) by 2015.

A wide array of strategies have been proposed to achieve this goal for a 30% reduction in the extent of the Gulf of Mexico hypoxic zone [10, 11]. These include agricultural fertilizer and manure best management practices (BMPs), tile-drainage management strategies, alternative cropping systems, creating and restoring wetlands and riparian buffers, reducing point sources, and diverting rivers for flood control in the Mississippi River delta region. The overall largest reductions in nitrate-N loads to the Gulf of Mexico are anticipated to arise from implementation of agricultural BMPs in the Midwestern Corn Belt [10]. The Upper Mississippi River Sub-Basin Hypoxia Nutrient Committee (UMRSHNC) sponsored a workshop in September, 2005 to review the effectiveness of agricultural BMPs at reducing nutrient loads in the Upper Mississippi River basin [12]. This chapter builds on results from that review, and summarizes the effectiveness of various agricultural BMPs to reduce nitrate-N losses from tile drained areas of the Midwestern Corn

Belt. These BMPs can be classified into three general categories: hydrologic modifications, nutrient management practices and landscape diversification.

2 Hydrologic BMPs

Large regions of the Upper Midwest are managed with artificial drainage, including subsurface tile drains and surface ditches. Subsurface tile drainage is used to lower the depth of the shallow water table in poorly drained soils in order to enhance soil trafficability and crop productivity. Nitrate-N arising from fertilizer and manure applications to row crops, or from mineralization of soil organic matter, is prone to discharge through subsurface tile drains [13, 14].

A general strategy for reducing nitrate-N losses in subsurface drainage is to alter depth and spacing of tile drainage systems (Table 1) to reduce tile discharge or nitrate-N concentration (or both). As tile spacing increases, or as the depth of tile drains becomes shallower, tile discharge generally decreases. In addition, nitrate-N concentrations may also decrease in response to prolonged periods of soil saturation and enhanced denitrification.

2.1 Tile drain depth and spacing effects on nitrate-N losses

In Indiana, Kladivko *et al.* [14] conducted a 15-year study of nitrate-N losses in subsurface drainage for drain spacings of 5 to 20 m. From 1986–1988, nitrate-N losses averaged 50 and 27 kg/ha, respectively (a 46% reduction), for spacings of 5 and 20 m. A continuous crop of corn was grown during this time, and N fertilizer was applied in spring at a rate of 285 kg/ha. From 1997–1999, nitrate-N losses averaged 16 and 13 kg/ha, respectively (a 19% reduction), for spacings of

Table 1: Summary of reductions in nitrate-N losses under various subsurface tile-drain management strategies.

Type of study	Reference	Drain parameters (m)	N fertilizer rate (kg/ha)	% Reduction in nitrate-N loss
Tile spacing	[14]	20 vs. 5	285	46%
	"	"	177	19%
	[16]	100 vs. 27	123	78%
Tile depth	[16]	0.9 vs. 1.5	123	59%
	[15]	0.9 vs. 1.2	138	15%
Controlled drainage	[20]	0.6–0.9 vs. 1.1	200	39%
	[19]	0.5 vs. 0.75	241	62–96%
	[18]	0.3 vs. 0.6	170	49%

5 and 20 m. During this time, N fertilizer was applied in spring to a corn–soybean rotation with a fall cover crop at a rate of 177 kg/ha. For this location, nitrate-N losses respond more to changes in fertilizer and cropping system management than to changes in drain spacing.

In Minnesota, Sands *et al.* [15] conducted a field experiment with a corn–soybean rotation designed to evaluate flow and nitrate-N losses from subsurface tile drains installed at depths of 0.9 or 1.2 m, and at spacings of roughly 10 or 20 m. They found an 18% reduction in annual flow and a 15% reduction in nitrate-N losses for the 0.9 m depth in comparison with the 1.2 m depth, without significant differences in nitrate-N concentrations. These results show that reductions in nitrate-N losses were largely attributed to reduced drainage flows at shallower tile depths.

Nangia *et al.* [16] used the Agricultural Drainage and Pesticide Transport (ADAPT) model to investigate the influence of subsurface tile drain depth and spacing on discharges of water and nitrate-N from tile drains under a corn–soybean rotation using a 50-year record of climatic conditions in southern Minnesota. The ADAPT model was calibrated and validated using a ten-year dataset for flow and nitrate-N losses from a 21 ha corn–soybean field in southern Minnesota. Baseline conditions for simulations included a tile spacing of 27 m, a tile depth of 1.2 m and a fall application of 123 kg/ha N fertilizer. For a subsurface tile depth of 1.2 m, increasing the tile spacing from 27 to 100 m reduced nitrate-N losses from 43.1 to 9.5 kg/ha, a reduction of 78%. Reductions in nitrate-N losses are also possible by decreasing depth of tile drains, at a spacing of 27 m, reducing tile depth from 1.5 m to 0.9 m reduced nitrate-N losses from 43.1 to 17.5 kg/ha, a reduction of 59%.

2.2 Controlled drainage effects on nitrate-N losses

Controlled drainage is an alternative to physically installing tile drains at shallower depths or wider spacings, and is cost effective on flat uniform fields with slopes of 1% or less [17]. Controlled drainage involves placing a water control structure near the outlet of the drainage system. Stop logs or float mechanisms in the control structure can be raised or lowered to regulate the level of the water table. Water is only discharged when the water table rises above the height of the stop log or float. In typical operation, tile drains are prevented from discharging water for the majority of the year from June through March, unless heavy rainfall occurs. From March to May, stop logs are lowered in order to drain the soil of excess water in preparation for vehicle traffic needed for tillage, fertilization and planting operations. Reductions in average nitrate concentrations from controlled drainage systems (due to denitrification) in comparison with conventional drainage systems range from 25–76% [18–20].

3 Nutrient management BMPs

Nutrient management BMPs attempt to improve water quality by reducing the inputs of nitrogen to the cropping system from fertilizer or manure, delaying the timing of fertilizer application to better match crop uptake, or adding nitrification inhibitors (Table 2).

Corn crops (in a corn–soybean rotation) in the upper Midwestern region are recommended to receive nitrogen fertilizer applications ranging from 115 kg/ha in Minnesota to 143 kg/ha in Iowa to 176 kg/ha in Illinois [21]. In much of Minnesota, corn crops typically receive from 15 to 60 kg/ha more N fertilizer than recommended, leading to significant losses of nitrate in tile drainage systems [8]. N fertilizer guidelines are typically based on economically optimum nitrogen rates determined from the ratio of corn prices to N fertilizer prices [22]. Environmental considerations are not factored into these guidelines.

3.1 N fertilizer rate effects on nitrate-N losses

Losses of nitrate-N in tile drainage typically decrease as the rate of fertilizer application decreases. Buzicky *et al.* [23] conducted a four-year study showing that, for continuous corn plots in Minnesota, nitrate-N losses in tile drainage were reduced from 29 to 21 kg/ha (a 28% reduction) when rates of spring-applied N fertilizer were reduced from 202 to 134 kg/ha. Nangia *et al.* [24] used the ADAPT model to simulate nitrate-N losses for a nearby field site in a corn-soybean rotation using a 50-year climatic record. Simulation results showed that decreasing N application rates from

Table 2: Summary of reductions in nitrate-N losses under various nutrient management BMPs.

Type of study	Reference	N fertilizer rate (kg/ha)	% Reduction
Fall N fertilizer application	[23]	134 vs. 202	28%
	[25]	112 vs. 182	10%
	[29]	136 vs. 180	17%
	[27]	143 vs. 174 (also switched from fall to spring application)	30%
	[16]	135 vs. 180	12%
	[24]	134 vs. 179	15%
	[25]	140 vs. 180	14%
	"	130 vs. 169	11%
Fall N vs. spring N application	[32]	NA	58%
	[8]	134	36%
	[25]	135	34%
	[16]	110	6%
Nitrification inhibitor	[33]	150	18%

180 to 135 kg/ha resulted in a 12% reduction in nitrate-N losses. Using the same model at the watershed scale where the field-scale site was located, Nangia *et al.* [24] showed that reducing fall N applications from 179.2 kg/ha to 134.5 kg/ha resulted in a 15% reduction in nitrate-N losses (from 28.2 to 24 kg/ha).

The extent of reductions in nitrate-N losses achieved at the watershed scale by reducing N fertilizer application rates depends on several factors, including the extent of subsurface tile drainage in the watershed and the extent of fertilizer N applications relative to nonfertilizer sources of N (manure and soil organic matter). Gowda *et al.* [25] simulated nitrate-N losses in two Minnesota watersheds with contrasting extents of subsurface tile drainage. Tile drainage is installed in 30% of the agricultural land in Sand Creek, whereas Bevens Creek has 50% of the land in tile drainage. Not surprisingly, annual nitrate-N losses in Sand Creek averaged 6.7 kg/ha, while nitrate-N losses in Bevens Creek averaged 19 kg/ha. Reducing N fertilizer application rates from 180 kg/ha to 140 kg/ha in Sand Creek caused a reduction in nitrate-N losses from 7.8 to 6.7 kg/ha (a 14% reduction). Reducing N fertilizer application rates from 169 to 130 kg/ha in Bevens Creek caused a reduction in nitrate-N losses from 21.3 to 19 kg/ha (an 11% reduction). Finally, Gowda *et al.* [26] studied nitrate-N losses in a Minnesota watershed with extensive subsurface tile drainage and animal agriculture operations and high soil organic matter content. Nitrate-N losses averaged 28.7 kg/ha, and these losses were reduced by less than 1 kg/ha (<3%) when N fertilizer applications were decreased by 20%.

Jaynes *et al.* [27] obtained a 30% reduction in nitrate-N concentrations by reducing N fertilizer applications and switching from fall to spring N applications in a portion of the Walnut Creek watershed located in Iowa. Gowda *et al.* [28] used simulation modeling for the same watershed to show that a 40% reduction in N fertilizer rates would only result in a 10% reduction in N loads at the watershed mouth. Baksh *et al.* [29] used the RZWQM for a single field in the same watershed, and showed that reducing the N fertilizer rate from 180 kg/ha to 136 kg/ha (a 30% reduction) decreased nitrate-N losses in tile drainage from 54 kg/ha to 45 kg/ha (a 17% reduction). These results show that for this study area in Iowa, reductions in fertilizer rate have a greater impact on nitrate-N losses at the field scale than at the watershed scale, and that changes in fertilizer application timing are more important than changes in the rate of application at the watershed scale.

Petrolia and Gowda [30] used the ADAPT model combined with econometric modeling to evaluate the economic implications of reduced nitrogen fertilizer application rates combined with land retirement to achieve a 30% reduction in nitrate-N losses from the Cottonwood River watershed in Minnesota. They compared targeted versus nontargeted strategies. When targeting the artificially drained soils that had the greatest potential to produce nitrate-N losses in tile drainage, the targeted strategy cost producers roughly $31/ac to achieve a watershed wide reduction in nitrate-N losses of 30%. Uniform implementation of fertilizer-management BMPs and land-retirement cost producers roughly $57/ac to achieve the same level of reduction. In addition to the cost differences, the targeted strategy required 25% less land area to be treated than with the uniform strategy.

3.2 N fertilizer application timing effects on nitrate-N losses

Fall applications of nitrogen fertilizer are more prone to losses through subsurface tile drainage systems than spring applications [8]. Fall-applied anhydrous ammonia is converted to nitrate through nitrification if soil temperatures are warm enough [31]. Spring applications are less prone to nitrification than fall applications due to cool soil temperatures and rapid uptake of soil N by the growing crop. Nitrate losses at the plot scale in subsurface drainage effluent were reduced by 36% in Minnesota [8] and by 58% in Illinois [32] when N fertilizer was applied in spring rather than in fall. At the watershed scale, similar reductions occur. Gowda *et al.* [25] showed a 34% reduction in nitrate-N losses for two Minnesota watersheds when N fertilizer was applied in spring rather than fall.

Nitrogen fertilizers will continue to be applied in fall as long as anhydrous ammonia prices are relatively low due to the relatively short number of days soils are trafficable in the spring due to snowmelt and rainfall on wet soils. Nitrification inhibitors are sometimes added to N fertilizer to slow the conversion of ammonium to nitrate. The use of N-serve as a nitrification inhibitor with fall-applied N fertilizer reduced nitrate-N concentrations in tile drainage water by 18% in Minnesota [33].

Many studies are underway to identify strategies for in-season correction of nitrogen deficiencies in rainfed corn. These strategies involve small starter applications of N at the time of planting, followed by a second sidedress application during June. Identification of nitrogen stress involves either the late-spring nitrate test [LSNT; 34], or timely identification of N deficiencies using sensors [35–37] or remote sensing [38–40]. These methods rely on the fact that N deficient crops are not as green as crops with sufficient N availability. Additional research on these strategies is needed to further develop the methodologies and evaluate their impact on crop productivity and environmental quality.

3.3 Impacts of manure N application rate on nitrate-N losses

Large concentrations of hog, cattle and poultry production exist in Midwestern states. Manure produced by these operations is generally stored until fall, when it is land applied to fields that will be planted to corn in the spring. Manure has traditionally been viewed as a waste to be disposed of on the land, but as nitrogen fertilizer prices increase, manure is increasingly being viewed as a valuable source of crop nutrients.

When manure is treated as a waste product there is a high likelihood that it will not be applied at rates that are agronomically reasonable. In Minnesota, from 10–20% of agricultural land receives applications of both N fertilizer and manure. Rates of N applied on these fields can far exceed those required for optimum crop production. This can lead to significant leaching of excess nitrogen to subsurface tile drains.

In general, there is no difference between the rates of N loss to subsurface tile drains from fertilizer versus manure if both are applied at equivalent rates of N [41].

Rates of N loss increase with the rate of applied N from manure [42]. Therefore, it is important to base manure applications on a combination of soil test results, crop-yield expectations and nutrient content of manure based on manure samples.

4 Landscape diversification

Landscape diversification attempts to reduce nitrate-N losses (Table 3) by broadening the cropping systems and environmental services provided on the landscape. Diversification takes many forms, including changes in cropping system, planting of cover crops or riparian buffer strips, agroforestry, prairie restoration or wetland restoration. Cropping system BMPs attempt to alter the sequence of crops using rotations with perennial crops, riparian buffer strips or crops that are efficient at removing residual nitrogen from the soil profile (cover crops). Perennial crops such as alfalfa can be grown in five-year rotations with corn and soybean. Since alfalfa is a legume, it typically receives N fertilizer applications only in the first year of growth during establishment and the remaining two years of the rotation do not receive any fertilizer.

Table 3: Summary of reductions in nitrate-N losses under various landscape diversification practices.

Practice category	Reference	Comparisons made	% Reduction
Fall planted cover crops	[14]	Winter wheat vs. no winter wheat	<60% (N rate changed too)
	[48]	Planting date Sept. 15 vs. Oct. 15	11–30%
	[46]	Rye vs. no rye	13%
	[47]	Rye vs. no rye	60%
Alternative cropping systems	[13]	Alfalfa vs. corn-soybean	97%
	[45]	Alfalfa vs. corn-soybean	51–63%
	[45]	Grassland and cover crops vs. corn-soybean	62–74%
	[49]	Short rotation woody crops vs. corn-soybean	5–15%
Wetland restoration	[53]	Mesocosm study	40–90%
	[51]	Constructed wetlands	19–59%

4.1 Impacts of alternative cropping systems on nitrate-N losses

In comparison with a corn–soybean rotation, the five-year corn–soybean–alfalfa–alfalfa–alfalfa rotation receives much lower rates of N fertilizer application. Randall *et al.* [13] showed that annual nitrate-N losses in Minnesota subsurface tile drainage averaged 50.7 kg/ha for a corn–soybean rotation, but only 1.7 kg/ha for a continuous crop of alfalfa (a 97% reduction). The same study showed negligible nitrate-N losses in a mixed crop of grass and alfalfa typical of the composition used in Conservation Reserve Program (CRP) plantings. Alfalfa and grass are deep-rooted plants whose annual evapotranspiration rates exceed those for a corn–soybean rotation. As a result of enhanced water uptake, nutrient uptake from the soil is also enhanced and tile drainage losses are reduced.

Chung *et al.* [43] attempted to use the EPIC model to simulate impacts of cropping-system management on tile drain losses of nitrate-N from the study of Randall *et al.* [13]. In general, nitrate-N loads were greatly underpredicted by EPIC for all cropping systems when default values were used for SCS runoff curve numbers. EPIC does not explicitly simulate effects of subsurface tile drainage, drainage effects are inferred from nitrate-N leaching through the soil profile to the depth of tile drains. Nitrate-N loss estimates could be improved by calibrating EPIC SCS runoff curve number using measured data. The poor performance of EPIC in tile drained agricultural systems calls into question its continued use (see for example, Ribaudo *et al.* [44]) for evaluation of strategies to achieve nitrate-N reductions in Midwestern agriculture.

Boody *et al.* [45] used the ADAPT model to study environmental benefits of cropping-system diversification for two agricultural watersheds in Minnesota. They simulated nitrate-N loads for each watershed in response to several scenarios, including A) current land use and land management, B) implementation of nutrient best management practices, C) replacing a portion of existing row crop land with alfalfa and wetlands, and D) planting cover crops and grassland. With BMPs (B), a reduction in nitrate-N loads ranging from 18–37% was achieved in comparison with current conditions (A). With additional alfalfa plantings and wetland restoration (C), a reduction in nitrate-N loads ranging from 51–63% was achieved. With cover crops and grassland, nitrate-N load reductions ranging from 62–74% were achieved. Boody *et al.* [45] concluded that to reach goals for reducing the area of the hypoxic zone in the Gulf of Mexico, additional diversification of the Midwestern agricultural landscape was needed in addition to nutrient management BMPs.

4.2 Impacts of cover crops on nitrate-N losses

Winter cover crops are planted shortly before or soon after harvest of a crop in fall and are killed in spring using cultivation or herbicides. Cover crops remove water and nitrogen from the soil profile after the primary crop is removed. Examples of cover crops include rye, small grains and clover. Reductions in nitrate-N losses from subsurface drains where cover crops were grown ranged from 13% in

Minnesota [46], to roughly 60% in Iowa [47]. Kladivko *et al.* [14] showed that annual nitrate-N losses from a subsurface drained soil in Indiana with low organic matter content decreased over a 15-year period from 38 kg/ha to 15 kg/ha (a 60% reduction) as a result of switching from continuous corn to a corn–soybean rotation with a fall cover crop of winter wheat. During the 15-year study, spring-applied N fertilizer rates were reduced from 285 kg/ha to 177 kg/ha, so the observed reductions are partly due to changes in fertilizer management and partly due to planting a cover crop. The major limitation to growing cover crops in the Midwest is related to poor fall establishment and poor winter hardiness. Feyereisen *et al.* [48] showed that nitrate-N losses in drainage water could be reduced by 30 or 11% if rye were planted as a cover crop in Minnesota on September 15 or October 15, respectively, and the rye was killed by May 15.

4.3 Impacts of riparian buffer strips and wetlands on nitrate-N losses

In the upper Midwestern region, riparian buffer strips are not as efficient at removing nitrate-N as perennial crops or cover crops. This is because of the prevalence of subsurface drains, typically installed at depths ranging from 1 to 1.5 meters, whose discharge is routed through the subsurface without ever flowing through the riparian buffer strip. However, nitrate-N in shallow groundwater (which is a small proportion of discharge from the landscape) can be removed at efficiencies ranging from 48–85% after flowing through the root zone of riparian buffer strips [49]. Updegraff *et al.* [50] used the ADAPT model to show that replacing corn–soybean crops in critical areas of Minnesota with short-rotation woody crops such as hybrid poplars was effective at reducing nitrate losses at the watershed scale. Simulations by Updegraff *et al.* [50] considered conversion of corn–soybean rotations to short rotation woody crops only on steeper, well-drained soils. For conversion levels of 10, 20 and 30%, nitrate-N losses to the mouth of the watershed were reduced by 4.7, 12.6 and 15%, respectively. Reductions were partially explained by lower rates of N fertilizer used in short-rotation woody crops in comparison with rates used on corn.

Another strategy for landscape diversification is to remove subsurface tile drains altogether in order to restore wetlands on the landscape. Wetlands store water in landscape depressions, reducing the volume of water delivered to surface waters. Wetlands also allow denitrification of nitrates, and if vegetation exists in the wetland, nitrates may also be taken up by the plants as they grow. Annual nitrate loads for landscapes with restored wetlands can be reduced by 20–60% [51–53]. The extent of nitrate removal by wetlands depends on factors such as the ratio of runoff volume from upland contributing areas to wetland storage volume, residence time of water in the wetland, temperature, wetland vegetation biomass and loading rates of nitrogen to the wetland. Removal efficiency decreases as the ratio of upland contributing area increases, as wetland storage volume decreases, as residence time decreases, as temperature decreases, as vegetation biomass decreases, or as nitrate loading rates increase.

5 Impacts of climate change

The effectiveness of many nitrogen BMPs depends on precipitation patterns. Mean annual precipitation in the Upper Midwestern region has increased by as much as 30–40% over the last 50 years [54]. Nangia *et al.* [24] showed that nitrate-N losses in a Minnesota corn–soybean rotation are a linear function of annual precipitation. As precipitation increases, nitrate-N losses also increase. In dry years, reductions in N fertilizer application will have little impact on nitrate-N losses from tile drainage due to limited tile discharge. The greatest reductions in nitrate-N losses through fertilizer management will be achieved in wetter years. During years with normal precipitation (73.7 cm), nitrate-N losses were reduced from about 31.7 kg/ha to about 24.3 kg/ha when N fertilizer application rates were reduced from 179.3 to 112 kg/ha. During the driest years (40 cm), nitrate-N losses were reduced from about 10 kg/ha to roughly 8 kg/ha when comparing the same N fertilizer application rates. As precipitation increases in the Upper Midwestern Corn Belt region, nitrate-N losses in tile drainage can be expected to increase, even if no changes are made in N fertilizer management.

References

[1] UNEP, The state of the marine environment: Trends and processes. *United Nations Environmental Program/Global Programme of Action.* http://www.gpa.unep.org/document_lib/en/pdf/report2-webversion_(global_soe).pdf, 2006.

[2] Larson, J., Dead zones increasing in world's coastal waters. http://www.earth-policy.org/Updates/Update41.htm, 2004.

[3] Bouwman, A.F., Van Drecht, G., Knoop, J.M., Beusen, A.H.W. & Meinard, C.R., Exploring changes in river nitrogen export to the world's rivers. *Global Biogeochemical Cycles,* **19**, pp. GB1002, 2005.

[4] Helly, J.J. & Levin, L.A., Global distribution of naturally occurring marine hypoxia on continental margins. *Deep Sea Research I*, **59**, pp. 1159–1168, 2004.

[5] Rabalais, N.N., Turner, R.E., Justic, D., Dortch, Q. & Wiseman, W.J., Topic 1 Report: Characterization of hypoxia. *NOAA Coastal Ocean Program Decision Analysis Series No. 15*, NOAA. Silver Spring, MD, 1999.

[6] Hogue, C., Dead zone. *Chemical Engineering News*, **84(40)**, pp. 40–42, 2006.

[7] Goolsby, D.A., Battaglin, W.A., Lawrence, G.B., Artz, R.S., Aulenbach, B.T., Hooper, R.P., Keeney, D.R. & Stensland, G.T., Topic 3 Report: Flux and sources of nutrients in the Mississippi-Atchafalaya River Basins. *NOAA Coastal Ocean Program Decision Analysis Series No. 17*, NOAA, Silver Spring, MD, 1999.

[8] Randall, G.W. & Mulla, D.J., Nitrate nitrogen in surface waters as influenced by climatic conditions and agricultural practices. *Journal Environmental Quality*, **30**, pp. 337–344, 2001.

[9] Mississippi River/Gulf of Mexico Watershed Nutrient Task Force. Action plan for reducing, mitigating and controlling hypoxia in the northern Gulf of Mexico. Washington, DC, 2001.

[10] Mitsch, W.J., Day Jr., J.W., Gilliam, J.W., Groffman, P.M., Hey, D.L., Randall, G.W. & Wang, N., Reducing nitrogen loading to the Gulf of Mexico from the Mississippi River Basin: Strategies to counter a persistent ecological problem. *BioScience*, **51(5)**, pp. 373–388, 2001.

[11] Dinnes, D.L., Karlen, D.L., Haynes, D.B., Caspar, T.C., Hatfield, J.L., Colvin, T.S. & Cambardella, C.A., Nitrogen management strategies to reduce nitrate leaching in tile drained Midwestern soils. *Agronomy Journal*, **94**, pp. 153–171, 2002.

[12] UMRSHNC, Gulf Hypoxia and local water quality concerns workshop. Ames, IA, 2005. http://www.umrshnc.org/index.php?option=com_content&task=view&id=19&Itemid=34

[13] Randall, G.W., Huggins, D.R., Russelle, M.P., Fuchs, D.J., Nelson, W.W. & Anderson, J.L., Nitrate losses through subsurface tile drainage in CRP, alfalfa and row crop systems. *Journal Environmental Quality*, **26**, pp. 1240–1247, 1997.

[14] Kladivko, E.J., Frankenberger, J.R., Jaynes, D.B., Meek, D.W., Jenkinson, B.J., & Fausey, N.R., Nitrate leaching to subsurface drains as affected by drain spacing and changes in crop production system. *Journal Environmental Quality*, **33**, pp. 1803–1813, 2004.

[15] Sands, G.R., Song, I., Busman, L.M., & Hansen, B., Water quality benefits of "shallow" subsurface drainage systems. *American Society Agricultural Biosystems Engineers Proceedings 2006 Annual International Meeting,* Publ. #062317, 2006.

[16] Nangia, V., Gowda, P.H., Mulla, D.J. & Sands, G.R., Modeling nitrate-nitrogen losses in response to changes in tile drain depth or spacing. International Annual Meeting *American Society Agricultural Engineers,* Paper No. 052022, 2005.

[17] Cooke, R.A., Sands, G.R. & Brown, L.C., Drainage water management: A practice for reducing nitrate loads from subsurface drainage systems. *Proc. Workshop on Gulf Hypoxia and Local Water Quality Concerns*. Upper Mississippi River Sub-Basin Hypoxia Nutrient Committee, Ames, IA, 2005.

[18] Drury, C.F., Tan, C.S., Gaynor, J.D., Oloya, T.O. & Welacky, T.W., Influence of controlled drainage-subirrigation on surface and tile drainage nitrate loss. *Journal Environmental Quality*, **25**, 317–324, 1996.

[19] Lalonde, V., Madramootoo, C.A., Trenholm, L. & Broughton, R.S., Effects of controlled drainage on nitrate concentrations in subsurface drain discharge. *Agricultural Water Management*, **29**, pp. 187–199, 1996.

[20] Kalita, P.K. & Kanwar, R.S., Effect of water table management practices on the transport of nitrate-N to shallow groundwater. *Transactions American Society Agricultural Engineers*, **36**, pp. 413–422, 1993.

[21] Sawyer, J.E. & Randall, G.W., Nitrogen rates. *Proc. Workshop on Gulf Hypoxia and Local Water Quality Concerns*. Upper Mississippi River Sub-Basin Hypoxia Nutrient Committee, Ames, IA, 2005.

[22] Nafziger, E.D., Sawyer, J.E. & Hoeft, R.G., Formulating N recommendations for corn in the corn belt using recent data. *Proc. North Central Industry Soil Fertility Conf. Des.* Moines, IA. PPI, Brookings, SD, 2004.

[23] Buzicky, G.C., Randall, G.W., Hauck, R.D. & Caldwell, A.C., Fertilizer N losses from a tile drained Mollisol as influenced by rate and time of 15-N depleted fertilizer application. *Agronomy Abstracts*, ASA-CSSA-SSSA Madison, WI, pp. 213, 1983.

[24] Nangia, V., Gowda, P., Mulla, D., & Kuehner, K., Evaluation of predicted long-term water quality trends to changes in N fertilizer management for a cold climate. International Annual Meeting *American Society Agricultural Engineers,* Paper No. 052226, 2005.

[25] Gowda, P.H., Dalzell, B.J. & Mulla, D.J., Model based nitrate TMDLs for two agricultural watersheds in southern Minnesota. *Journal American Water Resources Association*, **43**, pp. 256–263, 2006a.

[26] Gowda, P.H. & Mulla, D.J., Modeling alternative agricultural management practices in High Island Creek watershed in south-central Minnesota. *Journal Environmental Hydrology*, **14** (paper 13), pp. 1–15, 2006b.

[27] Jaynes, D.B., Dinnes, D.L., Meek, D.W., Karlen, D.L., Cambardella, C.A. & Colvin, T.S., Using the late spring nitrate test to reduce nitrate loss within a watershed. *Journal Environmental Quality*, **33**, 669–677, 2004a.

[28] Gowda, P.H., Mulla, D.J. & Jaynes, D.B., Modeling long-term nitrogen losses in Walnut Creek watershed, Iowa. *Agronomy Abstracts*, SSSA, Madison, WI, 2002.

[29] Baksh, A., Hatfield, J.L., Kanwar, R.S., Ma, L. & Ahuja, L.R., Simulating nitrate drainage losses from a Walnut Creek watershed field. *Journal Environmental Quality*, **33**, 114–123, 2004.

[30] Petrolia, D.R. & Gowda, P.H., Missing the boat: Midwest farm drainage and Gulf of Mexico hypoxia. *Reviews Agricultural Economics*, **28**, pp. 240–253, 2006.

[31] Schmidt, E.L., Nitrification in soil. Ch. 7. In: (F.J. Stevenson, ed.), *Nitrogen in Agricultural Soils*, Agronomy Monograph 22, ASA-SSSA-CSSA, Madison, WI, 1982.

[32] Smicklas, K.D. & Moore, A.S., Fertilizer nitrogen management to optimize water quality. In: (R.G. Hoeft, ed.), *Illinois Fertilizer Conf. Proc.*, pp. 117–124, Peoria, IL, 1999.

[33] Randall, G.W., Vetsch, J.A. & Huffman, J.R., Nitrate losses in subsurface drainage from a corn–soybean rotation as affected by time of nitrogen application and use of nitrapyrin. *Journal Environmental Quality*, **32**, pp. 1764–1722, 2003.

[34] Blackmer, A.M., Voss, R.D. & Mallarino, A.P., Nitrogen fertilizer recommendations for corn in Iowa. *Iowa State Coop. Ext.*, Iowa State Univ., **PM-1714**, Ames, IA, 1997.

[35] Schepers, J.S., Francis, D.D., Vigil, N. & Below, F.E., Comparison of corn leaf nitrogen concentration and chlorophyll meter readings. *Communication Soil Science and Plant Analysis*, **23(7&8)**, pp. 2173–2187, 1992.

[36] Raun, W.R., Solie, J.B., Stone, M.L., Martin, K.L., Freeman, K.W., Mullen, R.W., Zhang, H., Schepers, J.S., & Johnson, G.V., Optical sensor-based algorithm for crop nitrogen fertilization. *Communications in Soil Science and Plant Analysis*, **36**, pp. 2759–2781, 2005.

[37] Inman, D., Khosla, R. & Mayfield, R., On-the-go active remote sensing for efficient crop nitrogen management. *Sensor Review*, **25(3)**, 209–214, 2005.

[38] Daughtry, C.S.T., Walthall, C.L., Kim, M.S., Brown de Colstoun, E. & McMurtrey III, J.E., Estimating corn leaf chlorophyll concentration from leaf and canopy reflectance. *Remote Sensing of Environment*, **74**, pp. 229–239, 2000.

[39] Bausch, W.C., & Diker, K., Innovative remote sensing techniques to increase nitrogen use efficiency of corn. *Communications in Soil Science and Plant Analysis*, **32(7&8)**, pp. 1371–1390, 2001.

[40] Osborne, S.L., Schepers, J.S., & Schlemmer, M.R., Using multi-spectral imagery to evaluate corn grown under nitrogen and drought stressed conditions. *Journal Plant Nutrition*, **27(11)**, pp. 1917–1929, 2004.

[41] Randall, G.W., Iragavarapu, T.K., & Schmitt, M.A., Nutrient and pathogen losses in subsurface drainage water from dairy manure and urea applied for corn. *Journal Environmental Quality*, **29**, 1244–1252, 2000.

[42] Kanwar, R.S., Karlen, D.L., Cambardella, C. & Cruse, R.M., Swine manure and N-management systems: Impact on groundwater quality. *Clean Water, Clean Environment, 21st Century Conf. Proc. American Society Agricultural Engineers*, St. Joseph, MI, 1995.

[43] Chung, S.W., Gassman, P.W., Huggins, D.L. & Randall, G.W., EPIC tile flow and nitrate loss predictions for three Minnesota cropping systems. *Journal Environmental Quality*, **30**, pp. 822–830, 2001.

[44] Ribaudo, M.O., Heimlich, R., Classen, R. & Peters, M., Least cost management of nonpoint source pollution: Source reduction versus interception strategies for controlling nitrogen loss in the Mississippi River basin. *Ecological Economics*, **37**, pp. 183–197, 2001.

[45] Boody, G., Vondracek, B., Andow, D.A., Krinke, M., Westra, J., Zimmerman, J. & Welle, P., Multifunctional agriculture in the United States. *BioScience*, **55**, pp. 27–38, 2005.

[46] Strock, J.S., Porter, P.M. & Russelle, M., Cover cropping to reduce nitrate loss through subsurface drainage in the northern U.S. corn belt. *Journal Environmental Quality*, **33**, pp. 1010–1016, 2004.

[47] Jaynes, D.B., Kaspar, R.C., Moorman, T.B. & Parkin, T.B., Potential methods for reducing nitrate losses in artificially drained fields. In: (R. Cooke, ed.), *Proc. 8th Intl. Drainage Symp., American Society Agricultural Engineers*, St. Joseph, MI, pp. 59–69, 2004b.

[48] Feyereisen, G.W., Wilson, B.N., Sands, G.R., Strock, J.S. & Porter, P.M., Potential for a rye cover crop to reduce nitrate loss in southwestern Minnesota. *Agronomy Journal*, **98**, 1416–1426, 2006.

[49] Simpkins, W.W., Wineland, R.R., Andress, R.J., Johnston, D.A., Caron, G.C., Isenhart, T.M. & Schultz, R.C., Hydrogeological constraints on riparian buffers for reduction of diffuse pollution: Examples from the Bear Creek watershed in Iowa, USA. *Water Science Technology*, **45**, pp. 61–68, 2002.

[50] Updegraff, K., Gowda, P. & Mulla, D.J., Watershed scale modeling of the water quality effects of cropland conversion to short rotation woody crops. *Renewable Agriculture Food Systems*, **19(2)**, pp. 1–11, 2004.

[51] Xue, Y., Kovacic, D.A., David, M.B., Gentry, L.E., Mulvaney, R.L. & Lindau, C.W., In situ measurements of denitrification in constructed wetlands. *Journal Environmental Quality*, **28**, pp. 263–269, 1999.

[52] Mitsch, W.J. & Gosselink, J.G., *Wetlands*, 3rd edn. John Wiley & Sons: New York, 2000.

[53] Crumpton, W.G., Potential of wetlands to reduce agricultural nutrient export to water resources in the corn belt. *Proc. Workshop on Gulf Hypoxia and Local Water Quality Concerns*, Upper Mississippi River Sub-Basin Hypoxia Nutrient Committee, Ames, IA, 2005.

[54] Seeley, M.W., *The Minnesota Weather Almanac,* MHS Press, 2006.

CHAPTER 7

Effects of changing land use on nutrient loads and water quality in a Southeastern US Blackwater River Estuary

J.R. White[1], J. Hendrickson[2] & J.L. Conkle[1]
[1]*Department of Oceanography and Coastal Sciences, Louisiana State University, USA.*
[2]*St.Johns River Water Management District, Palatka, FL USA.*

Abstract

Nutrient enrichment in the coastal zone is a world-wide phenomenon, driven by consistent increases in population and land development. In order to counteract the effects of eutrophication, it is important to use historical records to first determine baseline conditions within the affected water body. Then the sometimes difficult and expensive process of constructing a nutrient budget must be completed, which will account for the within-basin point and nonpoint-source nutrient loads. Once this has been achieved, environmental managers will have the information to effectively target point source reductions through regulation and nonpoint-source reduction through best management practices. We take the reader through this process using the Lower St. Johns River Estuary, Florida as a model system to show degradation of water quality over time concomitant with increases in population and changes in land use, culminating in a determination of both the anthropogenic nitrogen and phosphorus loads.

1 Introduction

Continual increases in population density in the coastal zone have had a dramatic effect on both the land use and consequent resultant water quality of many estuarine systems. Changes in land use not only lead to increased nutrient loading, but also shifts to increasing bioavailable nutrient forms. Eventually, the nutrient status of the water body increases to a point where it is difficult for water-resource managers to mitigate expressions of eutrophic conditions including algal blooms, low dissolved oxygen (DO) and resulting fish kills. We have used the Lower St. Johns River Estuary, Florida to demonstrate how coupled population increase and land-use change can affect water quality. Land-use change can affect the overall nutrient loads but also the proportion of bioavailable nutrients. Finally, we have constructed a nitrogen and phosphorus budget for the basin and compare current loads with a historical on (estimated predevelopment) to determine the anthropogenic nutrient load to the system.

The St. Johns River (SJR) is one of the largest blackwater rivers of the southeast U.S. A blackwater river has water whose color ranges from clear tea to coffee. The river is located in northeast Florida and drains about 1/5th of the state, encompassing a 9562 square mile drainage area. The river is slow moving, with a slope of only 1.4 in/mile [1], and is essentially at sea level for its final 125 mi. The lower St. Johns is the estuarine portion of the river, formed at the confluence of the middle St. Johns and the Ocklawaha River, and encompassing a 2750 square mile area (Fig. 1). Within this reach, the St. Johns River is slightly more that 100 miles long and has a water surface area, including tributary mouths below head of tide, of 85,000 acres. Major centers of population within the LSJRB include Palatka, a city of 10,700 at the southern entrance to the basin; Green Cove Springs, a city of 4700 at the midpoint; and the Orange Park, Middleburg, and Jacksonville metropolitan area, with a population of over 1 million, in the northern portion of the basin [2]. The LSJR is a sixth-order, blackwater river estuary, and, along its length, it exhibits characteristics associated with riverine, lake, and estuarine aquatic environments [3].

The LSJR is divided into three ecological zones based on salinity. The three zones are as follows: 1) a predominantly freshwater, tidal lake-like zone that extends from the city of Palatka north to the mouth of Black Creek; 2) an alternately freshwater and marine, oligohaline zone extending from Black Creek northward to the Fuller Warren Bridge in Jacksonville; and 3) a predominantly marine and much narrower zone downstream from I-95 to the mouth [4]. The slow moving, lacustrine nature of the river facilitates phytoplankton primary production, and spring and summer algal blooms in this nutrient-rich river often exhibit chlorophyll *a* concentrations exceeding 100 μg/L.

The southern portion of the lower basin is largely rural, with predominant land uses in forestry and row-crop agriculture. The northern portion of the basin is distinguished by the heavily urbanized cities of Jacksonville, Orange Park and Middleburg. Roughly three quarters (64–82 per cent) of the basin's highly developed land uses (medium and high residential, high-intensity commercial and industrial) drain to the oligohaline and mesohaline lower St. Johns. In contrast, 62 to 98% of the basin's agricultural land uses drain to the fresh tidal reach.

1.1 Water-quality problems

Due to a long history of development within the basin, and its associated water-quality problems dating back to at least 1947 [5], the LSJR was one of the State's originally identified Surface Water Improvement and Management Act [6] water bodies for restoration, and has been identified as a high-priority water body for the

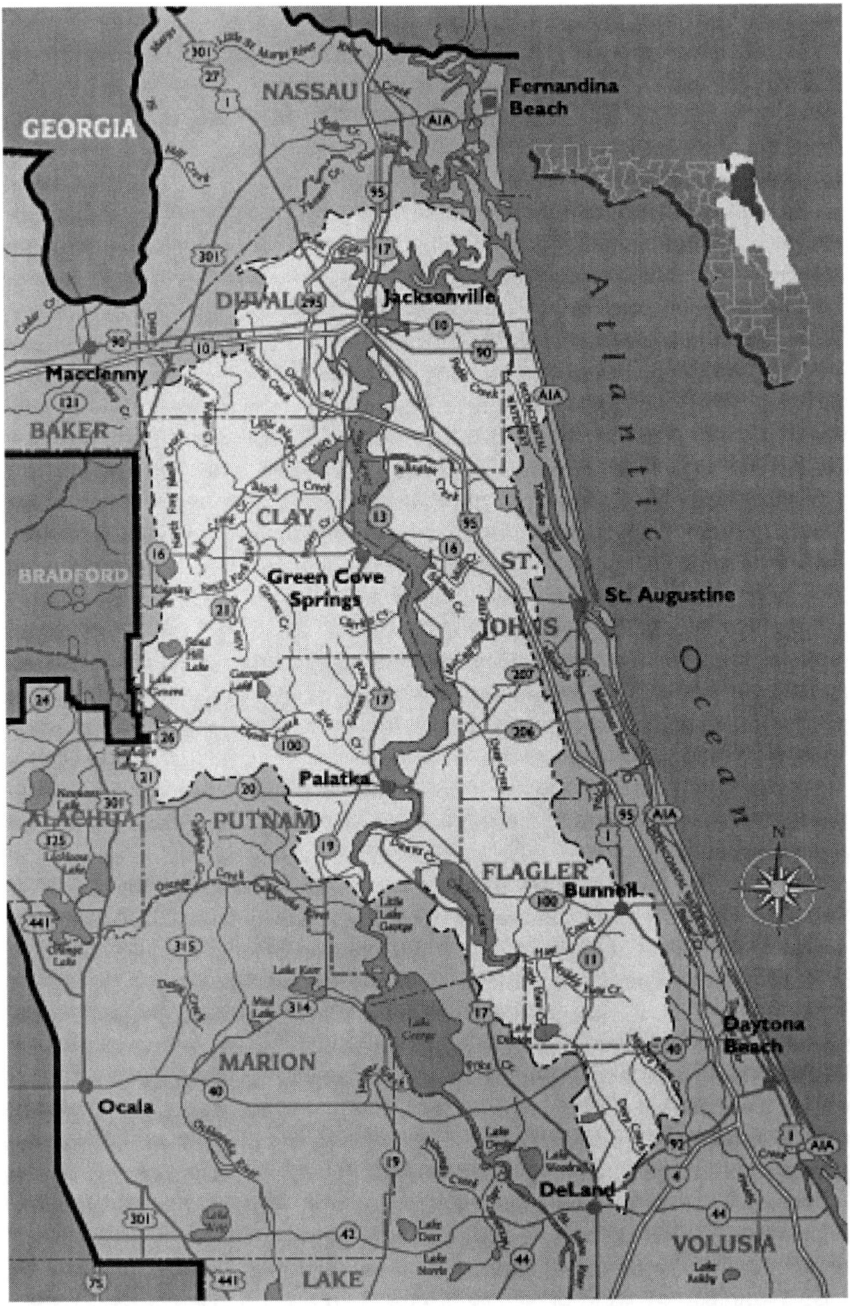

Figure 1: The Lower St. Johns River Basin showing population centers. The river flows through a series of connected lakes from south to the ocean outlet at Jacksonville, FL.

establishment of total maximum daily loads (TMDL) and pollutant-load-reduction goals (PLRG) to limit nutrient pollution.

Comprehensive external nutrient load assessments have been performed twice previously for the LSJR. In 1976, the firm of Atlantis Scientific [7], under authorization of the 1972 Clean Water Act Section 315, undertook a computation of the external load and concluded that point source comprised the majority of this load. Hendrickson and Konwinski [4] also computed the external load to the river for 1993–94, and concluded that nitrogen and phosphorus were 2.5 and 6 times greater than natural background respectively, with augmented nutrient loads (that load above natural background) approximately evenly split between point and nonpoint sources.

Accelerated eutrophication due to nutrient enrichment of estuaries represents one of the most significant water-quality problems faced by near-coastal waters world-wide [8]. Investigation into the effects of nutrient enrichment and elevated levels of algal biomass in the LSJR has, however, revealed noteworthy departures from the eutrophication paradigm established for other estuaries. The LSJR is, first, a blackwater river, with large amounts of allochthonous organic carbon and, at times, substantial light limitation to primary production imparted by colored dissolved organic matter. Also, in most river-mouth estuaries, algal production peaks in the broad, shallow zones that exist in the mesohaline reach, and is typically marine diatom-dominated. In contrast, the LSJR, due to its unique morphology, exhibits maximum algal standing stock in the upper, freshwater and oligohaline reaches. This morphology also results in a very different pattern in the zone of oxygen depletion; the most often cited manifestation of estuarine eutrophication [9]. Instead of the mesohaline zone vertical oxygen stratification typical of most eutrophic estuaries, the LSJR "dead zone" exhibits a stronger longitudinal character, with strong vertical mixing and top to bottom, long duration, and low oxygen levels.

Because of its broad, shallow shape and low flow – long residence time co-occurrence with spring and summer enhanced algal growth – the LSJR can exhibit prodigious amount of seasonal algal productivity and biomass. River flow, tide, and wind currents continually advect, disperse and diffuse algae, nutrients and salinity, with the result that the manifestations of nutrient loads frequently occur far from the source of entry. Changes in nutrient status, salinity, water-column light availability and water temperature bring about changing environments that seldom favor one particular algal group, so a succession of plankton communities is typically observed through the spring and summer growth seasons. Environmental stressors can at times be abrupt enough to lead to mass death in the algal community, leading to a sequence of water-quality characteristics with important implications for nutrient recycling, community change, and water-quality effects.

In addition to the elevated chlorophyll *a* values (algal blooms) and low DO levels, a number of widespread water-quality problems have been identified throughout the river that are indicative of an imbalance in the flora and fauna of the LSJR [10]. These problems include the following: a) fish kills; b) submersed aquatic shoreline vegetation covered in algal mats; c) excessive epiphyte growth further blocking light from submerged aquatic vegetation, d) anecdotal accounts of shoreline vegetation losses and reduced recreational fishing quality; e) river sediment conditions indicative of low benthic animal diversity; f) excessive organic matter sedimentation and prolonged anoxia; and g) the presence of potentially toxic dinoflagellates such as the *Pfiesteria*-like *Crytoperidiniopsoids* [11, 12], and *Prorocentrum minimum* [3], often co-occurring with fish kills or ulcerative disease syndrome in fish. All of these problems are connected by a common thread – they indicate accelerated eutrophication in an estuarine environment.

Numerous other studies have identified either high nutrient concentrations or eutrophic conditions [13–15] in the LSJR. In their assessment of nutrient loads to the LSJR and their potential effects, Hendrickson and Konwinski [4] determined the following: A combination of point- and nonpoint-source pollution has increased within-basin nutrient load to the LSJR 2.5 times over natural background for total nitrogen (TN) and 6 times higher for total phosphorus (TP). The areal nutrient loading is at 9.7 and 2.1 kilograms of nitrogen and phosphorus per hectare of watershed contributing area per year in the LSJRB, is one of the highest reported from studies in the southeastern United States.

It is clear that the LSJR receives high nutrient loads and is nutrient enriched, and that it exhibits the symptoms of estuarine eutrophication. While nutrient enrichment is not the only problem leading to impaired water quality in the LSJR, it is probably the most widespread and multifaceted.

2 Long-term water-quality trends

Few water-quality data exist for the St. Johns River prior to substantial levels of development. It is difficult to ascertain the point at which the St. Johns River Basin became developed to a degree that it exerted a significant effect on water quality. However, the sharp increase in the State's population that began subsequent to World War II appears to correspond to the earliest reports of water quality degradation (Fig. 2). Population density within the Basin upstream of Jacksonville remained relatively low until 1940. Between 1940 and 1950, the population within the basin increased by 39% [16].

By the time of the earliest, regular surface water quality monitoring efforts, beginning in the late 1960s, water quality in the middle St. Johns River appears to have already declined. Moody attributed this decline in Lake George to the occurrence of regular, severe algal blooms [17]. The principle factors leading to these algal blooms were believed to be: 1) upstream development and its concomitant nutrient enrichment through point- and nonpoint-source pollution, and 2) aquatic weed spraying to eliminate floating water hyacinth. There were clearly higher concentrations of blue-green algae in 1967 [17] than reported in an earlier study in 1939–40 (by Pierce [18]). Two water-quality sample events for Lake George, which were collected in July of 1967 and December of 1969 reported that in July, the sample contained 2.28 mg/L TN and 0.13 mg/L TP, while the December sample contained 1.3 mg/L TN and again, 0.13 mg/L TP.

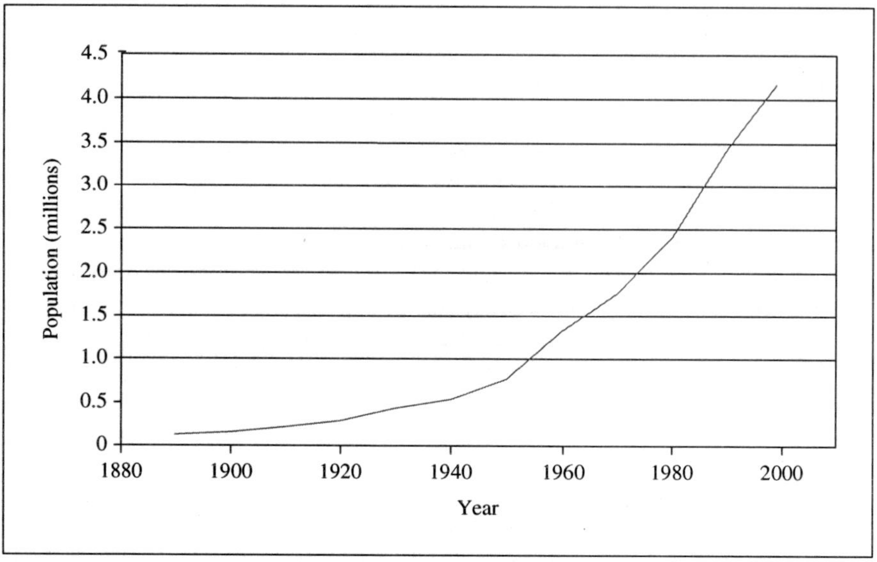

Figure 2: Population growth within the 14 Counties of the St. Johns River Basin, 1890–2000. Data from Dietrich [19] and University of Florida [20].

Three data sets have been identified that can help in understanding the nutrient status and hence open water ecology of the St. Johns prior to substantial development. The first of these is a study conducted by Pierce [18] reporting on several aspects of the water quality and plankton in the St. Johns River at locations in the Ocklawaha River mouth, in the river upstream and downstream of the Ocklawaha from September 1939 to November 1940. The second is the work of Odum [21] characterizing the phosphorus concentrations of waters of the State. The third study was published by the Florida State Board of Health in 1948 to determine the effect of untreated waste discharge in and around the City of Jacksonville.

The objective of the Florida State Board of Health study was to determine the effect of the discharge of untreated sewage in the vicinity of Jacksonville, and sampling stations were established in the St. Johns River near the city of Green Cove Springs to characterize upstream, background conditions. Two surveys were performed, the first conducted in May and June of 1945, and a second conducted from September 1945 to May of 1946. Though nutrient analysis was not performed, biological oxygen demand (BOD) and DO were examined and may be used to infer trophic status. Many of the reports' sampling stations were located in Jacksonville, which at the time was already significantly impacted, principally from the discharge of raw sewage. However, comparing the report's BOD data from the upstream, "un-impacted" site near the Shands Bridge to present-day concentrations at the same location suggests that water-column biodegradable organic matter has increased over time. The May–June survey produced what was referred to as a "highest average" (statistical methods were not explained in the report) of 0.83 mg/L; in comparison, the May–June mean BOD concentration at Shands Bridge from 1996 to 2000 was 2.14 mg/L. The September. 1945 to May 1946 survey produced a highest average concentration of 0.82 mg/L, while the 1996–2000 average is 1.43 mg/L for DO.

Under present-day conditions in the LSJR, algal biomass accounts for the majority of labile organic carbon in the river, and the relationship between BOD and chlorophyll *a* is highly significant, with chlorophyll *a* concentrations explaining over 70% of the variation in BOD (Chl-a = 14.39*(BOD) – 4.11; $R^2 = 0.71$). Based on this relationship, the 0.83 mg/L BOD measured in 1945 corresponds to a chlorophyll *a* of about 8 mg/m^3. Converting the present-day mean river color for this location to refractory organic nitrogen (assumes color has not changed; in reality, river color probably has declined somewhat due to basin development), and adding in the nitrogen content of algal biomass at 8 mg/m^3 chlorophyll *a*, a mean total organic nitrogen content of 0.48 mg/L can be calculated. With the inclusion of inorganic nitrogen, it would be expected from these BOD data that total nitrogen was in the neighborhood of 0.6 mg/L, comparable to the reconstructed historic Buffalo Bluff mean TN concentration of 0.687 mg/L.

The Odum [21] report to the Florida Geological Survey extensively surveyed orthophosphate and TP in surface waters around the State of Florida. Samples were collected at one time from many different locations, so annual trends cannot be inferred. These data for locations in the St. Johns River and its contributing streams are listed in Table 1. These data must be viewed selectively for the potential of anthropogenic nutrient contamination. For locations that likely still represented unimpacted reaches of the lower St. Johns River in 1952 (Lake George and Crescent Lake), these data suggest a concentration of TP of around 0.04 mg/L. Upstream of Lake Monroe (Brevard and Orange counties), the Odum data suggests a St. Johns River that was remarkably low in phosphorus.

Table 1: Total phosphorus concentrations determined for selected locations in St. Johns River Basin in 1952. Data from Odum [21].

Location	Date	TP, mg/L
Black Creek, Route 17	Aug. 9, 1952	0.04
Deep Creek, Hastings, Route 207	Jul. 14, 1952	0.54
Crescent Lake, Andalusia	Jul. 19, 1952	0.033
Doctor's Lake, Route 17	Aug. 9, 1952	0.065
Lake George at Silver Glen Springs	Aug. 14, 1952	0.044
Lake Monroe, Sanford	Jun. 23, 1952	0.18
Ortega River, Route 21	Aug. 9, 1952	0.044
St. Johns R., Crows Bluff, Volusia Co.	Sep. 3, 1952	0.117
St. Johns R., Palatka	Jul. 19, 1952	0.061
St. Johns R., Route 192 (Brevard Co.)	Jun. 23, 1952	0.007
St. Johns R., Route 50 (Orange Co.)	Jun. 23, 1952	0.015
St. Johns R., Green Cove Springs	Jul. 16, 1952	0.119

The data of Pierce [18], due to the length of his study and the comparatively large suite of measurements, provide compelling evidence of a river that was dramatically lower in nutrients and algal biomass. In 1939–40, Pierce reported blue-green algae (of the genera *Anabaena*, *Raphidiopsis* and *Microcystis*) ranging from too few to count for most months, to 36,000 cells/ml in August of 1940. In comparison, the annual mean peak blue-green cell count exiting Lake George for 1997–2000 [22] was significantly higher at 518,893 cells/ml. The Pierce [18] study also suggests a shift in the dominance of phytoplankton groups, with diatoms (primarily the genera *Coscinodiscus* and *Melosira*) formerly making up a much greater relative portion of the plankton.

The Pierce [18] study also provides data on nitrogen forms throughout the year (analysis for phosphorus was not performed) Due to some differences in methodology and uncertainties regarding sample handling and preservation techniques, only TN concentrations are compared. Pierce reported mean annual TN as 0.41 mg/L in Little Lake George (upstream of the Ocklawaha mouth), and 0.37 mg/L at Welaka (downstream of the Ocklawaha), values that are roughly one quarter of present-day concentrations. The Pierce TN numbers are similar to the present day estimated mean concentration refractory total organic nitrogen at Buffalo Bluff, of 0.46 mg/L.

Unaccounted for in the water-column measurements of these studies is the sequestration of nutrient in water hyacinth *(Eichornia crassipes).* Water hyacinth, introduced to the St. Johns shortly before 1900, quickly spread through the river, and anecdotal accounts prior to 1940 indicate widespread coverage. Annual reports to Congress on the progress of hyacinth control in the St. Johns [23] indicate that between 3000 to 13,000 acres of hyacinth were removed annually, suggesting that at least 5 to 10% (based on the sum of lake surface areas of the St. Johns from Lake Winder through Little Lake George) of the rivers water surface area typically may have been covered.

These studies suggest substantially lower water-column nitrogen and algal biomass, and marginally lower phosphorus concentrations prevailed in the predevelopment St. Johns River. If the concentrations of TN of 0.6 mg/L and TP of 0.04 mg/L are accepted as representative of water column (including floating macrophytes) natural background levels, then nutrients flowing into the lower St. Johns River from the upper and middle St. Johns and Ocklawaha Rivers today appear to be elevated between 1.5 to 4 times above predevelopment conditions.

Persistent, low concentrations of dissolved oxygen in the meso/polyhaline reach of the LSJR are a well-documented but poorly understood phenomena. First studied intensively in the 1950s, these occurrences were attributed to the discharge of untreated domestic sewage and industrial waste [24]. Despite large improvements in point-source treatment, the occurrence and severity of these episodes has remained unchanged over time.

3 Nutrient sources within the Basin

3.1 Point sources

Domestic wastewater facilities that discharge to surface waters are concentrated along the St. Johns River from Green Cove Springs to its mouth north of Jacksonville, and farther south near Palatka (Fig. 1). The largest domestic treated wastewater dischargers in the basin are the wastewater treatment facilities associated with the city of Jacksonville in the northern (downstream) end of the basin, including the Buckman Street, Arlington East, JEA District II, Southwest District, and Mandarin wastewater treatment facilities. Several of these facilities participate in reuse programs, and most are actively seeking ways to either include or improve nutrient-removal treatment [4, 25].

All domestic wastewater facilities discharging to the St. Johns River are required, at a minimum, to monitor for conventional pollutants such as total suspended solids (TSS), carbonaceous biological oxygen demand ($CBOD_5$), and fecal coliforms bacteria [25]. While most permits do not include nutrient effluent limits, nutrients must be monitored in many systems because of their potential negative effects on surface water, including their role in the formation of nuisance and harmful algal blooms.

Large industrial dischargers in the basin include power plants, pulp and paper mills, chemical plants, and manufacturing plants. The majority of industrial plants send their process wastewater through pretreatment facilities to publicly owned treatment works (POTWs). Facilities with significant nutrient discharges to the main stem of the LSJR include the Georgia-Pacific Corporation (which produces bleached and unbleached pulp and paper), Stone Container (which changed from a pulp and paper mill to a recycling mill in the 1990s, reducing the volume of discharge), and Anheuser-Busch (a brewery). Remaining discharges include nonprocess wastewater such as cooling water, softener regenerate, and boiler blowdowns, which do not contribute a significant nutrient load.

3.2 Nonpoint sources

Nonpoint sources of nutrient loading to the LSJR include septic tanks, marinas, silviculture, row-crop agriculture, dairies, stormwater from urban development and tributaries (including Black Creek, Dunns Creek, Deep Creek, Rice Creek, Julington Creek, Trout Creek, Sixmile Creek, Governors Creek, Clarkes Creek, Cedar Creek, Camp Branch, Mill Branch, and Dog Branch). Unlike traditional point-source effluent loads, nonpoint-source loads enter at so many locations and exhibit such large temporal variation that a direct monitoring approach is infeasible, except for the largest, most significant inputs. Those largest inputs are the upstream boundary of the LSJR at Buffalo Bluff, Dunns Creek, and the downstream boundary at the Atlantic Ocean.

4 Population trends

Between 1940 and 1950, the population within the basin increased by 39% to approximately 0.75 million people [16] (Fig. 2). Since 1950, the population has increased by an average of 0.7 million per decade rising to the year 2000 total of approximately 4.2 million. Such a dramatic increase in population is going to invariably lead to changes in land use as well as increased nutrient runoff associated with point and nonpoint discharges.

5 Land use and effects on water quality

Since 1973, there has been a pattern of shifting land use from more rural to more urban land use (Fig. 3). There has been a steady decline in forested/undeveloped land which comprised the majority of the land use in the basin at 73%. By 2005, this land-use category has declined to 59% of the basin in 1973. The land-use categories of both urban and agricultural lands made up similar percentages (13–14%) (Fig. 3). However, due to the increase in population, the urban land use has increased dramatically and comprised 32% in 2005. Approximately 5% of the land was conversion from agriculture to urban land use with the remainder coming from the forested/undeveloped category. In tracking land-use changes based on the 1973 survey, we can see that% urban has steadily increased to 120% more with no indications that this conversion has peaked (Fig. 4). It is likely that this trend will continue.

Land development influences the delivery of water-quality constituents to surface waters in two fundamental ways. Through fertilization, lawn maintenance, manure spreading, septic-tank operation, vehicular use, etc., nutrients and other pollutants are added to the land surface or to shallow groundwater in excess of natural land-cover conditions (i.e. native forest, wetland). Unlike the situation that tends to predominate on developed lands, natural land covers are highly conservative of

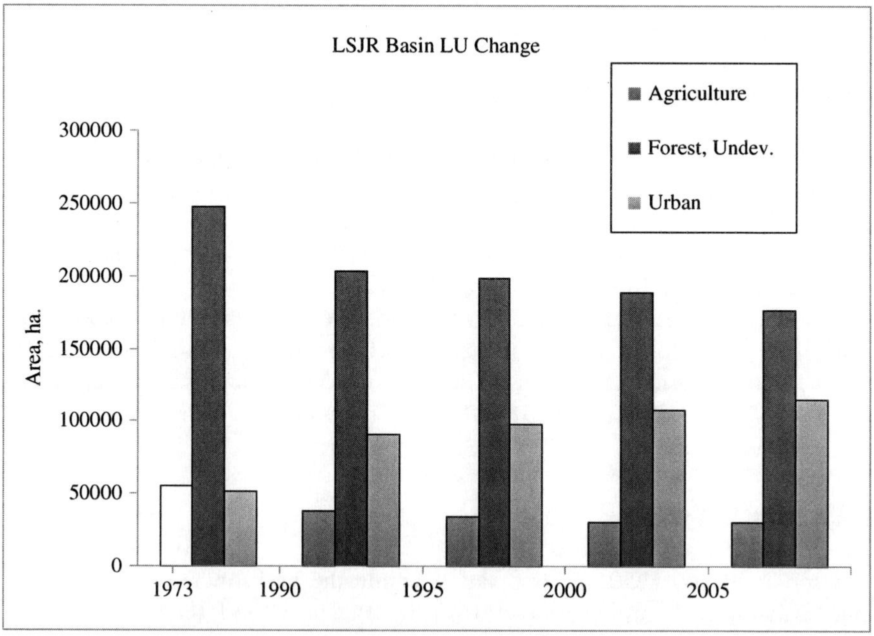

Figure 3: Land use (LU) changes from 1973–2005 for the Lower St. Johns River Estuary from 1973–2005.

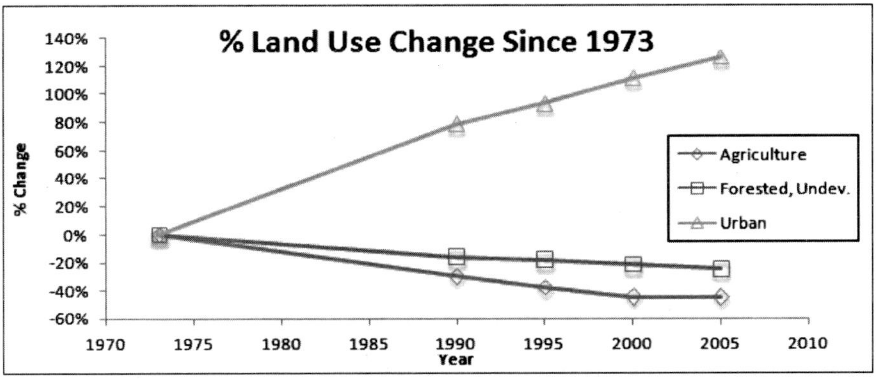

Figure 4: Percentage land-use change for agriculture, urban and forested/undeveloped. Percentage increase is based on the 1973 land-use assessment.

essential growth nutrients, and thus labile nutrient forms tend to be retained within these terrestrial ecosystems. In addition, the creation of impervious surfaces, drainage development, and the destruction of near-stream wetlands increases the amount of rainfall that ultimately ends up as runoff, thus increasing the pollutant-exporting capability in developed landscapes. Thus, the process of nonpoint-source pollution has both chemical and hydrologic components.

The effects of changing land use on both total and bioavailable nutrients is best demonstrated in a comparison of TN, which includes total inorganic N, labile organic N and refractory organic N components. Consider the basin comparison between natural forested land and mixed urban/commercial/residential watersheds. The difference may be a concentration increase of 214% from natural to urban land use (Fig. 5). However, the relative proportion of the total N has also changed such that when you pool the bioavailable pools (total inorganic + labile organic N), you have a net increase of 663% in the bioavailable pool [26]. This pool of N is what leads to expression of eutrophic conditions in coastal areas. Due to the higher relative amounts of labile nutrients in developed landscapes, deleterious nutrient load often exceeds that which would be inferred by absolute increases in nutrients alone.

6 Determination of a nitrogen and phosphorus nutrient budget

There are several components of the nutrient load that need to be assessed to determine the overall nutrient budget (the basin point-source load, the basin nonpoint-source load, and the upstream nutrient load). This is a vital first step in any eutrophic coastal area, especially for water management to determine the best course of action in mitigating nutrient loads. Once the source loads are identified, there will need to be a determination as to which loads can be most readily decreased in a cost-effective manner to yield the greatest benefit on the resultant water quality.

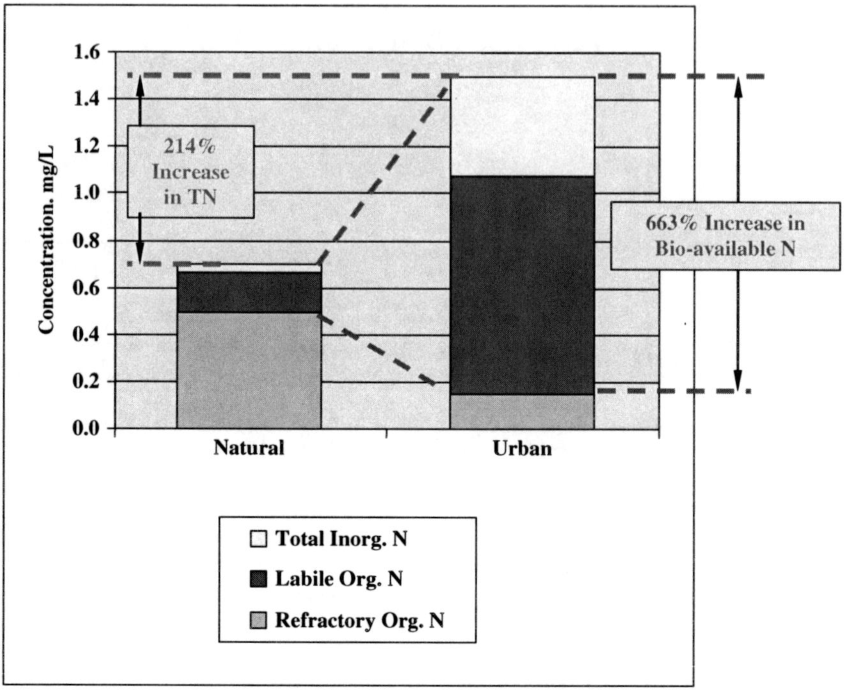

Figure 5: Comparison of total and bioavailable nitrogen forms in runoff from natural forested and mixed urban/commercial/residential watersheds.

While nitrogen is the primary nutrient of concern in coastal eutrophication, the morphology and hydrodynamics of the Lower St. Johns River estuary are such that there are regions of saline, fluctuating saline-oligohaline, and tidal-influenced freshwater zones. Therefore, phosphorus, which is generally limiting in freshwater systems, is likely an important contributor to the eutrophication status of the estuary and should therefore be accounted for in the nutrient budget.

6.1 Point sources

For the Lower St. Johns River Estuary, the point-source load estimation was performed using six separate data sets to gain available information on concentration, flow, point of discharge and service area. These data sets included 1) hard copy monthly operating report files maintained at the Florida Department of Environmental Protection (FDEP) Northeast District Office; 2) National Pollutant Discharge Elimination System (NPDES) electronic files obtained from FDEP Tallahassee; 3) Discharge quality data maintained by the Jacksonville Electric Authority; 4) Fifth-year synoptic surveys performed by FDEP or by contractor as part of permit renewal process; 5) a special 2-year sampling program conducted jointly by FDEP and Saint Johns River Water Management District (SJRWMD), and 6) a GIS data base of locational information.

6.2 Nonpoint sources

Unlike point-source effluent loads, nonpoint-source loads enter at so many locations and exhibit such large temporal variation that a direct monitoring approach is infeasible except for the largest, most significant inputs. At all other nonpoint entry points, statistical watershed modeling was relied upon to complete the external load budget. The watershed modeling approach used utilizes the relationship between land-use development and alteration in water quality and quantity to perform a spatial extrapolation of whole-basin nonpoint-source load. The formulation of this statistical model has its roots in the spreadsheet watershed load screening model, referred to as the pollution-load-screening model, which utilizes a computer-driven geographic information system framework to calculate constituent loads as the product of water-quality concentration associated with certain land-use practices, and runoff water volume associated with those same practices. The model's nonpoint-source pollutant export concentrations are specific to one of 15 different land-use classes. Water quantity is determined through a hybrid of the Soil Conservation Service (SCS) curve number method, and is the product of rain volumes and a coefficient (referred to as the runoff coefficient, or RC, with values ranging from 0 to 0.9) relating the propensity of various land-use and soil hydrologic group combinations to generate runoff. The computational approach of the PLSM is similar to that of the surface-water-management model (SWMM) screening level tool.

6.3 Upstream load

The upstream load is composed of the three large tributaries that make up the lower St. Johns: the middle St. Johns River, the Ocklawaha River, and Dunns Creek. These three tributaries make up approximately 61, 21, and 18 per cent of the long-term annual mean river discharge at Palatka. Because of autochthonous production in upstream lakes, the upstream load differs greatly from watershed loads that enter within the LSJR basin.

Because of the large amount of the entire LSJR flow that enters upstream (roughly 60%), a direct monitoring approach, featuring continuous measurement of discharge, and biweekly collection of water-quality samples, has been used to determine its constituent load. This is in contrast to the aforementioned watershed modeling approach that has been used to develop the downstream tributary load.

6.4 Nutrient budget

Compiling all available historical data and results of recent nutrient assessment, we can begin to understand the Lower St. Johns River Estuary nitrogen and phosphorus

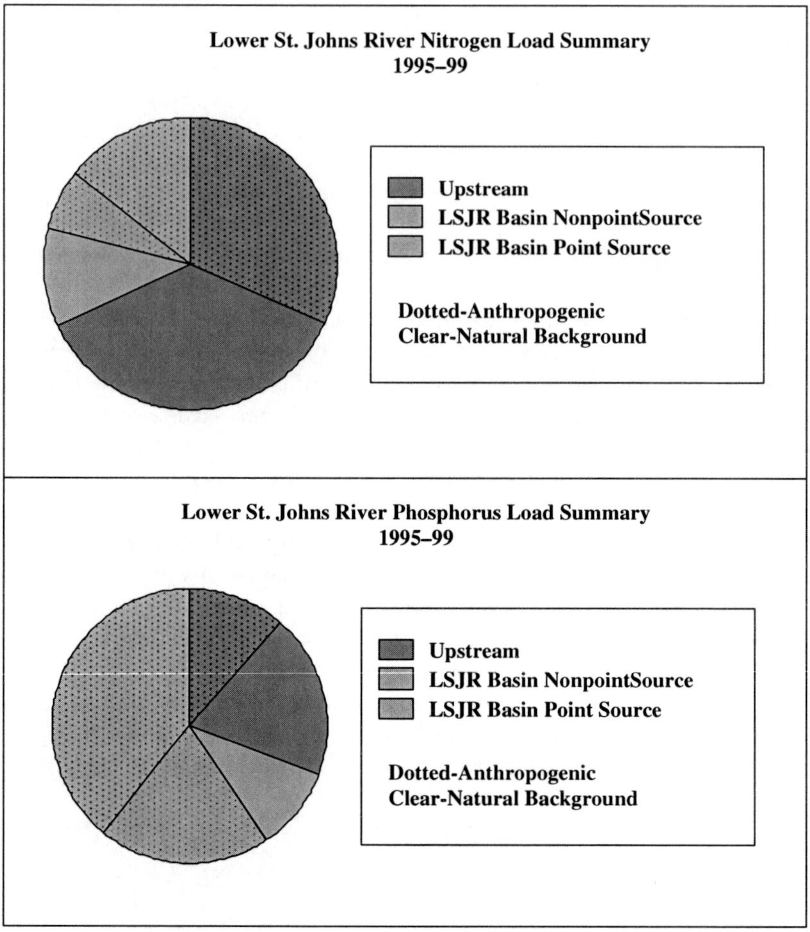

Figure 6: Nitrogen and phosphorus budgets for the Lower St. Johns River Estuary including the within basin point- and nonpoint-source loads and the upstream contributions.

budgets. For nitrogen, the largest component enters from upstream at 69% (Fig. 6). The within-basin contributions are split between point source (24%) and nonpoint-source (17%). From modeling efforts and historical data sets, we have been able to reconstruct the historical N loads. It is estimated that 50% of the upstream load is anthropogenic, due to human affects on the system. All the point-source additions are anthropogenic and approximately 55% of the nonpoint-source is anthropogenic. Consequently, 58% of the total nitrogen load to the Lower St. Johns River estuary is due to the additive effect of the human population and associated land-use changes to support that population.

The phosphorus budget has a different overall proportion when compared to the nitrogen budget. Only 31% of the phosphorus budget enters the basin from upstream. Another 30% includes the nonpoint-source contribution and the final 39% makes up the point-source discharges (Fig. 6). Clearly, reducing phosphorus concentrations may prove to be an easier task as municipalities generally have greater control and restrictions on point-source pollution.

The anthropogenic component of the phosphorus load includes 100% of the point sources, 33% of the nonpoint-source additions and 39% of the upstream load. Overall, the population and land-use changes have more than doubled the phosphorus load to the Lower St. John River estuary.

6.5 The internal or sediment load

The sediments of the Lower St. Johns River Estuary have become enriched with N and P as organic material associated with repeated algal blooms has settled to the bottom. Consequently, over time, there will be a release of N and P through mineralization or decomposition of this organic material that will continue to be a source of bioavailable nutrients to the water column. In a recent study, it was determined that the potential for sediment released NH_4-N was 2066 Mg/yr and 330 Mg/yr for dissolved reactive P [27]. Based on previously published nutrient load calculations [4], the sediment can potentially produce 28% of the total N loading and 21% of the total P loading from the watershed. Since this internal load is entirely in the bioavailable form, this internal load might exert an even greater influence on the nutrient dynamics of the estuary over time.

7 Effects of oceanic dilution on water quality

There have been numerous initiatives over time to counteract the changes in land use, high nutrient concentrations and the eventual expressions of eutrophication in the Lower St. John River Estuary including tighter permit restrictions on point source for both nitrogen and phosphorus and best management practices for agricultural lands in order to intercept nutrient runoff. When examining data on the total N and P of the estuary at Jacksonville, Fl, we observe no discernable increase in either N or P over the past 20 years, suggesting that initiatives to reduce loading have been successful in spite of the ever-present changes in land use and increases in population over that time (Fig. 7). The data show dramatic variability due primarily to the episodic nature of the rainfall patterns, where approximately 75% of the precipitation falls within about 5 months, which will affect both the upstream and nonpoint-source nutrient loads.

However, evidence suggests that this period of record is a particularly dry one with precipitation well below normal. Consequently, as an estuary, the ocean can exert significant influence due to the higher tidal amplitude and the very slow slope of the land and river. With low freshwater flows, a greater dilution by seawater can

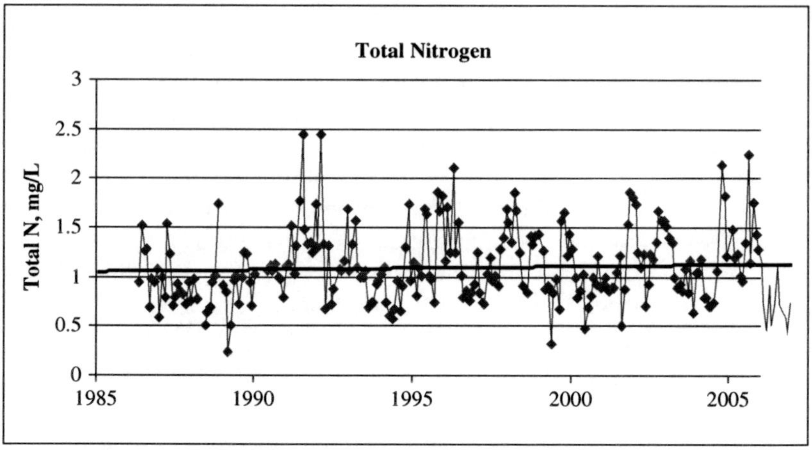

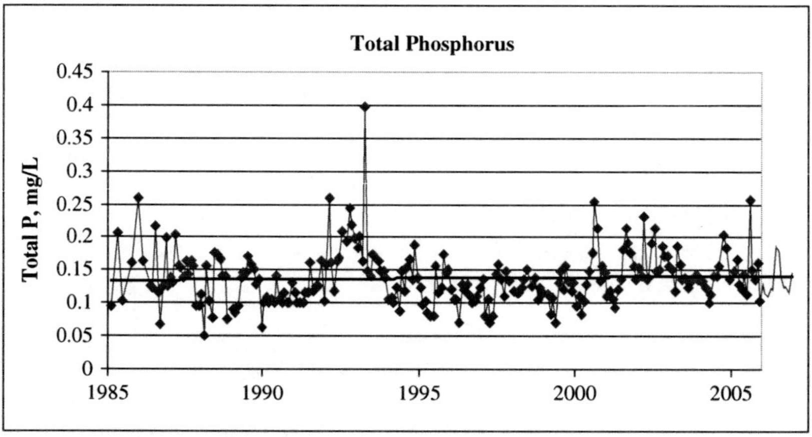

Figure 7: Total nitrogen and phosphorus for water samples collected at Jacksonville, Florida from the Lower St. Johns River estuary.

occur that can confound interpretation of the effectiveness of nutrient mitigation protocols on the nutrient status of the river.

In order to investigate this particular possibility, we have attempted to determine the effect of dilution of the river water by seawater over this 20-year timeframe. Using a fairly stable ocean concentration of N and P of the oceanic end member of the water and determining the amount of mixing within the estuary based on the salinity over time, we have been able to determine the concentration of the freshwater runoff from the basin into the river. In fact, when we subtract out the ocean mixing end member and plot the predicted freshwater end member for both N and P, we can see a more predictable pattern of increasing concentrations associated with the aforementioned changes in population and land use (Fig. 8). This trend suggests

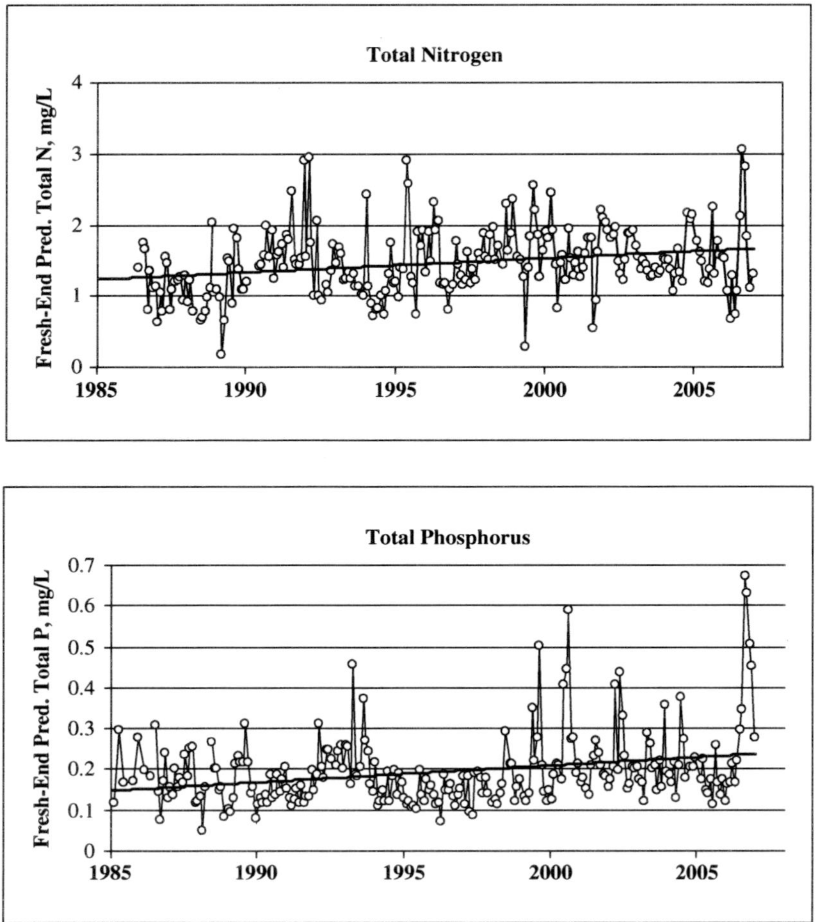

Figure 8: Calculated freshwater total nitrogen and phosphorus concentration for water samples collected at Jacksonville, Florida from the Lower St. Johns River estuary. The *y*-axis is the freshwater end-member predicted concentration for the watershed point- and nonpoint-source contributions subtracting out for dilution by the ocean tidal pulses and upstream contributions.

that despite best efforts to reduce nutrient loading to the estuary, the N and P concentrations of the runoff continues to increase.

8 Conclusions

The ever-increasing population in the coastal zone has had a marked effect on estuarine water quality around the globe. A concentration of people and associated land-use changes lead to anthropogenic point- and nonpoint-source discharges of nutrient into adjacent surface waters. These aquatic systems respond very quickly to these nutrient additions with similar expressions of eutrophication found in the St. Johns River Estuary including, low DO, increased algal blooms and associated fishkills.

In order to restore water quality back to some preimpact level, quantifying the contributions of the various nutrient sources in the coastal zone is an important step. This can be a difficult, time-consuming and expensive process. However, it is a critical first step in attempting to reverse the effects of eutrophication. Once both the source and forms of nutrients are known, efforts can be focused on those sources over which we have the greatest control and can make the largest impact. Specific land uses can be managed to reduce the increasingly bioavailable nutrient load. Additionally, as the population continues to increase and land use is impacted in the coastal zone in the future, the ability to identify and target specific nutrient loads may prove valuable in maintaining or improving water quality in the face of significant population pressure.

References

[1] Toth, D.J., Volume 1 of the Lower St. Johns River Basin Reconnaissance: Hydrogeology. Technical Report SJ93-7. *St. Johns River Water Management District,* Palatka, FL, 1993.

[2] Floyd, S.S., Irwin, E.M. & Evans, D.A., *Florida statistical abstract. University of Florida, Bureau of Economic and Business Research*, University Presses of Florida: Gainesville, 1997.

[3] Phlips, E.J., Cichra, M., Aldridge, F.J., Jembeck, J., Hendrickson, J. & Brody, R., Light availability and variations in phytoplankton standing crops in a nutrient-rich blackwater river. *Limnology and Oceanography,* **45(4),** pp. 916–929, 2000.

[4] Hendrickson, J. & Konwinski, J., Seasonal nutrient import-export budgets for the Lower St. Johns River, Florida. *St. Johns River Water Management District,* Palatka, FL, 1998.

[5] Florida State Board of Health, Report on the St. Johns River pollution survey. Bureau of Sanitary Engineering, Jacksonville, FL, pp. 59, 1948.

[6] Florida Statues, Title XXVIII, Natural Resources, Conservation, Reclamation, and Use. Chapter 373.451, *Water Resources*, 2007.

[7] Atlantis Scientific, Environmental Impact Assessment, Water Quality Analysis, St. Johns River (Florida). Prepared for the National Commission on Water Quality, NCWQ 75/81, 1976.

[8] Council, N.R., Clean coastal waters: Understanding and reducing the effects of nutrient pollution, National Academy Press: 2000.

[9] Officer, C.B., Biggs, R.W., Taft, J.L., Cronin, L.E., Tyler, M.A. & Boynton, W.R., Chesapeake bay anoxia: origin, development and significance. *Sciences,* **223**, pp. 22–27, 1984.

[10] Florida Department of Environmental Protection, Basin Status Report for the Lower St. Johns River Basin. Bureau of Watershed Management, 2002.

[11] Burkholder, J.M. & Glasgow, H.B., *Pfiesteria piscida* and other Pfiesteria-like dinoflagellates: Behavior, impact and environmental controls. *Limnology and Oceanography,* **42(5, part 2),** pp. 1052–1075, 1997.

[12] Burkholder, J.M. & Glasgow, H.B., Trophic controls on stage transformation of a toxic ambush-predator dinoflagellates. *Journal of Eukaryote Microbiology,* **44**, pp. 200–205, 1997.

[13] Bricker, S.B., Clement, C.G., Pirhalla, D.E., Orlando, S.P. & Farrow, D.R.G., National Estuarine Eutrophication Assessment: Effects of nutrient enrichment in the nation's estuaries. National Ocean Service, National Oceanic and Atmospheric Administrations, pp. 71, 1999.

[14] EPA, Nutrient Criteria Technical Guidance Manual: Estuarine and Coastal Marine Waters. Office of Water, EPA 822-B-01-033, 2001.

[15] Janicki, A. & Morrison, G., Developing a Trophic State Index for Florida's estuaries. Submitted to the Florida Department of Environmental Protection, Janicke Environmental Inc. (St. Petersburg, FL), 2000.

[16] Health, F.S.B.O., One River, One Plan – St. Johns River Basin: Report on water pollution control. Water Pollution Series #27. Tallahassee, FL, 1951.

[17] Moody, H., Factors in the decline of the fishery of the St. Johns River, Florida. Florida Game and Fresh Water Fish Commission Technical Report, 1970.

[18] Pierce, E.L., An annual cycle of plankton and chemistry of four aquatic habitats in Northern Florida. University of Florida Biological Science Series, Gainesville, *FL*, pp. 67, 1947.

[19] Dietrich, T.S., The urbanization of Florida's population: A historical perspective on county growth 1830–1970. Bureau of Economic and Business Research, University of Florida, Gainesville, FL, 1978.

[20] University of Florida. Florida estimates of population. Bureau of Economic and Business Research, University of Florida, Gainesville, FL, 2000.

[21] Odum, H.T., Dissolved phosphorus in Florida waters. Report to the Florida Geological Survey. Report of Investigations, No. 9. *Tallahassee, FL,*1953.

[22] Phlips, E.J. & Cichra, M.F., Plankton communities of the Lower St. Johns River. Annual Report to the St. Johns River Water Management District, Contract #97W165. University of Florida Department of Fisheries and Aquatic Sciences, Gainesville, FL., 2001.

[23] United State Army Corps of Engineers, Annual Reports to Congress. 1930–1967.

[24] Pyatt, E.E., On determining the pollutant distribution in tidal estuaries. *U.S. Geological Survey Water Supply Paper 1586-F. U.S. Government Printing Office*, Washington. DC, pp. 58, 1964.

[25] Florida Department of Environmental Protection. Health of the River: A survey of factors affecting water quality in the lower St. Johns River Basin. Prepared by the St. Johns River Task Force, Florida Department of Environmental Protection, Northeast District. Jacksonville, FL, 1997.

[26] Hendrickson, J.C., Trahan, N., Gordon, E. & Ouyang, Y. Estimating the relevance of organic carbon, nitrogen and phosphorus loads to a blackwater river estuary. *Journal of American Water Resources Association* **43(1),** pp. 264–279, 2007.

[27] Malecki, L.M., White, J.R. & Reddy, K.R., Nitrogen and Phosphorus Flux Rates from Sediments in the Lower St. Johns River Estuary. *Journal of Environmental Quality*, **33,** pp. 1545–1555, 2004.

CHAPTER 8

Effects of land-use changes and groundwater pumping on saltwater intrusion in coastal watersheds

Ahmet Dogan[1] & Ali Fares[2]
[1]*Department of Civil Engineering, Süleyman Demirel University, Isparta, Turkey.*
[2]*College of Tropical Agriculture and Human Resources, University of Hawaii-Manoa, Honolulu, HI, USA.*

Abstract

This chapter describes the occurrence of saltwater/freshwater interface in coastal aquifers. It has a brief description of the physics, hydraulics, and mathematics behind the theory of saltwater intrusion and the saltwater/freshwater interface. It also includes a section on numerical modeling techniques and the available computer models for saltwater-intrusion problems in coastal aquifers. Anthropogenic effects on saltwater intrusion such as changes in land use and landscape vegetation cover, groundwater pumping patterns, and the amount of pumping are also discussed. Finally, remedies, control and management of saltwater intrusion such as implementation of recharge wells, creation of artificial recharge basins, and construction of barriers are discussed. This chapter is expected to be useful to agricultural communities of coastal plains whose irrigation water is solely dependent on groundwater resources. It may also be of interest to individuals working in coastal-watershed and land-management businesses and governmental or local decision makers by addressing the problem of saltwater intrusion into coastal aquifers and its possible remediation and control techniques.

1 Introduction

Coastal aquifers are the main freshwater water supply sources for most urbanized coastal areas around the world. These aquifers are highly sensitive to disturbances such as land-use changes and groundwater pumping. Inappropriate management of these aquifers may limit their use as sources of freshwater due to saltwater intrusion. Saltwater intrusion and water-quality degradation have become two of the major constraints in groundwater management problems of coastal aquifers. Saltwater intrusion degrades water quality of production wells and consequently the wells have to be abandoned. To avoid saltwater-intrusion problems, adequate natural recharge of these aquifers should be maintained. Pumping rates and production-well locations are the two key parameters that can be optimized for adequate groundwater management.

Under a natural condition, pressure balance exists on both sides of the saltwater/freshwater interface. This natural condition turns into an environmental problem if this pressure balance is changed in favor of saltwater. This pressure imbalance is mainly the results of two major human activities: excessive groundwater pumping; and reduction of groundwater recharge as a result of land-use/land-cover changes. Long-term droughts and the subsequent reduction of recharge will also cause saltwater intrusion. During groundwater pumping from a coastal aquifer, freshwater levels drop; consequently the saltwater intrusion can move further inland toward the pumping wells. Eventually, the saltwater degrades the water quality of drinking and irrigation waters. To prevent saltwater intrusion, more and more countries adopt extensive monitoring schemes and numerical models to assess sustainable groundwater use that will protect their coastal groundwater resources.

This chapter presents our current knowledge of the saltwater-intrusion process in the coastal aquifers. The basic physical theory behind the saltwater intrusion, various conceptual modeling approaches, and preventive measures against the saltwater-intrusion problems have been discussed by focusing on the environmental and ecological concerns in coastal aquifers. Several numerical models and their application studies have also been discussed. Numerous studies have shown the effectiveness of numerical and optimization models in simulating the saltwater intrusion and the measures taken against saltwater intrusion before field implementation of costly measures. This chapter focuses on anthropogenic effects such as changes in land use and groundwater pumping and natural effects such as tides on saltwater intrusion in a cause-and-effect relationship context.

2 Concept of saltwater intrusion in coastal aquifers

All coastal aquifers are subject to saltwater intrusion, which is defined as flowing of saltwater from the sea toward inland freshwater aquifers by pushing the salt water/freshwater interface landward and/or upward. The density difference between freshwater and saltwater causes this inland flow from high-density seawater to low-density freshwater. If a denser fluid column coincides with less-dense fluid column, the former starts flowing towards the latter column because of the pressure difference between the two fluid columns until a new pressure balance is achieved. Pressure at the bottom of a water column is the product of specific weight and water height. If these water columns were connected at the bottom, then the saltwater column starts flowing towards the freshwater column because the pressure beneath the saltwater column is larger than that of the freshwater column of the same height. This phenomenon precisely explains the physics behind the natural saltwater intrusion in coastal aquifers. Naturally farther inland, the higher freshwater levels trigger flow of freshwater seaward, due to its higher potential. These two reverse flow types, i.e. from higher density to lower density and from higher potential to lower potential, complete the picture of the saltwater-intrusion process. At a coastal boundary, freshwater flows towards sea in the upper part of the coastal aquifer while saltwater flows toward inland in the lower part of the coastal aquifer. As a result, a freshwater lens forms on the top of the saltwater zone in the form of cone shape or a wedge towards inland (Fig. 1).

In coastal aquifers, the freshwater body overlies the saltwater body because the unit weight of freshwater is less than that of saltwater. A boundary surface exists between fresh and saltwater bodies known as the "saltwater/freshwater interface" The thickness of this interface is relatively constant along which freshwater and saltwater are mixed by hydrodynamic dispersion, which results from molecular diffusion and mechanical dispersion. This boundary is a transitional zone of varying salinity (Fig. 1). In aquifers, groundwater flow is dominantly laminar, thus the thickness of the saltwater/freshwater interface is small compared to the thickness of the aquifer. Therefore, in some analyses, an abrupt, sharp, well-defined, distinct interface in the aquifer cross section is assumed. Two fluids on either side of this sharp interface are assumed to be immiscible. If a coastal aquifer consists of two or more distinct aquifers, each aquifer will have its own interface. This interface is assumed to be static and, practically, there is no flow across the interface due to pressure balance on both its sides. Under these conditions, a steady-state seaward freshwater flow occurs discharging freshwater to the sea as a result of a constant hydraulic gradient. In unconfined aquifers, freshwater discharges to the sea across the sea floor. However, in confined aquifers, freshwater flows out to the sea by

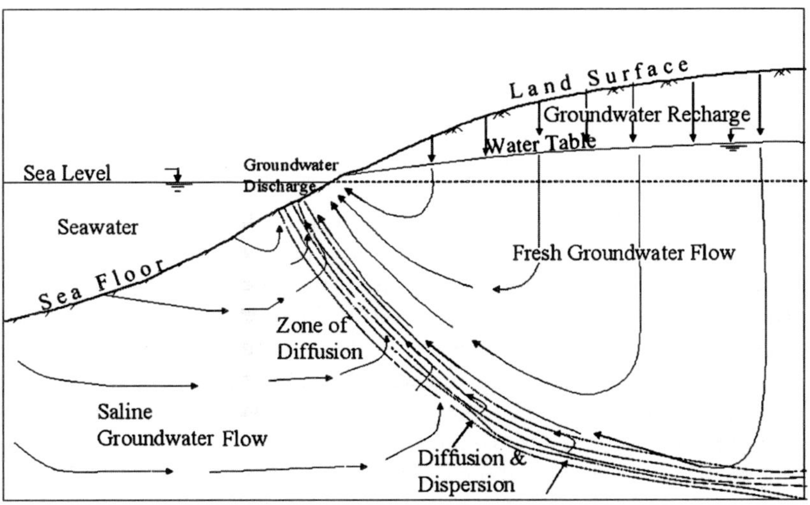

Figure 1: A schematic view of saltwater intrusion.

upward leakage through an overlying layer or by flowing to the sea directly. A wedge-shaped denser body of saltwater will develop beneath the lighter freshwater in each aquifer forming an interface between freshwater and saltwater.

Anthropogenic effects such as excessive groundwater withdrawals and/or severe reduction of aquifer recharge, distort the saltwater/freshwater interface into irregular patterns and sizes depending on location, manner and degree of those effects. Although groundwater pumping is the main cause of saltwater intrusion along the coasts, lowering of water table by drainage canals also leads to saltwater intrusion [1]. Generally, the shape and extent of saltwater-intruded zones depend on several factors such as the magnitudes of freshwater flow rates from the aquifer to the sea, the total rate of groundwater withdrawal compared to total freshwater recharge to the aquifer, the distance of pumping wells and drainage canals from the coast, rainfall intensities and frequencies, land-cover type over natural recharge areas, rates of evaporation, physical characteristics of aquifer materials, presence of confining units that may prevent saltwater from moving toward or within the aquifer, and tidal effects.

In addition to lateral intrusion from the sea, saltwater intrusion can occur through several processes including upward intrusion from deeper, more saline zones of an aquifer system, and by downward intrusion from brackish surface waters. "Saltwater encroachment" and "saltwater intrusion" have been used in the literature to refer to lateral and vertical movements of saltwater, respectively [2]. Another specific term "saltwater upconing" was also used to describe the movement of saltwater from a deeper saltwater zone upward into the freshwater zone in response to pumping at a well or well field [2].

Saltwater intrusion is a contamination source of freshwater resources when concentrations of dissolved solids exceed drinking and/or irrigation water standards. The timing and degree of saltwater intrusion varies widely depending upon the hydrogeological setting of aquifers. The time required for saltwater to move through the aquifer and reach a pumping well can be quite long, therefore many years may pass before saltwater intrusion is detected at a particular location.

3 Hydraulic approaches to treatment of saltwater intrusion

The first saltwater-intrusion problem was reported by Braithwaite [3] that was caused by overpumping many production wells in London and Liverpool, England. He attributed the degradation of water quality of these wells to the lowering of the groundwater levels to below the sea level. It is now well known that saltwater intrusion occurs even if the water-table level is higher than that of the saltwater. The first analytical solution developed to study saltwater intrusion was formulated by Ghyben [4] and Herzberg [5] independently. Thus, it is called the Ghyben–Herzberg relationship. Their analytical solutions approximate the saltwater-intrusion behavior based on a number of simplifying assumptions, which, most of the time, are not applicable to real conditions. Although analytical solutions oversimplify the real problems, they have been used as tools for first-cut engineering analyses in preliminary hydrogeologic investigations. They may serve also as teaching tools as well as verification tools for numerical models. Cheng and Ouazar [6] provided a number of analytical solutions to saltwater-intrusion problems that are of historical and practical importance. There are some key textbooks dealing with saltwater intrusion in the literature including Bear [7, 8], Todd [9], and Bear and Verruijt [10]. Advances in numerical solution techniques and computing capabilities enabled groundwater modelers to solve more general partial differential equations that realistically describe groundwater problems such as saltwater intrusion.

In a coastal aquifer there are three distinct zones; the freshwater zone, the mixing or diffusion zone, and the saltwater zone. The freshwater zone lies over the saltwater zone nearby the coastline. The salt concentration in the diffusion zone, where hydrodynamic dispersion exists, varies gradually from that of the saltwater to that of the freshwater. The thickness of the diffusion zone may change from 1 m to more than 100 m from system to system based on the hydrodynamic dispersion process [9]. The presence of seawater intrusion and tidal variations in coastal aquifers cause the solutes to move upward towards the coastline when it approaches the saltwater interface and then exits around the coastline [11]. Tidal variations of sea level cause groundwater fluctuation as well as oscillation of the saltwater/freshwater interface. Any stress applied within the freshwater region such as changes in discharge (i.e. pumping) or recharge causes movement of the freshwater–saltwater interface.

Saltwater–freshwater systems in coastal aquifers can be conceptualized by two basic hydraulic approaches depending on the relative thickness of the transition zone between freshwater and saltwater. The first approach is the freshwater-saltwater sharp-interface approach [12–14] and the second is the variable-density approach with solute transport (or hydrodynamic dispersion). The former approach ignores the mixing zone and considers the transition zone to be negligible compared to the aquifer thickness. In this approach, the saltwater-intrusion phenomenon is modeled as a two-region fluid flow separated by a sharp interface. However, hydrodynamic dispersion often causes a mixing zone across the interface between saltwater and freshwater. If the thickness of the mixing zone expands to a considerable extent than the sharp-interface approach may not be valid. In this case, the latter approach needs to be used to model the saltwater-intrusion problem using the density-dependent miscible flow and transport approaches. The latter approach assumes a mixing or transition zone and a solute-transport mechanism to model the hydrodynamic dispersion process [15–17]. Because of the improvements of computer technology and numerical solution techniques, this method has been used widely. It also represents the physical system more realistically compared to the sharp-interface approach. Several simulation models based on this approach were developed [18–20].

The appropriateness of either of these approaches and their method of analysis depends on the characteristics of the aquifer system under investigation and the problems of which solutions are being sought. The most important issue in modeling saltwater-intrusion problems is the proper conceptualization of the hydrogeologic system and definition of boundary conditions. Volker and Rushton [21] compared steady-state solutions using disperse- and sharp-interface approaches and showed that as the coefficient of dispersion decreases the solutions of those two approaches converge. Approaches and challenges for studying variable-density groundwater flow are summarized by Simmons *et al.* [22], Diersch and Kolditz [23], and Simmons [24].

3.1 Sharp-interface approach

In the sharp-interface conceptualization, the saltwater intrusion system can be formulated using two immiscible flow fields, namely freshwater and saltwater flow fields. The interface between these two flow fields acts as a common boundary (Fig. 2). Along this boundary, the two flow fields are coupled through the interfacial boundary condition of continuity of flux and pressure. In three dimensions, this boundary condition is highly nonlinear [7]. However, assuming horizontal flow and integrating the flow equations over the vertical dimension reduces the problem to a two-dimensional problem. The sharp-interface approach does not give information about the nature of the transition zone but simulates the regional flow dynamics of the groundwater system and the response of the interface to applied stresses.

The Ghyben–Herzberg [4, 5] relation between saltwater and freshwater states that, assuming hydrostatic condition in a homogeneous unconfined aquifer, the weight of a unit column of freshwater extending from the water table to the sharp interface is balanced by a unit column of saltwater extending from sea level to the same depth as the sharp interface [25]. As seen in Fig. 2, the Ghyben–Herzberg relationship can be written as:

$$z = \frac{\rho_f}{\rho_s - \rho_f} h \tag{1}$$

where ρ_f and ρ_s are fresh and saltwater densities (ML^{-3}), respectively, z is the depth of sharp interface from sea level for a given point (L), h is the distance

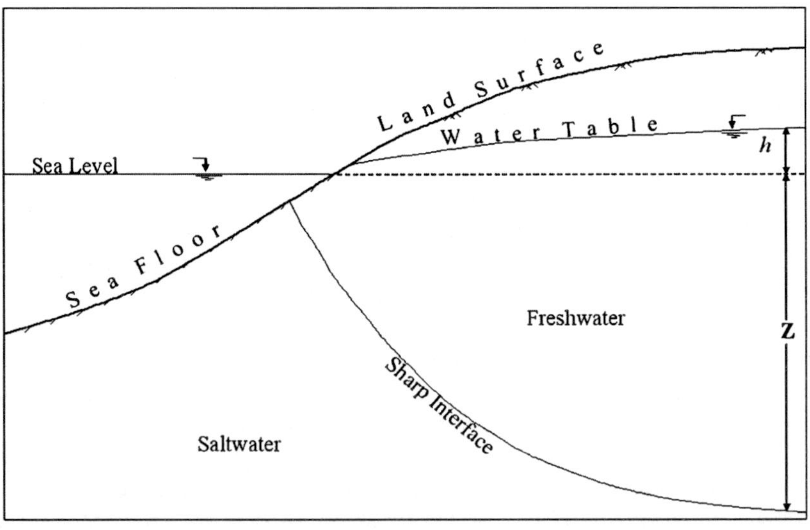

Figure 2: Conceptualization of saltwater intrusion based on the sharp-interface approach.

between water table and sea level (L). At about 20 °C, for $\rho_f = 1.0$ g cm^{-3} and $\rho_s = 1.025$ g cm^{-3}, Eq. (1) becomes $z_s = 40$ h. This means that if the water table in an unconfined coastal aquifer is lowered 1 m, the freshwater–saltwater interface will rise 40 m. The main drawback of the Ghyben–Herzberg relationship is that it assumes a hydrostatic condition for the saltwater phase, which is not valid in many real cases so that this relationship generally underestimates the depth to the interface [25]. This equation is reasonable away from the shoreline for mainly horizontal flow.

Hubbert [26] provided a more realistic relation describing steady-state groundwater flow in both fresh and saltwater zones. He formulated the following equation:

$$z = \frac{\rho_f}{\rho_s - \rho_f} h_f - \frac{\rho_s}{\rho_f - \rho_s} h_s, \tag{2}$$

where z is the elevation of a point on the interface at which head is measured (L), h_f and h_s are the freshwater and saltwater heads (L), respectively. Hubbert's relation must hold true at the interface to ensure the continuity of pressure field at the interface. For the stationary saltwater condition without any saltwater head gradient, if h_s is taken as zero by choosing the sea level as the datum, Eq. (2) becomes identical to Eq. (1), which is the Ghyben–Herzberg relationship.

When the saltwater is not static, then the freshwater and saltwater flow systems on both sides of the freshwater–saltwater interface need to be reformulated and solved simultaneously. In that case, the sharp interface is considered as a common boundary between those flow systems and it must satisfy Hubbert's [26] relation, Eq. (2). The saltwater and freshwater flow systems can be defined with the following equations [27]:

$$S_f \frac{\partial h_f}{\partial t} + \nabla.\vec{q}_f - Q_f = 0, \tag{3a}$$

$$S_s \frac{\partial h_s}{\partial t} + \nabla.\vec{q}_s - Q_s = 0, \tag{3b}$$

where h is the head (L), t is time (T), $\vec{q}$ is the specific discharge (LT^{-1}) determined by Darcy's law for constant density fluid as $\vec{q} = -\vec{K}.\nabla h$, K is the hydraulic conductivity (LT^{-1}), S is the specific storage (L^{-1}), Q is the source sink term (T^{-1}), $\nabla = (\partial/\partial x)\vec{i} + (\partial/\partial y)\vec{j} + (\partial/\partial z)\vec{k}$, and subscripts f and s refer to freshwater and saltwater, respectively. Equations (3a) and (3b), are parabolic partial differential equtions, they must be solved simultaneously for the freshwater head (h_f) and saltwater head (h_s). Once the freshwater and saltwater head distributions are known, the interface elevation at any x-y location in the aquifer can be obtained using Eq. (2). In regions away from the interface, only one type of fluid exists and the flow field can be described by a single equation without the interface [28]. The sharp-interface approach was used by some researchers, i.e. Wilson and Da Costa [29], and Essaid [13, 14].

In regional applications, this approach has advantages because of its relative simplicity, ease of development and use, and less intense data requirements. In addition to overlooking the dispersion zone, its main disadvantage for practical purposes is that it can not evaluate chloride concentration of individual production wells. Overall, this approach is useful in developing regional water-management plans.

3.2 Variable-density and dispersion approach

In reality, and as described earlier, the region between the freshwater and saltwater zones is not a sharp interface but instead it changes gradually over a finite distance, and is known as the zone of diffusion, zone of dispersion, or the zone of mixing (Fig. 3). Unlike the sharp-interface approach, this approach assumes saltwater and freshwater as miscible fluids. Water in this approach transports a solute (salt) that influences its density and viscosity; therefore, partial differential equations governing the groundwater flow system are used to simulate variable-density flow and solute transport with a relationship defining the density as a function of solute concentration. Thus, this approach requires coupling density-dependent groundwater flow and solute-transport equations. In this approach, the flow field is modeled using a single fluid with variable density, which is a function of concentration and pressure applied on the fluid.

The mass-balance equation for a variable-density fluid can be derived by combining the continuity equation, in which the storage term is written as a function of

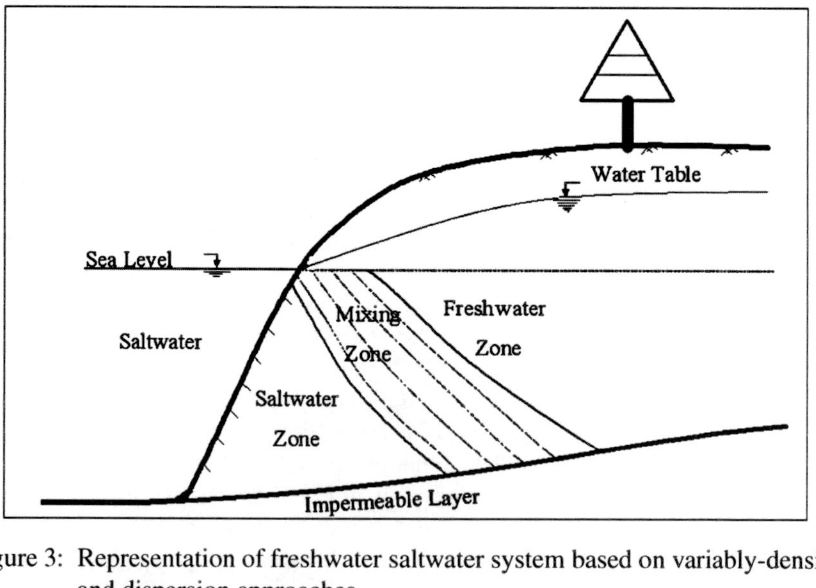

Figure 3: Representation of freshwater saltwater system based on variably-density and dispersion approaches.

change in pressure and concentration, with a pressure-based density-dependent form of Darcy's law as follows [30]:

$$\left(\rho s_{\mathrm{op}}\right)\frac{\partial P}{\partial t}+\left(n\frac{\partial \rho}{\partial t}\frac{\partial C}{\partial t}\right)=+\nabla.\left[\left(\frac{\rho k}{\mu}\right).(\nabla P+\rho g\nabla z)\right]+Q \tag{4}$$

where ρ is fluid density (ML^{-3}), S_{op} is specific pressure storage (LT^2M^{-1}), n is porosity (dimensionless), C is the mass-based solute concentration (MM^{-1}), t is time (T), k is the intrinsic permeability of the solid matrix (L^2), μ is the fluid viscosity ($ML^{-1}T^{-1}$), P is the pressure applied on the fluid ($ML^{-1}T^{-2}$), g is the gravitational acceleration vector (LT^{-2}), z is the upward coordinate direction (L), and Q is fluid mass source ($ML^{-3}T^{-1}$).

In the same way, the mass balance for a solute stored in solution can be expressed as

$$\frac{\partial(n\rho C)}{\partial t}=+\nabla.\left[n\rho D.\nabla C\right]-\nabla.(n\rho\vec{v}C)+QC^{*} \tag{5}$$

where D is the dispersion tensor that also includes molecular diffusivity of solutes (L^2T^{-1}), v is the fluid velocity (which is the specific discharge, q, divided by porosity, n) (LT^{-1}), and C^* is the solute concentration of the fluid sources (MM^{-1}).

The variable-density approach requires the simultaneous solution of the flow equation (Eq. (4)) and the transport equation (Eq. (5)) to obtain the pressure distribution and concentration distribution in the aquifer ranging from pure freshwater to pure saltwater. The equations are nonlinear in nature and require an iterative solution technique. Since there are three unknowns (P, C, ρ) and only two equations (Eqs. (4) and (5)), another equation is required that can relate fluid density to solute concentration. Such an equation of state relating density to concentration was given by Voss and Provost [31]:

$$\rho=\rho(C)=\rho_{\mathrm{o}}+\frac{\partial \rho}{\partial C}(C-C_{\mathrm{o}}), \tag{6}$$

where ρ_o [M/L^3] is reference fluid density at reference solute concentration C_o [M_s /M]. Usually, $C_o = 0$, and the reference density is taken as that of pure water. The factor $\partial\rho/\partial C$ is a constant value of density change with concentration. For example, for mixtures of freshwater and seawater at 20 °C, when C is the mass fraction of total dissolved solids, $C_o = 0$, and $\rho_o = 998.2$ [kg/m^3], then the factor, $\partial\rho/\partial C$, is approximately 700 [kg/m^3] [31].

In large regional simulations, a steady-state solute distribution can be assumed, in which case the concentration defining the density of fluid is known and will not change over the period of the analysis. Based on this assumption, there is no need to solve the transport equation, and the solution of the density-dependent flow equation (Eq. (4)) is sufficient to describe the flow system [27]. This approach has been used by several researchers, i.e. Weisss [32], and Kontis and Mandle [33].

The advantage of the variable-density method is that the response of the system to stresses influencing the thicknesses of the transition zone can be better analyzed and understood. Density-dependent flow and transport conceptualization represents the actual physical system more accurately than the sharp-interface approach. Also, the concentration distribution of solutes is simulated, and estimates of solute concentrations in individual wells are obtained. However, more input parameters are required, and the method is computationally more complex.

3.2.1 Equivalent freshwater heads approach

Equivalent freshwater head refers to a height of freshwater column that exerts the same pressure applied by saltwater at a given point. With the equivalent freshwater head approach, density effects can be converted into equivalent freshwater heads by which the density-dependent flow model turns into a constant-density flow model. Weiss [32] was one of the first to reformulate the groundwater flow equation in equivalent freshwater head form. Langevin and Guo [34] presented a methodology to couple a constant-density groundwater flow code with a solute transport code to simulate variable-density groundwater flow and solute transport in three dimensions using the equivalent freshwater head concept. They coupled a popular groundwater flow model, MODular three-dimensional finite-difference groundwater FLOW model (MODFLOW) [35, 36], and a popular transport model, Modular 3-D Transport model with Multi-Species structure (MT3DMS) [37], on a platform called the SEAWAT computer program [38].

Many existing models used for density-dependent groundwater simulation formulate the groundwater flow equations in terms of pressure. Langevin *et al.* [38] reformulated the flow equation in terms of freshwater heads to be able to use MODFLOW's flow equation routines. The equivalent freshwater head, h_f [L], can be defined as:

$$h_{\mathrm{f}} = \frac{P}{\rho_{\mathrm{f}} g} + z, \tag{7}$$

where z is the elevation of the point at which head is measured (L) and P is the fluid pressure at the point of measurement ($ML^{-1}T^{-2}$). The equivalent freshwater head formulation leads to a system of variable-density flow equations that can be solved relatively easily using the existing constant density groundwater flow equations in MODFLOW. The final form of the flow equation in terms of equivalent freshwater head is [39]:

$$\rho S_{\mathrm{sf}} \frac{\partial h_{\mathrm{f}}}{\partial t} + n \frac{\partial \rho}{\partial t} \frac{\partial C}{\partial t} = +\nabla . \left[\rho K_{\mathrm{f}} \left(\nabla h_{\mathrm{f}} + \frac{\rho - \rho_{\mathrm{f}}}{\rho_{\mathrm{f}}} \nabla z \right) \right] + Q \tag{8}$$

where S_{sf} is the freshwater specific storage [L^{-1}] defined as the volume of water released from storage per unit volume per unit decline of freshwater head, and C is the concentration of solute mass per unit volume of fluid [ML^{-3}]. For a constant-density system, Eq. (8) reduces to the flow equation solved by MODFLOW.

Motz [40] numerically investigated hydraulic heads in the freshwater part of saltwater/freshwater interface using MODFLOW and SEAWAT. He demonstrated that it is possible to represent the effects of the saltwater/freshwater interface in

MODFLOW simulations by specifying the boundary conditions as equivalent freshwater heads at the coastal boundary over the full thickness of the aquifer. This approach does not give any insight into the location or thickness of the interface and saltwater concentration at the interface, but it can be used as a guide during the construction of new regional groundwater flow models to ensure correct calculation of heads in the freshwater part of the aquifer.

The general form of the solute transport equation used in SEAWAT, which is solved in MT3DMS, is identical to Eq. (5) without the density term:

$$\frac{\partial(nC)}{\partial t} = +\nabla.[nD.\nabla C] - \nabla.(\vec{q}C) + QC^{*}. \tag{9}$$

For conditions with large spatial density gradients, the ∇C term in Eq. (9) should be formulated as $\rho\nabla(C/\rho)$ [41]. For most practical applications with moderate density variations, Zheng and Bennett [41] suggest that Eq. (9) represents a suitable approximation of the solute mass balance.

Equations (8) and (9) are coupled to model saltwater intrusion in variable-density groundwater systems. Fluid density is a function of solute concentration, transport is dependent on the flow field, and the storage term in the transient-flow equation incorporates changes in concentration. Concentrations resulting from the solution of Eq. (9) are used by an equation of state to calculate fluid density using the following linearized relationship between fluid density and solute concentrations similar to Eq. (6):

$$\rho = \rho_{\mathrm{f}} + \frac{\partial \rho}{\partial C} C. \tag{10}$$

This equation does not include the dependence of fluid density on temperature or pressure, and thus it should not be used for other than isothermal systems with an incompressible fluid. For other conditions Diersch and Kolditz [23] provided a summary of more rigorous forms of the equations of state. The partial differential equations presented in Eqs. (8) to (10) are simultaneously solved by numerical methods to obtain head and solute concentration distribution in the system.

4 Numerical models and case studies

Numerical modeling helps in analyzing coastal aquifer systems and provides quantitative insights as to their best management. It is one of the essential tools that have been used to understand groundwater flow and saltwater movement in coastal aquifers. It generally lies at the heart of any planning or research process of coastal aquifers. It provides a means to analyze complex systems of groundwater flow and saltwater movement in coastal aquifers. Such analysis is often impractical or impossible to do by analytical models or field studies alone. Field studies are costly and analytical models have limited practical use for aquifers with heterogeneous complex geometry; therefore use of numerical models is generally a necessity before additional more detailed field studies. Moreover, models provide significant insights into the potential mechanism of intrusion and clear guidance on the need for additional data, and the type and location of data needed [42]. The exclusion of seawater intrusion in numerical modeling results in an underestimation of solute mass rate exiting around the shoreline and unrealistic migration paths under the seabed [11]. Models are one of the best and most effective tools to design hydrogeologic investigations, locate and select well depths, and answer different "what if" questions. Models are also good tools for understanding the sensitivity of coastal aquifer systems to changes in their hydrologic components and anthropogenic effects such as land-use changes and groundwater exploitations. Without numerical models, testing different management scenarios and system analysis would be almost impossible with field experiments due to their prohibitive cost.

Developments of numerical models for saltwater-intrusion problems showed a parallel progress with groundwater-solute transport and groundwater–surface water interaction models over the years. Simmons [24] summarized the current research challenges and future possibilities in variable-density groundwater flow modeling. He illustrated the widespread importance, diversity and interest in applications of variable-density flow phenomena in groundwater hydrology. These applications include seawater intrusion, freshwater/saline-water interfaces and saltwater upconing in coastal aquifers, subterranean groundwater discharge, dense contaminant plume migration, and density-driven transport in the vadose zone. Sorek and Pinder [43] provided a survey of 15 computer codes that simulate saltwater-intrusion problems. Several authors have written groundwater modeling textbooks, including Anderson and Woessner [44] and Konikow and Reilly [45].

A critical step in any numerical hydrologic investigation is to select the appropriate model, which depends on setting clear modeling objectives. According to Maimone *et al.* [42], practical coastal aquifer modeling studies may have one or more of the following objectives: (1) determining the causes of existing saltwater intrusion and the mechanism that caused it, i.e. lateral intrusion, upconing, or downward leakage; (2) estimating the location of the interface; (3) evaluating the stability of interface in response to pumping; (4) determining the potential for intrusion based on current pumping or projected pumping; (5) estimating expected time of impact for specific well locations based on various pumping scenarios; and (6) testing various approaches to stopping or reversing intrusion, or assessing strategies for sustainable use of coastal aquifers as viable water supplies even with ongoing intrusion.

The sharp-interface approach has been used successfully in many regional modeling studies to analyze the long-term stability of coastal wells. This modeling approach can provide insight into horizontal movement of saltwater under the influence of both sea-level rise and coastal pumping. It can help estimate the optimum rate of long-term viable pumping without amplifying saltwater intrusion further inland. Models based on this approach have been used as primary planning tools in Florida, New York [42], and Hawaii [28].

In analyzing upconing of saltwater (Fig. 4), it is important to calculate the maximum sustainable pumping rate that avoids saltwater upconing, or to calculate the

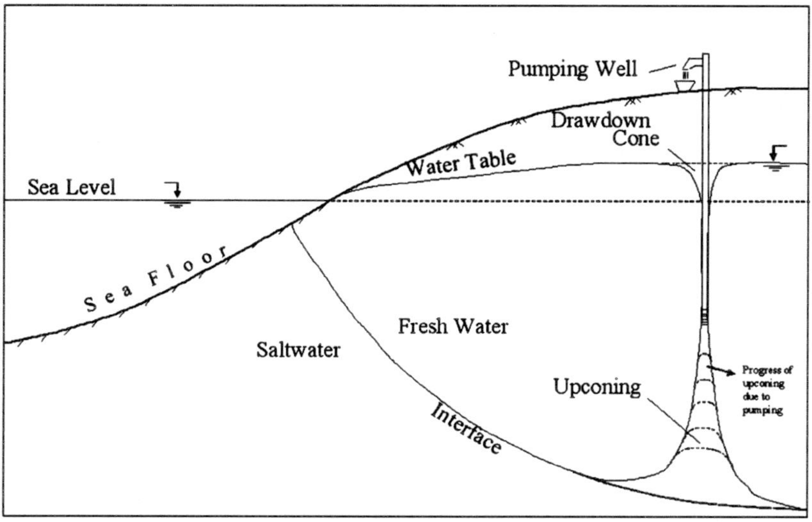

Figure 4: A schematic view of saltwater upconing under a production well.

expected levels of salt in the wells. In this case, an analytical model can be a solution to the saltwater upconing problems [46]. However, numerical models including sharp-interface models, density-dependent groundwater flow models, and coupled flow and transport models are more useful in simulating this situation in more complex hydrogeologic environment. In upconing problems, the density-dependent groundwater flow models, or coupled flow and transport models may have to be used if the salt concentration gradients underneath the freshwater zone need to be quantified and mapped.

Saltwater-intrusion models often do not have sufficient data to provide appropriate calibration and verification; therefore most of the models have been tested against standard analytical solutions to verify the correctness of the numerical approximations. This type of verification procedure is called benchmarking. Verifying a density-dependent flow model by testing it against data from either a laboratory-scale experimental or a field-scale case study is a difficult task because availability of these types of data sets is limited [47]. Simpson and Clement [48] proposed and used the term benchmarking as a way that the numerical algorithm can reproduce the prior history of a well-defined problem. It also refers to model testing against standard problems and/or well-controlled field and laboratory studies that have been sufficiently tested and are widely accepted by model developers. The most popular benchmarking problems are the Henry's saltwater-intrusion problem [15], for which an analytical solution exists, and the Elder's salt-convection problem [49], for which laboratory and numerical data are available.

Three commonly used models and their example application in a few case studies are briefly discussed below. These models are: 1) USGS SHARP model, a quasithree-dimensional, finite-difference model to simulate freshwater and saltwater flow separated by a sharp-interface model [13, 14], 2) the USGS finite-element variable-density flow and transport simulation code, SUTRA, [30, 31], and 3) SEAWAT model [38], which is a platform coupling the groundwater flow model, MODFLOW [35, 36], and the transport model MT3DMS [37].

SHARP is a quasithree-dimensional finite-difference model developed by Essaid [13, 14] to simulate freshwater and saltwater flow separated by a sharp interface in layered coastal aquifer systems. The model is quasithree-dimensional because each aquifer is represented by a layer in which flow is assumed to be horizontal. This model is a good example of the sharp-interface approach in modeling saltwater intrusion. The main assumption in this model is that the width of the freshwater–saltwater transition zone is small relative to the thickness of the aquifer so that it can be assumed that freshwater and saltwater are separated by a sharp interface and they do not mix. This modeling approach, in conjunction with vertical integration of the aquifer flow equations, can be used in regional-scale studies of coastal areas. This approach does not give information concerning the nature of the transition zone but does reproduce the regional flow dynamics of the system and the response of the interface to applied stresses.

SHARP was used to simulate the coastal aquifer of the Soquel-Aptos basin, Santa Cruz County, California. The topography of the Soquel-Aptos basin, Santa Cruz County, California, ranges from very steep valley slopes and angular land forms to nearly flat marine-terraced, sea cliffs and narrow beaches along Monterey Bay. The region is a populated urban area with increasing demand for freshwater. The principle aquifer is a layered aquifer with variable thickness. Saltwater has not intruded onshore and the position of the interface offshore was not known. An analysis was performed by Essaid [14] to estimate the amount of freshwater flow through the system, the position of the saltwater interface offshore, the quantity of discharge to the sea that must be maintained to keep the interface at or near the shore, and the rate at which the interface will move due to onshore development.

Initial conditions were obtained by simulating predevelopment conditions by letting the system reach the steady-state condition. Transient conditions were simulated for the period of 1930 to 1985. Historical well-pumping values were inputted to the model according to the time they were tapped to the aquifer and went into production. Prior to the predevelopment, the recharge to the system was $0.50\,m^3\,s^{-1}$, of which $0.47\,m^3\,s^{-1}$ discharged onshore to creeks and to the overlying Aromas Sand. Only $0.03\,m^3\,s^{-1}$ of the water flowed offshore to the ocean as fresh groundwater discharge. This small proportion of recharge was sufficient to maintain the freshwater/saltwater interface position offshore. With the development and increase in groundwater pumping, the 1981 onshore and offshore discharge rates decreased to $0.43\,m^3\,s^{-1}$ and $0.01\,m^3\,s^{-1}$, respectively. The 1930 to 1985 simulations predicted almost no movement of the interface despite significant changes in the groundwater flow system. Essaid [14] concluded that interface response was quite slow and took place over a long time. The slow response of the saltwater zone was a result of the considerably low horizontal and vertical hydraulic conductivities that impede the flow of saltwater into the interface zone.

SUTRA simulates fluid movement and transport of either energy or dissolved substances in a subsurface environment. The original versions of SUTRA [30, 31] are capable or simulating variable-density flow and transport of either heat or one dissolved species through variably to fully saturated porous media. Hughes and Sanford [50] modified SUTRA to simulate multispecies transport in which species may or may not affect fluid density and viscosity. SUTRA solves the two- or three-dimensional form of density-dependent saturated or unsaturated groundwater flow, and solute- and energy-transport equations using two-dimensional finite-element and three-dimensional finite-difference methods. Solute-transport simulation with SUTRA may be used for cross-sectional modeling of saltwater intrusion in aquifers at near-well or regional scales, with either dispersed or relatively sharp transition zones between the freshwater and saltwater. SUTRA provides, as the primary output, fluid pressures and either solute concentrations or temperatures as functions of time and space. SUTRA requires specification of pressures rather than hydraulic heads in saltwater-intrusion simulations.

Gingerich and Voss [51] applied SUTRA to model groundwater flow and solute transport in the Pearl Harbor aquifer, southern Oahu, Hawaii. They showed that the readjustment of the freshwater–saltwater transition zone would take a long time following changes in pumping, irrigation, or recharge in the aquifer system. They claimed that the Ghyben–Herzberg estimate of the freshwater/saltwater interface depth is not a good predictor of the depth of potable water. Their simulations showed that the transition zone moved upward and landward compared to the predevelopment period of year 1880.

Hunt *et al.* [52] and Voss [53] used SUTRA in cross section to numerically evaluate Oahu's southern coastal aquifer hydraulics and saltwater intrusion by analyzing and tracking the movement of the freshwater–saltwater transition zone. The purposes of this analysis were to give Oahu water managers reliable scientific data to help them decide on water allocation, managing Oahu's aquifers, and to quantify the amount of groundwater that would be safely produced from each aquifer. Oahu has high rainfall, highly permeable aquifers, and a coastal semiconfining layer overlaying many of the aquifers. The major aquifers on Oahu are composed of hundreds of thin lava flows that were extruded onto the land surface forming dike-type vertical aquifers. The layers, generally several meters thick, form a matrix of thin overlapping tubular units commonly tens to hundreds of meters wide and up to 30 km long that dip about 5 to 10 degrees from the mountainous recharge areas to the ocean. The coastal confining unit keeps heads high at the coast and creates a very thick freshwater lens. The combination of high recharge, high permeability and impeded discharge provides a rich freshwater supply for both drinking and irrigation. Other than anthropogenic contamination, saltwater exists at the bottom of each of Oahu's coastal aquifers and represents a potential threat by means of saltwater upconing.

This Southern Oahu's aquifer system was analyzed as a 2D cross section that represents the basal aquifer and caprock containing the basal freshwater lens, saltwater–freshwater transition zone, and deeper saltwater. Recharge to the aquifer enters through the water table, and through the upstream boundary of the aquifer representing inflow from dike compartments and the central plateau. Withdrawals occur near the coast. Recharge and withdrawal rates were changed with time according to historical data. The seaward boundary was held as zero pressure at the hydrostatic saltwater pressure. The rest of the boundaries were assumed to be impermeable. The transition zone was analyzed to predict the water quality of freshwater for eight different scenarios for the period of 1980–2080. These scenarios have different withdrawal and/or recharge rates [54]. The limiting concentration of potable water was defined as 2% saltwater concentration.

Some scenarios predicted sustainable potable water supplies, while others resulted in significant saltwater intrusion. Up to about 75% of the assumed recharge can be withdrawn before any significant saltwater intrusion affecting the water quality of production wells. According to this analysis, the key parameter controlling the saltwater intrusion was the net system discharge, not the percentage of recharge pumped. It was also found that the maximum safe withdrawals vary depending on the actual long-term recharge rate.

SEAWAT was developed by combining MT3DMS and a modified version of MODFLOW into a single program that solves the coupled variable-density groundwater flow and solute-transport equations [39]. MODFLOW was modified to solve the variable-density flow equation by reformulating the matrix equations in terms of fluid mass rather than fluid volume and by including the appropriate density terms. Fluid density is assumed to be solely a function of the concentration of dissolved constituents; the effects of temperature on fluid density are not considered. Temporally and spatially varying salt concentrations are simulated by SEAWAT using routines from the MT3DMS program. SEAWAT couples the groundwater flow equation with the solute-transport equation. The basic assumptions in SEAWAT development are the followings; Darcy's law is valid; the diffusive approach to dispersive transport based on Fick's law can be applied; isothermal conditions prevail; the porous medium is fully saturated; and a single, fully miscible liquid phase of very low compressibility is assumed.

Dausman and Langevin [55] constructed a variable-density groundwater flow model of Broward County, Florida using SEAWAT. In that study, SEAWAT was used to evaluate the relationship between water-level fluctuations and saltwater intrusion. The model was representative of many locations in Broward County that contain a well field, a control structure, a canal, the Intracoastal Waterway, and the Atlantic Ocean. The model was used to simulate short-term and long-term movements of the saltwater interface resulting from changes in rainfall, well-field withdrawals, sea-level rise, and upstream canal stages.

Long-term simulations, i.e. periods greater than 10 years, revealed that the upstream canal stage controls the movement and location of the saltwater interface. If the upstream canal stage is decreased by 30 cm, the saltwater interface takes 50 years to move inland and stabilize. If the upstream canal stage is then increased by 30 cm, the saltwater interface takes 90 years to move seaward and stabilize. If sea-level rises about 48 cm over the next 100 years as predicted, then inland movement of the saltwater interface may cause well-field contamination.

These results show that the upstream canal stage substantially affects the long-term position of the saltwater interface in this surficial aquifer system. The saltwater interface moves faster inland than seaward as a result of changes in the upstream canal stage. The saltwater-intrusion problem in the Florida Biscayne aquifer does not seem to be severe if the well-field withdrawal is increased for short-term drought problems based on the assumption that well-field withdrawals will decrease once the drought is over. Sea-level rise may be a potential problem to the water supply in Broward County because of the movement of the saltwater interface inland toward well fields.

The number of SEAWAT applications is increasing rapidly because SEAWAT is taking advantage of the widely used and accepted groundwater models: MODFLOW and MT3DMS models. Among these applications are: a study of movement of the saltwater interface in Broward County, Florida [55]; a second study that investigated the freshwater–saltwater interaction and the effects of pumping and sea-level change in the Lower Cape Cod Aquifer System, Massachusetts [56]; a third study that assessed the natural and the anthropogenic impacts on freshwater-lens morphology on Dog Island and St. George Island, Florida [57]; and lastly a study of salt transport in the Okavango Delta [58].

5 Land-use changes and groundwater pumping

Land-use changes on coastal aquifers are direct results of population increase and urban and agricultural developments. Increasing demand for water requires more groundwater exploitation from coastal aquifers. Population growth on coastal areas creates a chain reaction by triggering land-use changes and exploitation of more groundwater resources and deteriorating the coastal ecology and groundwater resources. As population increases, lands used for urban and agriculture increase and consequently demand for freshwater supplies increases. These stresses change the hydrologic and hydrogeologic characteristics of the coastal aquifers. They would have pronounced adverse effects on freshwater and saltwater quality, if proper precautions and preventative measures are not taken.

The effects of land-use changes are primarily seen on land-cover type, i.e. a forest area can be turned into an urban or agricultural area, or a barren land can be turned into a golf course. Obviously, changes in land use influence water infiltration and recharge to groundwater, and probably may change the microclimate of a region. As described before, according to the Ghyben–Herzberg relationship, there is a strong linear relationship between elevation of the freshwater table and the depth of the saltwater/freshwater interface such that 1-m drop in an unconfined coastal aquifer is compensated by about a 40-m rise of the interface. The key parameter is the height of the freshwater table above sea level or the quantity of fresh groundwater flow towards the sea. For example, if the land use of an area changes from forest to urban, then recharge rates in the area reduce due to increase in the proportions of impervious areas. Consequently, the water table drops, which causes the saltwater interface to move inland. On the other hand, if the change in land use increases the soil permeability, such as under golf courses, this would result in an increase of the recharge rates to groundwater, increasing groundwater flow to the sea and/or raising the water-table depth. As a consequence, saltwater intrusion will decrease.

Agricultural and urbanization activities on coastal zones might have a variety of effects on their aquifers depending on the hydrogeology of these areas, type of agricultural practices, and the degree of urbanization. If agricultural land is drained by lowering the water table, then saltwater intrudes inland by responding to low water-table levels. In coastal areas, urban and agricultural land uses might have parallel environmental conflicts. For example, in coastal urban areas, the underlying aquifer system must serve as a source of water supply, in which water levels drop due to pumping, while the aquifer water levels must be maintained at certain elevations to prevent saltwater intrusion. Changes in natural groundwater flow patterns and the associated reduction in groundwater flow toward coastal bays alter salinity and affect the local marine ecology. It can be concluded that land-use changes and groundwater pumping directly affect saltwater intrusion in the coastal aquifer systems as well as in coastal bays. Therefore, a careful water- and land-use management plan should be applied on coastal zones.

The Everglades in south Florida is a very good example for land-use changes and corresponding saltwater-interface response to those changes. A highly controlled water-management system has evolved during the 1900s largely to provide drained land for a rapidly expanding population in the Everglades. Draining of Everglades's wetland areas during the last 75 years has provided the opportunity for westward expansion of agricultural, mining, and urban activities, which changed the land-use and water-use characteristics of the region substantially [59]. Urban and agricultural growth and land-use change can greatly impact the ecological health and stability of coastal areas. A review of 100 years of land-use and population changes in the Everglades illustrated impacts of those changes on water resources in a coastal area where urban areas are growing rapidly and replacing agricultural areas [59]. Some declines in water levels can be directly attributed to municipal groundwater withdrawals; however, water-level declines over wider areas were a direct result of canal drainage, i.e. dropping the water-table elevations. Landward movement of the saltwater interface has been an issue of local and regional concern since the 1940s. In decreasing importance, canal overdrainage, overpumping from wells located near the coast, and upconing of seawater are the primary sources of saltwater in the surficial aquifer system. Significant changes in land and water uses at the coastal zones contributed to the deteriorating conditions of the marine ecosystem in south Florida. Renken [59] showed that saltwater intrusion in the surficial aquifer system in south Florida was a direct consequence of water-management practices, concurrent agricultural and urban development, and natural drought conditions. These findings would be true for any other areas with similar conditions.

The seaward groundwater flow affects coastal ecosystems; it sustains the flow and aquatic habitats of coastal streams during periods when surface runoff is low. Groundwater discharge also helps to maintain water levels and water budgets of freshwater lakes, ponds, and wetlands near the coast. Dissolved chemicals in groundwater coming from agricultural lands affect the salinity and geochemical budgets of coastal ecosystems and play a role in the biological-species composition and productivity of these systems. Although coastal groundwater systems have been contaminated by many types of chemical constituents, the current concern has focused on the discharge of excess nutrients, particularly nitrogen, to coastal ecosystems. Nutrient contamination of coastal groundwater occurs as a consequence of land-use and water-use changes. Wastewater disposal from septic systems and pollutant and sediment runoff from agricultural and urban areas are the main sources of the nutrient contamination. Nutrient overenrichment can lead to excessive production of algal biomass, loss of important habitats such as seagrass beds and coral reefs, changes in marine biodiversity and distribution of species, and depletion of dissolved oxygen and associated die-offs of marine life [60].

Groundwater-pumping effects on saltwater intrusion should be analyzed in two different scales; the regional scale and individual well scale. In order for an aquifer to supply freshwater to wells, the regional system must be capable of providing the required quantity of water. However, even though the regional system may be in equilibrium, saltwater upconing beneath pumping wells can make these wells produce high salinity water. Although, the regional system may be capable of sustaining the rate of production, the drawdown around an individual pumping well may cause saltwater upconing. Reilly and Goodman [2] showed that in a well analysis in Truro, Cape Cod, MA, although the regional system was in equilibrium and capable of sustaining $4200\,m^3\,d^{-1}$ at a particular well field based on regional estimates, the actual well was not capable of this production because extensive drawdown occurred around the well causing saltwater upconing. Estimates from local analysis showed that the maximum permissible withdrawal from a single well should not exceed $1800\,m^3\,d^{-1}$; thus, the withdrawal rate for a well should be limited, and additional wells should be installed if more capacity is required. To reduce the risk of upconing while producing water from freshwater lenses, horizontal shafts (sometimes called a Maui shaft) can be used. The horizontal shafts can produce large volumes of freshwater by skimming water from near the top of the freshwater lens [61].

The total amount of groundwater extraction, locations of wells, and individual well pumping rates should be taken into account all together in analyses of saltwater intrusion. This kind of analysis requires simultaneous use of optimization and numerical groundwater models that are capable of simulating density-dependent flow and solute transport. Optimization and management models should be able to optimize pumping rates, well locations, and number of wells. However, the nonlinearity in the variable-density groundwater flow brings difficulties in optimization models due to computational time and burden. Therefore, the complex optimization models with the constraint of saltwater-intrusion problems have yet to become very practical. Nonetheless, saltwater intrusion into wells can also be dealt with in simpler and indirect approaches, i.e. by constraining drawdown at a number of control points, or by minimizing the overall intruded saltwater volume in the whole aquifer. A typical analysis of temporal and spatial variations of coastal saltwater intrusion generally follow the following steps; 1) identification of the principal factors that control the extent of saltwater intrusion, 2) evaluation of long-term trends in groundwater-withdrawal rates, groundwater-level change, rainfall, and increases in chloride concentration; and 3) determination of causal relations between the position of the saltwater interface, water-management practices, and the expansion of agricultural and urban areas.

6 Tidal effects and sea-level rise on saltwater intrusion in coastal aquifers

Tidal activity in the oceans may force the saltwater to intrude further inland and creates thicker interface compared to the cases without tidal effect. Tidal fluctuations also change the configuration of the interface because of the changes in the flow pattern and the velocity of groundwater near the shoreline. If the depth of an aquifer is much larger than tidal amplitudes, the tidal fluctuation does not have much effect on how far the seawater intrudes into the aquifer but a significant change is observed in the configuration of salt-concentration contours [62]. This change is more pronounced at the top of the aquifer than at the bottom, and is caused by the infiltration of saltwater into the top of the aquifer during the high tides.

Tidal effects on coastal aquifers have been subject to numerous recent studies [63–68]. Inouchi *et al.* [68] presented an approximate analytical solution and a numerical model for analyzing seawater intrusion in a confined aquifer including the effects of tides. Nielsen [64] reported the first analytical investigation on the slope effects. Essink [69] studied the impact of sea-level rise on saltwater intrusion in the Netherland. From the year 1990 to 2100, an average of 49 cm of global mean sea-level rise is estimated, which will accelerate the salinization process in the aquifers and shift the mixing zone between freshwater and saltwater further inland. This situation will seriously impact every coastal aquifer by easing the upconing process where groundwater is heavily exploited [69]. Ataie-Ashtiania *et al.* [62] investigated the effects of tidal fluctuations on seawater intrusion and groundwater dynamics in an unconfined aquifer. They reported that the effects of tidal fluctuations are more significant for a sloping beach than for a vertical face with more pronounced saltwater intrusion. In the case of the sloping beach, unlike the vertical face beaches, the saltwater can easily move inland over the beach at the high-tide stage and then infiltrate vertically through the top of the aquifer. They also stated that neglecting the effects of tidal fluctuations underestimates the saltwater-intrusion impact on groundwater quality near the shore, because the large-amplitude tidal fluctuations force the seawater to intrude further inland and also create a thicker interface in shallow coastal aquifers. Kim *et al.* [70] conducted field studies in a multilayered coastal aquifer in the eastern part of Jeju Island, Korea, to observe the tidal effects on seawater intrusion. They reported that tidal effects on the groundwater level reached up to 3 km inland from the coastline. They found a zone where freshwater and saltwater moved alternately in opposite directions, as influenced by tidal fluctuations.

Although the tidal influence on groundwater dynamics has been studied extensively, the effects of tides on the fate of chemicals in the aquifer have not been investigated adequately [71]. These effects should be quantified in order to determine the pathways of land-originated nutrients and contaminants entering seawater, and provide useful information for improving strategies for sustainable coastal resources management and development. Tide-induced groundwater fluctuations and oceanic oscillation affect the chemical transport and transformation in the aquifer near the shore. Li *et al.* [71] reported that a subsurface estuary may play an important role in determining the subsurface chemical fluxes to coastal waters. They used MODFLOW and PHT3D [72] to model contaminant transport and biodegradation in coastal aquifers affected by tidal oscillations. They used toluene as a representative biodegradable contaminant and oxygen as the electron acceptor. Their simulation results demonstrated that tidal fluctuations lead to the formation of an oxygen-rich zone in the near-shore aquifer area. Aerobic bacterial activity sustained by high oxygen concentration in this active zone degrades the contaminants. These effects may have significant implications for the beach environment [71]. El-Kadi [73] developed a model for hydrocarbon biodegradation in tidal aquifers. He modified SUTRA [30, 31] to simulate multispecies fate and transport and combined with a bacterial growth submodel. He applied the model to a hypothetical tidal aquifer. He reported that tides cause additional mixing for nutrients and oxygen-enhancing degradation in the unsaturated zone of the coastal aquifer. He also developed a quantitative approach to assessing the bioremediation in tidal aquifers.

7 Control and management of saltwater intrusion

The first step in control and management of saltwater intrusion is to collect sufficient data to adequately understand the costal aquifer system and its associated problems. Existing data on water levels and salinity in coastal wells should be reviewed. A data base consisting of the present situation of production wells and their current and historical pumping rates, recharge estimates, aquifer hydrogeologic parameters, and estimated position of the interface should be obtained. Once the available data and information have been collected and reviewed, a conceptual model of the mechanism of intrusion is hypothesized [42].

The second step is to use an appropriate numerical model to gain a deeper understanding of intrusion and to test the hypothetical conceptual model. Modeling lies at the heart of control and management processes. Measures considered for preventing intrusion must be tested using the models. It is recommended that a saltwater-intrusion model be developed before any field implementation of those measures is carried out considering their high cost [42]. The developed and calibrated intrusion model can provide a clear picture of the problem and potential threats. Once the problem of saltwater intrusion is clearly identified, potential means of mitigating intrusion can be investigated. The desired state after restoration, in terms of sustainable rates of withdrawals and the groundwater tables and piezometric levels should be determined by water managers. Each option considered to be a solution to the problem must be tested using the numerical model to determine its effectiveness and applicability.

Maimone *et al.* [42] gave examples of some of the potential measures against saltwater intrusion as follows:

1. Enhancing aquifer recharge to increase freshwater heads to resist saltwater intrusion. Aquifer recharge can be enhanced by spreading any available surface water, or treated waste water, or by capturing surface runoff in recharge basins and allowing them to infiltrate.
2. Lowering the demand for water to reduce pumping from the aquifer. This can be done by convincing or educating the public not to waste water, or increasing water prices, or supplying treated waste waters for irrigation and other nonpotable water uses. In some cases, legal action to forbid water uses for certain times and for certain activities (i.e. car washing) may be helpful to reduce water demand.
3. Creating hydraulic injection barriers to prevent intrusion into unaffected portion of the aquifer by injecting treated waste waters to form a narrow zone in which the freshwater gradient is toward the sea.
4. Tapping alternative aquifers that are located either below or above the impacted aquifer. This relieves the pumping stress on the impacted aquifer.
5. Relocating the wells to areas of higher freshwater heads or areas less susceptible to intrusion. Relocation can also be used to spread out the pumping cone of depression and reducing the potential localized intrusion of upconing.
6. Modifying pumping rates or pumping schedules to allow the well heads to recover.
7. Restricting the pumping rates or the placement of new wells.
8. Replacing deep wells with horizontal wells for skimming freshwater, i.e. Maui-type horizontal shafts.
9. Constructing physical barriers such as slurry walls or sheet piles that can be applied in shallow intrusion situations in small-scale projects.
10. Extracting saltwater using scavenger wells while freshwater pumping continues to stabilize or lower the upconing and increase the storage capacity of freshwater zone.
11. Conjunctive use of surface water and groundwater to offset the excessive reliance on groundwater.
12. The natural recharge can be increased by proper land use (natural vegetation and choice of crops), land-tillage practices, the installation of check dams, retention and detention basins for flood control, and weirs in surface waters, so as to raise the water levels therein and to divert water to adjacent spreading grounds. The main principle of these measures is to hold up the water as long and as much as possible in order to give it more time for infiltration, rather than to let it run off directly. Most of these measures are also favorable in erosion and flood-control terms, but the quality of the water infiltrating in urban areas may be doubtful.

The most important aspect in coastal-aquifer management is the selection of alternative solutions, all of which have some kind of complex tradeoff. The ideal selection of the best alternative requires running saltwater-intrusion and decision-support models. All aspects of decision-making criteria, including economic, environmental, social, technical, and political considerations, should be involved in the selection process of the alternative solutions. Groundwater development and restoration of deteriorated fresh groundwater resources in coastal aquifer systems requires integrated water management of surface water and groundwater, both in terms of water quantity and water quality. Integrated use of numerical models with a geographic information system (GIS) for the input data and presentation of the model output will be helpful in the decision-making procedure. If there is no comprehensive model and/or computer capacity is available, or if there is not enough knowledge or experience with such models, simpler models, such as those ignoring dispersion and assuming sharp interfaces and steady-state conditions, can be used. Those simpler models can be surprisingly helpful for preliminary analysis.

Saltwater barrier projects in Los Angeles County, California were among the good examples showing the complex challenges to be faced in managing a coastal aquifer system. The Central and West Coast groundwater basins are two coastal aquifer systems located adjacent to the Pacific Ocean in California. Severe groundwater exploitation from these basins from the early 1900s to the late 1950s caused saltwater intrusion taking coastal wells out of use, and threatening the sustainability of water supply to the area. Groundwater-management agencies took major steps to halt the intrusion and control the overdraft including construction of freshwater injection wells along the coast, limiting the annual amount of groundwater pumping by legislation, purchasing water from alternative sources for making up the annual and accumulated overdrafts and for use in artificial replenishment and injection wells [74].

The first injection well was tested in the early 1950s by the Los Angeles County Flood Control District using an abandoned water well in Manhattan Beach. The test was successful in creating a freshwater mound and reversing the gradient back to seaward [74]. With the success of the first test, a larger test was carried out including 9 recharge wells, spaced 152.4 m apart, and 54 observation wells from February 1953 to June 1954. This test successfully created a pressure ridge along the injection line, reversing the previous landward gradient to a seaward gradient, which stopped saltwater intrusion [75]. Three major saltwater barrier projects were successfully implemented in California. The West Coast project started in 1953 and had 153 injection and 276 observation wells along a 14.5-km stretch as of 2003. The Alamitos Gap (Los Angeles County) project started in 1964 and had 44 injection and 4 extraction wells, and 239 observation wells along a 3.5-km stretch as of 2003. Dominguez Gap (Los Angeles County) project started in 1969 and had 94 injection and 232 observation wells along a 6.9-km stretch as of 2003 [76]. In all of these projects, potable waters purchased from the Colorado River and north California were used to ensure the water quality of groundwater reserves.

Because of the high maintenance and operation costs of the injection wells, the search for new alternatives is underway in the saltwater-barrier projects in Los Angeles County. The cost of injection water was steadily increasing since 1960, from 1.7 cent m^{-3} to a maximum of 42.8 cent m^{-3} in 2001. In the 2001 water year, a total of 37.48 million m^3 water was injected into barriers at a cost of nearly \$15 million. Johnson and Whitaker [76] reported that nine alternative saltwater barriers were identified by URS Greiner and Woodward-Clyde [77], including slurry walls, deep-soil mixing, grout curtains, jet grouting, *in-situ* vitrification, channel lining, rubber dams, nitrogen-gas injection, and biological barrier walls. After a through economic and technical analysis using numerical models and optimization models, the nitrogen-gas injection and deep-soil-mixing alternatives were recommended over traditional injection wells. Pilot testing of nitrogen-gas injection alternative is currently underway and preliminary designs for deep-soil mixing were completed [76].

This project showed that injection wells have been successfully used to both control saltwater intrusion and to replenish the overexploited aquifers in California since the 1950s. However, the rising cost of the injection water requires other alternatives to injection wells to be sought. Saltwater-intrusion models in conjunction with optimization models were utilized to maximize the groundwater production while minimizing injection requirements with the constriction of drinking-water quality standards. Based on these results, new alternatives were found to stop saltwater intrusion. Reichard and Johnson [78] applied simulation-optimization methods to the West Coast Basin of coastal Los Angeles to obtain useful quantitative guidance for controlling seawater intrusion. Their goal was to determine the most cost-effective way to raise water levels along the coast, either by increasing injection or reducing pumpage through substitution of delivery of surface water, so as to better control seawater intrusion. For the base-case optimization analysis, assuming constant groundwater demand, substitute delivery was determined to be most cost effective. These studies showed the complexity of the coastal-aquifer management due to saltwater intrusion, and they also showed the necessity of saltwater-intrusion and optimization models in decision making and management processes in coastal aquifers.

8 Summary and conclusion

The current understanding of the saltwater-intrusion process, interaction between saltwater–freshwater environments, and their hydraulic and hydrogeologic characteristics were presented in this chapter. In the saltwater-intrusion process, heavier saltwater moves inland until it is balanced by the freshwater along the saltwater/freshwater interface. Pressure balance occurs at both sides of this interface. Excessive groundwater pumping, changes in land uses, reducing the natural recharge rates, and other human activities reducing the freshwater heads in coastal aquifers change this balance in favor of saltwater. In many aquifers, the occurrence and movement of saltwater have been changed by groundwater pumping. Human activities not only cause saltwater intrusion but also cause adverse impacts on coastal ecosystems. Overexploitation of groundwater resources reduces the seaward flow of freshwater to the coastal ecosystem, increasing the salt concentrations and reducing dissolved oxygen and some nutrients carried by groundwater flow in estuaries. Agricultural practices may also influence the salinity and geochemical budgets of coastal ecosystems that may greatly affect the flora, fauna, and aquatic habitats in coastal ecosystem. Nutrient overenrichment can lead to excessive production of algal biomass, loss of important habitats such as seagrass beds and coral reefs, changes in marine biodiversity and distribution of species, and depletion of dissolved oxygen and associated die-offs of marine life.

Several numerical models were developed to simulate the saltwater-intrusion process, including SHARP, SUTRA, and SEAWAT. Numerical models simulate the saltwater/freshwater interface as either sharp or diffusive. The former approach assumes no mixing between saltwater and freshwater. The models that use this approach are called sharp-interface models. They couple saltwater and freshwater flow equations along the interface. In the regional scale, these models give successful results in terms of general movement of saltwater interface with less computational complexity. However, these models can not be used if saltwater concentrations need to be predicted, e.g. to asses water-quality problems in pumping wells. To simulate the transition zone (or mixing zone) between the saltwater and freshwater, one needs to use the latter type of models. These models simulate single fluid flow with variable density changing with salt concentration using a state equation describing the relationship between fluid density and its salt concentration. This approach is more realistic but it requires more computational efforts compared to the sharp-interface approach. There are three types of saltwater intrusion: lateral and upward saltwater intrusion (regional), downward leakage from brackish surface water (local), and saltwater upconing beneath a pumping well (local).

Water needs for increasing populations along coastal zones and economic developments produce a number of complex and unique challenges to hydrologists, water-resources managers, and public decision makers. Today, most of the coastal aquifers are generally under the threat of saltwater intrusion due to intensive groundwater pumping. A consensus on an effective management option by many stakeholders and other concerned parties must be developed to cope with the potential threat of saltwater intrusion. This effective management strategy must be supported by suitable numerical models and field investigations. The remedial or preventive measures against saltwater intrusion can be one or a combination of the followings: i) increasing recharge naturally or artificially, ii) decreasing the groundwater extraction by demand management or alternative water supplies, iii) creating barriers (hydraulic or physical) to stop or reverse inland saltwater movement, iv) relocation of pumping wells or changing the schedule of pumping, and v) implementing conjunctive use of surface and groundwater resources.

The time scale of observing the benefits of management decisions may be relatively long, but if these critical coastal resources are adversely impacted by overexploitation then consequent saltwater intrusion will be very costly and last long term. Remediation measures for these impacted coastal freshwater resources will be very difficult and expensive to implement. In many cases, saltwater-impacted coastal aquifers cannot be restored to a viable freshwater condition and have to be abandoned.

References

[1] Kumar, C.P., Management of groundwater in saltwater ingress coastal aquifers. *Groundwater Modelling and Management*, ed. N.C. Ghosh & K.D., Sharma, Capital Publishing Company: New Delhi, pp. 540–560, 2006.

[2] Reilly, T.E. & Goodman, A.S., Analysis of saltwater upconing beneath a pumping well. *Journal of Hydrology*, **89(3–4)**, pp. 169–204, 1987.

[3] Braithwaite, F., On the infiltration of saltwater into the springs of wells under London and Liverpool. *Proc. Inst. Civil Engineers*, **14**, pp. 507–523, 1855.

[4] Ghyben, W.B., *Nota in verband met de voorgenomen putboring nabij Amsterdam*, Tijdschrift van Let Koninklijk: Inst. Van Ing., 1888.

[5] Herzberg, A., Die Wasserversorgung einiger nordseebader: J. Gasbeleucht. *Wasserversorg*, **44**, pp. 815–819, 1901.

[6] Cheng, A.H.D. & Quazar, D., Analytical solutions (Chapter 6). *Seawater Intrusion in Coastal Aquifers- Concepts, Methods and Practices*, eds J. Bear, A.H.D. Cheng, S. Sorek, D. Quazar & I. Herrera, Kluwer Publishers: Dordrecht, pp. 163–191, 1999.

[7] Bear J., *Hydraulics of groundwater*, McGraw Hill: New York, p. 567, 1979.

[8] Bear J., *Dynamics of Fluids in Porous Media*, American Elsevier: New York. p. 746, 1972

[9] Todd, D.K., *Groundwater Hydrology*, Wiley: New York, p. 535, 1980.

[10] Bear, A.V. & Verruijt, A., *Modeling Groundwater Flow and Pollution, Theory and Applications of Transport in Porous Media*; D. Reidel Publishing Co. p. 414, Boston, MA, 1987.

[11] Zhang Q., Volker, R.E. & Lockington, D.A., Influence of seaward boundary condition on contaminant transport in unconfined coastal aquifers. *Journal of Contaminant Hydrology*, **49(3)**, pp. 201–215, 2001.

[12] Henry, H.R., Salt intrusion into freshwater aquifers. *J. Geophys.Res.*, **64**, pp. 1911–1919, 1959.

[13] Essaid, H.I., The computer model SHARP, a Quasi three dimensional finite-difference model to simulate freshwater and saltwater flow in layered coastal aquifer systems. *USGS Water Resources Investigations Report 90-4130*, Reston, VA., 1990.

[14] Essaid, H.I., A multilayered sharp-interface model of coupled freshwater and saltwater flow in coastal systems: Model development and application. *Water Resources Research*, **26(7)**, pp. 1431–1454, 1990.

[15] Henry, H.R., Effects of dispersion on salt encroachment in coastal aquifers. *US Geological Survey Water-Supply Paper*, **1613-C**, pp. C71–C84, 1964.

[16] Pinder, G.F. & Cooper, H.H., A numerical technique for calculating the transient position of the saltwater front. *Water Resour. Res.*, **6(3)**, pp. 875–882, 1970.

[17] Huyakorn, P.S., Anderson, P.F., Mercer, J.W. & White, J.H.O., Saltwater intrusion in aquifers: development and testing of a three-dimensional finite element model. *Water Resour. Res.*, **23**, pp. 293–312, 1987.

[18] Galeati, G., Gambolati, G. & Neuman, S.P., Coupled and partially coupled Eulerian–Lagrangian model of freshwater-seawater mixing. *Water Resour. Res.*, **28**, pp. 147–165, 1992.

[19] Putti, M. & Paniconi, C., Picard and Newton Linearization for the coupled model of saltwater intrusion in aquifer. *Adv Water Resour*, **18(3)**, pp. 159–170, 1995.

[20] Das, A. & Datta, B., Optimization based solution of density dependent seawater intrusion in coastal aquifers. *J. Hydrol. Eng., Am. Soc. Civil Eng.*, **5**, pp. 82–89, 2000.

[21] Volker, R.E. & Rushton, K.R., An assessment of the importance of some parameters for seawater intrusion and a comparison of dispersive and sharp-interface modeling approaches. *J Hydrol*, **56**, pp. 239–50, 1982.

[22] Simmons, C.T., Fenstemaker, T.R. & Sharp, J.M., Jr., Variable density groundwater flow and solute transport in heterogeneous porous media: approaches, resolutions and future challenges. *J Contam. Hydrol.*, **52(1–4)**, pp. 245–275, 2001.

[23] Diersch, H.-J.G. & Kolditz, O., Variable-density flow and transport in porous media: approaches and challenges (Review paper). *Adv. Wat. Res.*, **25**, pp. 899–944, 2002.

[24] Simmons, C., Variable density groundwater flow: from current changes to future possibilities. *Hydrogeol. J.*, **13**, pp. 116–119, 2005.

[25] Freeze, R.A. & Cherry, J.A., *Groundwater*, Prentice-Hall: Englewood Cliffs, New Jersey, p. 604, 1979.

[26] Hubbert, M.K., The theory of groundwater motion. *Journal of Geology*, **48(8)**, pp. 785–944, 1940.

[27] Reilly, T., Analysis of groundwater systems in freshwater-saltwater environments (Chapter 18). *Regional Groundwater Quality*, ed. W.M. Alley, Van Nostrand Reinhold: New York, NY, pp. 443–469, 1993.

[28] Essaid, H.I., USGS SHARP Model (Chapter 8). *Saltwater Intrusion in Coastal Aquifers- Concepts, Methods and Practices*, eds J. Bear, A.H.D. Cheng, S. Sorek, D. Quazar & I. Herrera, Kluwer Academic Publishers: Dordrecht, The Netherlands, pp. 213–248, 1999.

[29] Wilson, J.L. & Da Costa, A.S., Finite element simulations of a saltwater/freshwater interface with indirect toe tracking. *Water Resources Research*, **18(4)**, pp. 1069–1080, 1982.

[30] Voss, C.I., SUTRA – A finite-element simulation model for saturated-unsaturated fluid density dependent groundwater flow with energy transport or chemically reactive single species solute transport. *USGS Water Resources Investigations Report 84-4369*, USGS, p. 409, 1984.

[31] Voss, C.I. & Provost, A.M., SUTRA, A model for saturated-unsaturated variable-density groundwater flow with solute or energy transport. *USGS Water-Resources Investigations Report 02-4231*, p. 250, 2002.

[32] Weiss, E., A model for the simulation of flow of variable-density groundwater in three-dimensions under steady state conditions. USGS: Reston, VA, *USGS Open File Report 82-352*, p. 59, 1982.

[33] Kontis, A.L. & Mandle, R.J., Modification of a three-dimensional groundwater flow model to account for variable water density and effects of multiaquifer wells. *Water Resources Investigations Report 87-4265*, USGS: Reston, VA, p. 78, 1988.

[34] Langevin, C.D. & Guo, W., MODFLOW/MT3DMS–Based simulation of variable-density ground water flow and transport. *Ground Water*, **44(3)**, pp. 339–351, 2006.

[35] McDonald, M.G. & Harbaugh, A.W., A modular three dimensional finite-difference groundwater flow model. *USGS Techniques of Water-Resources Investigations*, Book 6, Chapter A1, USGS, p. 586, 1988.

[36] Harbaugh, A.W., Banta, E.R., Hill, M.C. & McDonald, M.G., MODFLOW-2000, the US Geological Survey modular groundwater model: User guide to modularization concepts and the groundwater flow process. *USGS Open-File Report 00-92*. USGS, p. 121, 2000.

[37] Zheng, C. & Wang, P.P., MT3DMS, A Modular Three-Dimensional Multispecies Transport Model for Simulation of Advection, Dispersion and Chemical Reactions of Contaminants in Groundwater Systems. Vicksburg, Mississippi: Waterways Experiment Station, US Army Corps of Engineers, 1999.

[38] Langevin, C.D., Shoemaker, W.B. & Guo, W., MODFLOW-2000, the U.S. Geological Survey Modular Ground-Water Model – Documentation of the SEAWAT-2000 version with the variable-density flow process (VDF) and the integrated MT3DMS Transport Process (IMT). *USGS Open-File Report 03-426*. USGS, p. 43, 2003.

[39] Guo, W. & Langevin, C.D., User's Guide to SEAWAT: A Computer Program for Simulation of Three-Dimensional Variable-Density Ground-Water Flow. *Techniques of Water-Resources Investigations Book 6 Chapter A7*, p. 77, 2002.

[40] Motz, L.H., Representing the Saltwater-Freshwater Interface in Regional Groundwater Flow Models. Groundwater and Saline Intrusion: Selected Papers from the 18th Saltwater Intrusion Meeting, SWIM 2004, Cartegena, Spain, May 31-June 2, 2004, Hydrogeology and Groundwater Series, No. 15, *Geological Survey of Spain*, eds L. Araguas, E. Custodio & M. Manzano, pp. 33–39, 2005.

[41] Zheng, C. & Bennett, G.D., *Applied Contaminant Transport Modeling*, 2nd edn. JohnWiley & Sons Inc: New York, p. 621, 2002.

[42] Maimone, M., Harley, B., Fitzgerald, R., Moe, H., Hossain, R. & Heywood, B., Coastal aquifer planning elements (Chapter 1). *Coastal aquifer management; monitoring, modeling, and case studies*, ed. A.H.D. Cheng & Ouazar, D. Lewis Publishers: Boca Raton, London, New York, and Washington D.C. pp. 1–28, 2004.

[43] Sorek, S. & Pinder, G.F., Survey of computer codes and case histories. *Saltwater Intrusion in Coastal Aquifers – Concepts, Methods, and Practices*, ed. J. Bear, *et al.*, Kluwer Academic Publisher: AA Dordrecht, The Netherlands, pp. 399–461, 1999.

[44] Anderson, M.P. & Woessner, W.W., *Applied Groundwater Modeling – Simulation of Flow and Advective Transport,* Academic Press: San Diego, California, p. 381, 1992.

[45] Konikow, L.F. & Reilly, T.E., Groundwater modeling. *The Handbook of Groundwater Engineering*, ed. J.W. Delleur, CRC Press: Boca Raton, pp. 20-1-20-40, 1999.

[46] Motz, L.H., Saltwater upconing in an aquifer overlain by a leaky confining bed. *Groundwater*, **30(2)**, pp.192–198, 1992.

[47] Simmons, C.T., Narayan, K.A. & Wooding, R.A., On a test case for density-dependent flow and solute transport models: the salt lake problem. *Water Resour. Res.*, (**35**), pp. 3607–3620, 1999.

[48] Simpson, M.J. & Clement, T.P., Theoretical analysis of the worthiness of Henry and Elder problems as benchmarks of density-dependent groundwater flow models. *Advances in Water Resources*, **26**, pp. 17–31, 2003.

[49] Elder, J.W., Transient convection in a porous medium. *Journal of Fluid Mechanics*, **27(3)**, pp. 609–623, 1967.

[50] Hughes, J.D. & Sanford, W.D., SUTRA – MS A Version of SUTRA Modified to Simulate Heat and Multiple-Solute Transport. *USGS Open File Report 2004-1207*, p. 141, 2004.

[51] Gingerich, S.B. & Voss C.I., Three-dimensional variable-density flow simulation of a coastal aquifer in southern Oahu, Hawaii, USA. *Hydrogeology Journal*, **13(2)**, pp.436–450, 2005.

[52] Hunt, C.D., Ewart, C.G. & Voss, C.I., Region 2 – Hawaiian Islands. *Hydrogeology: The Geology of North America*, eds W. Back, J.S. Rosenshein & P.R. Seaber, Vol. 0-2, Geolog. Soc. America: Boulder Colorado, pp. 255–262, 1988.

[53] Voss, C.I., USGS SUTRA Code – history, practical use, and application in Hawaii (Chapter 9). *Saltwater intrusion in coastal aquifers- concepts, methods and practices*, ed. J. Bear, A.H.-D. Cheng, S. Sorek, D. Ouazar & I. Herrera, Kluwer Academic Publishers: Dordrecht, The Netherlands, pp. 249–314, 1999.

[54] Voss, C.I. & Souza, W.R., Variable density flow and solute transport simulation of regional aquifers containing a narrow freshwater–saltwater transition zone. *Water Resources Research*, **23(10)**, pp. 1851–1866, 1987.

[55] Dausman, A.M. & Langevin, C.D., Movement of the saltwater interface in the Surficial Aquifer System in response to hydrologic stresses and water-management practices. Broward County, Florida: *U.S. Geological Survey Scientific Investigations Report 2004-5256*, Reston, VA. 2005.

[56] Masterson, J.P., Simulated Interaction Between Freshwater and Saltwater and Effects of Ground-Water Pumping and Sea-Level Change, Lower Cape Cod Aquifer System, Massachusetts. *USGS Scientific Investigations Report 2004-5014*, p. 78, 2004.

[57] Schneider, J.C. & Kruse, S.E., Assessing selected natural and anthropogenic impacts on freshwater lens morphology on small barrier Islands: Dog Island and St. George Island, Florida, USA, *Hydrogeology Journal*, **14(1–2)**, pp. 131–145, 2006.

[58] Zimmermann, S., Bauer, P., Held, R., Kinzelbach, W. & Walthe, J.H., Salt transport on islands in the Okavango Delta: Numerical investigations. *Advances in Water Resources*, **29(1)**, pp. 11–29, 2006.

[59] Renken, R.A., Synthesis on the Impact of 20th Century Water-Management and Land-Use Practices on the Coastal Hydrology of Southeastern Florida. *U.S. Geological Survey Open-File Report 00-449*, pp. 41–42, 2000.

[60] National Research Council, *Clean Coastal Waters - Understanding and Reducing the Effects of Nutrient Pollution*, National Academy Press: Washington, DC, p. 405, 2000.

[61] Gingerich, S.B. & Oki D.S., Ground Water in Hawaii. *USGS Fact Sheet 126-00*, p. 6, 2000.

[62] Ataie-Ashtiani, B., Volker, R.E. & Lockingtonb, D.A., Tidal effects on seawater intrusion in unconfined aquifers. *Journal of Hydrology,* **216,** pp. 17–31, 1999.

[63] Underwood, M.R., Peterson, F.L. & Voss, C.L., Groundwater tens dynamics of atoll islands. *Water Resour. Res,* **28 (11)**, pp. 2889–2902, 1992.

[64] Nielsen, P., Tidal dynamics of the watertable in beaches.*Water Resour. Res.* **26(9)**, pp. 2127–2134, 1990.

[65] Nielsen, P., Aseervatham, R., Fenton, J.D. & Perrochet, P., Groundwater waves in aquifers of intermediate depths. Adv. *Water Resour.,* **20,** pp. 37–43, 1997.

[66] Li, L., Barry, D.A., Stagnitti, F., Parlange, J.-Y. & Jeng, D.S., Beach water table fluctuations dueto spring-neap tides, *Adv. Water Resour.*, **23**, pp. 817–824, 2000.

[67] Jeng, D.S., Li, L. & Barry, D.A., Analytical solution for tidal propagationin a coupled semi-confined/phreatic coastal aquifer, *Adv. Water Resour.*, **25(5)**, pp. 577–584, 2002.

[68] Inouchi, K. Kishi, Y. & Kakinuma, T., The motion of coastal groundwater in response to the tide. *Journal of Hydrology,* **115**, pp. 165–191, 1990.

[69] Essink, O.G.H.P., Impact of sea-level rise in the Netherlands (Chapter 14). *Saltwater intrusion in coastal aquifers- concepts, methods and practices*, ed. J. Bear, A.H.-D. Cheng, S. Sorek, D. Ouazar & I. Herrera, Kluwer Academic Publishers: Dordrecht, The Netherlands, pp. 507–530, 1999.

[70] Kim K.-Y., Seong, H., Kim T., Park, K.H., Woo, N.C., Park, Y.S., Koh, G.W. & Park, W.B., Tidal effects on variations of fresh–saltwater interface and groundwater flow in a multilayered coastal aquifer on a volcanic island (Jeju Island, Korea). *Journal of Hydrology,* **14**, pp. 462–472, 2006.

[71] Li, L., Barry, D.A., Jeng, D.-S. & Prommer, H., Tidal dynamics of groundwater flow and contaminant transport in coastal aquifers (Chapter 6). *Coastal Aquifer Management; Monitoring, Modeling, and Case Studies*, ed. A.H.-D. Cheng, & D, Ouazar, Lewis Publishers: Boca Raton, London, New York, and Washington DC, pp. 115–141, 2004.

[72] Promer, H., PHT3D, A Reactive Multicomponent Transport Model for Saturated Porous Media. User's Manual Version 1.0 Contaminated Land Assessment and Remediation Research Center. The University of Edinburgh, UK, pp. 141, 2002.

[73] El-Kadi, A.I., Modeling hydrocarbon biodegradationin tidal aquifers with water-saturation and heat inhibition effects. *J. Contam. Hydrology*, **51**, pp. 97–125, 2001.

[74] Lipshie, S.R. & Larson, R.A., The west coast basin, Dominguez Gap, and Alamitos seawater – intrusion barrier system, Los Angeles and Orange Counties, California. *AEG News*, **38(4)**, pp. 25–2, 1995.

[75] California Department of Water Resources, Seawater Intrusion in California, Appendix B by the Los Angeles County Flood Control District. *Bulletin* **No. 63**, 1957.

[76] Johnson, T.A. & Whitaker, R., Saltwater intrusion in the coastal aquifers of Los Angeles County, California (Chapter 2). *Coastal aquifer management; monitoring, modeling, and case studies*, ed. A.H.-D. Cheng & D., Ouazar, Lewis Publishers: Boca Raton, London, New York, and Washington DC, pp. 29–48, 2004.

[77] URS Greiner and Woodward-Clyde, Final Report: Alternative Seawater Barrier Feasibility Study, 1999.

[78] Reichard, E.G. & Johnson, T.A., Assessment of regional management strategies for controlling seawater intrusion. *J. Water Res. Plan. and Mngmt*, **131(4)**, pp. 280–291, 2005.

CHAPTER 9

Restoration and protection plan for the Nawiliwili Watershed, Kauai, Hawaii, USA

Aly I. El-Kadi[1, 2], Monica Mira[1], James E.T. Moncur[1, 3] & Roger S. Fujioka[1]
[1]*Water Resources Research Center,*
[2]*Department of Geology and Geophysics and*
[3]*Department of Economics, University of Hawaii, USA.*

Abstract

This study dealt with developing a restoration and protection plan for the Nawiliwili Watershed, Kauai, Hawaii, USA. The proposed plan covers the nine elements required by the U.S. Environmental Protection Agency for watershed-based plans that are developed or implemented with Section 319 funds to address the requirements of Section 303(d) of the federal Clean Water Act for listed waters. The elements include identification of the causes that will need to be controlled to achieve contaminant-load reductions; an estimate of the load reductions expected for the management measures described in the plan; a description of the nonpoint-source (NPS) management measures that will need to be implemented to achieve the load reductions; an estimate of the amounts of technical and financial assistance, associated costs, and resources needed; an information and education component for the public; a schedule for implementing NPS management measures identified in the plan; a description of interim measurable milestones for determining whether NPS management measures are being implemented; a set of criteria that can be used to determine whether loading reductions are being achieved; and a monitoring component to evaluate the effectiveness of the implementation efforts over time. Limitations of the approaches adopted, especially due to data limitations, are stressed where appropriate.

1 Introduction

As required by the U.S. Environmental Protection Agency (USEPA), watershed-based plans that are developed or implemented with Section 319 funds to address 303(d) listed waters must include at least the nine elements listed below:

1. An identification of the causes and sources that need to be controlled to achieve contaminant-load reductions
2. A description of the nonpoint-source (NPS) management measures that need to be implemented to achieve the load reductions and an identification of the critical areas in which those measures need to be implemented
3. An estimate of the load reductions expected for the suggested management measures
4. An estimate of the amounts of technical and financial assistance needed, associated costs, and the sources and authorities that will be relied upon, to implement this plan
5. An information/education component to enhance public understanding of the project and encourage the public's early and continued participation in selecting and designing the NPS management measures that will be implemented
6. A schedule that is reasonably expeditious for implementing NPS management measures identified in this plan
7. A description of interim measurable milestones for determining whether NPS management measures or other control actions are being implemented
8. A set of criteria that can be used to determine whether loading reductions are being achieved over time and whether substantial progress is being made toward attaining water-quality standards
9. A monitoring component to evaluate the effectiveness of the implementation efforts over time, as measured against the criteria established as described above.

The objective of this study was to assess the status of the Nawiliwili Watershed on Kauai, Hawaii, USA, and to develop a plan for its restoration and future protection. A three-phase study was conducted to achieve this objective. Phase 1 was concerned with validating and documenting existing environmental data [1]. Phase 2 was aimed at identifying current sources of pollution and contamination in the watershed [2]. Finally, Phase 3 dealt with developing a restoration and protection plan for the watershed, which is covered in this chapter. Additional details of the plan are provided in the publication by El-Kadi *et al.* [3].

Examples of similar studies include those for the Gallows Run Watershed in Pennsylvania [4] and the Breton Bay Watershed in Maryland [5]. The current study differs from others in that it deals with a tropical watershed that is also a tourist destination. As should be expected, water-flow conditions and water-quality problems are greatly influenced by land use, climate condition, and hydrological features of the watershed. Although typical water-quality problems exist, such as those due to nutrients and bacteria, approaches for restoration and protection are dependent on the watershed's site-specific condition. For example, the uniqueness of tropical problems is clearly evident in unsuitability of the US mainland bacterial water-quality indicators for tropical environments [6].

Elements of the Nawiliwili Watershed restoration and protection plan, in the order listed above, are covered in the following sections. In addition to the mandated NPS management measures for contaminant-load reductions, Section 3 covers strategies that complement restoration measures.

2 Nawiliwili watershed assessment

2.1 The watershed

Nawiliwili Bay is fed by three main streams: Huleia Stream, Puali Stream, and Nawiliwili Stream (Fig. 1). Major changes have occurred in the Nawiliwili Watershed throughout geologic and historical times, including uses of land, harbor construction, industrialization, and development. Several hundred years ago the flood plain of the Huleia River was simply used for growing taro. Water that was diverted to flood the taro patches ultimately returned to the same watershed. Many taro fields were eventually converted to rice fields around 1860 and soon after, the sugar plantations came in. After World War II, some sugarcane fields were converted to ranches. More drastic changes took place around the 1930s when Nawiliwili Bay was dredged and the land was reclaimed, or filled in, to build the harbor and the breakwater wall. As time passed, Lihue developed into a commercial district. The Kauai Surf Hotel was built and subsequently grew, evolved, and changed to the Kauai Westin Hotel, and then to the Kauai Marriott Hotel. During the time of the hotel expansion, two streams that flowed into Kalapaki Bay were diverted under the hotel, and now they discharge into Nawiliwili Stream. A sewage treatment facility and golf course were placed near the hotel at Kauai Lagoons.

In accordance with section 303(d) of the federal Clean Water Act, Nawiliwili Bay, Nawiliwili Stream, and Huleia Stream are currently listed by the Hawaii Department of Health (HDOH) as water bodies in which water quality is impaired by excessive turbidity. Nawiliwili Bay is also listed as being impaired by excessive nutrients, enterococci, and chlorophyll a. Many of the streams in the Nawiliwili Watershed drain into and can cause pollution conditions in Nawiliwili Bay. Streams, harbor sites, and beaches are used by residents and by tourists for kayaking, swimming, and other recreational activities. In this regard, Kalapaki Beach is a primary swimming area in Nawiliwili Bay.

2.2 Water-quality problems and sources of contaminants

To assess quality problems of the watershed, ten primary sites and four alternative sites were chosen for sampling (Fig. 2). The water-quality parameters measured were turbidity, salinity, temperature, nitrate and phosphate, and fecal indicator bacteria. Data collected over about a year were used to determine point- and nonpoint-source contributions of nutrients and bacteria. They were also used to determine whether the water within the watershed met water-quality standards for the above parameters or if it represented a health hazard. Hydrological models were also used in assessing contributions of point and nonpoint sources of contamination in the watershed.

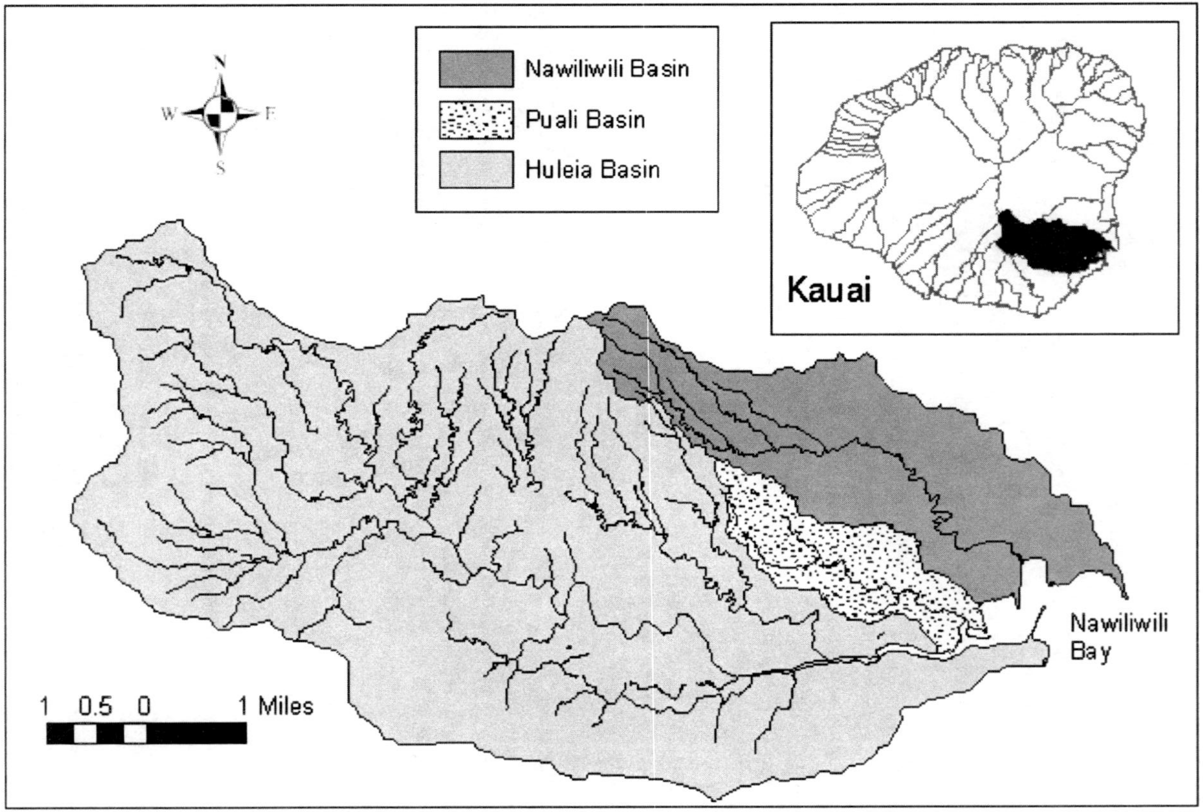

Figure 1: The Nawiliwili Watershed, Kauai with its main perennial streams and their respective basins. (Source: The Internet site of the Hawaii Statewide GIS Program, HDBEDT [7].)

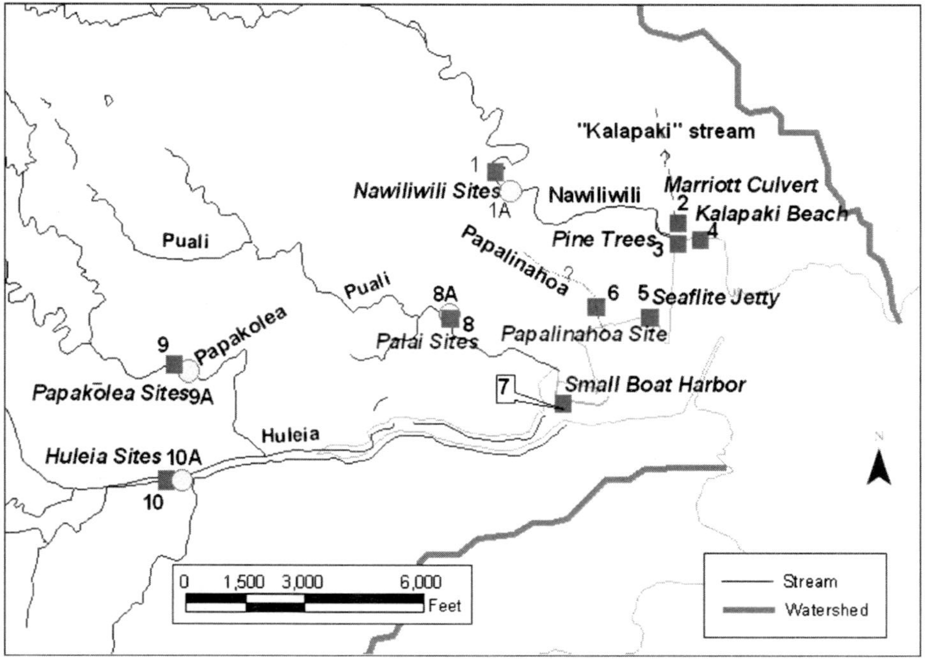

Figure 2: Nawiliwili Watershed sampling sites. Squares represent primary sites, labeled 1 through 10. Circles represent alternative sites, labeled 1A, 8A, 9A, and 10A. Question marks indicate approximate location of undocumented streams. (Source of original maps: HDBEDT [7].)

Based on the assessment, it was concluded that Nawiliwili Bay is subjected to accidental sewage, chemical, fertilizer, and oil/gasoline spills. Sediment sources include agricultural lands, construction sites, channels, a quarry, and urban areas. Nutrients originate from agricultural lands, golf courses, cesspools, forested areas, urban areas, and wastewater treatment spills. Bacterial contamination originates from cesspools, forested areas, urban areas, and wastewater treatment spills. In addition, there is a chance that other chemicals from various sources are also present. The focus for this study, however, is on nutrient and sediment contamination.

2.3 Severity of water-quality problem

An assessment of turbidity data indicated that five of ten sites in Fig. 2 have a 60% probability for turbidity to be at or to exceed a suggested standard of 5 nephelometric turbidity units. The Papakolea Stream site has the highest sediment level and the Nawiliwili Stream site the second highest. These results emphasize the significance of sediment load originating from agricultural land-use districts that feed the two sites. Here it is important to distinguish between lands that are merely within the agricultural land-use district (which may be unused and unmanaged) and lands that are currently used for agricultural production. Up-to-date maps are not available; however, our visits to the area actually confirm that the area upstream from the Papakolea sampling site is mostly unmanaged agricultural land (Fig. 3). The banks of Papakolea Stream are very steep, causing erosion problems. In the stream is a large accumulation of sediment, which might be contributing to the high bacterial count at this site.

Qualitative assessment indicated erosion problems in the Nawiliwili Watershed. The website of the National Resources Conservation Service (NRCS [8]) provides estimates for the loss from fields on Kauai in the range of 6.7 to 11 metric tons/ha/yr. El-Kadi *et al.* [3] used the Universal Soil Loss Equation within a detailed modeling scheme of streamflow output from the watershed (see Section 3). However, the model seems to underestimate sediment loss in comparison to field observations and to NRCS's estimates. Field studies are needed for accurate estimates of sediment losses.

High turbidity seems to be triggered when the daily rainfall rate ranges from 5.08 to 7.62 cm. The high sediment load after a major storm occurred on May 13, 2002 demonstrated the vulnerability of the watershed to rainfall, in terms of sediment loads.

There is an 80% probability for phosphate concentrations at the ten primary sites to be at or to exceed the level of 0.01 mg/l suggested by this study. Seven of the ten sites have an 80% probability for nitrate concentrations to be at or to exceed the suggested level of 0.1 mg/l.

For nitrate, Nawiliwili Stream had the highest load, followed by Puali and Papakolea Streams. The nitrate load per unit hectare was highest at Puali Stream and second highest at Nawiliwili Stream. The Huleia Stream and Nawiliwili Stream sites received the highest loads of phosphate, but the Nawiliwili site alone ranked

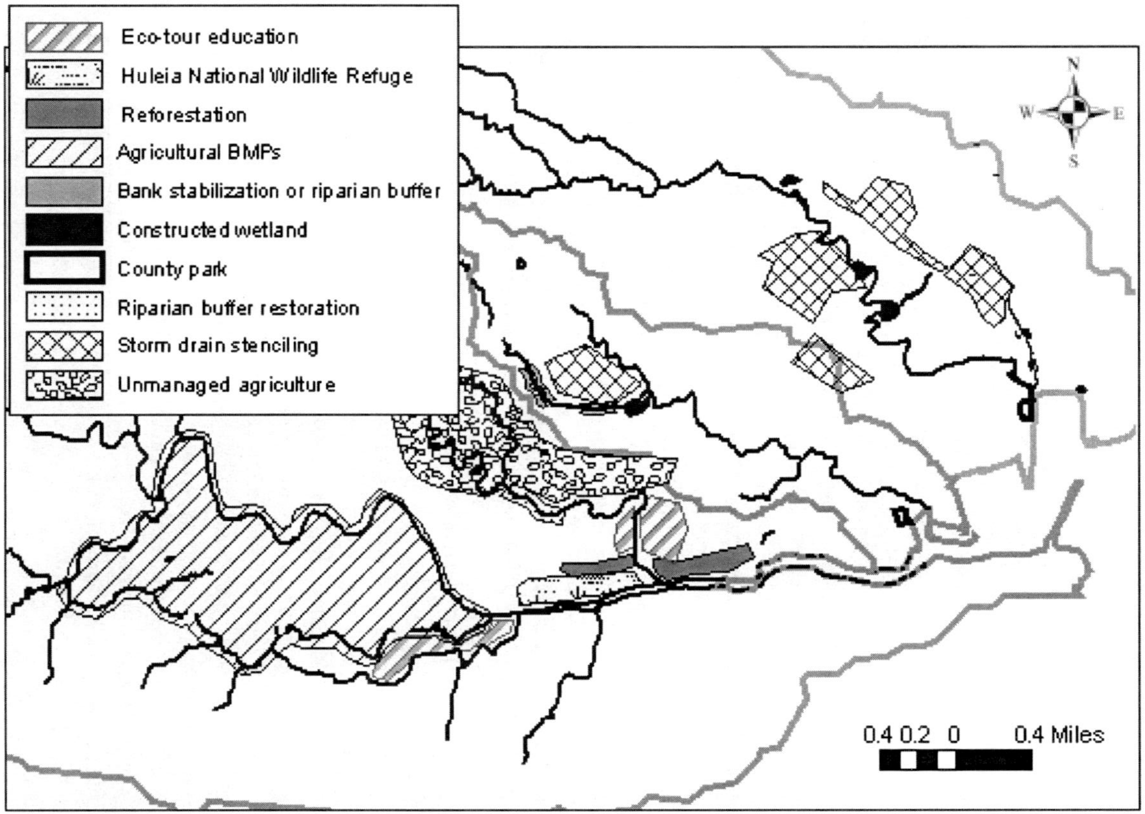

Figure 3: Approximate locations of management measures suggested for the restoration of the Nawiliwili Watershed. (Source of original maps: HDBEDT [7].)

highest in terms of phosphate load per unit hectare. Estimated loads to Nawiliwili Bay were on the order of 5.4 and 1.8 metric tons/yr for nitrate and phosphate, respectively. El-Kadi *et al.* [3] presented a detailed modeling scheme of streamflow output from the watershed. For total nitrate and phosphate, the study estimated 39 and 10.8 metric tons/yr, respectively, which are higher than the values described above. The higher values are most likely more accurate, considering the level of modeling adopted. However, model limitations, including the absence of data for calibration, still put severe restrictions on such estimates.

Most water samples obtained from stream sites greatly exceeded the current USEPA recreational water-quality standards for fecal coliform (200 CFU/100 ml) and enterococci (33 CFU/100 ml). At six sites (two on Nawiliwili Stream and one each at Marriott Culvert, Pine Trees, Papakolea Stream, and Papalinahoa Stream), the geometric mean concentration was above 1000 CFU/100 ml for fecal coliform. At seven sites (the same six sites plus Puali Stream), the geometric mean concentration for enterococci was at or exceeded 1000 CFU/100 ml. Site 1 (Nawiliwili Stream) had the highest geometric mean concentrations for both fecal coliform (71740 CFU/100 ml) and enterococci (2914 CFU/100 ml). Site 2 (Marriott Culvert) had geometric mean concentrations of 4915 CFU/100 ml for fecal coliform and 1939 CFU/100 ml for enterococci.

At Site 4 (Kalapaki Beach), the geometric mean concentrations of fecal coliform (11 CFU/100 ml) and enterococci (14 CFU/100 ml) were below current USEPA standards. On the other hand, the concentrations of enterococci exceeded the state of Hawaii's standard of 7 CFU/100 ml for marine waters. The sample collected on May 14, 2002 was characterized by elevated concentrations of fecal coliform (22,400 CFU/100 ml) and enterococci (14,800 CFU/100 ml). The water-quality data from Site 5 (Seaflite Jetty) revealed elevated concentrations of fecal coliform (372 and 404 CFU/100 ml) and enterococci (232 and 348 CFU/100 ml) only for two rainy-day events.

Exceeding USEPA standards for bacterial indicators should be carefully assessed because it appears that the sources of fecal bacteria on Kauai are environmental in nature and not necessarily from sewage sources. Overland and subsurface flows wash the fecal bacteria from the soil into streams. It thus seems that it is not possible to use such indicators as evidence for reducing levels of bacteria through management decisions. Moreover, it has been concluded that the concentrations of fecal indicator bacteria are not related to health risks from sewage contamination [9]. Thus, more reliable fecal indicators, such as *Clostridium perfringens* and FRNA coliphages, are needed. The U.S. Food and Drug Administration [10] indicates that *C. perfringens* might not be a good indicator of sewage discharge. However, Fujioka and Shizumura [6] showed that *C. perfringens* is better than fecal bacteria as an indicator of sewage contamination in streams in Hawaii. HDOH is using Fujioka's *C. perfringens* water-quality standards to determine when waters are contaminated with sewage. Until completely reliable indicators are identified, it seems sanitary surveys and bacterial source tracking may provide the only definitive answers for assessing sewage contamination. However, the difficulty is that sanitary surveys alone cannot detect underground leakage from a sewage source.

Nearly all of the sampling sites contained low numbers of *C. perfringens*, indicating that the streams in the Nawiliwili Watershed are not being directly contaminated with sewage discharge. The only exception was the Pine Trees site, where the recommended standard of 50 CFU *C. perfringens*/100 ml was exceeded.

Papakolea Stream had the highest concentration of FRNA coliphages, and this was taken as evidence of cesspool contamination. Since some water samples from Nawiliwili, Papalinahoa, and Puali Streams contained elevated levels of FRNA coliphages and low levels of *C. perfringens*, we concluded that these streams are being occasionally contaminated with cesspool wastes.

3 Strategies and actions for improving water quality in the Nawiliwili Watershed

3.1 Managing stormwater runoff and quality

Unabated urban stormwater runoff that is discharged at rapid velocities and high volumes was identified as a priority concern in the Nawiliwili Watershed. Increasing the degree of urbanization can cause irreversible damage to aquatic habitat and biota [11]. Impacts on habitat quality, which in turn affect aquatic biota, begin to occur at relatively low levels of urbanization characterized by 8% to 12% impervious cover [12].

Best management practices (BMPs) can be used to integrate stormwater-management practices with provisions for habitat protection, water quality, and water quantity (e.g. ASCE [13]). Other management practices include better site-design practices (BSDPs), policy changes, and education and outreach programs. BSDPs differ from BMPs regarding their potential to prevent pollutants and runoff volumes from reaching waterways, whereas BMPs only help treat runoff and reduce pollutant loads. BSDPs can protect a watershed by conserving its natural features and resources, by using low-impact site-design techniques, by reducing impervious land cover, or by utilizing natural features for stormwater management (see Georgia Stormwater Management Manual [14]).

Extensive technologies are available for stormwater BMPs and BSDPs. However, only certain designs may be appropriate or feasible for use in the Nawiliwili Watershed, considering that many of these practices are designed for implementation at the inception of land development. For highly urbanized areas, existing stormwater-management practices need to be retrofitted and hence can be more costly. This problem is more severe in the Nawiliwili Stream Basin.

Example BMPs include natural drainage swales, detention basins, ponds, and constructed wetlands. Suggested specific management measures for the Nawiliwili Watershed are shown in Fig. 3. The use of a specific BMP depends on the type of land use. In many cases it may be necessary to design a creative BMP to meet the requirements of a specific site. A future study should be conducted to determine the feasibility, cost, and effectiveness of each BMP through site evaluations and through a series of demonstrations or test projects.

Among suggested BMPs, constructed wetlands provide multiple benefits. Besides being one of the most reliable BMPs for their ability to treat pollutants and improve water quality, constructed wetlands can also control runoff volume by storing it in a shallow basin. By controlling the runoff volume, stream bank erosion caused by runoff from peak storms can be reduced. Constructed wetlands can therefore improve the downstream habitat. By enhancing diverse vegetation they can provide additional wildlife habitat and aesthetic values in urban areas. Constructed wetlands, which can be easily integrated into the landscape, can be used as an educational tool, such as a community-adopted project. Except for the early years, constructed wetlands require minimal maintenance and little energy inputs, and they are generally less expensive than conventional systems [15].

There are, of course, some drawbacks to constructed wetlands. It is possible that invasive species, whose spread is difficult to control, can become established in wetlands. Also, wetlands may increase the temperature of the water that is returned to natural systems, causing potential harm to sensitive fish species [16]. Ample groundwater or another source of base flow is required to sustain the wetland vegetation. Additionally, appropriate vegetation for Hawaii environments will need to be determined.

The biggest drawback to constructed wetlands is that they require large areas of open space and are therefore subject to land availability. Due to this constraint, they are not always feasible solutions in already-developed, densely populated urban areas.

Although a more complete survey of the watershed needs to be conducted, we have already identified some potential sites for constructed wetlands (see Fig. 3). It should also be realized that the site locations are merely suggestions and that the actual selection and construction are beyond the scope of this study. Appropriate negotiation with owners is needed, leading preferably to cooperative partnership.

3.2 Preventing soil erosion and sedimentation from agricultural lands

A large portion of the Nawiliwili Watershed is comprised of agricultural land. Some of this land is being actively farmed and ranched, while other parts are just lying fallow. All-terrain-vehicle (ATV) riding and eco-tours have become growing businesses on agricultural land. Such activities exacerbate erosion by dislodging sediment from roads with 4-wheeling vehicles. Intensive agriculture, such as sugarcane, causes periodic disturbances of soil during harvesting and tilling activities. On the other hand, ranching and permanent crop agriculture may improve water quality by reducing the frequency of these disturbances. However, soil from agricultural roads is subject to erosion and can end up in streams and waterways if not properly managed.

The following sections discuss restoration projects that address erosion from agricultural roads as well as solutions to sedimentation caused by cattle. The NRCS office on Kauai offers an extensive list of BMPs available to farmers, ranchers, and others needing soil-conservation advice. These resources are also available as an online BMP encyclopedia (see for example, National Watershed Manual, National Soil Survey Handbook [17]).

3.2.1 Promoting videotapes produced by soil- and water-conservation districts

In a partnership with Grove Farm and Kauai Coffee Company, the East and West Kauai Soil and Water Conservation Districts (SWCDs) recently completed a project, "Demonstration and Training in Critical Area Stabilization Techniques on Agricultural Roads and Unprotected Waterways," and produced five educational videotapes that explain different methods of erosion control for agricultural roads on Kauai. The videotapes also document demonstration BMP projects that were implemented. In each scenario, a conservation problem was presented with several options for a solution. The solution that was chosen for a site was implemented and lessons learned from the project were then shared. Many of the projects took place in the Nawiliwili Watershed on Grove Farm land. BMPs were tailored for the area, and locally available materials were recycled for use in some of the projects.

3.2.2 Expanding use of conveyor-belt water bars to prevent erosion

A simple project that was demonstrated in the videos described above involved the use of a protruding water bar that was adapted specifically for the Nawiliwili Watershed. The BMP, previously used on forest roads on the mainland, called for the use of cut tires. For Kauai, a conveyor belt from the old sugar mill was substituted for the tires. Recycling old materials is not only environmentally friendly but also cost effective. The belt was nailed to a pressure-treated Albesia board provided by a local tree farmer. The protruding water bar was then placed at a 35° to 45° angle to the road to ensure the swift movement of water from the road. Once the water moves off the road, it can go into a roadside ditch, filter strip, or other outlet. Since the water spends less time on the road, there is less chance for sheet and rill erosion.

3.2.3 Locating water troughs for cattle away from streams

Another practice to reduce sediment inputs to streams is to place water troughs about 30 m (100 ft) away from streams. It has been shown that this practice can keep cattle out of streams up to 90% of the time. Keeping cattle out of streams could also reduce bank erosion and cattle-related bacterial problems. A pilot project should be conducted by one of the ranchers in the Nawiliwili Watershed to determine the effectiveness of this activity.

3.2.4 Developing a "working farm" to demonstrate BMP implementation

A partnership could be formed between NRCS and a local farmer or rancher who has successfully implemented BMPs on his/her property. This person's property would function like a classroom or laboratory where tours or training workshops could be held to demonstrate BMP implementation and their ability to reduce erosion and sediment loads. Any lessons learned or design modifications made for local applications can be offered as a part of the tour or workshop.

3.2.5 Providing solutions for ATV riding and eco-tour erosion

ATV riding and eco-tours take place on privately owned land, and thus their activities are somewhat unregulated. Tour companies need to be educated about impacts that can be associated with their activities and also about preventing such impacts.

The SWCD videotapes are appropriate educational tools for this purpose. The tapes recommend efforts to ensure that only clean water is discharged into streams. Erosion control mats could be used while establishing vegetation to further reduce erosion. The use of filter strips or a small sediment basin lined with geotextile mats could further reduce the amount of sediment reaching impacted streams. The videotapes also discuss ways to stabilize steep road banks. In addition, a self-monitoring program may need to be developed. Designing and implementing an appropriate program will ultimately make tour companies responsible for their actions.

3.3 Updating land-use maps

Updated information is crucial for the development of future plans for the watershed and the implementation of BMPs. For example, it is not possible to distinguish between active agricultural lands and unused or unmanaged lands that are merely within an agricultural land-use district. Obviously, pollutant loads associated with each use can differ, leading to possible errors in identifying sources of such pollutants and in defining solutions for reducing pollutant loads.

3.4 Promoting water recycling and conservation practices

The East and West Kauai SWCDs offer awards to individuals or groups for adopting innovative water conservation practices to irrigate crops without the use of county water. One of the winners, Gary Ueunten, uses rainwater runoff that flows by gravity from his roof to irrigate his two-hectare farm. This method can also be used for landscape irrigation or in fish culturing.

In Hawaii, a combination of a growing population and limited water resources is reducing the availability and quality of drinking-water supplies. There are also environmental problems and financial costs resulting from the disposal of wastewater. Hence increasing the safe use of recycled water can address all these problems. HDOH has long been an advocate for water reuse as long as it does not compromise public health and our valuable water resources [18].

An example of water reuse in the Nawiliwili Watershed concerns Kauai Lagoons, which uses up to 1.2 mgd of R-2 water from the Lihue Waste Water Treatment Plant for golf-course irrigation. According to the Kauai Division of Wastewater, the current average reuse is about 1 mgd. Six injection wells are currently being drilled for emergency overages. Until the drilling is completed, no comment about what currently happens to spills will be made by the division. However, cases have been documented for spills that have found their way to Nawiliwili Bay via streams and diversions.

Obviously, there is a need to increase the use of recycled water. Such uses will have to conform to guidelines for the treatment and use of recycled water [18]. Restrictions include the suitability of the type of recycled water for specific uses.

For the Nawiliwili Watershed, specific actions to encourage recycling include (1) promoting and supporting current recycling efforts, (2) offering incentives such as awards and tax or other types of credits, (3) producing and distributing a pamphlet that offers water-recycling projects and ideas, and (4) launching an educational campaign to use gray water for landscape irrigation. Some of these ideas could be incorporated into educational opportunities for architects, plumbers, contractors, home-builders, and do-it-yourselfers.

3.5 Enforcing and revising current water-quality policies and regulations

As mentioned earlier, stormwater runoff is a major contributing factor in polluting the waterways of the Nawiliwili Watershed. Stormwater-management manuals were designed to reflect policies, and an important factor in the effectiveness of such policies is enforcement. Many of Kauai's policies, laws, and manuals were outdated. The Storm Water Runoff System Manual for the County of Kauai is now available. There have been some positive revisions to the drainage manual, such as the requirement of detention basins for projects of a certain size. Unfortunately, however, many policies ignore new technologies that are currently available. For example, in subsection 3.2 of the Basic Principles of the Storm Water Runoff System Manual, the first statement is "a. Natural Drainage Ways shall be used for storm runoff drainage ways wherever possible." This seems to be a direct contradiction to the reasons stated for adopting the manual if the natural drainage way of choice (stream) continues to be used. The manual would be more consistent with its purpose if the term "natural drainage ways" was more clearly defined to exclude streams but include swales, buffer systems, and other "natural" options. Peak flows and velocities could be reduced substantially.

Because the revised Storm Water Runoff System Manual is designed to address new developments, no amendments address the correction of old problems, such as the contribution of pollutants entrained in stormwater runoff that is currently being discharged by the existing infrastructure. Moreover, because the manual is designed to maintain the current water quality of a watershed, it does not address the problem that many watersheds, including Nawiliwili, do not currently meet the water-quality standards and thus are placed on the 303(d) list by USEPA and HDOH.

It should also include a provision concerning reduction of impervious surface coverage, street design, open-space plans, native revegetation, zoning densities, and lot sizes. Basins can be designated as sensitive, impacted, moderately impacted, and intact; then the most appropriate goals can be assigned for each basin. Ordinances to adopt low-impact development standards and perhaps interim standards may be required as part of a policy revision. Another source for model ordinances that protect local resources is the USEPA [19], whose web site links to the Local Government Environmental Assistance Network.

Kauai's current policies may not reflect the changes that are necessary to make the Nawiliwili Watershed sustainable or to make the waterways comply with the Clean Water Act. The current practice of voluntary action on behalf of the county and landowners has proved to be inadequate to keep the water quality in the Nawiliwili Watershed from exceeding standards set forth by the state government.

Therefore, the recommendation offered by the inhabitants of the Nawiliwili Watershed is to find a mechanism for the enforcement of existing policies. It is also recommended to expand or change policies to include the implementation of more and better BMPs not only to maintain water quality but also to improve it. It is prudent for the County of Kauai to recognize and authorize watershed councils or neighborhood boards for participating in the development of natural resource conservation plans. Finally, there is a need for more coordinated effort between state and county agencies. In the past, the division of responsibility has led to serious problems [20].

3.6 Integrating the ahupuaa concept with modern watershed management

It has been recommended by the Ahupuaa Action Alliance, Hawaii, that the ahupuaa system [21] be adopted as the legal framework for planning and resource management in Hawaii. A holistic approach such as this would contribute to the overall sustainability of the watershed. The integration of the ahupuaa concept into modern watershed management is an attempt to reconnect man, nature, and government. The hope is that when one recognizes one's place in the watershed as a steward, it removes the attitude that the watershed is one's own resource to take from but the government's responsibility to protect. This attitude can then be replaced with the idea that the watershed is a resource for all of its residents as long as they make responsible decisions.

3.7 Controlling invasive and non-native species

Invasive plant and animal species pose a threat not only to all of Hawaii's watersheds and water resources, but also the tourism-based economy, agriculture, health, and quality of life. Habitat destruction and the introduction of alien species have been the predominant causes of biodiversity loss in Hawaii for over a century. More native species have been eliminated from Hawaii than anywhere else in the United States [22]. Native species comprise only a small portion the species composition in the Nawiliwili Watershed.

The red mangrove *Rhizophora mangle* is an invasive species that is actively spreading in the Nawiliwili Watershed. Huleia estuary and Alekoko Fishpond have been inundated with this mangrove. The rock walls of the fishpond are being torn apart by the mangrove roots, and the estuary itself seems to be shrinking in size as the mangrove closes in. The introduced mangrove appears to facilitate the establishment of opportunistic exotics such as the Samoan crab while concurrently enhancing local species richness [23]. Therefore, controlling the spread of mangrove species and removing some of their existing range are necessary for the protection of native species (David Smith, personal communication, 2003). In order to control the spread, it may be useful to develop public interest for mangrove uses such as firewood, mulch, and building or crafting material. In the meantime, floating booms or other devices could be used to trap propagules and control further spreading.

Finally, feral ungulates that are not native to the Nawiliwili Watershed, such as pigs, negatively impact water quality. In mauka areas of the watershed they destroy habitat by removing some of the remaining native vegetation, making way for more aggressive species. They also directly affect water quality by causing erosion and other sediment contributions by routing in soil near stream banks. Additionally, pigs are responsible for creating some health risks. A control plan for pigs may include allowing/encouraging more pig hunting in mauka sections of the watershed. Fencing is another option for controlling pigs. A priority site for pig fencing is the Huleia National Wildlife Refuge.

3.8 Encouraging collaboration among various agencies

Collaboration among government agencies, nongovernment organizations, and community members is critical for the success of watershed management. Building relationships and encouraging cooperation can have a synergistic effect by maximizing human resources that already have a role in watershed restoration, minimizing project overlap, and thereby maximizing cost effectiveness. Collaboration also encourages the flow of information. The implementation of restoration activities will require continued cooperation in order for the plan to be successful. More collaborative effort will lead to better solutions, to efficient use of resources, and to a greater sense of community responsibility.

Community involvement is also a critical component of watershed planning, decision making, and management. Community-based organizations, such as watershed councils and neighborhood boards, can provide additional support to government agencies in the planning and decision-making processes. Community participation and support could be more successful if community-based organizations were recognized by the County of Kauai and given a leading role in the planning process.

3.9 Developing a water budget for the watershed

A water budget is an accounting of all of the inflows, outflows, and changes of storage within a system. Inflows and outflows may include water from tributaries, ditches, irrigation diversions, and other inputs. Water that may be flowing out of a source watershed into an entirely different basin also needs to be accounted for. Irrigation systems and groundwater withdrawals change natural flow patterns. Developing water-use plans and assessing water quality are difficult if not impossible in the absence of a water budget.

Like many other Hawaii watersheds, the Nawiliwili Watershed has become a complex network of interconnected ditches, irrigations systems, diversions, flumes, and reservoirs. When the sugar industry was emerging, massive quantities of water were needed to irrigate the crop. In some regions, there was little water available locally, so it was brought in from elsewhere. This required the building of a complex network of irrigation systems that still exist today. However, very little sugarcane is still being grown, and yet water is still flowing through these systems.

There is a lack of knowledge regarding the condition of these systems and regarding the volumes of water diverted to other watersheds.

A water budget may be prepared by first consulting the Commission on Water Resource Management's (CWRM) database. A field survey would need to be conducted to verify the information contained in the database. Also, large landowners would need to be consulted to determine if all of the diversion works have been accounted for.

If there are no plans by large landowners for the water that is flowing in the irrigation systems, it may be determined that there is no longer a need to divert the water. If this is the case, then a petition may be filed with CWRM to amend or establish interim instream flow standards. The benefits of setting values for instream flows and returning some of the water may include recruitment of native fish, restoration of habitat, dilution of some contaminants, and increasing stream flushing capacity. Other stream uses include drinking or aquifer water recharge.

4 Expected load reductions due to management measures

El-Kadi *et al.* [3] provided baseline values and an assessment of expected load reductions for nutrients and sediment in the Nawiliwili Watershed, based on suggested remediation strategies or BMPs. The Generalized Watershed Loading Function (GWLF) model, developed by Haith and Shoemaker [24], was used in the calculations. Uncertainty of model results is stressed due to limitations of the model and the lack of appropriate data for model validation. The watershed was divided into subbasins and values were estimated for each. The baseline sediment loss was estimated in the range of 0.45 to 2.44 metric tons/ha/yr, for the watershed of a total area of about 9300 ha. Nutrients yields were estimated as follows:

Annual dissolved nitrogen yield = $2 \times 10^{-3} - 2.2 \times 10^{-2}$ metric tons/ha/yr
Total annual nitrogen yield = $3.1 \times 10^{-3} - 1.8 \times 10^{-2}$ metric tons/ha/yr
Annual dissolved phosphorus yield = $0 - 6.0 \times 10^{-2}$ metric tons/ha/yr
Total annual phosphorus yield = $9 \times 10^{-3} - 6.9 \times 10^{-2}$ metric tons/ha/yr

One of the highly recommended actions for this study concerns the cesspool systems. The model was used to estimate load reductions when cesspools are replaced by normally operating septic systems or sewer systems. The reduction in yields amounted to 25% and 16% for dissolved and total nitrogen, respectively, and 92% and 62% for dissolved and total phosphorus, respectively.

An attempt was also made to design a variable-width riparian buffer plan based on Wenger [25]. According to that study, it seems that riparian buffers are not an effective BMP for streams where the land slopes are greater than 25%, which is the case for a large portion of the Nawiliwili Watershed. Thus, it is expected that these buffers may not be an ideal BMP. Moreover, accessibility to some areas may not even be possible for the implementation of these measures.

Designed buffer widths ranged from 16 to 28 m. If an area has slopes greater than 25%, then the buffers should extend beyond these high-slopes zones to an area where the slope is less than 25% [25]. Such a design might not be suitable for parts of many of Hawaii's watersheds, so alternative buffering measures are needed. For buffers having a width greater than 15 m, as in this case, the removal efficiency for total suspended solids, total nitrogen, and total phosphorus can be conservatively estimated at greater than 50%.

5 Economic implications and management of the watershed plan

Economic impacts of the watershed conservation plan include potential expenditures by the agricultural, recreation and tourism, and household sectors. In addition, since all three types of users have direct and indirect connections to other sectors of the economy, these impacts will reverberate throughout the economy of Kauai (and, for that matter, potentially throughout the entire state). Identifying these impacts and estimating their linkages will require considerably greater resources than are available to this project, but this section will outline significant elements of the overall picture.

5.1 Preliminary considerations

Before delving into the costs of implementing the recommendations of the watershed plan, we note several considerations growing out of general economic principles. First and foremost is that "economic value" covers, in principle, any and all uses of water, including market as well as nonmarket uses and including costs incurred by decision makers themselves and costs imposed on third parties. Value inheres in nonfinancial considerations such as aesthetic enjoyment, even though we have only imperfect means of measuring such values (and the scope of this project does not call for applying even these imperfect techniques).

Second, the recommendations of this study, if carried out, will generate both benefits and costs in pursuit of enhanced water quality. The economic value of these benefits is measured, in principle, by the total "willingness to pay" (WTP) for them. Being largely nonmarket in character, benefits of improved water quality are very difficult to measure. As an example, water-based recreation in the Nawiliwili Watershed and in Nawiliwili Bay clearly has considerable economic value. The first study of the economic value of beach recreation in Hawaii was done in 1972 for Oahu [26, 27]. The results indicate that a day at the beach for Oahu residents is valued at $1.50 to $5.90 per person, depending on the particular beach. (Note that the 1972 dollar values have been adjusted to 2003 price levels, using the consumer price index.) Very little subsequent work has been done on this subject.

However, of special interest is the increase in the value of a beach visit due to improved water quality, rather than the overall value of such a visit. No studies on this subject have been done for Hawaii, but in other places, researchers have applied contingent valuation techniques to estimate WTP for water-quality improvement at local beaches. Russell [28] summarized several such studies done in the Philippines, Latin America, the United States, and the United Kingdom. While cross-country comparisons are probably inapt, residents surveyed in Uruguay, for example, reported WTP of $14 per household per year for an "improvement" in water quality at nearby beaches. A study in Rhode Island yielded WTP estimates of $80 to $187 per household. In all these cases WTP estimates are well below 1% of annual income. While no similar information is available for Hawaii or, more specifically, for Kauai, it is apparent that in a tourist-based economy, water-based recreation can be very highly valued. Nawiliwili Bay is heavily used by beach-goers, thus the benefits of water-quality improvement may well be substantial.

Another aspect of the value of water quality is the cost of health care associated with illness traceable to swimming or other water-based activities. Unfortunately, no epidemiological studies have been done for Nawiliwili on the health effects of existing contamination or the associated costs of medical care. Such studies are difficult and costly, but the lack of them could simply reflect the rarity or minor character of illnesses caused by water contamination and the lack of broad concern for such problems.

It should be kept in mind that any given level of water quality, or any program designed to achieve that level, is socially desirable only if the benefits derived from it exceed its costs. Unless the willingness of all members of society to pay for the specified level of water quality exceeds the value of resources necessary to achieve it, society would be better off applying those resources to some other goal.

Third, laws and regulations do not necessarily or precisely reflect benefits or costs, as broadly defined above. Likewise, there may be imbalances between those who fund water-quality improvements (e.g. farmers or taxpayers) and those who reap the benefits (e.g. tourists and environmentalists). Imperfections in the processes of lawmaking and regulation or the inability to measure benefits and costs means that the laws and regulations may or may not reflect the values of all those who underwrite the costs or those who receive the benefits of enhanced water quality. However, laws and regulations define the requirements that sponsors of this project must implement.

Finally, any changes in regulations growing from the desire for better water quality will have economic effects beyond the direct changes themselves. These effects could be studied through input–output models [29] or more sophisticated general equilibrium methods, although the requisite data are not available for present purposes.

5.2 Costs of remediation of septic tanks and sewer systems

The problem of cesspools contributing to pollution in the watershed could be alleviated by connecting residences to sewer systems or, where that alternative is prohibitively costly, by replacing cesspools with septic tanks. New technologies for septic systems should be explored. For example, EnvironEDGe Technologies, Inc. [30] markets septic tanks that, according to the company, are much superior to conventional tanks in terms of quality and protecting the environment. (Note that information is included for reference only. We do not endorse vendors or any products.)

Some idea of costs for these options is available from an environmental impact statement (EIS) [31] of wastewater treatment facilities in the Waimanalo area on Oahu, as well as from vendors and installers of septic tanks.

For sewer systems, installation costs vary greatly, depending on the density of lots and location of the house vis-à-vis existing sewer lines. The EIS cited above gives costs for two areas. One, labeled Sewer Improvement District (SID) #2, is densely developed with some 350 homes. It could connect to existing sewer lines by gravity feed. This would involve capital costs of $7,160,000 plus $110,000 for annual maintenance, or an annualized cost of about $2,100 per home. (Annualized cost assumes a 20-year life, 6% discount rate, and no salvage value. They are calculated using the formula $A = Pi/[1-(1 + i)^{-T}]$, where $A is the annual equivalent of a principal amount $P invested for *T* years at discount rate *i*.)

In addition to the high capital cost, this alternative has technical challenges (some pipes would be below the water table) and is opposed by a majority of the homeowners.

A nearby area (SIDs #3 and #5) is somewhat smaller with only 100 homes. However, this area would require construction of an additional pumping station, with associated higher operating costs. Here, the sewer connection cost was estimated at $6500 per household per year.

Alternatively, either area could meet wastewater objectives by installing or upgrading individual wastewater-treatment systems. The Waimanalo EIS estimated capital costs for septic tanks at between $1830 and $12,000 for the "typical" and "worst-case" scenarios, respectively. The 350 homes in SID #2 could be upgraded with individual systems at a capital cost of $640,000. Annualizing and adding annual maintenance costs would bring the total cost estimate to $162,000, or $460 per household per year. Outfitting the other area with individual wastewater-treatment systems would give a very similar estimated annual cost per household.

Another estimate of septic-tank costs resulted from Internet searches and subsequent phone calls to providers of such equipment. The website Water Tanks [32] gives prices for the tanks per se of between $500 and $1100. Installation costs were quoted at between $2000 and $6000, depending on soil type, and maintenance costs at about $75 per year, based on a three-year pump-out.

The Nawiliwili Watershed, in general, is a larger area that is more rural in character and that has less densely situated homes than the Waimanalo areas studied in the EIS. Outside urban areas, sewage systems are probably prohibitively expensive. They are also less necessary, to the extent that because of the low density in rural areas, septic tanks can be operated efficiently and effectively without an accumulation of residual contamination.

Information about tax map key parcels with cesspool systems in the Nawiliwili Watershed was obtained from the HDOH data. The total number of parcels is 470. Unfortunately, the data does not include the number or size of cesspools in each parcel. Using the price quotes listed above, the total cost of septic tanks would range from $235,000 to $517,000 for tanks, plus $940,000 to $2.82 million for installation. The average total cost would be about $2.3 million. Land owners/operators would be responsible for the annual maintenance cost. Considering Hawaii's above-average living expenses, the actual cost would most likely be higher than the averages given. Yet, the numbers provided here serve as a guideline for a more accurate analysis.

5.3 Costs of other recommended remediation efforts

Precise and reliable data on costs for recommended measures would require extensive engineering studies, but some general notions can be obtained from extant literature. Two reports are particularly useful: the Kailua Bay Advisory Council (KBAC) watershed-management plan [33] and the stormwater-pollution management practices report by the Texas Statewide Storm Water Quality Task Force [34].

The KBAC report identifies several segments of the Kailua watershed needing attention and gives cost estimates for various possible pollution-control and remediation projects. The cost figures in the report can serve as a guideline for future implementation of the current plan.

The Texas study (also adapted for use in the KBAC report) provides information on a wider variety of remediation measures. No absolute dollar figures are provided; instead, only a broad qualitative valuation is given. The valuations do not include land costs that, for both Oahu (as in the KBAC report) and Kauai, can be expected to differ substantially from that of Texas. For example, wet ponds have relatively high space requirements, so the judgment that they are "low" in cost probably does not apply in the Kauai context. That said, Table 1 includes some of the KBAC conclusions.

Table 2 provides a cost estimate of the main items dealing with the proposed restoration activities. We emphasize again that all estimates are provided as guidelines and that elaborate studies should be undertaken to finalize such estimates. Uncertainties are related mostly to the absence of site-specific data regarding such factors as land price and size of installations. A good approach would be to subcontract construction activities to qualified nonprofit and for-profit entities through bidding procedures.

5.4 Potential funding sources

In addition to making public funds available, the government can institute a wide variety of regulations, subsidies, and tax schemes. For example, for the Neuse River Basin and several other watersheds, North Carolina has a system of incentive and bonus payments to landowners [35]. With federal and state participation, the Conservation Reserve Enhancement Program provides payments to owners of agricultural lands for up to 15 years, as well as subsidies for BMP installation. Up to 100% of the installation costs can be covered, if the contract is a permanent one.

Among the available funding sources is HDOH's Environmental Planning Office, Funding Sources for Communities – Watershed Focus. Additional information can be obtained from the Environmental Planning Office, 919 Ala Moana Blvd., Room 312, Honolulu, HI 96814, phone (808) 586-4337.

The draft Kailua Waterways Improvement Plan, Volume II [33], contains a preliminary list of potential grant programs and funding sources that are available to support the implementation program. Potential funding amounts are given where available. Finally, cost-sharing and low-interest loans are effective incentives for restoration implementation. Examples include that of the Virginia agricultural incentives program [36].

Table 1: Stormwater best management practices.

	Land use			Feasibility and benefits					
Cover type or BMP	Residential	Commercial	Industrial/ hotspots	Space	Aesthetics	Habitat	Safety	Maintenance	Relative cost
Native vegetation, preserved or established	Y	Y	Y	H	H	H	H	H	H
Turf grass	S	S	S	L to H	L	L	H	H	L
Disconnect impervious areas	Y	Y	Y	M	M	M	H	M	M
Swales and channels	Y	Y	N	M	M	M	H	H	M
Bioretention/rain gardens	Y	S	N	M	H	H	H	L	M
Grass filter strips	S	Y	N	M	M	M	H	M	M
Wet ponds	S	S	IL	H	H	H	M	L	L
Constructed wetlands/ extension detention wetlands	S	S	IL	H	H	H	M	L	L
Pervious pavement	S	Y	N	–	M	M	H	M	H
Surface sand filters	S	Y	Y	L	M	L	H	M	L
Perimeter sand filters	S	Y	Y	L	M	L	H	M	L
Infiltration basins	Y	Y	N	H	M	M	H	M	M
Infiltration trenches	Y	Y	N	M	M	L	H	M	M
Dry detention basins	S	S	N	M	L	L	M	L	M
Parking lot detention	N	S	N	L	L	–	–	–	–

Source: Tetra Tech EM Inc. [33].
Y = yes, S = sometimes, N = no, IL = use with impermeable liner; H = high, M = moderate, and L = low.

5.5 Restoration and protection plan management

We propose the establishment of a new entity, the Nawiliwili Watershed Restoration Office, with a structure similar to that proposed for the Kailua Protection Plan (Tetra Tech Em Inc. [33]). The office will manage and supervise all restoration activities and also be responsible for integrating the efforts of watershed stakeholders who are concerned about the watershed's health and welfare. These stakeholders include watershed residents; landowners; commercial, industrial, and agricultural businesses; community groups; schools and academic institutions; and county, state, and federal agencies. The efforts of all parties need to be synchronized toward implementing a holistic watershed management program using the plan proposed here as a guiding framework.

Table 2: Cost estimate for the restoration activities of the Nawiliwili Watershed.

	Cost estimate (in thousands of US dollars)					
Action	Year 1	Year 2	Year 3	Year 4	Year 5	Total
WATERSHED RESTORATION CENTER						
Staff and office cost	\$100	\$105	\$110	\$116	\$122	\$553
Advisory group meetings	\$10	\$10.4	\$11	\$11.6	\$12	\$55
EDUCATION						
Education programs in schools	\$5	\$5.5	\$5.7	\$5.8	\$6	\$28
Native tree planting in Huleia National Wildlife Refuge	\$5	\$5.5	\$5.7	\$5.8	\$6	\$28
Educational research center for Alekoko Fishpond	\$5	\$5.5	\$5.7	\$5.8	\$6	\$28
Low-impact development Video/Workshop	\$7					\$7
Educational program for eco-tour operators	\$1	\$1	\$1	\$1.5	\$1.5	\$6
Educational plaques	\$0.4	\$0.6	\$0.6	\$0.7	\$0.7	\$3
Storm-drain stencilling project	\$0.1	\$0.1	\$0.1	\$0.2	\$0.2	\$0.7
Educational program for agricultural conservation	\$5	\$5.5	\$5.7	\$5.8	\$6	\$28
SOIL EROSION FROM AGRICULTURAL LAND						
Use of conveyor-belt water bars	\$1					\$1
Relocation of water troughs for cattle away from streams	\$20					\$20
Working farm for BMP implementation	\$180	\$20	\$20	\$20	\$20	\$260
Updating land-use maps	\$50					\$50
Water recycling and conservation practices	\$3	\$3	\$3	\$4	\$4	\$17
Solution for eco-tour and ATV erosion problems	\$1	\$1	\$1	\$1.5	\$1.5	\$6
CAPITAL IMPROVEMENTS						
Strategy and installation of basin inserts	\$5					\$5
Construction of storm-water wetlands	\$2100	\$20	\$21	\$22	\$23	\$2186
CONTROL OF NON-NATIVE/INVASIVE SPECIES						
Monitoring and control program for mangrove	\$10	\$5	\$5.5	\$5.5	\$6	\$32
Community workdays program	\$1	\$5	\$5.3	\$5.5	\$5.8	\$22.6
Plans to encourage hunting	\$1	\$1	\$1.1	\$1.2	\$1.2	\$5.5

(*Continued*)

Table 2: Continued

Action	cost estimate (in thousands of US dollars)					
	Year 1	Year 2	Year 3	Year 4	Year 5	Total
ELIMINATION OF CESS-POOL CONTAMINATION	$450	$473	$496	$521	$547	$2486
DEVELOPMENT OF A WATERSHED WATER BUDGET	$10					$10
LOW-IMPACT DEVELOPMENT STRATEGIES						
Basin designation	$10					$10
Zoning density	$10					$10
Tree, forest, and open space protection	$10					$10
Stormwater-management standards	$10					$10
Grading restrictions	$10					$10
Reducing impervious surface recover	$30					$30
Utilizing natural features for stormwater management	$20	$5	$5	$5	$5	$40
Education	$1					$1
Partnership with agencies/ communities	$0					$0
HABITAT RESTORATION AND PROTECTION	$10	$10	$11	$11.5	$12.5	$55,256
IMPROVEMENT OF HULEIA ESTUARY	$20	$10	$10			$40
Total	$3101.5	$692.5	$724.4	$750.4	$786.4	$6054.8

The office operation will be supervised by a core group that includes representatives from HDOH, Kauai County, Hawaii Department of Land and Natural Resources (HDLNR), and an advisory stakeholder group. Also, as suggested for the Kailua Protection Plan, partners should include other government and private entities, in particular, USEPA, U.S. Fish and Wildlife Service, U.S. Department of Agriculture, Hawaii Department of Transportation, academic institutions, and the private sector. Cooperation should be established with volunteer groups from the community, schools, and nongovernment organizations. Roles and responsibilities should be clearly set through interagency agreements defining the lead agencies or organizations, based on the nature of the implementation action.

6 Developing and implementing education and outreach programs

The purpose of a watershed education program is to increase awareness of watershed conditions and provide opportunities that allow community members to participate in the solution. Education is one of the most powerful tools to enhance environmental conditions and reduce pollutant loads. Ideally, education can lead to prevention. For education to be effective, it must begin at an early age. Simple concepts like stewardship can teach children that they are responsible for their actions within the watershed. Students who learn how watershed dynamics work may realize that making responsible decisions can improve the condition of the watershed.

Community education and outreach programs can be an effective way to spread awareness about watershed issues, but it is a difficult task to reach the adult generation. On a small island it would seem that the community would be more connected to the environment, like their ancestors were, but this is not necessarily the case. Absence of interest is based on the difficult economic conditions and the lack of time to think about their place in the watershed. Frustration with economic conditions has led to blaming government agencies for the exploitation of resources and the degradation of water quality. Education programs intend to change this attitude. It is time for community members to take control of their resources and be a part of the solution.

Education programs for children, as recommended above, can help in reaching the adult population as well. Children who bring new values home can inspire a change in their parents. In addition, adult education and outreach programs can be effective tools for community members who are willing to participate. The most effective education programs connect people with the resource through hands-on activities that allow education to lead to structured outcomes that turn knowledge into action [37].

Educational opportunities can be offered via local television programs, eco-tours, or community-participation projects such as beach cleanup days and storm-drain stenciling. A memorable experience can lead to the passing on of information, thus perpetuating the educational experience.

A watershed education program should not only increase general public awareness but should also include more technical education and support targeted groups such as landscapers, developers, contactors, eco-tour companies, small businesses in industrial areas, restaurants, and resorts. Businesses that participate in eco-friendly activities could use their "watershed-friendly" practices as part of their promotional material. The community in Bellevue, Washington, formed a group called the "Business Partners for Clean Water" [38]. Forming a similar group may be a good way to involve many of the small businesses that operate in the Nawiliwili Watershed.

Specific education programs can include expanding native tree planting in the Huleia National Wildlife Refuge, using Alekoko Fishpond as an educational research center, and promoting a videotape explaining and demonstrating the ahupuaa concept and low-impact development strategies, using educational plaques, and initiating storm-drain stenciling projects (see Fig. 3 for suggested locations of stenciling projects).

One of these programs concerns Alekoko Fishpond, which is one of the most important cultural resources of the Nawiliwili Watershed and the island of Kauai. Although the fishpond is currently privately owned, restoration is feasible if federal or local community funds can be obtained to purchase the fishpond and place it in a public land trust. After restoration, the pond can be used as an educational center and living laboratory to serve the schools, the community, and visitors through workshops on local history, science, and aquaculture. The restoration of Alekoko Fishpond would be the result of the community adopting ahupuaa principles, reconnecting themselves to their resources, and giving back to the community.

Another program is related to the effort by Don Heacock, a Kauai resident who leads boy scouts to annually plant native trees in the Huleia National Wildlife Refuge, an activity that provides a combination of education and BMP implementation. Native trees that are planted in the refuge have a better chance of survival due to the lack of goats that populate the land on the Kipu Kai side of Huleia River. With the cooperation of the Huleia National Wildlife Refuge and the U.S. Fish and Wildlife Service, Heacock hopes for more participation by every scout troop and school group in the Nawiliwili Watershed in future years. Since the refuge is set aside as public land for the restoration of native plant species and for providing a habitat for endangered water birds, it could feasibly serve to house and safeguard a native forest for years to come.

7 Priorities and schedule of plan implementation

7.1 Priorities

It was difficult to establish definitive priorities for the restoration activities due to minimal participation at community meetings. However, a few items were emphasized repeatedly. Two of these are the preparation of a water budget and the setting of instream flows. These actions may create the need to re-evaluate prior studies that may have overlooked the uncertainty of instream flows and the watershed budget.

In general, the project activities can be grouped according to issues dealing with nutrient- and sediment-load reduction, water-resource assessment, and education. The overlap of these groups is obvious. For example, preparation of a water budget, setting instream flows, and reducing sediment and nutrient loads are critical in restoring stream aquatic health. In addition, the effective design of BMPs greatly depends on accurate knowledge of the water budget of the watershed, including instream flows. Education is a cornerstone of the protection plan, which depends to a great extent on community and visitors' good will and participation. During discussions at the restoration plan meetings, the following activities were given high priority:

- Implementation of erosion control and BMPs, including those related to nutrient-load reduction
- Preparation of a water budget and setting of instream flows
- Development and implementation of a watershed curriculum
- Posting educational plaques

- Revision of National Pollutant Discharge Elimination System files
- Collaboration of state and county agencies

Our own assessment put the activities in the order listed above. However, the overall community consensus was that all of the identified restoration activities were equally important and that the priorities would be set forth according to cost and feasibility. Action is needed by government agencies to lead the community restoration activities. The implementation of one or more of the activities recommended in this report will give the community an opportunity to finally see the results of their input.

7.2 Schedule of plan implementation

A timeline for plan implementation should be subjected to extensive discussion and adjustment as community input is considered. Securing the necessary funds is the most critical and challenging part of the plan to restore the watershed. Meetings of the core group, institutional partners, and volunteer groups would be held to form an advisory committee to guide the process of generating proposals, with research topics that address this protection plan, for submission to various agencies. The next step would be to hire a coordinator and staff for the Nawiliwili Watershed Restoration Office to coordinate plan implementation. The hiring and manning of the office would take up to six months. The possibility of assigning the position to a person from a state organization, such as HDOH, should be explored to minimize the cost, especially regarding the overhead involved in setting up new office space. Tasks of the coordinator would include evaluating measures of success by reporting to the advisory committee as the projects get underway and are completed. Tasks would also include participating in project reviews and presenting project results to the community and to the project advisors.

It is possible that small projects, such as establishing buffer zones and replanting riparian vegetation, would start within six months of initial funding. Pamphlets and educational material could be completed within a few months; however, it may take longer to establish contacts and develop partnerships and to work on community education. Developing/producing videotapes and workshops may take up to a year. Implementing education programs in schools would take at least a year. An ATV road stabilization project would probably take up to two years to complete, depending on the number of other projects in progress and the level of funding. Large-scale projects, such as constructed wetlands and those related to policy changes, would take about five years due to the need for multiagency cooperation. Other complicating factors include the need for extensive development plans and potential diversity of funding resources. Restoring Alekoko Fishpond may extend over 10 years as a project with many phases, and eventually this project may have to spin off and become an entity in itself.

8 Measures for evaluating plan success

Major project milestones, tasks, and deliverables should be assessed regularly. Tasks include the development of public-education programs and public-participation activities, construction of demonstration projects, completion of monitoring activities, and establishing and implementing a monitoring program. In addition, and most important, measures of success should include the restoration program's impact on water quality. Specific objectives for measures of success include improvement of surface water quality based on analytical results; improvement of habitat quality and eradication of invasive species; physical stabilization of streams and other water bodies; and, ultimately, in cooperation with the state's total maximum daily load (TMDL) program, delisting of water bodies in the watershed under Section 303(d) of the Clean Water Act. Other important measures of progress include: establishment of an advisory group; preparation of funding applications and securing funds; creation of the Nawiliwili Watershed Restoration Office and hiring of staff; design and construction of demonstration projects and implementation of BMPs; creation and implementation of public outreach and education programs, including a training program; contracting with vendors; adherence to program schedules and budgets; and publication of research findings.

9 Plan evaluation

9.1 Criteria for success of load-reduction strategies

As stated earlier, measures of success should include assessing the restoration program's impact on water quality as manifested by load reduction of sediment and nutrients. Improvements of surface water quality should be evaluated through analytical results. A monitoring program should be the vehicle for assessing water quality on a regular basis. Influent and effluent from demonstration projects and BMPs should be used to assess their efficienacies. Monitoring results should be compared to state water-quality criteria and TMDLs, where applicable, to track the progress on meeting water-quality requirements and designated uses in the watershed. It is strongly recommended that a new round of sampling be conducted prior to implementation of any demonstration projects or BMPs, in order to provide a baseline by which to judge the progress on meeting water-quality targets. Periodic sampling – including quarterly, semiannual, and annual sampling events – would be carried out throughout the duration of the implementation program.

9.2 Revision of plan and program implementation

The proposed implementation program of the Nawiliwili Watershed Restoration and Protection Plan should be a living document that can be revised and refined as the program matures. Periodic updates and revisions to this program plan are anticipated in response to levels of funding, agency and public participation, future conditions and developments, and lessons learned. Additional demonstration projects may be proposed in the future, as development patterns evolve and new problems or approaches become apparent. The measure of success discussed in the previous section should be incorporated into plan revisions to present an accurate picture of current conditions. In addition, changes in supporting conditions – such as infrastructure upgrades, development patterns and infrastructure capacity, land-use patterns, water-quality regulations, and critical habitat areas – must be incorporated into the plan to support valid and effective recommendations for future work.

Reviews and revisions will be the responsibilities of the restoration office, with input and concurrence of the core group and partnering agencies. Revisions should take place on an as-needed basis. However, annual reviews need to be undertaken describing progress of various projects, as well as the needs of the community and watershed. Issues that require immediate changes can be addressed on an as-needed basis.

10 Monitoring plan

Monitoring should be an integral part of the restoration and protection plan for the Nawiliwili Watershed. The objective of monitoring is to assess existing conditions and to track the progress of the various restoration and protection strategies. The current watershed condition is described in Furness *et al.* [1] and El-Kadi *et al.* [2]. The design of specific remediation activities might require installing new monitoring equipment and collecting data from other sources, including USGS, HDOH, NRCS, USEPA, County of Kauai, and community groups. The following issues, adopted from Tetra Tech EM Inc. [33], should be carefully examined in order to provide a complete and effective monitoring plan.

10.1 Data management

Data obtained so far and those to be obtained during the restoration phase (from the monitoring program and from other sources) should be compiled into a project water-quality database maintained by staff of the Nawiliwili Watershed Restoration Office. Data should be systematically organized and categorized for analysis and comparison. The resulting database can be made available on the Internet to allow participating agencies and researchers to benefit from the data and to inform the public about the implementation program and the water quality in the watershed.

10.2 Water-quality sampling

A routine sampling program is an important component of the implementation program. The primary goal of the monitoring program is to fill in the gaps of existing knowledge and to collect a complete and regular record of surface-water and groundwater quality in the watershed. Sampling of shallow groundwater (e.g. springs, seeps, and agricultural wells) should also be initiated and done on a regular basis. Another important goal is to sample influent and effluent from demonstration projects and

BMPs to assess their effectiveness, in order to help select applicable practices for the watershed and to advance the state of knowledge about water-quality BMPs.

Monitoring results should be compared to state water-quality criteria and TMDLs, where applicable, to track the progress of meeting water-quality requirements and designated uses in the watershed. Monitoring should be carried out by professionals and volunteers. Using volunteers will help stretch scarce funding while giving the public a real stake in managing the program and the watershed. Another important benefit is the education of the public about watershed health and management.

10.3 Watershed assessment

It is recommended to assess the health of the watershed by examining conditions such as the physical stability of streams and wetlands, the quality of aquatic and riparian habitats, and the presence of exotic and invasive species. The assessment should also emphasize locating pollutant-source areas.

A watershed-assessment protocol would be developed from existing bioassessment protocols for Hawaii and other tropical and subtropical regions and from other stream and surface-water assessment methodologies. Assessments would be conducted at the beginning and end of the implementation program, at a minimum, and annually if resources permit. Assessment data would be entered into the data-management system described previously.

10.4 Quality assurance

All data collection and monitoring should be conducted under a rigorous quality-assurance program. The purpose of the program would be to ensure that all data, whether obtained from other sources or generated through the program, are of good quality and are useable for implementing and evaluating the restoration program. Data collection, sampling, and monitoring should be conducted according to a sampling and analysis plan and quality-assurance project plan (QAPP) developed specifically for the program. The QAPP would conform to state or USEPA quality-assurance guidelines as appropriate, depending on the requirements of the various grant and funding sources for the implementation program. The QAPP developed during Phases 1 and 2 of this study [1, 2] could also be used for the restoration phase.

Acknowledgements

This chapter covers some of the results of the project "Assessment and Protection Plan for the Nawiliwili Watershed," principal investigator Aly I. El-Kadi, and co-principal investigators Roger Babcock, Roger S. Fujioka, Clark C.K. Liu, Jacquelin N. Miller, James E.T. Moncur & Philip Moravcik – all with the Water Resources Research Center, University of Hawaii. This project is jointly funded by the U.S. Environmental Protection Agency, under Section 319(h) of the Clean Water Act, and the Hawaii State Department of Health, Clean Water Branch. Community members who have contributed are listed in the study reports [1–3]. Completion of this study would not have been possible without their input. Their help is greatly appreciated.

This is contributed paper CP-2008-06 of the UH Water Resources Research Center, Honolulu.

References

[1] Furness, M., El-Kadi, A.I., Fujioka, R.S. & Moravcik, P.S., *Assessment and Protection Plan for the Nawiliwili Watershed: Phase 1 – Validation and Documentation of Existing Environmental Data*, WRRC-2002-02, Water Resources Research Center, University of Hawaii: Honolulu, 2002.

[2] El-Kadi, A.I., Fujioka, R.S., Liu, C.C.K., Yoshida, K., Vithanage, G., Pan, Y. & Farmer, J., *Assessment and Protection Plan for the Nawiliwili Watershed: Phase 2 – Assessment of Contaminant Levels*, WRRC-2003-02, Water Resources Research Center, University of Hawaii: Honolulu, 2003.

[3] El-Kadi, A.I., Mira, M., Dhal, S. & Moncur, J.E.T., *Assessment and Protection for the Nawiliwili Watershed: Phase 3 – Restoration and Protection Plan*, WRRC-2004-05, Water Resources Research Center, University of Hawaii Honolulu, 2004.

[4] Genesee/Finger Lakes Regional Planning Council, *www.grwabucks.org/report.htm*

[5] The Center for Watershed Protection, *www.dnr.state.md.us/watersheds/surf/proj/wras.html*

[6] Fujioka, R.S. & Shizumura, L.K., *Clostridium perfringens,* a reliable indicator of stream water quality. *J. Water Pollut. Control Fed.*, **57**, pp. 986–992, 1985.

[7] HDBEDT, *www.hawaii.gov/dbedt/gis/*

[8] NRCS, *http://www.nrcs.usda.gov/technical/land/meta/m5058.html*

[9] Hardina, C.M. & Fujioka, R.S., Soil, the environmental source of *E. coli* and enterococci in Hawaii's streams. *Environ. Toxicol. Water Qual.*, **6**, pp. 185–195, 1991.

[10] USFDA, *http://vm.cfsan.fda.gov/~mow/chap11.htm*

[11] May, C.W., Welch, E.B., Horner, R.R., Karr, J.R. & Mar, B.W., Quality Indices for Urbanization Effects in Puget Sound Lowland Streams, Water Resources Series Technical Report No. 154, University of Washington. Ecology Publication No. 98-04, 1997.

[12] Haub, A. & Hoenig, L., *Aquatic Habitat Evaluation & Management Report*, City of Olympia, Olympia, Washington, 1999.

[13] American Society of Civil Engineers (ASCE), *A Guide for Best Management Practice (BMP) Selection in Urban Developed Areas*, ASCE Publications: Reston, Virginia, 2001.

[14] Atlanta Regional Commission and Georgia Department of Natural Resources Environmental Protection Division, Georgia Stormwater Management Manual, 2001.

[15] Ocean Arks International, *http://www.oceanarks.org/wetlands/*
[16] U.S. Environmental Protection Agency (USEPA), Storm Water Technology Fact Sheet Storm-water wetlands, EPA 832-F-99-025, Washington, DC, 2002.
[17] National Watershed Manual, National Soil Survey Handbook, *www.nrcs.usda.gov/technical/references*
[18] Hawaii State, Department of Health (HDOH), Guidelines for the Treatment and Use of Recycled Water, Wastewater Branch, Honolulu, Hawaii, 2002.
[19] USEPA, *http://www.epa.gov/owow/nps/ordinance*
[20] KRP Information Services, Water Quality Management Plan for the County of Kauai, Honolulu: Hawaii, 1993.
[21] Kaneshiro, K.Y., Chinn, P., Duin, K.N., Hood, A.P., Maly, K. & Wilcox, B.A., Hawaii mountain-to-sea ecosystems: Social-ecological microcosms for sustainability science and practice. *EcoHealth J.*, **2**, pp. 349–360, 2005.
[22] Kauai Invasive Species Committee (KISC), Action Plan, Kilauea: Hawaii, 2003.
[23] Demopoulos, A.W.J., *Mangrove Forest Ecosystems*, Department of Oceanography, University of Hawaii at Manoa: Honolulu, Hawaii, 2003.
[24] Haith, D.A. & Shoemaker, L.L., Generalized watershed loading functions for stream flow nutrients. *Water Resour. Bull.*, **23**, pp. 471–478, 1987.
[25] Wenger, S., A Review of the Scientific Literature on Riparian Buffer Width, Extent and Vegetation, Office of Public Service and Outreach, Institute of Ecology: Georgia, 1999.
[26] Moncur, J.E.T., The Value of Recreation Areas on Oahu. University of Hawaii Center for Governmental Development: Honolulu, Hawaii, 1972.
[27] Moncur, J.E.T., Estimating the value of alternative outdoor recreation facilities within a small area. *J. Leisure Res.*, **7(4),** pp. 301–311, 1975.
[28] Russell, C.S., *Applying Economics to the Environment*, Oxford University Press: New York, pp. 332–337, 2001.
[29] Hawaii State, Department of Business, Economic Development and Tourism (HDBEDT), The Hawaii Input-Output Study: 1997 Benchmark Report, Honolulu, Hawaii, 2002.
[30] EnvironEDGe Technologies, Inc., *http://www.environedge.com/*
[31] Hawaii Pacific Engineers, Inc., Draft Supplemental Environmental Impact Statement: Waimanalo Wastewater Facilities Plan, Koolaupoko, Oahu, Hawaii, 1998.
[32] Water Tanks, *http://www.watertanks.com/category/35/*
[33] Tetra Tech EM Inc., Draft Kailua Waterways Improvement Plan: A Watershed Approach for Improving Water Quality in the Kailua Waterways System, 2003.
[34] Texas Statewide Storm Water Quality Task Force, *http://www.txnpsbook.org/About.htm*
[35] Wossink, G.A.A. & Osmond, D.L., Cost analysis of mandated agricultural best management practices to control nitrogen losses in the Neuse River Basin, North Carolina. *J. Soil Water Conserv.*, **57**, pp. 213–220, 2002.
[36] Virginia, *http://vmirl.vmi.edu/ev2000/PPT/bayless.ppt*

[37] Center for Environmental Research and Service, Troy State University, Considerations for Stormwater and Urban Watershed Management, Developing a Program for Complying with Stormwater, Phase II MS4 Permit Requirements and Beyond, Department of Biological and Environmental Sciences, Troy State University: Troy, Alabama, 2000.
[38] City of Bellevue, Water-quality protection for Bellevue Businesses, City of Bellevue Utility Department, Bellevue, Washington, 1993.

CHAPTER 10

Estimating the benefits from restoring coastal ecosystems: a case study of Biscayne Bay, Florida

Donna J. Lee[1] & Anafrida B. Wenge[2]
[1]*ENTRIX Inc., Environmental and Natural Resource Management Consultants, USA.*
[2]*Institute of Food and Agricultural Sciences, College of Agriculture and Life Sciences, Food and Resource Economics Department, University of Florida, USA.*

Abstract

Biscayne Bay is a subtropical estuarine ecosystem in South Florida. The Bay serves as a nursery for Atlantic reef fish and provides critical habitat for a variety of species including 31 animals listed as endangered, threatened and "special concern." Noteworthy are the American crocodile, Atlantic green turtle, Atlantic hawksbill turtle, Cape Sable seaside sparrow, Florida panther, West Indian manatee, and Wood stork. Rapid development in the Miami-Dade area has substantially altered the Bay and its coastline. This study quantifies the economic and environmental benefit from a decade of Biscayne Bay restoration projects.

1 Introduction

United States' natural ecosystems have been invaded by approximately 5000 exotic plant species that compete with 17,000 native plant species for space and resources. Commercial agriculture and forest sectors Florida spend $265 million per year controlling invasive plants (Lee [1]). In natural areas, establishment of exotic plants has hampered coveted recreational activities and jeopardized the viability of many listed threatened and endangered species. State response has been public funding for controlling invasive plants in natural areas to the tune of $32 million per year. This study examines the value of restoring ecosystems damaged by exotic plants through a case study of Biscayne Bay Florida. Since 1987, more than $10.6 million has been spent on exotic-plant removal and coastal-habitat restoration in Biscayne Bay. Results show the net gain from restoring coastal ecosystems in Biscayne Bay Florida to be from $32 million to $36 million with an internal rate of return between 8.9% and 9.6%.

2 Cost of invasive plants in the US

United States' natural ecosystems have been invaded by approximately 5,000 exotic plant species that compete with 17,000 native plant species for space and resources. Many of the invaders were brought to the US to be grown for food, feed, fiber, and ornamental purposes. While most nonindigenous species are unable to survive in the wild, some possess characteristics that enable them to grow and reproduce rapidly unchecked by natural enemies and out compete native plants for space, sunlight, and nutrients. Often growing in monocultures, invasive plants diminish biodiversity, alter habitat, and eliminate natural food sources for native birds, reptiles, and mammals. In February 1999, President Bill Clinton signed Executive Order 13112 that allocated $28 million to combat invasive species.

Invasive plants are responsible for $25 billion in damages to US food and horticultural crops each year. The US food and horticulture industries each spend upwards of $9.5 billion controlling invasive plants. To control just a few of the plants (Purple loosestrife, Melaleuca, and invasive aquatics) invading natural systems, expenditures top $104 million each year. The total cost including damages and control expenditures of invasive plants in the agriculture and horticulture sectors is $34.5 billion (Pimentel [2]).

In Florida, exotic plants are common and numerous. In the past century, over 1300 exotic plant species have become established in the State; 124 species are destructive to natural areas and have been classified by the Florida Exotic Plant Pest Council as Category I and Category II invasive species· In upland ecosystems the ten "worst" invasive plants identified by the FL DEP are Melaleuca, Brazillian pepper, Lygodium spp., Chinese tallow, Australian pine, Cogon grass, Ardesia spp., Chinaberry, Air Potato, and Ligustrum. Category I invasive plants are believed to have been originally introduced for ornamental purposes 47% to 69%, agriculture 21%, and accidental 6% (Fox *et al.* [3]). It is worth noting that the horticulture industry brought to Florida an estimated 25,000 plant species from other countries (Pimentel, 2000) however, only a very small percentage of those have successfully invaded natural areas (FLDEP [4]). Because of their habitat destroying proclivity, invasive-plant management and habitat restoration are key components in statewide efforts to protect endangered species (FLEPPC [5]).

Private expenditure in Florida for controlling invasive plants in the agriculture and forest industries is $265 million per year (Lee [1]). State expenditure for prevention and control of invasive plants is $103 million per year (FLDEP [6]). Public funding to control invasive plants in natural areas is $32 million per year, $6.3 million for upland invasive plants and $25.7 million for freshwater aquatic invasive plants (FLDEP [4]).

In Biscayne Bay, Florida, federal, state and local agencies have joined forces to restore coastal upland habitats through the removal of exotic vegetation and the planting of native vegetation. In addition, historically altered wetlands are being restored through the removal of exotic vegetation and fill. Since year 2000, more than $11 million has been spent on coastal habitat restoration efforts in Biscayne Bay, Florida as shown in Table 1.

3 Restoring coastal ecosystems in Biscayne Bay: a case study

Biscayne Bay, Fla., is a 428-square-mile (1,109-square-kilometer) subtropical estuarine ecosystem which formed 3000 to 5000 years ago when sea level rose and flooded the natural limestone depression that is now south Florida. The bay supports a diverse flora and fauna and serves as a nursery for coral reef and marine ecosystems (USGS [7]). The unique environment of Biscayne Bay is home to 31 animal species that have been listed as endangered, threatened or "of special concern". Seven species listed as endangered by both the State and Federal governments are as follows: American crocodile, Atlantic green turtle, Atlantic hawksbill turtle, Cape Sable seaside sparrow, Wood stork, Florida panther, and West Indian manatee (Cantillo *et al.* [8]).

Throughout the last century, rapid population growth, urbanization, and development of the Miami-Dade County area have altered the Bay environment. Extensive dredging in the early 1900s reshaped the Bay and created navigation channels the largest being the Atlantic Intra-coastal Waterway (ICW). Remnants of ICW construction are a series of large spoil-fill islands that have become popular as recreational sites. Human activity, unstable shorelines, and overgrowth of exotic vegetation have accelerated island erosion and contributed to Biscayne Bay turbidity (Wanless *et al.* [9] and Milano [10]). Documented incidents linking exotic vegetation (Australian pine) and coastal-erosion problems around the world are highlighted in Box 1. In addition to erosion problems, invasive plants in Biscayne Bay have overtaken native habitats and altered coastal wetlands. Plants such as Australian pine (*Casuarina equisetifolia)* and Brazillian pepper (*Schinus terebinthifolius*) are among the worst for choking out native plants and destroying native animal habitats (SFWMD [11]).

To preserve and protect the natural setting, Biscayne National Monument was established in 1968 for the "education, inspiration, recreation and enjoyment of present and future generations a rare combination of terrestrial, marine, and amphibious life in a tropical setting of great natural beauty" according to Public Law 90–606. The monument was enlarged in 1980 and designated Biscayne National Park, recognized at the time as the largest marine park in the National Park System (USGS [7]).

Box 1. Australian pine: evidence of erosion and habitat alteration

Australian pine (*Casuarina spp*) is native to Australia, the South Pacific Islands, and Southeast Asia. Trees grow 5 to 10 feet per year and can reach heights up to 150 feet. With their towering height and thick shallow roots, Australian pines tip over in high winds. Gordon [12] cites evidence that outside its native environment, Australian pine contributes to shoreline erosion at least in part by crowding out native vegetation that possesses the deep root structures that are more conducive to sandy beach environs harsh wind and wave conditions.

In India, Australian pines were planted along a coastal dune belt on Sagar Island in the early 1990s as a vegetative windbreak and source of timber. By 1995, storm surges knocked down the trees and destroyed the entire dune belt. Alternative vegetation has been planted to re-establish the dune (Bandyopadhyay [13]).

Australian pine is an exotic invasive in the Andros Islands in the Bahamas and grows throughout the coastal zone. The tree is considered a threat to the stability of the coastal zone yet even along shoreline areas that have been significantly altered there has been no apparent effort to remove the invasive plant (Sealy *et al.* [14]). On San Salvador Island in the Bahamas, Australian pine was introduced in the late 1920s and became established during the 1950s and 1960s in conjunction with the installation US military bases. Because of their ability to destabilize dunes, Australian pine is considered a "clear and present danger" in the Bahamas (Rodgers [15]).

Australian pine was introduced to Florida in 1898. Trees were planted extensively as windbreaks along canals, around agricultural fields, beside roadways, and to provide shade near homes. The trees naturalized and spread across the southern region of the State and along both coasts. Across the state, Australian pine can be found on over 300,000 acres (Doren and Ferriter [16]). On public land Australian pine is under maintenance control on 3457 acres (FLDEP [4]). Cultivation and sale of the plant is now banned in Florida.

Removal of Australian pine and Agave from Dry Tortugas National Park in Florida began in 1992. With average spending of $20,000 per year plants were eradicated in 10 years (Doren and Ferriter [16], DOI [17]). Total cost for plant eradication in the Dry Tortugas is estimated to be $50,000 per acre over 40 acres.

Along a 2.5-mile stretch of beach in John U. Lloyds State Park, approximately 36,000 cubic yards of sand was being washed away each year due in part to the demise of native vegetation and extensive growth of Australian pine and other invasive plants. As part of a larger beach management effort, $300,000 was allocated to remove Australian pine and restore native plants.

(Continued)

Box 1. Continued

Uninhibited growth of exotic vegetation, loss of bird and marine habitats, and eroding shorelines along Biscayne prompted a major effort led by Miami-Dade County (Milano [18]). The Biscayne Bay Restoration and Enhancement Program coordinates activities that include removal of exotic vegetation, excavation of emplaced spoil fill, planting of native vegetation, stabilization of eroding shorelines and long-term success monitoring. A significant component of the program is the restoration and re-establishment of natural ecosystems to create suitable and stable habitats for native birds, fish, and mammals. Spending between 1990 and 2000 totaled $10.6 million (Milano [10]).

Between 1987 and 2000, $10.6 million has been spent to implement ecosystem restoration and enhancement in Biscayne Bay. As of 2000, ten large scale (exceeding one hectare in size) wetland restoration projects and twelve island restoration projects have been completed. The projects were funded by Miami-Dade County Department of Environmental Resources Management (DERM) in conjunction with State and Federal partners including; Florida Department of Environmental Protection, FDEP; South Florida Water Management District, SFWMD; Biscayne Bay Environmental Enhancement Trust Fund, BBEETF; Miami-Dade Seaport Department; and Florida Inland Navigation District, FIND.

An important component in all of the projects was removal of exotic vegetation. The dominant invasive plants removed were Australian pine (*Casuarina equisetifolia*), Brazillian pepper (*Schinus terebinthifolius*), Burma reed (*Neyraudia reynaudiana*), Inkberry (*Scaevola taccad*a), and Seaside mahoe (*Thespesia populne*a). Native vegetation was replanted at most sites to recreate one or more of the following types of coastal communities: coastal strand and maritime hammock; mangrove wetland and freshwater wetland; and dune.

On most of the island projects, lime rock boulders and filter fabric were installed to stabilize eroding shorelines. A stretch of mature red mangroves along the Oleta River State Park was also reinforced with natural lime rock boulders to reduce erosion and protect the mangroves. Additional activities included creating a high salt marsh, recreating a historical riverbed, building numerous osprey nesting platforms, installing boat-docking facilities on two islands, and planting native species on a university campus. Detail on the 22 restoration projects appears in Table 1.

4 Description of Biscayne Bay restoration costs

4.1 Wetland project costs

Cost of clearing infested acreage and removing invasive plants ranged from $4600 to $5200 per acre. Average cost for removing invasive plants was $4900 per acre for wetland projects. Excavation of fill cost an average of $5 per cubic yard of material removed. For each project, 10% of the total budget went to "mobilization" costs.

Table 1: Biscayne Bay restoration and enhancement projects.

No.	Project	Remove		Construct		Restore				Total cost
		Invasive plants	Dredge spoils	Shore stable	Tidal connect	Coastal	Mangrove	Dune	Dune	
		acs	*cu yds*	*lf*	*no.*	*acs*	*acs*	*acs*	*lf*	
1	Bear Cut Preserve	21.5	41,600	0	3	0.0	10.0	0.0		$800,000
2	Bill Baggs Cape FL State Park	85.0	630,000	0	2	0.0	85.0	0.0		$2,800,000
3	Biscayne Bay Vista Campus	2.0	10,000	0	1	0.0	0.0	0.0	0	$140,000
4	North Virginia Key Dune	7.0	23,000	0	2	0.0	4.0	2.0		$320,000
5	National Bulk Carrier Site	140.0	0	0	0	0.0	0.0	0.0		$300,000
6	Oleta River State Park – Phase I	13.0	55,000	0	1	0.0	13.0	0.0		$300,000
7	– Phase II	45.0	58,000	0	2	0.0	28.5	0.0		$1,500,000
8	– Mangrove stabilization	0.0	0	1900	1	0.0	0.0	0.0		$430,000
9	Highland Oaks Wetlands	8.2	0	0	0	0.0	14.2	0.0		$260,000
10	Chicken Key Bird Rookery	7.0	33,000	0	1	0.0	3.7	0.0	1200	$600,000
11	Dinner Key Island	4.0	0	500	0	4.0	0.0	0.0		$275,000
12	Flagler Monument Island	4.5	0	900	0	4.0	0.0	0.0	750	$220,000
13	Teachers Island	3.7	0	320	0	0.0	0.0	3.7	0	$140,000
14	Morningside Island	4.0	0	500	0	0.0	0.2	4.0		$290,000
15	Mangrove Islands	2.8	0	0	0	0.0	2.8	0.0		$5,000
16	Legion Island	7.0	0	500	0	5.0	0.0	0.0		$150,000
17	Pelican Island	5.0	0	1200	0	4.5	14.0	0.0	150	$300,000
18	Quayside Island	5.0	0	2000	0	4.5	0.3	0.0	360	$590,000
19	Helkers Island	4.0	0	880	0	4.0	0.8	0.0	480	$300,000
20	Crescent Islands	2.0	0	800	0	0.4	0.5	0.0	600	$200,000
21	Little Sandspur Island	1.0	0	620	0	0.0	0.5	1.0	220	$170,000
22	Sand Spur Island	15.0	0	3002	0	14.0	2.5	2.1	0	$531,000

Expenditures for habitat restoration activities ranged from $0 to $2,800,000 per project. Additional information about Biscayne Bay wetland project costs can be found in Table 2 as projects numbered 1 through 10.

4.2 Island project costs

Cost of clearing infested acreage and removing invasive plants ranged from, $5200 to $9,000 per acre. Average cost for removing invasive plants was $7100 per acre for island projects. No excavation activities were reported for the island projects. For each project, 10% of the total budget was spent on "mobilization" costs. Expenditures for habitat restoration activities ranged from $5000 to $590,000 per project. Additional information on Biscayne Bay island project costs can be found in Table 2 as projects numbered 11 through 22.

4.3 Total project cost

Total expenditures for Biscayne Bay restoration projects as reported through year 2000 was $10,621,000. Inflating values using the Turner building cost index (Turner Corp. [19]) and the consumer price index (US Department of Labor [20]) yields the present value of total expenditures in 2006 dollars of $16,187,682. Expenditures in years 1995 through 2000 were inflated with the Turner building cost index. Expenditures in years 1990 through 1994 were inflated using the CPI because Turner values were not available for those years.

4.4 Estimated maintenance cost

Long-term maintenance for each restoration site was estimated at 10% of average restoration cost. For wetland and island projects, the annual cost was $155,337 per year.

5 Assessing the benefits from restoring Biscayne Bay

5.1 Environmental valuation methods

Methods such as hedonic price, travel cost, and random utility models use information from market traded goods to infer environmental values indirectly. The property-value method and marginal-productivity method are other means of utilizing available market data to indirectly ascertain environmental service values. The contingent-valuation method obtains values directly by querying subjects regarding their willingness to pay for environmental services. The benefit-transfer method draws on findings from previous valuation studies in one locale to make inferences about the environmental values in another locale. Benefit transfer is the method applied in this study. For more explanation about valuing coastal environmental resources, the reader is referred to Letson and Milon [21].

Table 2: Estimated allocation of total project costs by category.

	SITE	Date	Remove: Invasive plants	Remove: Fill	Mobilization	Restoration and other	Total cost	Year 2006: PV total cost
	Unit cost 1–10		*$4900 /ac*	*$5.525/c yd*	*10%*			
1	Bear Cut Preserve	1996	$105,350	$229,840	$80,000	$384,810	$800,000	$1,199,238
2	Bill Baggs Cape FL State Park	1999	$391,000	$1,921,500	$280,000	$207,500	$2,800,000	$3,865,965
3	Bay Vista Campus	1995	$9800	$55,250	$14,000	$60,950	$140,000	$218,178
4	Virginia Key Dune	1999	$34,300	$127,075	$32,000	$126,625	$320,000	$441,825
5	National Bulk Carrier Site	1994	$270,000	$0	$30,000	$0	$300,000	$479,878
6	Oleta River State Park – Phase I	1990	$59,800	$167,750	$30,000	$42,450	$300,000	$578,676
7	– Phase II	1999	$220,500	$320,450	$150,000	$809,050	$1,500,000	$2,071,053
8	– Mangrove	1990	$0	$0	$43,000	$387,000	$430,000	$829,436
9	Highland Oaks Wetlands	1999	$40,180	$0	$26,000	$193,820	$260,000	$358,982
10	Chicken Key Bird Rookery	1994	$34,300	$182,325	$60,000	$323,375	$600,000	$959,756
	Unit cost 11–22		*$7100 /ac*	*None*	*10%*			
11	Dinner Key Island	2000	$28,400	$0	$27,500	$219,100	$275,000	$363,739
12	Flagler Monument Island	1994	$31,950	$0	$22,000	$166,050	$220,000	$351,911
13	Teachers Island	2000	$26,270	$0	$14,000	$99,730	$140,000	$185,176
14	Morningside Island	1997	$28,400	$0	$29,000	$232,600	$290,000	$415,719
15	Mangrove Islands	1996	$2500	$0	$500	$2000	$5000	$7495
16	Legion Island	2000	$49,700	$0	$15,000	$85,300	$150,000	$198,403
17	Pelican Island	1992	$35,500	$0	$30,000	$234,500	$300,000	$545,266
18	Quayside Island	1993	$35,500	$0	$59,000	$495,500	$590,000	$1,045,788
19	Helkers Island	1993	$28,400	$0	$30,000	$241,600	$300,000	$531,757
20	Crescent Islands	1992	$14,200	$0	$20,000	$165,800	$200,000	$363,510
21	Little Sandspur Island	1999	$7100	$0	$17,000	$145,900	$170,000	$234,719
22	Sand Spur Island	1993	$106,500	$0	$53,100	$371,400	$531,000	$941,209
	Total cost		$1,589,650	$3,004,190	$1,062,100	$4,965,060	$10,621,000	$16,187,682
	% Total cost		15.0%	28.3%	10.0%	46.7%	100.0%	

5.2 Coastal ecosystem values from previous studies

The value of Florida saltwater marshland was estimated by Bell [22]. Bell applied the property-value method and the marginal productivity value method. Based on historic costs of acquiring wetlands, Bell assessed Florida wetland values at $2879 per acre. Based on the contribution to commercial and recreational fisheries, Bell arrived at $3337 per acre as the value of Florida salt water marshes.

Table 3: Value of coastal ecosystem services from previous studies.

	Site	Value	Year	2006[b]	Source
Coastal wetlands [a]		*$/acre*		*$/acre*	
All ecosystem services	Global	$4043	1994	$5763	Costanza *et al.* [23]
Ecosystem services excluding direct human uses	Global	$3518	1994	***$5016***	Costanza *et al.* [23]
Commercial and recreational fisheries	Florida	$3337	1998	$4227	Bell [22]
Wetland acquisition cost	Florida	$2879	1998	$3647	Bell [22]
Direct human uses [c]	Global	$524	1994	***$747***	Costanza *et al.* [23]
Coastal estuaries		*$/acre*		*$/acre*	
All ecosystem services	Global	$9240	1994	$13,174	Costanza *et al.* [23]
Ecosystem services excluding direct human uses	Global	$8853	1994	***$12,622***	Costanza *et al.* [23]
Direct human uses [d]	Florida	$540	1995	***$748***	Milon [21]
Direct human uses [c]	Global	$387	1994	$552	Costanza *et al.* [23]
Disturbance regulation [e]	Global	$229	1994	***$326***[f]	Costanza *et al.* [23]

[a] Coastal wetlands includes saltwater marsh, tidal marsh, and mangroves.

[b] An interest rate of 3% was used to inflate benefit values.

[c] Direct human uses include recreation, food, and cultural uses.

[d] Direct human uses includes fishing, swimming, boating, nature watching, water sports, hunting, passive uses, and commercial shell fishing.

[e] Disturbance regulation implies habitat protection from storms and other aspects of environmental variability primarily from vegetation structure.

[f] Value in 2006 dollars converts to $0.122 per linear foot and $642 per linear mile assuming a linear foot and a linear mile are ½ acre wide, average width of restored areas equal to 104 ft or 0.0198 mile.

Milon [21] estimated the value of the Indian River Lagoon a coastal estuary located in South Florida. Milon applied the property-value method, marginal-productivity method, contingent-valuation method, and the travel-cost method. Based on property values, human uses (including fishing, swimming, boating, nature watching, water sports, and hunting), passive uses, and commercial shell fishing, Milon estimated an ecosystem value of $724 million per year. Unlike the Indian River Lagoon, the restored areas in Biscayne Bay do not include residential properties, so we subtracted the $33 million in property values leaving a total of $691 million. The Indian River Lagoon covers an area of 2000 square miles or 1.28 million acres. With this information we arrived at a unit value of coastal estuary of $540 per acre in 1995 dollars.

Costanza *et al.* [23] compiled findings from a myriad of studies to provide a reference on the values for a wide range of ecosystem types and the services they provide. This approach would be best classified as benefit transfer. The team of authors categorized 11 aggregate ecosystem types and defined 17 classes of services. Among the aggregate ecosystems types were coastal estuaries and coastal wetlands (categorized as tidal marsh and mangroves). Among the 17 service classes we chose three: "recreation", "food production", and "cultural" to capture direct human use values and allocated the remainder to indirect human use values and nonuse values. With this aggregation, the coastal estuary ecosystem value to direct human use is $377 per acre and the value to indirect human use and nonuse is $8863 per acre in 1994 dollars. The coastal wetland ecosystem value to direct human use is $459 per acre and the value to indirect human use and nonuse is $3,584 per acre in 1994 dollars.

Benefit values from Bell [22], Milon [21], and Costanza *et al.* [23] are shown in Table 3. Selected values from Table 3 (shown in bold italics) were used in the benefit transfer analysis for Biscayne Bay and are displayed in Table 4.

6 Applying benefits transfer to Biscayne Bay restoration

Indirect and direct use benefit from Biscayne Bay restoration projects was estimated based on the number of acres restored, the type of ecosystem restored, and the estimated ecosystem benefit per acre for each type of ecosystem for Florida in 2006 dollars.

The indirect and nonuse ecosystem service values (U_j) from each project j is defined (eqn (1)) as a function of u_i the value per acre for indirect and nonuse services for restored ecosystem i (where i = shoreline, coastal, dune and mangrove) and E_{ij} the number of acres of ecosystem type i restored under project j.

$$U_j = \sum_{i=1}^{4} u_i E_{ij}. \tag{1}$$

The direct use ecosystem service values (V_j) from each project j is defined (eqn 2) as a function of v_i the value per acre direct use services for restored ecosystem i

Table 4: Value of coastal ecosystem services for Biscayne Bay restoration.

Ecosystem services	Shore	Coastal	Dune	Mangrove
Indirect human use and nonuse	*$326*	*$12,622*	*$5016*	*$5016*
Direct human use	0	*$748*	*$747*	*$747*
All services	*$326*	*$13,370*	*$5763*	*$5763*
Description	*Construction activities to stabilize shoreline and reduce erosion*	*Restore native vegetation to create and enhance coastal strand and maritime hammock ecosystem*	*Restore native vegetation to create and enhance dune ecosystem*	*Restore native vegetation to create and enhance mangrove and fresh wetland ecosystem*

(where i = shoreline, coastal, dune and mangrove) and E_{ij} the number of acres of ecosystem type i restored under project j.

$$V_j = \sum_{i=1}^{4} v_i E_{ij}. \tag{2}$$

The wetlands projects (numbered 1 through 10) restore areas with human uses that include fishing, kayaking, bird watching, and snorkeling. The indirect human use values yields the following benefit estimate (b_I)

$$b_I = \sum_{j=1}^{10} U_i. \tag{3}$$

The combined direct and indirect human use values for wetlands projects 1 through 10, yields the benefit estimate (b_{II})

$$b_{II} = \sum_{j=1}^{10} (U_j + V_j). \tag{4}$$

Island projects (numbered 11 through 22) restored areas that are popular recreation sites. The benefits from both direct and indirect human use values are given by (b_{III})

$$b_{III} = \sum_{j=11}^{22} (U_j + V_j). \tag{5}$$

The "base" estimate of the ecosystem service benefit value from projects 1 through 22 sums the values b_I and b_{III} in eqns (3) and (5)

$$A = b_I + b_{III}. \tag{6}$$

The "upper end" estimate of ecosystem service benefit value from projects 1 through 22 sums the values b_{II} and b_{III} in eqns (4) and (5)

$$A = b_{II} + b_{III}. \tag{7}$$

Per acre ecosystem service values (u_i and v_i) are given in Table 4. The sizes of the restored ecosystems (E_i) are displayed in Table 1. Using these values and eqns (1) through (7) ecosystem benefits were estimated and appear as a worksheet in Table 5.

7 Net benefits from the Biscayne Bay restoration projects

The annual benefit from Biscayne Bay restoration projects (eqns (6a) and (6b)) is estimated to be from $1,590,808 to $1,712,790. Long-term maintenance at 10% of average restoration cost comes to $155,337 per year. Restoration benefit adjusted for maintenance costs is $1,433,471 to $1,555,454 per year. The total

Table 5: Ecosystem services benefits worksheet summary.

		Ecosystem services		
		Indirect use and nonuse values	Direct human use values	All values
		\$/year	\$/year	\$/year
1	Bear Cut Preserve	\$50,158	\$7471	\$57,629
2	Bill Baggs Cape FL State Park	\$426,343	\$63,504	\$489,847
3	Bay Vista Campus	\$0	\$0	\$0
4	Virginia Key Dune	\$30,095	\$4483	\$34,577
5	National Bulk Carrier Site	\$0	\$0	\$0
6	Oleta River State Park – Phase I	\$65,205	\$9712	\$74,918
7	– Phase II	\$142,950	\$21,292	\$164,243
8	– Mangrove	\$1486	\$0	\$1486
9	Highland Oaks Wetlands	\$71,224	\$10,609	\$81,833
10	Chicken Key Bird Rookery	\$32,978	\$4912	\$37,890
	Subtotal 1–10 (b_I)	\$820,440	\$121,983 (b_{II})	\$942,422
11	Dinner Key Island	\$50,880	\$2990	\$53,870
12	Flagler Monument Island	\$60,205	\$4332	\$64,538
13	Teachers Island	\$18,809	\$2764	\$21,573
14	Morningside Island	\$21,457	\$3138	\$24,595
15	Mangrove Islands	\$14,044	\$2092	\$16,136
16	Legion Island	\$63,503	\$3738	\$67,240
17	Pelican Island	\$129,763	\$14,092	\$143,854
18	Quayside Island	\$64,396	\$4262	\$68,658
19	Helkers Island	\$60,807	\$4424	\$65,232
20	Crescent Islands	\$14,761	\$1709	\$16,470
21	Little Sandspur Island	\$10,652	\$1514	\$12,167
22	Sand Spur Island	\$202,133	\$13,902	\$216,035
	Subtotal 11–22	\$711,411	\$58,957 (b_{II})	\$770,368
	Estimated annual benefit	Base estimate: $A = b_I + b_{III} =$ \$1,590,808 Upper estimate: $A = b_{II} + b_{III} =$ \$1,712,790		

benefit (TB) can be calculated as follows. Where A is the annual benefit and r is the annual interest rate, the present value of the ecosystem services benefit stream in perpetuity is

$$\mathrm{TB} = \frac{A}{r}. \tag{8}$$

At an annual interest rate of $r = 0.03$, total benefit from Biscayne Bay restoration projects (eqn (8)) is estimated to be from \$47,782,372 to \$51,848,459.

The present value of net benefits (NB) is defined to be total benefit (TB) minus total cost (TC):

$$\mathrm{NB}=\mathrm{TB}-\mathrm{TC}. \tag{9}$$

The present value of the total cost of the restoration projects completed between 1990 and 2000 is \$16,187,682. The present value of net benefits (eqn. (8)) from Biscayne Bay restoration projects is \$31,594,690 to \$35,660,776.

Another indicator of the net gain from public expenditure is the internal rate of return (IRR) that is computed as follows:

$$\mathrm{IRR} = \frac{A}{\mathrm{TC}}. \tag{10}$$

For the 22 restoration projects, the internal rate of return (eqn 9) is estimated to be between 0.0886 and 0.0961 or between 8.9% and 9.6%.

8 Summary

Invasive plants in natural areas are notorious for their ability to alter ecosystems. In Florida coastal areas, invasive species have replaced native plants, dislodged native animals, and radically altered hydrologic processes such as tidal flows, dune replacement, and shoreline erosion rates. This study applied the benefit-transfer method to assess the value of restored ecosystems in Biscayne Bay Florida. Results indicate the present value of a perpetual stream of ecosystem service benefits from the restoration projects is between \$48 million and \$52 million. The cost of the projects was \$16 million in 2006 dollars. Thus the net benefit from the restoration projects is estimated to be \$32 million to \$36 million. The internal rate of return from restoration expenditures is 8.9% to 9.6%.

Acknowledgements

Gary R. Milano from the Department of Environmental Resources and Management, Miami-Dade County provided invaluable information regarding the details of the Biscayne Restoration projects and a helpful review of the manuscript. Doria R. Gordon of The Nature Conservancy and Department of Botany at the University of Florida graciously lent her expertise in reviewing and improving the manuscript.

Alan Hodges of the Food and Resource Economics Department at the University of Florida provided useful comments on condensed version of this manuscript.

References

[1] Lee, D.J., The private cost of upland invasive plants in Florida, U. Florida, Institute of Food and Agricultural Sciences, Florida's Invasive Species Working Group website, http://www.iswgfla.org/, 2005.

[2] Pimentel, D., *Biological invasions,* CRC Press: Boca Raton, FL, 2002.

[3] Fox, A.M., Gordon, D.R. & Stocker, R.K., Challenges of reaching consensus on assessing which non-native plants are invasive in natural areas, *Hort. Sci.,* **38(1)**, pp. 11–13, Feb. 2003.

[4] FLDEP, Upland invasive exotic plant management program, Florida Department of Environmental Protection, Bureau of Invasive Plant Management, *Annual Report FY 2002-3*, 2004.

[5] FLEPPC, List of Florida's Invasive Species, Florida Exotic Plant Pest Council web site http://www.fleppc.org, 2006.

[6] FLDEP, Florida Department of Environmental Protection, Bureau of Invasive Plant Management website, http://www.dep.state.fl.us/lands/invaspec/2nd levpgs/Research.htm, 2006.

[7] USGS, South Florida Ecosystem History Project, US Geological Survey website http://sofia.usgs.gov/flaecohist/biscaynebay.html, 2006.

[8] Cantillo, A.Y., Hale, K. Collins, E. Pikula, L. & Caballero, R., Biscayne Bay: Environmental History And Annotated Bibliography, US DOC, NOAA, *Technical memorandum* NOS NCCOS CCMA 145, 2000.

[9] Wanless, H.R., Cottrell, D., Parkinson, R. & Burton, E., Sources and circulation of turbidity, Biscayne Bay, FL, *Final report to Florida Sea Grant College Program* and Dade County, p. 499, 1984.

[10] Milano, G.R., Island restoration and enhancement in Biscayne Bay, *Proceedings of the 26th Annual Conference on Ecosystem Restoration and Creation*, ed., P.J. Cannizarro, Hillsborough Community College: Tampa, FL, pp. 1–17, 2000.

[11] SFWMD, South Florida Water Management District website http://www.sfwmd.gov/org/wrp/wrp_ce/projects/bb/goal3.html, 2006.

[12] Gordon, D.R., Effects of invasive, non-indigenous species on ecosystem processes: lessons from Florida. *Ecological Applications*, **8(4)**, pp. 975–989, Nov. 1998.

[13] Bandyopadhyay, S., Natural environmental hazards and their management: a case study of Sagar Island, *India Singapore Journal of Tropical Geography*, **18(1)**, pp. 20–45, 1997.

[14] Sealy, K.S., Flowers, L., Nero, V.L., Semon, K.L. & Bradley, K., The State of the Coast: Report on Andros and South Andros, henge.bio.miami.edu/coastalecology/THE STATE OF THE COAST.pdf, 2007.

[15] Rodgers, J.C. III., The distribution of Casuarinas on San Salvador Island, The Bahamas. *Southeastern Geographer*, **45(2)**, pp. 222–238. 2005

[16] Doren, R.F. & Ferriter, A., *Weeds Won't Wait!, An Assessment Of Invasive Exotic Plants In Florida,* A Report to the South Florida Ecosystem Restoration Task Force and Working Group, South Florida Ecosystem Restoration Task Force, Miami, Florida. 305/348.1665, 2002.

[17] Dept of Interior (DOI), Dry Tortugas National Park, *2000 Annual Report*, National Park Service, 2000.

[18] Milano, G.R., Restoration of coastal wetlands in southeastern Florida. *Wetland Journal,* **11(2)**, pp. 15–24, 1999.

[19] Turner Corp., Turner Corporation Building Cost Index, http://www.turnerconstruction.com/corporate/content.asp?d=20, 2006.

[20] US Department of Labor. Bureau of Labor Statistics, Consumer Price Indexes, www.bls.gov/cpi, 2006.

[21] Milon, J.W., Natural resource valuation of Indian River Lagoon, (Chapter 8). *Florida Coastal Environmental Resources: A Guide To Economic Valuation And Impact Analysis*, eds D. Letson, & J.W. Milon, Florida Sea Grant College Program: SGR, p. 124, 2002.

[22] Bell, F.W., The economic value of saltwater marsh in Florida's commercial fisheries, (Chapter 5) *Florida Coastal Environmental Resources: A Guide To Economic Valuation And Impact Analysis*, eds D. Letson, & J.W. Milon, Florida Sea Grant College Program: SGR, p. 124, 2002.

[23] Costanza, R., d'Arge, R., deGroot, R., Farber, S., Grasso, M., Hannon, B., Limburg, K., Naeem, S., O'Neill, R.V., Paruelo, J., Raskin, R.G, Sutton, P. & van den Belt, M., The value of the world's ecosystem services and natural capital, *Nature*, **387(15)**, pp. 283–260, May 1997.

CHAPTER 11

The economic value of watershed conservation

Brooks Kaiser[1,3], Basharat Pitafi[2], James Roumasset[3] & Kimberly Burnett[3]
[1]*Department of Economics, Gettysburg College, Gettysburg, PA, USA.*
[2]*Department of Economics, Southern Illinois University, Carbondale, IL, USA.*
[3]*Department of Economics, University of Hawaii, Honolulu, HI, USA.*

Abstract

Watershed conservation creates benefits within and beyond the management area of interest. The magnitude of the benefits also depends considerably on economic policies accompanying conservation measures. Direct benefits come from biodiversity protection, improved recharge, and the improved quality of ground and surface water. Additionally, the health of a watershed has profound implications on near-shore resources, including beaches and coral reefs. In the case of the Koolau Watershed on the island of Oahu, Hawaii, total economic benefits are found to be well above costs. The anticipated cost of watershed conservation into the indefinite future has a present value of $43.2 million using a discount rate of 1%. The benefits of watershed conservation stemming from groundwater recharge alone vary widely depending on the assessment of increased recharge but may be more than $900 million provided that conservation is accompanied by pricing reform. Benefits to near-shore resources (including the avoidance of beach closures and reef sedimentation) range from $4.2 million to $22.0 million even before accounting for flood and sewage-spill damages in March, 2006.

1 Introduction

Watershed conservation creates benefits within and beyond the management area of interest. Direct benefits are those realized in the watershed itself, such as improved water quality and quantity, and biodiversity protection. Additionally, the health of a watershed has profound implications on near-shore resources below its reaches, including beaches and coral reefs. This chapter reviews the major benefits of watershed conservation and discusses the economic value of these activities.

Within the watershed itself, conservation activities assist in protecting important groundwater sources by maintaining healthy canopies, native plant cover, reducing erosion, and fending off nuisance plant and animal species that degrade the system. These activities are a vital part of overall watershed maintenance and are reviewed in the introduction to this volume. Systems that are often excluded from the discussion of watershed management are those downstream from the area of interest. Improved watershed health leads to less runoff to ecosystems below and therefore cleaner and more valuable beaches and reefs.

Perhaps the most direct benefit from tropical watershed conservation is that of aquifer recharge. The Hawaiian island of Oahu gets about 90% of its freshwater supply from groundwater. Alternative production techniques such as desalination are costly and postponement of their need is a valuable policy goal. Postponement can occur through the supply side by maintaining or potentially enhancing forest quality. The Koolau conservation district is a 97,760–acre area along the Koolau mountain range running the entire windward (Eastern) side of the island (see Fig. 1). There are hundreds of inches of rain each year in some locations, and the general trend is for high levels of rainfall along the crest of the range declining with elevation. The form and composition of the forest in large part determines how much of the water will run off, how much sediment it will carry, and how much will recharge the aquifers from which Oahu draws its water supply.

Kaiser *et al.* [1] calculate that the net present value lost from a decrease in recharge of 41 MGD to the Pearl Harbor aquifer may be between \$1.42 billion and \$2.63 billion dollars, depending on the assumptions made regarding the social discount rate. If the current levels of groundwater recharge are maintained, the scarcity rents will rise from \$0.6 per thousand gallons in year 2000 to nearly \$3 in year 2072. On the other hand, if the forest damage reduces the recharge by 41 MGD, the scarcity rents in year 2000 jump to over \$1 per thousand gallons and rise to nearly \$3 by 2057. They also calculate that if recharge to the aquifer from the Koolaus ceased altogether, the reduction of inflow to the Pearl Harbor aquifer would be approximately 133 MGD. The lost net present value from such a disruption would amount to \$4.57 to \$8.52 billion.

Biodiversity and aesthetics are among the other direct benefits provided by tropical watersheds. In the Koolau mountains, these forest benefits have been valued between \$2626 and \$7860 million in net present value terms, composed of species habitat benefits (\$487 to \$1434 million), biodiversity benefits (\$660,000 to \$5.5 million), subsistence benefits (\$34.7 to \$131 million), hunting-related benefits (\$62.8 to \$237 million), aesthetic values (\$1040 to \$3070 million), commercial harvests (\$0.6 to \$2.4 million), and ecotourism (\$1000 to \$2980 million) (Kaiser *et al.* [1]).

Tropical watershed conservation also has important indirect benefits to offsite resources. Runoff from forests may damage nearshore resources such as beaches, reefs, and marine resources. For example, runoff from forest watersheds can contribute to reef sedimentation and nutrient loading in nearshore waters (e.g. Mackenzie

Figure 1: Pearl Harbor Aquifer, Oahu, Hawaii. (Source: United States Geological Survey).

[2]), and streams transport toxins to reefs, marine fisheries and recreational areas, resulting in eutrophication and algae growth that smothers out the (hard) scleratinian corals (Dollar & Tribble [3]; Jokiel [4]). Conservation of forests that reduces runoff, therefore, provides benefits for nearshore resources.

2 Direct benefits of watershed conservation: the Pearl Harbor aquifer

Figure 1 shows the Pearl Harbor coastal aquifer. The aquifer is recharged from the Koolau mountain watershed, which rises up to 1000 meters with annual rainfall reaching 4000 millimeters at high elevations (Giambelluca *et al.* [5]). In the early 1900s, the State removed destructive cattle and feral ungulates (mainly pigs and goats) and replanted vegetation in degraded areas around the watershed. These watershed conservation efforts have helped to protect aquifer recharge.

Currently, anthropogenic pressures and biological changes that work to reduce forest cover and increase runoff, at the expense of recharge, threaten the watershed. Existing feral pig and goat populations destroy vegetative cover and compact the soil, likewise increasing runoff and decreasing recharge. Urban encroachment and increased recreational use of watersheds, e.g. hiking and off-road vehicular traffic have similar effects. A growing threat stems from a possible shift in the forest structure from the spread of an invasive weedy tree, *Miconia calvescens*. Miconia has a shallow root structure, contributing to landslides, and can virtually eliminate the understory, thereby decreasing infiltration and increasing runoff. Miconia currently covers 70% of previously native forest on two Tahitian islands (Meyer [6]) and has already spread to nearly 150 thousand acres of forestland in the State of Hawaii. In the Koolau watershed, Miconia is present and the island's invasive species committee is actively engaged in suppressing this population. It is believed there are no longer any adult trees in the watershed but there are many juveniles and the seed bank is expected to last at least 14 years (Meyer, personal communication, 2007).

Groundwater in the Pearl Harbor aquifer, as in many other coastal areas, is commonly modeled as a Ghyben–Herzberg lens where freshwater floats on a saltwater layer underneath (Mink [7]; Shade and Nichols [8]; Oki [9]; Oki *et al.* [10]), as shown in Fig. 2. The freshwater from the lens leaks into the ocean at a rate that varies directly with the head level. If extraction is more than the net recharge (net of inflow and leakage), the lens contracts and the head level falls, requiring greater lift, and therefore increasing costs for further extraction.

Water is extracted and distributed by the Honolulu Board of Water Supply (HBWS), an agency of the City and County of Honolulu. The first extraction of groundwater on Oahu occurred in 1879, following surface-water shortages in the 1870s. By 1917 the civil water authority of the day (Honolulu Water Commission) was issuing statements of concern about sustainability of the resource (HBWS [11]). Current HBWS estimates indicate if growth in demand continues at its current pace then new sources of will be needed by 2020 (Kaiser *et al.* [1]), however there remains political reluctance to raise prices or use other rationing mechanisms.

Users in the area are divided into domestic, commercial, and agricultural categories. Most of the water use is in the first two categories with agriculture being close to 2% of the total water use (Malla [12], Table 4.2). The HBWS charges cost-recovery prices for the water supplied to the users. This amounts to pricing water at the average cost of extraction and distribution, and ignores the opportunity cost of resource drawdown (marginal user cost). A potential alternative but much more

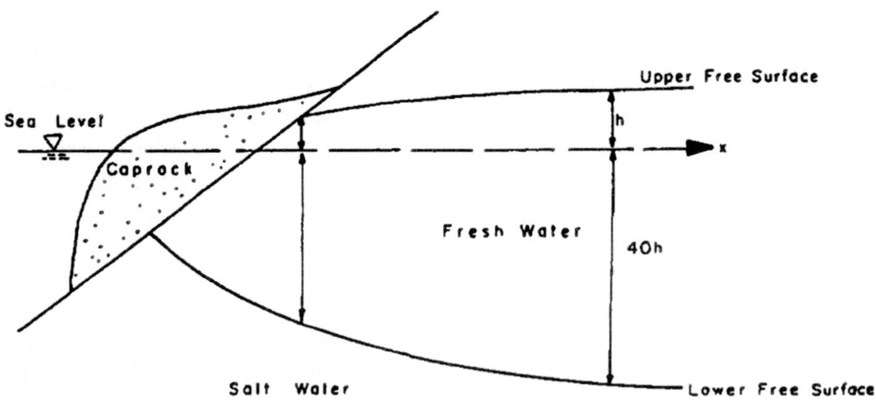

Figure 2: Ghyben–Herzberg freshwater lens (after Mink [7]).

expensive source of freshwater, which can be used when the groundwater is depleted, is desalination of seawater.

Pitafi and Roumasset [13] have modeled this aquifer to estimate the value of watershed recharge service. They set up a regional hydrologic-economic model to optimize groundwater use, along the lines of Krulce *et al.* [14], which extends previous models (e.g. Brown and Deacon [15, Cummings and Burt [16], Moncur and Pollock [17], among others) by allowing recharge to vary continuously with the head level. Water is extracted from a coastal groundwater aquifer that is recharged from a watershed and leaks into the ocean from its ocean boundary depending on the aquifer head level. They find that gains from watershed conservation that prevent a 1% loss of recharge are about $43 million in present value terms. If the watershed conservation prevents a 10% loss of recharge, then the gains are over $546 million. These gains are obtained with the status-quo pricing policy, under which average cost pricing is used. However, if the groundwater is being efficiently priced, so that prices equal the marginal user costs across individuals and time, the gains increase dramatically, as discussed in the final section of this chapter.

3 Indirect benefits of watershed conservation: near-shore resources

Near-shore resource quality is dependent not only on direct use but also on the quality of upland watershed. Recent efforts at economic valuation of Hawaii's coral reef assets have generated values up to $10 billion dollars in present value, or $360 million per year (Cesar [18]). Too often, the role of the watershed in protecting these valuable assets is underplayed or ignored in policy decisions. We investigate the economic implications of expanding near-shore resource policy to include explicit upland conservation efforts. The intent of this section is to promote integrated management of marine and terrestrial environmental resources for the state of Hawaii in particular and to define explicitly the interconnectedness among said resources to examine the potential benefits from such integrated policy efforts.

Near-shore resources are directly damaged by tourists and residents, e.g. through water pollution and alteration of marine ecosystems. Less obviously but not necessarily less importantly, beaches, reefs, and marine resources may also be damaged by runoff from upland areas. For example, runoff from forested watersheds can contribute to reef sedimentation and nutrient loading in near-shore waters, and channelized streams transport toxins to reefs, marine fisheries, and recreational areas. Short-term turbidity episodes may not kill reefs, but they will promote eutrophication and algal growth over coral substrate growth. The reduction in calcifying reef builders may result in smaller, more slowly growing reefs that do not support as much ecosystem diversity or long-run biomass production. Long-run ecosystem service production will decline, including beach and sand production, habitat quality for recreational divers and snorkelers, and fish nurseries. Since

Hawaii's reefs are young, at the edge of the reef-supporting climatic zone, and without great diversity and resilience, they are particularly fragile.

Lacking an approach to integrated resource management, planners tend to impose restrictions on direct use, e.g. by closing beaches, limiting the number of visitors, or restricting access. By neglecting the indirect sources of resource degradation, however, these restrictions are partially effective at best. Moreover restricting access to the object of tourism has the dual effects of shifting tourism to other, possibly ecologically fragile, venues and of reducing overall tourism demand. The latter has the additional effect of reducing the economic tax base and rendering conservation projects even more difficult to fund. With lower conservation budgets, user restrictions tend to be used more and more as a conservation device, resulting in a vicious cycle of resource degradation and diminishing fiscal capacity.

A more promising approach to near-shore resource conservation is to consider both direct and indirect sources of ecological damage in formulating an integrated conservation strategy. In particular, reducing runoff by conserving upland watersheds can be an important component of near-shore resource conservation. In addition to protecting near-shore resources, it reduces the need for user restrictions and concomitant erosion of economic activity and fiscal capacity. Upland conservation also increases groundwater recharge, thus maintaining the value of groundwater aquifers and further avoiding the drain on the economy and fiscal capacity that would accompany the required increases in water prices.

This section provides a preliminary estimate of the benefits of runoff mitigation, with particular emphasis on the enhanced value of near-shore resources. By demonstrating both the importance and the difficulty of integrated resource management, the section will also provide a prima facie basis for a broader multidisciplinary program of research to examine the detailed linkages between terrestrial and marine resources and between environmental resources and the economy.

Using Geographical Information Systems (GIS) software to frame the problem, we superimpose the reef structure, the watersheds that begin in the Koolau conservation district, the island's roads and trails, and known *Miconia calvescens* plants on a landsat photo of Oahu. (For those who are visually oriented, a color illustration that summarizes the complexity of the problem is available online (see Fig. 1.1 in Kaiser *et al.* [19]).) It is apparent that particularly in the southern portion of the district, changes in forest quality may be affecting downslope reef resources. *Miconia calvescens* is an invasive species of concern because it is believed to reduce forest cover and increase runoff and sedimentation. Roads and trails are longstanding sources of concern for increasing runoff and sedimentation in forested watersheds.

In order to value the benefit of watershed conservation on near-shore resource, we begin by establishing runoff as the linkage between forest quality and beach and reef quality. We then ascertain how forest quality affects runoff, and then explore the relationship between runoff and near-shore resources. Because none of these relationships are understood with certainty, we take several approaches to quantifying these relationships. We then use these relationships to examine two management scenarios: first, status quo management of the Koolaus, and second, management under a fully funded KMWPP. We obtain estimates of damages to beach values and reefs under each scenario. The difference in these will be the benefit of upland conservation on these near-shore resources.

4 Watershed health and runoff

4.1 Summary results from survey of experts

4.1.1 Water quality

We surveyed watershed and near-shore resource experts at length to determine the current state of the watershed and the threats it faces for the purpose of water quality. We use these results to generate the expected damages to the watershed and near-shore resources.

The near-shore resource experts were asked to indicate their opinions regarding the current state of runoff and sedimentation affecting ocean resources on the windward and leeward sides of the Koolaus.

There is some controversy over the connection between human interaction in the watershed and near-shore resource quality, though most respondents believe that runoff and sedimentation rates are not at their natural background levels. About one quarter of respondents feel that all watersheds are impaired with toxics and other waste products, and over 50% believe there are elevated levels of runoff and sedimentation from all areas except North, while at the same time between 5 and 25% indicate human intervention has reduced runoff and sedimentation, particularly to Kaneohe bay.

There was somewhat greater consensus about what effects runoff and sedimentation might have. Overall, natural background sedimentation was described as allowing high water quality, as were sedimentation rates below background. Sedimentation rates only 10–20% above background reduced confidence both in water quality and in the ability to determine water quality.

The forestry questionnaire also asked about the respondents' understanding of reef impacts from changes in forest quality. The forest experts were provided with a characterization of the watershed's basic connection to near-shore resources, along with some background information regarding construction of the H3 freeway and its impact (discharge, turbidity, and fecal coliform levels) on Halawa stream in the leeward Koolaus. This summary was given so the respondents could consider the extent to which runoff and sedimentation may change reef composition and the value of other beneficial assets.

Under the assumption of natural background sedimentation, 32% of the respondents believed that water quality would be high. Given sedimentation rates 10–20% below or 100–150% above background, 30% of respondents answered in the same way, with an assumption of high water quality. However, the majority of the respondents were uncertain on each and every condition described.

Overall, there appears to be more consensus that turbidity will rise with increased sedimentation than that biological water quality will fall. Since turbidity is the biggest threat to reef corals, and there is more certainty regarding its impacts, we concentrate on establishing connections between the conservation district and sedimentation rather than between the conservation district and biological properties of water quality.

4.1.2 Reef growth

The near-shore resource experts responded to questions about reef growth. Hawaii's reefs, like its islands, are young and grow relatively slowly due to the types of corals present, which do not, for example, include any of the more rapidly growing, branching *acropora* species. Long-run growth rates for Hawaiian reefs keep pace with island subsidence except in periods of rapid increases in sea level (Moore [20]). Over half the respondents (55–65% depending on the portion of the watershed) felt that sedimentation levels on the leeward side of the Koolaus and on the southern end of the windward Koolaus were elevated by human intervention, while a third also felt that the northern portion of the windward Koolaus were currently experiencing elevated levels of runoff and sedimentation. However, 5–15% of the experts believed that human intervention had reduced the flow of runoff and sedimentation to the ocean on the leeward side and 0–25% believed this for the windward side. We then asked what Hawaiian reef quality would look like under these different possibilities. Their responses indicated that if sedimentation were at its natural background rate, 55% of the experts believe reef growth would be normal for temperature substrate, etc., and 9% say reef growth is deterred but without significant impact. If sedimentation is at rates 10–20% (100–150%) above its natural background, 14% (5%) think reef growth is normal, 27% (5%) think growth is deterred but without significant impact, 9% (36%) think growth is deterred to the detriment of fisheries, and 5% (45%) think growth is deterred to the detriment of recreation, such as diving and snorkeling. In contrast, if sedimentation rates are 10–20% (100–150%) below natural background, 41% (36%) say reef growth is normal, 5% (0%) say growth is deterred to the detriment of recreation, 9% (5%) say growth is enhance but without significant impact, 5% (18%) say growth is enhanced with gain to fisheries, and 9% (14%) say reef growth is enhanced with gains to recreation. For each alternative, an average of 28% of respondents were uncertain.

4.2 Econometric relationships between watershed health and runoff

The survey exercise above was our first approach to determining the relationship between the health of the watershed and runoff to near-shore resources. Now we attempt to determine the connection between forest quality, runoff, and sedimentation affecting the Koolau conservation district for ourselves. To do this, we construct raster data of land cover, soils, and relevant hydrological properties for the district and then analyze how changes in the land cover, simulating forest quality changes from invasive species or increased usage, will affect runoff and sedimentation.

4.2.1 Existing upland quality and connection to runoff

We base our understanding of water balance in Oahu on infiltration and rainfall data obtained from the University of Hawaii and the County of Honolulu, respectively. Figure 3 illustrates the difference between rainfall and infiltration, which at constant soil moisture storage is the sum of runoff plus evapotranspiration. Note that negative values are possible. These reflect highly irrigated lands.

We regress these spatial estimates of runoff with evapotranspiration on solar radiation levels, elevation, slope, soil characteristics, and proximity to roads to determine how runoff changes with respect to these characteristics for each of the land-cover types delineated by NOAA's CCAP (Coastal Change Assessment Program) project for Oahu. In this way, we interpolate water-balance equations by land cover and can use these equations to predict changes in runoff from changes in land cover. (For the visually oriented, Fig. 7.2 in Kaiser *et al.* [19] (online) shows these land classifications.) Lands in the Koolau Conservation District fall almost entirely into the evergreen forest classification or the scrub/shrub classification.

Once we have connected Oahu's land cover to its hydrological properties spatially, we can describe each land area in terms of its water-balance characteristics. Oahu is divided into cells where each cell covers 40,000 square meters, and each cell is associated with land-cover data and other characteristics, including rainfall, estimated recharge (El-Kadi [21]), soil type, roads, trails, and streams. Data not otherwise attributed is from the Hawaii Office of Planning. Table 1 provides summary data by select land classifications for variables affecting water

Figure 3: Runoff and evapotranspiration on Oahu.

Table 1: Land-use classification and water balance.

	Evergreen forest	Scrub/ shrub	Low-intensity developed	High-intensity developed	Cultivated land
Rain (mm/yr)	2550 (1417)	1913 (1353)	1066 (440)	864 (293)	1041 (284)
Elevation (ft)	907 (565)	973 (668)	266 (278)	159 (175)	536 (342)
Slope (degrees)	26.5 (22)	36.1 (25)	7.1 (12)	3.4 (9)	6.9 (9)
Class A Soils (portion)	0.01 (0.1)	0 (0)	0.06 (0.23)	0.04 (0.20)	0 (0)
Class B Soils (portion)	0.57 (0.49)	0.47 (0.50)	0.48 (0.50)	0.31 (0.46)	0.83 (0.38)
Class D Soils (portion)	0.42 (0.49)	0.52 (0.50)	0.34 (0.47)	0.47 (0.50)	0.09 (0.29)
Pineapple (portion)	0 (0)	0 (0)	0.04 (0.19)	0.01 (0.11)	0.34 (0.47)
Recharge (mm/yr)	1054 (857)	734 (807)	243 (282)	104 (191)	312 (289)
Proximity to Road (m)	2833 (2001)	2242 (1717)	421 (566)	296 (396)	708 (545)
Proximity to trail (m)	3249 (2575)	4889 (4445)	6154 (4074)	5925 (3593)	8512 (3047)
Proximity to stream (m)	262 (474)	619 (1078)	970 (1347)	1256 (1388)	700 (744)
Solar radiation (cal)	366 (52)	381 (51)	419 (47)	438 (43)	423 (30)
Runoff + ET (mm/yr)	1496 (675)	1179 (669)	823 (355)	760 (232)	729 (399)
N. Obs.	8000	15924	4013	2094	2488
Runoff + ET as% of rain	59%	62%	77%	92%	70%

balance; for example the average rainfall is 2550 mm/yr in areas that are evergreen forest but only 864 mm/yr in areas that are high-intensity developed. The final row compares the percentage of rainfall that becomes runoff or evapotranspiration for that land classification. We see that this ranges from a low of 59% in evergreen forest areas to a high of 92% in high-intensity developed areas.

Land Use and Land-cover data is taken from NOAA's Coastal Change Analysis Program (CCAP) for Oahu in 2000, which is derived from Landsat satellite imagery with 30–m resolution. The project tracks 22 land covers, 14 of which are present on Oahu, and four of which we discuss below. Evergreen Forest and Scrub/Shrub are the current land covers in the Koolau Conservation district, while low- and high-intensity developed areas describe alternative land covers for all or parts of the area.

Evergreen forest describes areas dominated by trees generally greater than 5 meters tall and greater than 20 percent of total vegetation cover. More than 75 percent of the tree species maintain their leaves all year, and canopy is never without green foliage. Scrub/shrub describes areas dominated by shrubs less than 5 meters tall with shrub canopy typically greater than 20 percent of total vegetation. This includes tree shrubs and young trees in early successional stages as well as trees stunted by environmental conditions. Low-intensity developed land includes areas with a mixture of constructed materials and vegetation. Impervious surfaces account for 21 to 49 percent of total cover. High-intensity developed land includes highly developed areas where people reside or work in high numbers. Impervious surfaces account for 80 to 100 percent of the total cover.

Rain, recharge and runoff with evapotranspiration are measured in millimeters per year, elevation in feet, slope in degrees, and solar radiation in calories per cm^2/day. Class A soils are the most absorptive, best draining soils, and Class D are the least. Pineapples have reduced transpiration rates so an indicator variable is used for lands in pineapple cultivation. These variables are measured as the fraction of that land cover with the associated soil type or pineapple cultivation. Road, trail, and stream proximity are measured by cell distances (meters) computed by Arcview's spatial analyst.

4.2.2 Estimation of runoff impacts by land cover

We regress the sum of runoff and evapotranspiration (RunET) on land cover, elevation, slope, soil classification, solar radiation, pineapples if applicable, and proximity to roads and streams. We try two scenarios: one where the effects of the nonland-cover variables are held constant across land-use types, and the second where these coefficients are also allowed to vary by land cover. Results are in Table 2.

We see from the positive, significant coefficient on elevation that rainfall, and therefore runoff, is indeed orographic, meaning that it follows the elevation of the mountains. We include fogdrip, an indicator variable for cells above 615 m, to account for cloud forest. Slope is surprisingly negative, though it is fairly correlated with elevation (0.48), solar radiation (–0.38), and Koolau (0.30), which may affect its influence. Solar radiation is negative as expected; more sunshine dries out the land. The least absorptive, Class D soils, are positive and the coefficient is sizeable and significant with the exception of when they are on high-intensity developed land. This corroborates the opinions of experts that high-intensity developed land has little recharge due to much of the surface area being impermeable (e.g. paved, roofed). The indicator variable for land in the Koolau conservation district is positive and significant for the land-use types that have a presence there; this may be due to the higher rainfall in the area but may also stem from the correlation with slope.

Table 2: Regression of runoff + evapotranspiration on land characteristics.

Variable	I (all)	II (Evergreen forest)	III (Scrub/shrub)	IV (Low-intensity developed)	V (High-intensity developed)	VI (Cultivated land)
Elevation	0.15*	0.44*	0.04*	0.07*	0.09*	0.31*
	(0.006)	(.01)	(0.008)	(0.02)	(0.03)	(0.03)
Slope	–3.00*	–2.44*	–2.63*	–1.84*	–1.38*	–2.7*
	(0.11)	(0.20)	(0.14)	(0.38)	(0.47)	(0.60)
Solar Rad	–3.03*	–2.84*	–3.59*	–1.50*	–0.79*	–2.12*
	(0.06)	(0.11)	(0.09)	(0.14)	(0.16)	(0.23)
D Soils	125*	241*	83*	39*	–4.2	195*
	(4.6)	(9.4)	(6.7)	(9.4)	(7.4)	(19)
Koolau	269.5*	75.0*	293.0*	–9.7	–69	–46
	(6.9)	(11.7)	(10.8)	(27)	(44)	(136)
Pineapple	20.8	43	73.5*	–210.7*	–393.8*	28.2*
	(12.9)	(64)	(34.2)	(22)	(32.8)	(17.1)
Road	0.09*	0.05*	0.12*	0.11*	0.06*	–0.08*
	(0.002)	(0.003)	(0.003)	(0.008)	(0.009)	(0.01)
Trail	–0.01*	–0.03*	–0.006*	0.001	0.00	0.001
	(0.001)	(0.002)	(0.0009)	(0.002)	(0.00)	(0.003)
Stream	–0.01*	–0.01	–0.02*	–0.02*	–0.02*	0.13*
	(0.002)	(0.009)	(0.003)	(0.004)	(0.004)	(0.008)
Evergreen	–					

Scrub/shrub	−30*					
	(5.5)					
Low dev	108*					
	(8.6)					
High dev	128*					
	(11)					
Cultivated	23*					
	(10)					
Fogdrip	127.0*	560.2*	54.8*			
	(12.03)	(26.4)	(14.7)			
East (by cell)	2.44*	4.98*	2.30*	2.81*	2.77*	3.81*
	(0.05)	(0.16)	(0.07)	(0.14)	(0.19)	(0.20)
South (by cell)	−2.58*	−2.85*	−2.56*	−3.29*	−3.43*	−7.67*
	(0.06)	(0.12)	(0.08)	(0.15)	(0.22)	(0.19)
Const	1974.0*	1445.2*	2210.0*	1471.7*	1233.0*	1886.2*
	(28.6)	(65.9)	(42.6)	(71.4)	(87.0)	(114)
R sq. adj.	0.68	0.70	0.67	0.49	0.56	0.57
N. Obs.	32610	8000	15924	4013	2094	2488

Pineapple has a large significant negative effect in low- and high-intensity developed land but a positive and smaller effect on cultivated land. It may be that compared to irrigated sugar, pineapple's reduced transpiration does not outweigh this contribution to water balance. By increasing runoff and evapotranspiration on net, distance from a road has the opposite effect one would anticipate, though in the case of cultivated land the increased distance decreases runoff + ET. Distances from trails and streams have small, negative effects.

In the regression holding the impacts of other variables constant, we see that compared to evergreen forest, scrub/shrub appears to have slightly lower runoff, while cultivated land, low-intensity developed, and high-intensity developed have increasingly higher runoff, respectively. These results are expected from the raw percentages of runoff + ET calculated in Table 1 with the exception of scrub/shrub, which has a slightly higher runoff percentage (62%) than evergreen forest (59%). In the individual regressions, scrub/shrub also has a much higher constant value, though certainly for elevations above 615 m the gap is significantly smaller. Elevation itself seems to matter much less and solar radiation considerably more for scrub/shrub.

4.2.3 Predicted changes in runoff if land cover changes

Table 3 below shows how runoff and evapotranspiration would change if existing land use/land cover changed to another land use/land cover.

The numbers in the table represent the proportions of precipitation that are estimated in regression analysis as runoff or evapotranspiration in the water balance under a given land cover for an area that has the characteristics in terms of elevation, slope, east and north positions, soil hydrology properties, pineapple plantings, distance to road, distance to trail, distance to stream, solar radiation, Koolau conservation status, and fogdrip potential of existing land uses. Thus, the diagonal shows the estimated current level of runoff + ET from each land cover.

Table 3: Predicted Runoff + ET from land cover change.

	Predictions Means			
Coeff:	Evergreen	Scrub/shrub	Low-intensity developed	High-intensity developed
Evergreen	62.6	64.3	76.8	82.7
Scrub/shrub	65.4	68.1	80.6	86.4
Low-intensity developed	66.8	59.8	79.4	86.3
High-intensity developed	59.4	56.8	80.5	89.9

Means for each land-cover category are used to generate estimated runoff + ET using coefficients from each regression. Predicted runoff + ET are about 3% higher than the means from the raw data. Evergreen forest has the lowest predicted runoff-ET (62.6%) and high-intensity developed has highest (89.9%). Predictions match average data fairly well. Evergreen forest accounts for about 1/3 of Koolaus, with scrub/shrub as 2/3. Transforming the Koolaus into low-intensity developed land would increase runoff + ET by about 15%. The potential expanse of such a change is shown in Figures 7.3 and 7.5 in Kaiser *et al.* [19].

Figure 7.5 in the report shows transformation of the Koolaus from currently forested conditions to low-intensity developed. Forest quality deterioration is likely to mimic effects of impervious surfaces from pig damages that compact soils, canopy reductions that increase erosion and runoff, or urban development and other human access. While this figure shows a total transformation of the formerly forested area, each GIS grid cell can be treated individually in analysis if desired.

5 Runoff and near-shore resources

5.1 Marine pollution due to runoff from conservation district

Marine pollution in the form of suspended sediments and associated particulate nutrients in Oahu's waters arrive mainly through storm events (Hoover [22]). Hoover has studied storm runoff at sites on Oahu that have conservation land use and compared them to those that have agricultural and urban land uses as well. Using only a few sites, he has insufficient data to identify quantitatively land use impacts, and attributes control of specific discharge to rainfall, while noting that at one site at least (Opaeula) discharge is much lower than would be expected if rainfall were the dominant factor at the site. Table 4 summarizes data for the four main sites. Figure 4.1 in Kaiser *et al.* [19] shows how the four sites differ significantly in characteristics other than land use. Kahana, while conservation land, is quite steep, while Opaeula (in the Paukauila system) is much longer and passes through a long flat portion of agricultural land. Heeia and Kaneohe offer the most similar geophysical properties and differences between those two areas might suggest more clearly the role of urbanization separate from other factors.

We see that Heeia has more of its sediments discharged during storms than Kaneohe, and slightly less of its water discharge.

Hoover's analysis also provides regression parameters for annual specific sediment yields from the study sites. Though the relationships between yield and runoff are quite different at each site, exponential functions fit the data fairly well for each watershed. Table 5 shows the regression results from Hoover [22] for the equation Yield = A*Runoff^{B}. Yield is measured in kg/ha/yr while Runoff is in cm/yr.

There are no clear trends associated with land use, though it is possible that greater conservation land increases the curvature of the relationship between yield and runoff, especially considering that Heeia's geophysical properties resemble

Table 4: Discharge from Hawaiian streams (Hoover [22]).

Site	Con-servation % of total	Agri-cultural % of total	Urban % of total	Runoff % of rainfall	Base water discharge % of total	Storm water discharge % of total	Base suspended sediment discharge, % of total	Storm suspended sediment discharge, % of total
Kahana	100	0	0	76	65	35	1.4	99
Heeia	100	0	0	40	90	9.6	7.8	92
Opaeula	56	44	0.2	31 (upper watershed)	43	57	3.6	97
Kaneohe	84	0	16	41	84	16	32	68

Table 5: Estimated annual sediment yield as a function of runoff in Hawaiian streams.

Site	*A*	*B*	*R*-sq	Yield (kg/ha/cm)
Kahana	6.3×10^{-6}	3.4	0.97	4.1
Heeia	5.1×10^{-5}	3.8	0.72	43
Opaeula	2.3	1.5	0.84	10
Kaneohe	5.5×10^{-2}	2.0	0.60	8.3

Kaheohe's more than Kahana's, though the estimated parameters are closer to Kahana's.

There is evidence of a relationship between land use and the types of nutrients contained in the sediments. Conservation land sites (Kahana and Heeia) exhibit low, constant values for dissolved nitrogen and phosphorus during low-flow conditions (Hoover [22], 175), but low-flow concentrations are more variable and higher for the agricultural and urban sites. The agricultural sites have the greater variability. Particulate organic nitrogen and particulate organic carbon appear to be comparatively depleted and less variable at urban sites and urban-site sediments contain more nonsoil detritus both from urbanization and because increased channelization increases the transport of sediments in general. Urban sites also have elevated particulate inorganic phosphorus, though the reason is unclear. Phosphate (PO_4) levels are significantly elevated for urban areas, particularly in storm periods and compared to global figures on similar watersheds. (Hoover [22], 185). There are elevated levels of nitrate (NO_3) and ammonia (NH_3) in agricultural base flows and urban storm flows.

Hoover does not track the contaminants in runoff that generate beach closures (fecal coliform, e. coli) or those that Hawaii is considering using as alternate mechanisms (described below). We use USGS data from 1988 to 1998 for Kaneohe to test for a relationship between fecal coliform (col/100 ml) and discharge (cfs). Observations were taken on 34 occasions between Dec. 1988 and Feb. 1998. Fecal coliform levels ranged from a low of 870 col/100 ml (Sept 1989) to a high of 20,000 col/100 ml reached in both Dec. 1991 and Feb. 1995. Discharges on these dates were 17 cfs, 19 cfs, and 25 cfs, respectively, while the low discharge was 8 cfs in Oct. 96 (fec. col. 5200 col/100 ml) and the high was 44 cfs in Apr. 89 (fec. col. 5400). The correlation between fecal coliform and discharge is 0.25.

Regression analysis using several different functional forms and attempting to account for seasonal or annual fluctuations failed to illuminate a highly significant relationship between fecal coliform levels and discharge. The clearest relationship between fecal coliform and discharge came from a log–log functional form, indicating that it is the rates not the levels of each that are most consistently related. Results using this functional form are shown in Table 6. Model I includes only discharge as an explanatory variable for fecal coliform levels. Model II adds time of day of the discharge. Model III includes discharge and the month of the year.

Table 6: Influence of discharge on fecal coliform levels, Kaneohe Bay. Dependent variable = ln(fecal coliform), std. errors in parentheses.

Variable	I	II	III	IV	V
ln(discharge)	0.48 (0.33)	0.51 (0.32)	0.49 (0.33)	0.50 (0.33)	0.52 (0.33)
Time of day		−0.001 (0.0009)			−0.001 (0.001)
Month			0.013 (0.04)		
Summer dummy				0.12 (0.32)	0.13 (0.32)
Constant	6.68* (0.88)	7.89* (1.32)	6.58* (0.94)	6.61* (0.91)	7.81* (1.35)
Nobs	34	34	34	34	34
Adj. R-sq.	0.03	0.05	0.01	0.01	

Model IV includes discharge and treats summer as distinct from other periods of the year. Model V includes time of day and the summer distinction.

Though coefficients are only significant for the constant, they are stable across regressions, with a 1% increase in discharge increasing fecal coliform by 0.5%. Thus, for expositional purposes, and until better scientific data is available, we will assume that if discharge increases 15% from forest quality deterioration that fecal coliform levels and related beach closures will increase 7.5%.

5.2 Beach-closure conditions

Beach closures for health standards are determined by the HI State Department of Health (DOH) based on Environmental Protection Agency (EPA) guidelines. For the state of Hawaii, waters within 1000 feet of shore must meet a standard where the geometric mean of enterococcus counts is less than or equal to 7 colonies per 100 ml. The DOH has added an additional test for *Clostridia perfringens*, which must have a median value less than or equal to 5, in an attempt to more closely monitor sewage concerns. Humid tropical soils are capable of supporting enterococcus, E. coli, and fecal coliform bacteria without sewage and the viruses these bacteria generally indicate. Under these conditions, a correlation between enterococcus levels and swimming-related illnesses has not been established for Hawaii, though the 7 colonies/100 ml standard was set based on temperate climate studies that suggested this figure corresponded to 10 mild illnesses for every 1000 bathers.

Hawaii's tropical climate fosters the production of other water-borne illnesses that are not tested for and therefore will not induce beach closures. These include approximately half of the 100–200 cases of leptospirosis in the United States each year.

Leptospirosis is particularly of concern in any discussion of tropical watershed because it is transmitted to humans through freshwater and mud containing urine of animals carrying the bacteria. *Staphylococcus aureus* is another concern for Hawaii, as it appears to survive in soils and salt water and is increased with the presence of swimmers.

Beach days lost to runoff are therefore an underestimate of the actual water quality problems in Hawaii stemming from watershed forest quality. Beach closures and warnings are relatively rare on Oahu. From 1994 to 2005, individual beaches have been posted as closed or with warnings a total of 414 days in (DOH [23]; DOH [24]). Of these, a total of 225 days of closures were attributed directly to heavy rains causing the deteriorating sewage system to overflow. Though the sewage and storm-drain systems are independent, with separate pipes and operations, Hawaii's storm waters are clearly infiltrating the sewer system. During rainy periods, over 200 million gallons per day of water goes through the largest sewage treatment plant (Sand Island), rather than the normal 70 million gallons per day, causing breaks in the pressurized lines and returning contaminated water to the untreated storm drains (Knutson [25]). Thus, 54% of beach days lost in the past decade or so have been attributed to heavy rains. Using daily average attendance at the beaches affected, there were approximately 266,000 lost beach days for individuals at these beaches over the period (HI Databook [26]; HI Databook [27]), from a total of approximately 205 million beach days (46,423 beach goers per day) taken by tourists averaging 110,845 per day and residents numbering approximately 1,437,000 total. If we consider 82% of beach days taken by tourists and 18% by residents, tourists account for 38,067 beach visits per day and residents for 8356 visits per day. Average visits per year per resident are then estimated to be 2.12 days and average visits per year per tourist are estimated to be 0.47. If 0.13% of these days are currently lost to runoff-related beach values, we generate an estimate of current damages in lost economic value (using the Bell and Leeworthy demand curves detailed in Kaiser *et al.* [19]) equal to per person losses of $0.02 per tourist and $0.07 per resident, for estimated total damages of $207,800.

The current health of the Koolaus, for the purpose of beach quality, appears healthy in as much as we can measure beach quality by beach closures. However, the period of 1994–2003 exhibited very low annual precipitation levels, with annual precipitation averaging 27% below longer run levels (Hawaii Data Book [28]). Heavy rains play an increasing and much more significant role in beach postings from 2004 forward, and have clearly demonstrated that the sewer capacity is currently inadequate to handle higher than average rainfall. Rainfall in 2004 was almost double the long-run average (as measured at Honolulu International Airport). There were 75 beach closure postings, 64 of which were attributed to heavy rains. In 2005, there were 143 beach postings, all of which were attributed to heavy rains (DOH [29]).

Between February and March of 2006, the island of Oahu experienced 44 consecutive days of heavy rain. State agencies blame this extraordinary amount of rainfall for causing the state's largest sewage spill on March 24, 2006. More than 48 million gallons of untreated sewage flowed into the ocean forcing Honolulu city officials to post signs warning tourists on Waikiki Beach to stay out of the water.

The spill was significant in that it was the first time in history Hawaii Department of Health closed the state's most valuable tourist destination due to contaminated water. DOH closed significant portions of Waikiki Beach for approximately 12 days (Several health-related damages can likely be attributed to the sewage spill, although we do not do so here. There were at least three severe incidences (one resulting in loss of life) that have been blamed on bacteria in the water. Signs were not posted at these beaches until 5 days after the spill.), for a loss of over 330,000 beach days, and warning signs were posted at other beaches around the state for over two months.

5.3 Lost value to beaches from change

If runoff were to increase 15%, we would expect to see an increase in beach closures as more sediments and associated pathogens are washed to the shore. As discussed above, we estimate that a 15% increase in runoff would generate an expected 7.5% increase in beach closures due to fecal coliform increases. The relatively low damages described in section 5.2 ($207,800 total) increase to $0.07 per tourist and $0.14 per resident from an immediate 7.5% decrease in beach days, for a total of $679,000 in damages though lost economic value expected per year ($472,000 tourist/$207,000 resident). These sums clearly demonstrate a direct connection between forest quality and water quality. The more likely case where degradation occurs more slowly over 20 years is discussed below, as is the effect of proposed mitigating activities.

5.4 Lost value to reefs from change

Using Hoover's estimates in conjunction with Cesar's presentation of the relationship between sedimentation and coral cover, we find that a 15% increase in runoff is likely to have only a negligible impact on coral cover regardless of the characteristics of the watershed. For each of the three watersheds in Hoover's (2002) study, we calculate that daily sedimentation rates are quite low and that a 15% increase in runoff will lead to increases in sedimentation but they will not impact coral cover. Table 7 summarizes.

If these three sites are a reasonable average of the variety of conditions on Oahu, we see that coral cover might only decline 0.007%. If this were the loss of coral

Table 7: Change in coral cover from change in runoff and sedimentation rate.

	Kahana	Heeia	Kaneohe
Sedimentation rate (mg/cm^2/day)	0.070	0.037	0.063
Sedimentation with 15% increase in runoff	0.111	0.056	0.073
Percent coral cover	23.786	23.796	23.788
Percent coral cover, 15% increase in runoff	23.773	23.790	23.785

cover for the entire island of Oahu, with 504 square kilometers of reef, the loss would be worth a minimum of $105,840 per year (at $0.03 per square meter), and a maximum of $331.3 million (at $93.91 per square meter). Since runoff from the Koolaus does not reach all shores, we calculate the percentage of reef that may be impacted by streams and runoff from the Koolaus.

We predict that 79% of Oahu's reefs, or 398.16 square kilometers, are positioned so that runoff from the Koolaus affects them. The change in consumer surplus generated by the damages to this reef from a one percent increase in runoff (in its current state) from Oahu to the state of Hawaii is calculated to be between $26,200 and $32,000 per year. An instantaneous fifteen percent increase in runoff would generate $393,000 dollars per year in damages. This figure will increase if reefs are deteriorating with time.

6 Likelihood of forest damages

6.1 Threats to watershed health

The forest canopy and the forest understory are believed to play different ecological roles in the tropical forested watershed. Loss of understory is expected to be particularly worrisome with respect to near-shore resources as the bare soil will lead to increased erosion and sedimentation of streams and near-shore resources. Loss of canopy is more likely to affect long-term groundwater recharge and also to reduce the efficiency of the forest's role in mitigating the speed with which runoff occurs. The two forest levels are treated separately in our discussion. Discussions with experts led us to ask specifically about the following threats to forest quality:

1. Feral pigs
2. Other feral animals
3. Invasive plants
4. Miconia calvescens, a specific invasive plant
5. Vehicular (ATV or 4 × 4) access
6. Fire.

These categories then became the focus of survey questions of a broader set of experts, discussed below.

Feral pigs are separated from other feral animals because there is a known and active feral-pig population in the district that creates benefits for hunters as well as costs to the ecosystem. Other feral animals have less expected direct benefit and policy implications may differ due to these differences.

Miconia calvescens is separated from other invasive plants because it is credited with the ability to change watershed characteristics due to its particular biological properties, whereas other invasive plants may simply change the forest from one shade of green to another.

Motorized vehicle recreation can be a significant source of damage in the watershed, causing erosion and facilitating the colonization of weeds by exposing bare soil. Trail use is another concern since humans can damage vegetation directly through trampling, and indirectly by providing the ignition source for fire or introducing weeds. Humans can also increase the likelihood of plant pest introductions, and have been implicated as a major culprit in the spread of such major pests as *Clidemia hirta*. In terms of volume, hiking is the highest impact human activity in the conservation district and has the potential to be detrimental. The close proximity of the Koolau Mountains to Waikiki allows many of the trails within the watershed area to be readily accessed, subjecting them to high visitor loads.

Fires result in the removal of vegetation cover, loss of the soil-anchor attribute of root masses, and exposure of bare mineral soil (KWPMP [30]). This combination subjects burned areas to high levels of erosion. A large fire may also reduce evapotranspiration and significantly increase runoff. Additionally, heat levels can reduce permeability of soils and reduce recharge levels.

6.2 Results of survey of watershed experts

In our survey of forested watershed experts, we requested the experts to evaluate the level of seriousness of the specific threats described above.

Tables 8–11 show the distribution of responses to the question of how serious the forested watershed experts believe each of seven threats, identified as possible issues for the Koolaus, are to the forest quality. Since threats are expected to be incremental and slow, the scale for choices is nonlinear.

From our results, it appears that most experts feel Miconia and invasive plants pose a most serious threat to the canopy forest cover (46% and 41% agreement,

Table 8: Feral pigs and other feral animals.

	Feral pigs		Other feral animals	
Response	Threat to canopy	Threat to understory	Threat to canopy	Threat to understory
Not a serious threat (0–5% chance of quality degradation)	0.00	0.00	5.41	5.41
Not serious (5–10%)	8.11	2.70	8.11	2.70
Mild threat (10–20%)	13.51	0.00	8.11	8.11
Mild-moderate threat (20–50%)	13.51	2.70	27.03	8.11
Moderate-serious threat (50–80%)	24.32	13.51	27.03	43.24
Serious threat (80–100%)	29.73	72.97	5.41	16.22
I don't know enough to answer	10.81	8.11	16.22	16.22
Left blank	0.00	0.00	2.70	0.00
Num. answering question	37	37	36	37

Table 9: Miconia and other invasive plants.

	Miconia		Invasive plants	
Response	Threat to canopy	Threat to understory	Threat to canopy	Threat to understory
Not a serious threat (0–5% chance of quality degradation)	0.00	0.00	0.00	0.00
Not serious (5–10%)	5.41	0.00	2.70	0.00
Mild threat (10–20%)	5.41	2.70	8.11	2.70
Mild-moderate threat (20–50%)	5.41	5.41	8.11	2.70
Moderate-serious threat (50–80%)	21.62	13.51	32.43	21.62
Serious threat (80–100%)	45.95	70.27	40.54	64.86
I don't know enough to answer	10.81	8.11	8.11	8.11
Left blank	5.41	0.00	0.00	0.00
Num. answering question	35	37	37	37

Table 10: Vehicular and trail access.

	Vehicular access		Trail access	
Response	Threat to canopy	Threat to understory	Threat to canopy	Threat to understory
Not a serious threat (0–5% chance of quality degradation)	5.41	5.41	10.81	10.81
Not serious (5–10%)	16.22	10.81	32.43	24.32
Mild threat (10–20%)	18.92	13.51	27.03	32.43
Mild-moderate threat (20–50%)	21.62	21.62	13.51	13.51
Moderate-serious threat (50–80%)	8.11	13.51	2.70	8.11
Serious threat (80–100%)	10.81	18.92	2.70	0.00
I don't know enough to answer	18.92	13.51	10.81	10.81
Left blank	0.00	2.70	0.00	0.00
Num. answering question	37	36	37	37

Table 11: Fire.

Response	Fire	
	Threat to canopy	Threat to understory
Not a serious threat (0–5% chance of quality degradation)	2.70	2.70
Not serious (5–10%)	16.22	18.92
Mild threat (10–20%)	21.62	10.81
Mild-moderate threat (20–50%)	18.92	18.92
Moderate-serious threat (50–80%)	13.51	16.22
Serious threat (80–100%)	16.22	21.62
I don't know enough to answer	10.81	10.81
Left blank	0.00	0.00
Num. answering question	37	37

respectively). Miconia is thought to be a slightly greater threat than invasive plants in general, indicating that while they do believe that Miconia changes watershed characteristics to promote runoff, they also believe other invasive plants in Hawaii share these qualities either because they are indicators of forest disruption or because features that make them successful invaders, including shallow root systems that expand quickly and aggressively and the reduction in biomass expected with monotypic stands.

Feral pigs are another concern in reducing canopy forest cover, as 24% of respondents said they possess a moderate to serious threat, and 30% said they pose a serious threat. The respondents also classified fire as an intermediate level threat to the forest canopy, with 16% calling it serious, 14% calling it moderate to serious, 19% labeling it as mild to moderate, and 22% calling it mild to moderate. Of lesser concern to the canopy is vehicular access, other feral animals, and trail access (in decreasing order).

With respect to the understory forest cover, experts indicated feral pigs and then Miconia have the highest threat levels and both were identified by a significant majority as serious threats (73% and 70% agreement, respectively). Invasive plants in general were another serious concern to the understory, with 65% of respondents classifying this threat as serious. Similar to the canopy question, fire was identified as an intermediate threat, with vehicular access, other feral animals, and trail access again of much smaller concern.

6.3 Status-quo conservation-level impacts

The survey results indicate that without mitigation there is an expected range of 38% to 71.5% deterioration in overall forest quality. We consider a 38% quality deterioration to be approximated by a land-use change from forested watershed to low-intensity development as this is roughly equivalent to the impermeable cover in such land use. Similarly, we consider a 71.5% deterioration to be approximated by a mix of 1/2 low-intensity development and 1/2 high-intensity development. Thus, the increase in runoff under the status quo over the next 20 years is estimated at between 15% and 18%, for an annualized rate of deterioration of between 0.852% and 1.04%.

6.4 Expected outcomes of increased conservation

The survey also asked about the impacts of mitigation under an evolving watershed management plan currently under the auspices of the Koolau Mountain Watershed Partnership and supervised by the Department of Land and Natural Resources. (The plan is available online at http://www.state.hi.us/dlnr/dofaw/wmp/koolau/KMWPMP.PDF.) We find that full implementation of the actions in the watershed management plan should decrease the likelihood of quality deterioration in the Koolaus by an expected 72%. We calculate this figure in a multi-step process delineated below:

1. Create weighted averages of expected damages to canopy and understory for each threat (including a minimum threat using the bottom range of damages and a maximum threat using the top range);
2. Create weighted average of expected damages to canopy and understory for each threat for each of 14 conservation actions that are outlined in the Koolau Watershed Partnership Management Plan;
3. Calculate differences between mitigated and unmitigated expected damages for canopy and understory, generating an average difference for the minimum and the maximum damages for each mitigation action and each threat;
4. Calculate cumulative reduction in threats from implemented actions for each threat;
5. Average cumulative reduction in threats across threats.

Details of the calculations may be found in Kaiser *et al.* [19]. The resulting reductions in expected damages are shown in Table 12.

Table 12: Cumulative reduction in threats by threat.

Threat	Remaining threat to canopy after mitigation	Remaining threat to understory after mitigation
Feral pigs	21%	6%
Other feral animals	34%	32%
Miconia calvescens	5%	3%
Invasive plants in general	3%	1%
Vehicular access	29%	20%
Trail access	77%	77%
Fire	45%	39%
Average	31%	25%

Thus, the expected deterioration in forest quality with conservation will only range from 11% to 20% on average. This is estimated to be similar to a shift to 1/4 to 1/2 low-intensity developed land, and thus an increase in runoff of 3.75% to 7.5% over twenty years, or an annualized deterioration of 0.201% to 0.41%.

7 The value of integrated resource management

Using runoff as the connecting ecological factor between beach values, reef values, and forest quality, we determine the expected benefits of forest conservation to near-shore resources under the status-quo level of conservation (expenditures of approximately $300,000 per year) as well as under a scenario in which the Koolau Mountain Watershed Partnership Plan is fully funded with approximately $3 million to spend each year for 5 years followed by $375,000 per year in maintenance expenditures. Under the current conditions, we predict an annualized deterioration in forest quality that leads to an increase of runoff of 0.852% and 1.04%, while the increased spending is expected to reduce this deterioration to an annualized rate of between 0.201% and 0.41%.

Demand for beach days is estimated separately for tourists and residents, with status-quo levels of demand at approximately 2.12 resident days per year and 0.47 tourist days per year. We estimate that a one percent increase in runoff increases fecal coliform bacteria counts by 0.5% and will increase beach closure days similarly, given the extremely low threshold for acceptable levels of bacteria in Hawaii's waters and the relatively low number of beach closures that occur. Without conservation, the expected 15% increase in runoff would generate an expected 7.5% increase in beach closures due to fecal coliform increases.

Measuring the loss in economic value from closed beaches may be considered akin to measuring the burden of a tax where there are no revenues to the tax-assessing agency. We calculate that damages to lost beach days will have a present value of between $12.5 million and $15 million with no mitigation, and between $3.1 million and $6.3 million with conservation activities. The expected net benefit of mitigation is between $6.2 million and $11.9 million using a 1% discount rate over 20 years with mitigating activities.

Savings from mitigating activities under this assumption are between $9.1 million and $17.9 million, about 33% higher than using the marginal user's losses. This higher loss may more accurately reflect the opportunity cost of beach closures that affect the average beach user. Additionally, since 2004 and a return to average rainfall patterns after a decade of drought, beach closures have increased dramatically after heavy rains, and further increases in damages appear imminent. Thus, we expect damages to be increasing.

Demand for healthy reef is estimated per square meter and is based on amenity values, recreational values, and fisheries values. We estimate that a 1% increase in runoff will decrease coral cover by only 0.00047%. We estimate that approximately 79% of the 504 square kilometers of reef surrounding Oahu will be impacted by the increase in runoff from the Koolaus, and that 8% of reefs lie below predominantly upslope conservation land. Minimum damages from decreased coral cover associated with a 15% increase in runoff range from $393,000 to $1.156 million per year.

If the degradation takes place over 20 years, using 1% discount rates, the minimum damages without mitigation are estimated to have a present value between $478,000 and $583,000 using the values generated by the estimation of coral cover loss due to sedimentation increases. Mitigation decreases these damages to between $113,000 and $230,000 in present value. The savings from mitigation are thus between $240,000 and $470,000 in present value.

Using the values generated by the estimate of coral cover loss due to a change in the upland conservation district, we assume that 225,800 square meters of reef are lost over 20 years at an even rate of degradation of 11,290 square meters per year. The damages without mitigation are estimated to have a present value over 20 years (1% discount rate) of $3,384,000, while successful mitigation will leave the district intact and damages are estimated at zero. This more blunt estimation provides an upper bound for the expected damages as well as savings from mitigation efforts.

The present value of mitigation costs is estimated at $14.9 million above current expenditures over 20 years at a 1% discount rate. The present value of damages with no mitigation is estimated at between $13.0 million and $25.4 million, while the present value of damages with mitigation is estimated at between $3.4 million and $8.8 million. Thus, mitigation expenditures should reduce the present value of damages by between $4.2 million and $22.0 million.

Though it is therefore questionable as to whether the benefit:cost ratio would be greater than one for conservation costs versus near-shore benefits, two important points should be apparent. First, there are near-shore damages occurring due to upslope changes, and these damages are expected by forest and water-quality experts to increase over the next 20 years. Second, upslope conservation activities will decrease the probability of damages significantly over this time period, in addition to other benefits they convey directly to forest quality. These benefits should be included in any decision making for forest-conservation expenditures.

8 The value of improved pricing policy

8.1 The value of price reform

Watershed degradation can lead to reduced recharge of groundwater aquifers (Doolette and Magrath [31]; Mackay and Band [32]). In Section 6.3, a rate of runoff increase (recharge loss) from watershed deterioration is estimated at about 1% annually, and over the next 20 years about 15–18%. Conservation of watersheds can help to preserve the groundwater supplies by avoiding this loss of recharge (Kaiser and Roumasset [33]). But if groundwater is being misallocated due to underpricing, the benefits of watershed conservation will be diminished. In effect, the water saved by watershed conservation would be increasingly wasted by underpricing. One example of groundwater waste is the current policy on Oahu of pricing water at average extraction and distribution costs, resulting in under-pricing and excessive consumption. In contrast, efficient water allocation requires full marginal cost pricing, i.e charging marginal instead of average extraction costs, charging the actual distribution costs, and charging the opportunity costs of resource drawdown, including the higher extraction costs that drawdown implies (see e.g. Koundouri [34], for a recent survey; also see Nieswiadomy [35]; Agthe *et al.* [36]; Howe [37]).

Correcting overuse through pricing reform can provide substantial welfare gains (see e.g. Noel *et al.* [38]; Feinerman and Knapp [39]), and make watershed conservation more beneficial. Without watershed conservation, changing from the status-quo pricing mechanism to the efficient mechanism should generate up to $877.8 million in present-value savings. However, proposals to induce efficient use through pricing reforms often are found to be politically infeasible (Dinar and Wolf [40]; Postel [41]; Johansson [42]). This may lead policy makers to consider watershed conservation as an alternative to price reform.

To illustrate the futility of watershed conservation without pricing reform and also the limited gains of pricing reform without watershed conservation, we compare the gains of just one of the two reforms with both together. For transparency, we concentrate only on quantity issues and abstract from sedimentation and other water-quality issues. The decomposition exercise can be thought of in terms of sequencing reforms. What are the separate welfare gains of doing pricing reforms and then watershed conservation, and what are the decomposed gains if the order of the reforms is reversed?

8.2 Combining pricing reform and watershed conservation

In order to evaluate the available policy choices, Pitafi and Roumasset [13] compare watershed conservation versus pricing reform and compare doing either of the options alone versus both reforms together. They also analyze the case of immediate watershed conservation combined with delayed pricing reform versus the case of taking both measures immediately. The analysis is conducted using the Pearl Harbor aquifer on the island of Oahu in Hawaii. They model two pricing scenarios: efficiency and status-quo pricing, and three watershed conservation policies: no conservation, conservation that prevents a 1% loss of recharge, and conservation that prevents a 10% loss of recharge.

Figure 4 summarizes their welfare estimates under four different scenarios. In policy A, status-quo pricing is continued and lack of watershed conservation causes a 1% recharge loss. (In reality, the loss may be greater or smaller, may occur in the future rather than immediately, and/or may happen once or multiple times. Here, the assumption is that the net effect of all the losses from lack of watershed conservation is equal to that of one percent immediate loss of recharge. Analysis with 10% loss is also reported later in this section.) In policy B, efficient pricing is undertaken but again lack of watershed conservation causes a 1% recharge loss. In policy C, status-quo pricing is continued but watershed conservation prevents recharge loss. In policy D, efficient pricing is undertaken and again watershed conservation prevents recharge loss. Starting from policy A, the gains from pricing reform (moving to policy B) are

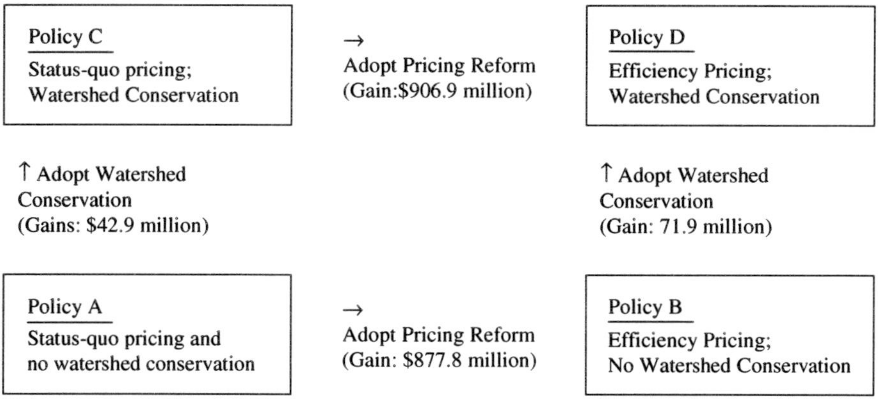

Figure 4: Present value of welfare gain from pricing reform and watershed conservation (preventing loss of 1% recharge; normal recharge is 281 mgd). Starting from policy A, the gains from pricing reform (moving to policy B) are about $878 million. In comparison, the gains from watershed conservation (moving to policy C) are about $43 million. Adopting watershed conservation after pricing reform (moving from policy B to D) provides much larger gains (about $72 million).

about $878 million. In comparison, the gains from watershed conservation (moving from policy A to policy C) are about $43 million.

In addition, since status-quo pricing involves overuse and wastage, a unit of recharge is more valuable at efficient prices than at status-quo prices. To see this, note that if we are in policy A, and move to policy C (adopt watershed conservation that prevents the loss of recharge), the welfare gain is about $43 million. Instead, if we are in policy B, and move to scenario D (adopt watershed conservation), the welfare gain is about $72 million. Watershed conservation is, therefore, more valuable under efficient pricing than under the status-quo pricing.

If the present value of conservation costs is greater than $43 million, but less than $72 million, then conservation would be warranted if and only if pricing reform is done first. This may indeed be the case. The present value of the cost of watershed conservation through the Koolau Mountain Watershed Partnership Plan described in Section 7 is about $50 million, at 1% discount rate.

The difference between the benefits of pricing reform and watershed conservation depends on the amount of recharge loss that is being prevented by watershed conservation. Figure 5 examines the welfare effects if lack of watershed conservation would cause a 10% loss of recharge. Once again, watershed conservation undertaken after pricing reform is more valuable than before the reform.

However, this time, the gain from pricing reform alone (B – A) is almost the same as the gain from watershed conservation (C – A). This is because watershed conservation is now providing a bigger service (preventing a 10% recharge loss). For even larger recharge losses prevented, gains from watershed conservation will be higher than the gains from pricing reform.

Policy C Status-quo pricing; Watershed Conservation	→ Adopt Pricing Reform (Gain: $906.9 million)	Policy D Efficiency Pricing; Watershed Conservation
↑ Adopt Watershed Conservation (Gains: $546.2 million)		↑ Adopt Watershed Conservation (Gain: $906.7million)
Policy A Status-quo pricing; No Watershed Conservation	→ Adopt Pricing Reform (Gain: $546.4 million)	Policy B Efficiency Pricing; No Watershed Conservation

Figure 5: Present value of welfare gain from pricing reform and watershed conservation (preventing loss of 10% recharge; normal recharge is 281 mgd). Starting from policy A, the gains from pricing reform (moving to policy B) are about $546 million. The gains from watershed conservation (moving to policy C) are also about the same. Adopting watershed conservation after pricing reform (moving from policy B to D) provides larger gains (about $907 million).

Gain $906.9 million	Gain $677.7 million	Gain $493.7 million
↑ Immediate pricing reform	↑ Pricing reform after 10 years	↑ Pricing reform after 20 years
Gain $42.9 million		
↑ Adopt watershed conservation		
Status-quo pricing and no watershed conservation*		

Figure 6: Welfare gains from water-management reforms (*1% loss of recharge scenario). Starting from status-quo pricing and no watershed conservation, adoption of watershed conservation provides modest gains. Immediate pricing reform after that provides the biggest gains. Delaying the pricing reform reduces its gains substantially.

Finally, delay in adopting pricing reform can substantially affect the resulting gains as shown in Fig. 6.

Thus, welfare gains from watershed conservation are small compared with those from pricing reform unless the recharge benefits of watershed conservation are particularly large. The benefits of conservation increase substantially in the presence of price reform. A conservation project that has positive net benefits at corrected prices may have benefits less than costs if price reform is not concurrently implemented. Relatedly, if watershed conservation is adopted and leads to delays in adopting pricing reform, substantial potential gains are lost.

9 Concluding remarks

The economic value derived from watershed conservation is a highly variable composite of economic, political, geophysical, and biological or ecological factors. We examine the interlinked ecological services provided by the forested watershed encompassing the Koolau mountains on the island of Oahu, Hawaii, to provide a framework for estimating this composite value. Efficient pricing acts to prolong the life of the resource by incorporating the dynamic scarcity of the asset and extends any benefits accrued from conservation activities.

Beyond improved pricing, watershed conservation involving mitigation of long-term threats has direct and indirect economic benefits. In particular, expenditures on watershed conservation that increase groundwater recharge will also protect downstream resources from damages due to sedimentation and runoff. Restricting activities, e.g. through beach closures, may mitigate damages but will impose unnecessary costs rather than provide long-term improvements in resource quality.

Forested watersheds like those of the Koolau mountains produce a host of valuable ecological services that range from groundwater quantity and surface-water quality to maintenance of biodiversity and recreational beach use. The more intensely the resources are used, the more valuable conservation becomes. This is due both to the direct demand for the ecological services and to the increased risk of losing the asset's productivity through inefficient, myopic, use.

The anticipated cost of watershed conservation in the Koolaus into the indefinite future has a present value of \$43.2 million using a discount rate of 1%. The benefits of watershed conservation stemming from groundwater recharge alone vary widely depending on the assessment of increased recharge but may be more than \$900 million provided that conservation is accompanied by pricing reform. Benefits to near-shore resources (beach closures and reef sedimentation) range from \$4.2 million to \$22.0 million, even before accounting for disasters such as the recent sewage spill into the Ala Wai Canal. Though extremely difficult to measure as a function of conservation plans targeted toward watershed conservation rather than species and biodiversity preservation, benefits will accrue in these categories as well and may be worth billions of dollars. This is particularly likely in the Koolaus because two of the largest watershed threats, feral ungulates and invasive plants, are also the greatest threats to biodiversity, and because Hawaii is home to so many endemic species.

To justify watershed protection in this case, we need not quantify their benefits thoroughly, because the direct economic benefits delineated above are well above the costs. Such win-win scenarios are expected to be more common in natural resource conservation than in other economic cases for two important reasons. First, since much natural resource use has evolved from open-access exploitation, improved policy mechanisms for use, particularly using pricing that accurately reflects the costs borne by both present and future users, can greatly extend the benefits through time of the resources. Secondly, since ecological services provided by natural resource assets are often integrated across end uses, activities that enhance the capital stock for the purpose of improving one set of ecological services may jointly provide enhancement of additional sets of ecological services valued by other users.

Acknowledgement

Thanks to Hawaii SeaGrant Agreement No. NA16RG2254 and to the University of Hawaii Water Resources Research Center for financial assistance.

References

[1] Kaiser, B., Krause, N. & Roumasset, J., *Environmental Valuation and the Hawaiian Economy.* Online publication at UHERO http://www.uhero.hawaii.edu/workingpaper/HawaiiEnviroEvaluation.pdf, 1999.

[2] Mackenzie, F., Nonpoint Source Pollutant Nutrient Fluxes, Cycling, and Management in Southern Kane'ohe Bay UNIHI-SEAGRANT-R-EL-26, 2001.

[3] Dollar, S.J. & Tribble, G.W., Recurrent Storm Disturbance and Recovery: A Long Term Study of Coral Communities in Hawaii. *Coral Reefs,* **12**, pp. 223–233, 1993.

[4] Jokiel, P.L., Hunter, C.L., Taguchi, S. & Waterai, L., Ecological Impact of a freshwater "reef-kill" in Kaneohe Bay, Oahu, Hawaii. *Coral Reefs,* **12**, pp. 177–185, 1993.

[5] Giambelluca, T. & Ridgley, M., Water Balance, Climate Change and Land-use Planning in the Pearl Harbor Basin, Hawai'i, *Water Resources Development,* **12(4)**, pp. 515–530, 1996.

[6] Meyer, J.-Y., Epidemiology of the invasion by *Miconia calvescens* and reasons for a spectacular success. Proceedings of the First Regional Conference on Miconia Control, Aug 26–29, 1997, Papeete, Tahiti, Gouvernement de Polynésie Française/ University of Hawaii at Manoa/Centre ORSTROM de Tahiti, 1998.

[7] Mink, J.F., State of the Groundwater Resources of Southern Oahu. Honolulu, HI: Board of Water Supply, City and County of Honolulu, 1980.

[8] Shade, P.J. & Nichols, W.D., Water budget and the effects of land-use changes on ground-water recharge, Oahu, Hawaii, US Geological Survey Professional Paper, 1412–C, 1996.

[9] Oki, D.S., Geohydrology of the central Oahu, Hawaii, ground-water flow system and numerical simulation of additional pumping. U.S. Geological Survey Water-Resources Investigations Report, 97–4276, 1998.

[10] Oki, D.S., Souza, W.R., Bolke, E.L. & Bauer, G.R., Numerical analysis of the hydrogeologic controls in a layered coastal aquifer system, Oahu, Hawaii, USA. Hydrogeology Journal, **6(2)**, pp. 243–263, 1998.

[11] Honolulu Board of Water Supply. *Water for Life*. (2006) Online publication: http://www.boardofwatersupply.com/files/Wfl_Website.pdf

[12] Malla, P.B., *The economics of urban water in Hawaii: Empirical analysis and case studies.* Ph.D. Dissertation, University of Hawaii, 1996.

[13] Pitafi, B. & Roumasset. J., Evaluating interdependent watershed conservation and groundwater-management reforms. *Journal of the American Water Resources Association*, **42(6)**, pp. 1441–1450, 2006.

[14] Krulce, D.L., Roumasset, J.A. & Wilson, T., Optimal management of a renewable and replaceable resource: The case of coastal groundwater. *American Journal of Agricultural Economics*, **79(4)**, pp. 1218–1228, 1997.

[15] Brown, G. & Deacon, R., Economic Optimization of a Single-Cell Aquifer. *Water Resources Research*, **8(3)**, pp. 557–564, 1972.

[16] Cummings, R.G. & Burt, O.R., The economics of production from natural resources: note. *American Economic Review*, **59(5)**, pp. 985–990, 1969.

[17] Moncur, J.E.T. & Pollock, R.L., Scarcity rents for water: a valuation and pricing model, *Land Economics,* **64(1)**, pp. 62–72, 1988.

[18] Cesar, H. & Beurkering, P., Economic Valuation of Hawaii's Coral Reefs, *NOAA Final Report (FY 2001–2002)* Online at http://www.hawaii.edu/ssri/hcri/rp/cesar/noaa_final_report_01–02/cesar_final_report-01–02.htm, 2002.

[19] Kaiser, B., Pitafi, B. & Roumasset, J., Mitigating Rrunoff as part of an integrated strategy for nearshore resource conservation. Final Report. Sea Grant (NA16RG2254) Project R-TR/5. http://homepage.mac.com/ondinebak, 2005.

[20] Moore, J.G., Ingram, B.L., Ludwig, K.R. & Clague, D.A., Coral ages and island subsidence, Hilo drill hole *Journal of Geophysical Research,* **101(B5):** 11,599–11,605, 1996.

[21] El-Kadi, A., Recharge Data, personal communication, 2003.

[22] Hoover, D., *Fluvial Nitrogen and Phosphorus in Hawaii: Storm Runoff, Land Use, and Impacts on Coastal Waters.* University of Hawaii Dissertation (August 2002). UMI # 3057363.

[23] Hawaii Department of Health. Beach Closure Data, 1994–2004. Personal Communication, Feb 24, 2004.

[24] Hawaii Department of Health. Beach Closure Data, 2004–2005. Personal Communication, May 22,2006.

[25] Knutson, C., Toxic Neighbors. Honolulu Magazine, September 2005.

[26] Hawaii State Department of Business, Economic Development and Tourism. 1999 *State of Hawaii Data Book.* Online at http://www3.hawaii.gov/DBEDT/index.cfm?section=statistics_and_economic_information516&siteID=1, 1999.

[27] Hawaii State Department of Business, Economic Development and Tourism. 2004 *State of Hawaii Data Book.* Online at http://www3.hawaii.gov/DBEDT/index.cfm?section=statistics_and_economic_information516&siteID=1, 2004.

[28] Hawaii State Department of Business, Economic Development and Tourism. 2003 *State of Hawaii Data Book.* Online at http://www3.hawaii.gov/DBEDT/index.cfm?section=statistics_and_economic_information516&siteID=1, 2003.

[29] Hawaii Department of Health. Beach Closure Data, 2004–2005. Personal Communication, April 12, 2005.

[30] Koolau Mountains Watershed Partnership. Koolau Mountains Watershed Partnership management plan. Available at: http://www.state.hi.us/dlnr/dofaw/wmp/koolau/KMWPMP.PDF, 2002.

[31] Doolette, J.B. & Magrath, W.B., Watershed Development in Asia; Strategies and Technologies. Paper 127, World Bank – Technical Papers, 1990.

[32] Mackay, D.S. & Band, L.E., Forest ecosystem processes at the watershed scale: dynamic coupling of distributed hydrology and canopy growth. *Hydrological Processes*, **11(9)**, pp. 1197–1217, 1997.

[33] Kaiser, B. & Roumasset, J., Valuing indirect ecosystem services: the case of tropical watersheds. *Environment and Development Economics*, **7(4)**, pp. 701–714, 2002.

[34] Koundouri, P., Current Issues in the Economics of Groundwater Resource Management. *Journal of Economic Surveys*, **18**, pp. 703–38, 2004.

[35] Nieswiadomy, M. The demand for irrigation water in the High Plains of Texas, 1957–1980. *American Journal of Agricultural Economics,* **67**, pp. 619– 626,1985.

[36] Agthe, D.E., Billings, R.B. & Buras, N., *Managing Urban Water Supply.* Kluwer Academic Publishers: The Netherlands, 2003.

[37] Howe, C., The functions, impacts and effectiveness of water pricing: Evidence from the United States and Canada. *International Journal of Water Resources Development,* **21(1)**, pp. 43–53, 2005.

[38] Noel, J.E., Gardner, B.D. & Moore, C.V., Optimal regional conjunctive water management. *American Journal of Agricultural Economics,* **62**, pp. 489– 498, 1980.

[39] Feinerman, E. & Knapp, K.C., Benefits from groundwater management: magnitude, sensitivity, and distribution. *American Journal of Agricultural Economics,* **65**, pp. 703–710, 1983.

[40] Dinar, A. & Wolf, A., Economic and political considerations in regional cooperation models. *Agricultural and Resource Economics Review*, **26**, pp. 7–22, 1997.

[41] Postel, S. *Pillar of Sand: Can Irrigation Miracle Last*? pp. 235–236. W.W. Norton & Company: New York, NY, 1999.

[42] Johansson, R.C., Pricing Irrigation Water: A Literature Survey. Policy Research Working Paper # WPS2449. Washington, DC.: The World Bank, 2000.

CHAPTER 12

Impact of best management practices in a coastal watershed

K.T. Morgan
Southwest Florida Research and Education Center,
Soil and Water Science Department,
University of Florida, USA.

Abstract

Draining of wetlands to adapt the coastal plain for agricultural and urban use has occurred in many locations throughout the United States. The Kissimmee River, Lake Okeechobee, and Everglades are part of a vast wetland system that historically extended over 200 miles from the Kissimmee chain of lakes, near Orlando, ending in the mangrove estuaries of Florida Bay, south of Miami. This nutrient-poor wetland system supported a diverse and large community of species across huge seasonal and interannual variation in rainfall. The combination of a subtropical climate and supply of potentially arable land has proven to make South Florida a desirable place to farm and live. With agricultural and urban development of the landscape, several areas of Lake Okeechobee and the Everglades have experienced increased nutrient loading, particularly phosphorus, resulting in shifts in the algae and plant communities found within lakes, marshes, and near-shore marine environments. Reducing this "phosphorus enrichment" is the primary goal of Lake Okeechobee and Everglades restoration efforts brought about by the Everglades Forever Act (EFA). Site-specific best management practices (BMPs) to reduce the quantity and improve the quality of runoff leaving agricultural lands have improved the water quality of associated wetlands. Another method used to reduce phosphorus for complying with the EFA includes the development of man-made wetlands, called stormwater-treatment areas (STAs) where phosphorus removal is achieved through the accumulation and burial of peat sediments. These restoration actions should reverse environmental impacts of increased P loading while maintaining the original goals of supporting agricultural production and urban development.

1 Introduction

Degradation of water quality in two watersheds in Central and South Florida began with the drainage of these watersheds starting in the 1880s and continuing through the 1960s. The drainage of wetlands and watersheds along the Kissimmee River, Lake Okeechobee, and the Everglades was done to make arable land available for agricultural and urban development. While these actions were taken with little or no regard to their environmental impact, currently some of these actions are now being reversed. However, much environmental damage has occurred and reclamation will be costly, both temporally and monetarily. But these reclamation efforts are essential for the sustainability of agricultural production and increased urban development in the region.

Past actions associated with draining wetlands to adapt the costal plain of south Florida for agricultural and urban use have been repeated in many locations throughout the United States and the world. The current reclamation efforts in the Okeechobee and Everglades watersheds can be used as a model for other impacted costal plains. The reader will see that agricultural best management practices have and will continue to improve water quality in the sensitive ecosystems of south Florida. This chapter will review the hydrology of the Kissimmee River and Everglades ecosystems, changes in land use and agricultural development in South Florida, and resulting water-quality and ecosystem changes. The chapter will conclude with legislative mandates to reclaim these watersheds and current improvements made in water quality in South Florida.

2 Hydrology of the Kissimmee River and Everglades ecosystems

The Kissimmee River and Everglades are part of a vast wetland system that historically extended over 200 miles from the Kissimmee chain of lakes, near Orlando, south through Lake Okeechobee into the freshwater marshes of the Everglades and ending in the mangrove estuaries of Florida Bay, south of Miami (Fig. 1). Water flow to the Everglades started with rainfall accumulating in central Florida and slowly flowing south through the meandering Kissimmee River to Lake Okeechobee. From Lake Okeechobee, the water continued moving slowly south, through the Everglades toward Florida Bay, between the Florida mainland and keys. During this slow progression through the Everglades, the water recharged the Biscayne Aquifer, supplied nutrients for vegetative growth, provided freshwater for fish, maintained the food chain for migratory birds, and reduced saltwater intrusion from both the Atlantic Ocean and Gulf of Mexico [2, 3].

The upland ecosystems along the Kissimmee River include pine (*Pinus* sp. L.) and hardwood forests containing a mixture of temperate and tropical species [34, 44]. Natural land cover includes saw palmetto (*Serenoa repens* Batr.) and pine flatwoods, hardwood forests, cypress sloughs, and small isolated wetlands with herbaceous vegetation. In contrast, The Everglades was an expanse of sawgrass

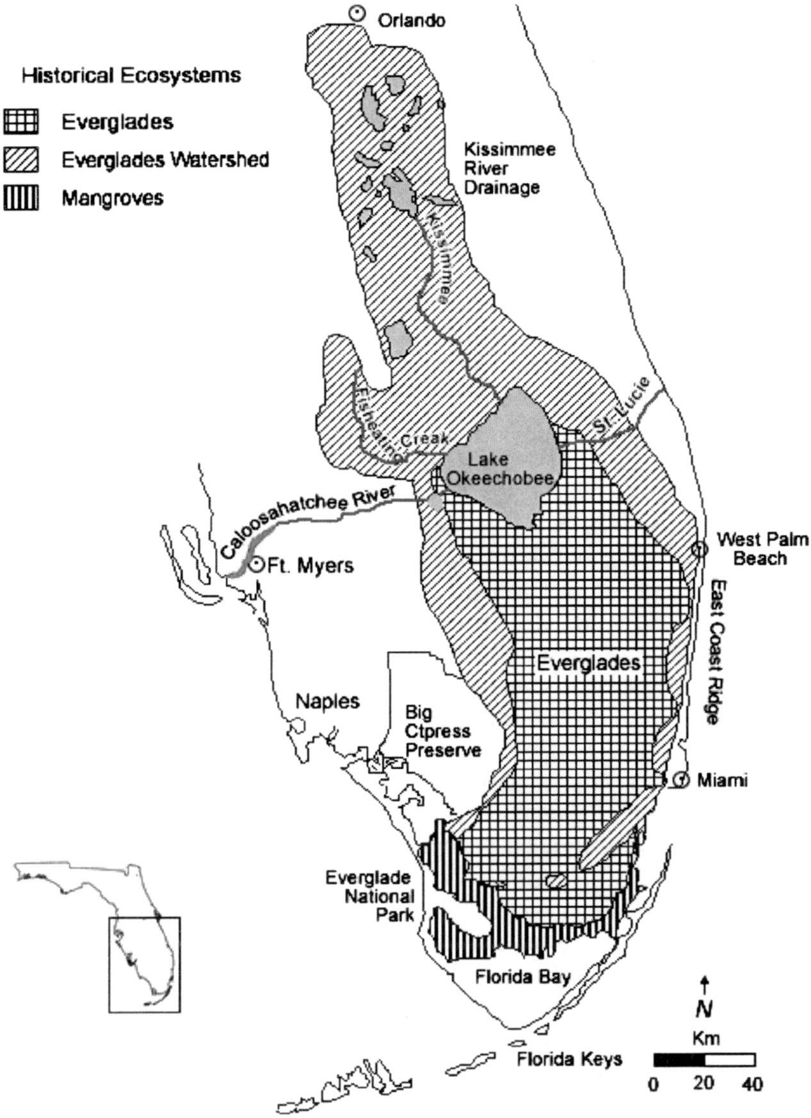

Figure 1: The historical watershed of the Everglades ecosystem.

(*Cladium jamaicense* Crantz) wetlands and was called the "River of Grass". The original Everglades extended south from Lake Okeechobee to Florida Bay and was bounded by the swamps of the Big Cypress on the west and Atlantic Ocean on the east [33].

Soils along the Kissimmee River developed over geological time primarily into Spodosols. Spodosols are characterized as having undergone extensive leaching of organic matter and aluminum from the surface horizons [15]. These soils develop in poorly drained areas with sandy, acidic parent material. The leached horizon is uncoated sands with low water- and nutrient-holding capacities. The most striking property of Spodosols is a black to reddish-brown horizon of amorphous accumulations of organic matter and aluminum at a depth of less than 1 m with low hydraulic conductivity, and high cation-exchange capacity [15]. Water falling on the Spodosols north of Lake Okeechobee readily penetrate the relatively short distance between the soil surface and accumulates above the spodic horizon, saturating the soil. Water flow is then predominately horizontal along the spodic horizon to any point of drainage carrying nutrients it contains. Any water percolating through the spodic horizon would leach nutrient to the nutrient-poor parent material below. At a depth of a few to several meters below the surface, this water enters the limestone aquifer system underlying the state of Florida.

Soils south of Lake Okeechobee developed into Histosols. In general, Histisols contain more than 50% organic matter in the upper 80 cm of soil depth [25]. The soils are composed of partially decomposed plant material that accumulates over geological time scales. Morris and Gilbert [24] stated that these soils formed in South Florida from sawgrass since the Holocene period (<10,000 years ago). As with Spodosols, vertical water movement in Histosols of south Florida is limited to a few meters by limestone deposits.

This nutrient-poor wetland system supported a diverse and large community of species across huge seasonal and interannual variation in rainfall [16]. The water level of Kissimmee River and Lake Okeechobee would fluctuate dramatically with rainfall patterns within their drainage basins. Total annual rainfall varies from less than 1000 mm to almost 2000 mm over a seven- to ten-year cycle [10]. Approximately three-quarters of the annual precipitation occurs during the four-month period from mid-June through September [10, 21, 43]. Water levels within the marshes and lakes of the system rose during the wet seasons and water slowly flowed south [11]. During wet years, flow rates of the Kissimmee River would increase dramatically contributing to substantial increases in lake levels of Lake Okeechobee. Lake Okeechobee would overflow its southern rim and contribute a broad sheet of water to the Everglades, increasing water depths and lengthening hydroperiods. Due to the nearly level topography of the Everglades, water within the marshes moved slowly and was maintained for longer periods of time, typically well into the following dry season. During the dry season, evaporation exceeded rainfall and water levels would recede [11]. The annual pulse of fresh water into the estuary of Florida Bay created a highly productive interface between the freshwater system and the Gulf of Mexico [23]. The ecosystem supported a huge array of animal species. Large flocks of wading birds nested along the interface between the freshwater Everglades and the estuaries [1, 13, 29, 30].

3 Changing land uses of South Florida

The combination of a subtropical climate and supply of potentially arable land has proven to make South Florida a desirable place for humans to live. The population of this region has increased from only a few tens of thousands in 1900 to over six million people in 2000 [47]. An increase of almost two million new residents is projected over the next 20 years. This increase in human populations resulted in major changes to the ecosystems of South Florida. Extensive drainage of wetlands and clearing of uplands began in the 1880s to create agricultural lands [19, 20].

Wetlands along the Kissimmee River and around Lake Okeechobee were drained and converted to agricultural lands for the raising of cattle and growing winter vegetables, sugar cane, rice, and sod [20]. The hydrologic connection between Lake Okeechobee and the Everglades was severed to develop agricultural lands and control flooding south of the lake. Water that had flowed from Lake Okeechobee to the Everglades was shunted east to the Atlantic Ocean through the St. Lucie Canal, and west to the Gulf of Mexico through the Caloosahatchee River. As a result, the volume of water flowing through The Everglades was decreased and the pattern of the water through the ecosystem was significantly altered [7].

More intensive land use is associated with the transition from open space to agriculture and residential uses, necessitating increased flood protection. These changes reduce the water available to natural areas. In order to provide increased flood protection in developed areas, water tables were lowered. This action reduced both the local storage of groundwater and the probability that local rainfall would recharge well fields, which places greater demand on the regional system for water.

4 Agricultural development in South Florida

A large portion (more than 380,000 ha) of the original Everglades that lay immediately south of Lake Okeechobee was drained and developed into the Everglades Agricultural Area (EAA). Along the Eastern edge of the Everglades about 1.2 million ha has been developed for urban use. To facilitate this development, construction of an extensive system of canals and levees was built to meet flood-control and water-storage needs of growing agricultural and urban development. Called the Central and Southern Florida Control Project, these structures were completed between 1920 and 1960 (Fig. 2). These activities resulted in alterations in the hydrology and water quality of both Lake Okeechobee and the Everglades. Alteration in the hydrology of South Florida included the depth, duration, and timing of annual flooding (hydropattern) and the amount of time each year that the ground is covered with water (hyproperiod) [2, 16, 19, 22]. Changes in the timing and flow of the water impact many aspects of the ecosystem, causing changes in water quality.

In a recent survey, the area of Histosols in south Florida is estimated to be 0.8 million ha [24]. Approximately 15% of the Histosols in the Everglades were found to be in agricultural production; the major crops in this area are sugarcane and winter vegetables. Decomposition of the organic matter comprising Histosols results from oxidation by aerobic soil micro-organisms when these soils are drained. During decomposition, these soils release nitrogen and phosphorus at rates approaching 1200 and 73 kg ha^{-1} yr^{-1}, respectively [9, 42]. As the organic matter of Histosols decompose, the soil depth decreases or subsides. The rate of subsidence in south Florida have been estimated to be approximately 2.5 cm yr^{-1} for the period 1924 to 1978 [37], resulting in as much as 175 cm reduction in soil

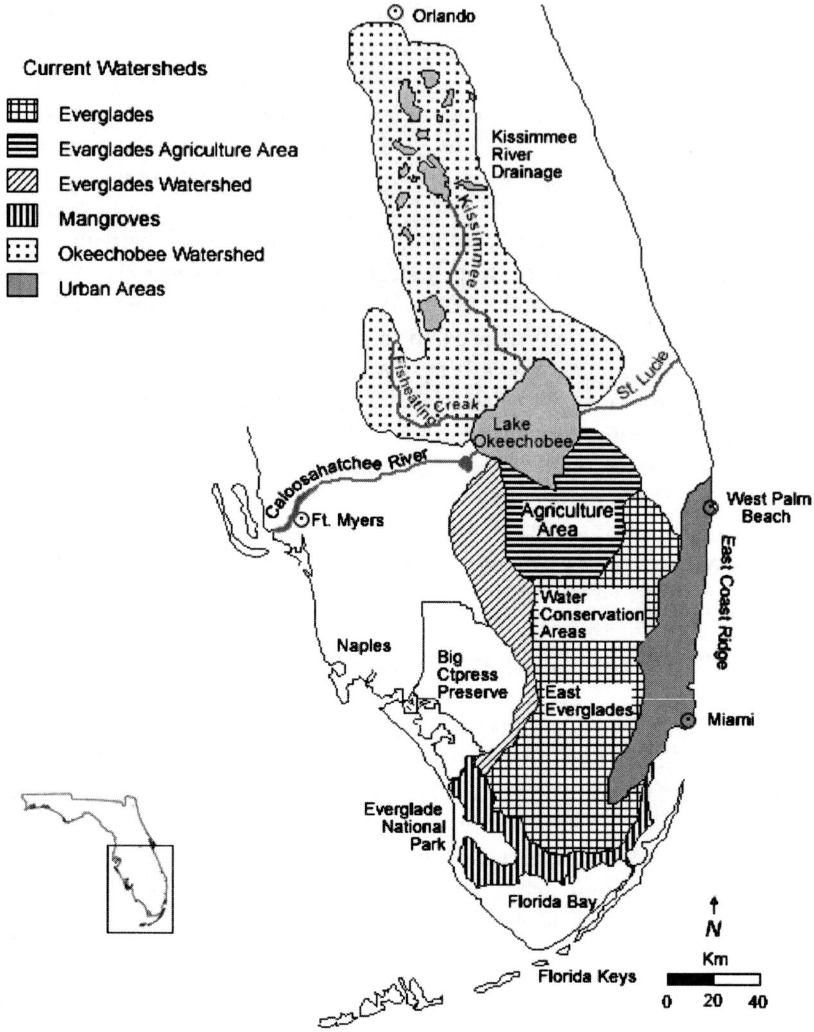

Figure 2: Alterations in the watershed of the Everglades ecosystem.

depth of some heavily cropped areas over the past 80 years. Improved agricultural practices, such as maintenance of higher water tables and flooding of fallow land, implemented in the late 1970s have reduced this rate to approximately 1.5 cm yr^{-1}. As a result, the Histosols in the EAA are becoming increasingly shallow, in some cases less than 20 cm thick. These soils overlie hard limestone rock and may not be able to sustain agricultural production in the near future.

5 Water-quality and ecosystem changes

Historically, the nutrient supply to the Kissimmee River and Everglades watersheds was provided primarily through rainfall, thus, the vegetative communities selected for species with low nutrient requirements [16, 32]. However, with agricultural and urban development of the landscape, several areas of Lake Okeechobee and the Everglades have experienced increased nutrient loading [7, 8, 12, 27]. Nutrient inflows into the Kissimmee River and the Everglades, particularly phosphorus, have been responsible for many of the changes to the ecology of the Lake Okeechobee and the Everglades.

The increased levels of both phosphorus and nitrogen in surface waters of south Florida have resulted in shifts in the algae and plant communities found within lakes, marshes, and near-shore marine environments. In the past, the marshes of the Everglades were characterized by a mixture of sawgrass and slough communities. However, increased nutrients, particularly phosphorus, have resulted in a shift to a community dominated by cattail (*Typha latifolia* L.) [2, 16]. Due to these changes in the plant communities, food chains have been altered [5].

Therefore, reducing this "phosphorus enrichment" is the primary goal of Lake Okeechobee and Everglades restoration efforts [14, 22]. The Florida legislature passed the Everglades Forever Act (EFA) in 1994 (Section 373.4592, Florida Statutes), which directs the State of Florida to develop a phosphorus criterion for the Everglades Protection Area. The criterion numerically interprets an existing narrative standard which states: "In no case shall nutrient concentrations of a body of water be altered so as to cause an imbalance in natural populations of aquatic flora or fauna". The EFA mandated the reduction of phosphorus through the reduction of loads and water treatment. Two plans, Lake Okeechobee Protection Plan (LOPP) and Comprehensive Everglades Restoration Plan (CERP), were developed to implement improved agricultural management practices to lower phosphorus through improved on-farm water-management techniques and to construct managed wetlands to reduce phosphorus through vegetative uptake and soil storage [46, 49, 50].

6 Lake Okeechobee protection plan

Lake Okeechobee is approximately 150,000 ha in size and has a drainage basin containing approximately 1.1 million ha. During the last century, agricultural and urban development in the watershed and the construction of the Central and South Florida Project for flood control have caused excessive nutrient inputs. By the late 1980s, Lake Okeechobee was highly polluted with phosphorus resulting from point- and nonpoint-source agricultural and urban discharges into the watersheds along the Kissimmee River [5, 7]. The lake also receives large internal loads of phosphorus from the underlying mud sediments, where decades of past nutrient loads have accumulated. Total phosphorus concentration in Lake Okeechobee has almost doubled since the 1970s and chlorophyll levels have significantly increased over the same period [39]. As a result, there have been serious changes in the ecosystem, the most visible one being an increased frequency of algal blooms.

Lake eutrophication was attributed primarily to phosphorus loads form agricultural runoff in its watershed. The majority of Florida's 4000 ha of cattle pasture are located in south and central Florida [26]. Much of what was once native subtropical wet prairie ecosystem in this region is now managed for grazing. Land-use changes within these ecosystems have resulted in dramatic changes in the wildlife habitat characteristics and the patterns of nutrient flow for upland, marsh, and lake ecosystems [30].

The main focus of the LOPP is to reduce inputs of phosphorus to the lake from the watershed. A key aspect of the restoration program is that it includes cooperation among landowners, local government, and other state and federal agencies. Agricultural landowners are encouraged to implement measures to reduce the amount of phosphorus migrating off their land.

Site-specific improved agricultural production practices or best management practices (BMPs) improving the quality of runoff leaving agricultural lands are particularly important aspects of the LOPP. The Florida Department of Agriculture and Consumer Services has developed a voluntary BMP implementation program for activities including dairies and cow-calf operations, and vegetables and citrus production [18]. Examples of BMPs that reduce runoff-water quantity are detention ponds and grassed swales. Practices that improve runoff-water quality are reduced fertilizer application rates, ditch sediment control, dairy waste lagoons, and livestock confinement.

7 Comprehensive Everglades restoration plan

The CERP is the largest of the two restoration plans and is funded, managed and implemented through an unprecedented 50–50 partnership between the sate of Florida and federal government. The purpose of the plan is to re-establish a more natural flow of water throughout South Florida, including the Everglades, as well as ensure reliable water supplies for agricultural and urban use and provide flood control. Restoring a more natural flow of water to the Everglades should result in a long-term, sustainable water supply for South Florida with improved water quality.

The EFA proposed a stringent 100 ppb standard for P in surface water for the entire Everglades with a compliance deadline of December 31, 2003. Landowners, including farmers, in the Okeechobee Basin, pay one-tenth of a mil ad-valorem tax (~$38 million annually) and farmers within the EAA and C-139 basin pay an Agriculture Privilege Tax (~$11.5 million annually, or ~$60 per ha). Florida is paying the full cost of the water-quality improvements required by the State under the EFA and 50 per cent of the cost to implement the CERP, with the federal government paying the other half. To date, the state of Florida has committed nearly $1.3 billion to clean up the Everglades (EFA) and nearly $1.5 billion to restore a more natural flow of water to the River of Grass (CERP).

8 Compliance with the Everglades Forever Act

In 1996, SFWMD determined that the best interim methods to reduce phosphorus for complying with EFA included improved farming practices and the development of man-made wetlands, called stormwater-treatment areas (STAs). Farmers can choose from a list of approved BMPs and must monitor off-farm discharge for both flow volume and P concentration. On-farm studies in the EAA have shown that a significant portion of the total P load in drainage water leaving farms is in the particulate form [4, 17, 31, 35, 36, 41]. Therefore, most of the BMPs target the reduction of particulate P and sediment loads in drainage water [6, 38]. A primary BMP is field leveling prior to planting to reduce sheet erosion and improve water management. Small raised berms can also be constructed parallel to field ditches and drainage canals. With this BMP, water that may pond on the field after a heavy rain must percolate vertically eliminating sheet flow to drainage ditches. Main canal and field ditch sumps can be designed to further reduce off-farm sediment discharge. These sumps are deeper and/or wider areas in canals or ditches upstream of culverts or discharge pumps to reduce water flow and allow heavier materials to settle to the bottom. Cleaning and maintenance of drainage ditches and sumps are also prescribed as BMPs. Ditch and sump cleaning removes any debris and vegetable matter from the drainage system that can contribute to sediment P loading. Ditch maintenance includes establishing vegetated banks for stabilization and use of weed and trash booms to block these materials from leaving the farm.

Evaluations conducted since 1996 have confirmed that STAs were the best interim step toward achieving the long-term water-quality and hydropattern-restoration goals for the Everglades [40]. These managed wetlands were constructed to remove water-borne nutrients through plant growth and the accumulation of decomposing plant material in a layer of peat. The goal was that by 2006, these technologies will reduce phosphorus in water entering the Everglades by 90 per cent from a decade ago [40].

An important aspect of stormwater-treatment area optimization research is to determine the function of phosphorus retention mechanisms that serve in phosphorus removal. Vegetation growth results in the rapid, short-term nutrient uptake from the soil and water column. These nutrients can be temporarily stored in the vegetative structures of the plants before being released back into the water column from sediments after plant death. The microflora community of this wetland system is highly productive, but is greatly impacted by P enrichment [45]. Periphyton (a community of algae, bacteria, and microfauna) proliferate at the soil surface of impacted wetlands and are highly efficient at reducing P from the water column by immobilization of the nutrient [28, 48]. A nutrient mass balance has been established by measuring the total phosphorus concentration of inflow water, precipitation, plant tissue, wetland sediment, microbial immobilization, and outflow water. This mass-balance model is used to determine the timing and impacts of STA maintenance practices.

Long-term phosphorus removal is achieved through the accumulation and burial of peat sediments [40]. Through burial, much of the phosphorus in the underlying peat deposits will be sequestered and functionally removed from the overlying water column. Sediment nutrient content and accumulation are currently being documented using feldspar a white, crystalline mineral not found within the areas natural sediments. Layers of feldspar are deposited in a small area of each test cell where, after settling, they create a distinct horizon layer [40]. Sediment cores are taken one year later and deposition rates are measured by the amount of sediment visible above the feldspar marker. The sediment profile is divided and analyzed for total phosphorus content. Results of this analysis help to determine the rate of sedimentation and the mass of phosphorus retained in the sediment.

9 Water-quality improvements

In 1997, it was determined that numerous dairies have runoff total phosphorus concentrations in excess of 1000 ppb, and that current monitoring data show some sites had concentrations above 7000 ppb [40]. Four basins in the Lake Okeechobee watershed (approximately 115,000 ha) contribute the highest phosphorus concentrations and loads (35% of total) to the lake. All four basins contain active dairy operations. In 1997, the Florida Department of Environmental Protection enacted the Dairy Rule (chapter 62–670, Florida Administrative Code) that required that all dairies within the Lake Okeechobee watershed and its tributaries implement BMPs for the purpose of reducing phosphorus inputs into the lake. Under the rule, each farm was required to develop a site-specific plan providing for the collection, storage and disposal of wastewater. However, minimum on-site retention of water during a 25-year, 24-hour storm event was mandated. The general concept behind the Dairy Rule was to achieve a nutrient balance by retaining phosphorus on-site for uptake by forage crops, followed by harvesting of hay or other crops for cattle feed [39, 46]. Nondairy sources of phosphorus in the Lake Okeechobee watershed are primarily from beef-cattle pastures. Although animal densities and runoff phosphorus concentrations associated with beef-cattle pastures are relatively low, the large area (approximately 190,000 ha) of this land use makes them a major contributor of phosphorus.

The overall net phosphorus import in the northern Lake Okeechobee watershed was 1888 tons per year in 2002 which is a 28% decrease compared with data obtained in 1991 [40]. In 2005 the total phosphorus loading to the lake was approximately 600 metric tons per year, for a total reduction of 77% since 1991.

Agricultural BMPs reduced phosphorus concentrations leaving the EAA by 22% between 1996 (140 ppb) and 1999 (109 ppb) (STA1, Table 1). Inflow rates into STA 1 increased after 1999 due to reflux from sediment accumulation in the system. Other STAs (STAs 2 and 6, Table 1) indicate a further reduction in total phosphorus concentrations entering STAs from the EAA of 51% from levels in 1995. Inflow concentrations entering these STAs are also increasing due to the same sediment phosphorus reflux. Outflows from the STAs are less than 35 ppb, well below the original 100 ppb goal established in the EFA.

10 Impacts of tropical weather events on water quality

South Florida experienced an extremely rare occurrence with a series of hurricanes in 2004 and 2005. The region was hit by three major hurricanes and a remnant of a fourth in less than seven weeks in 2004, followed by a fifth hurricane a year later. Collectively, the storms drove a cascade of water and nutrients across the region beginning in mid-August and lasting in many areas to the spring 2006. The Florida Department of Environmental Protection analyzed deviations from water-quality criteria for 2005 and reported that The Everglades water quality generally meets state numeric criteria, except total phosphorus (Table 2). Despite the 2004/2005 hurricane-related impacts, STA operations were able to reduce total phosphorus concentrations by 55 per cent during 2005. The operational STAs together treated

Table 1: Average annual total phosphorus levels for storm-treatment areas (STA) in the Comprehensive Everglades Restoration Plan 1996–2005.

	STA 1		STA 2		STA 5		STA 6	
Year	Inflow (ppb)	Outflow (ppb)	Inflow (ppb)	Outflow (ppb)	Inflow (ppb)	Outflow (ppb)	Inflow (ppb)	Outflow (ppb)
1995	140	27						
1996	112	28						
1997	98	21						
1998	122	22					56	17
1999	103	23					62	21
2000	147	31					71	16
2001	149	42			252	97	136	29
2002	148	44	78	16	251	83	68	17
2003	151	51	66	17	268	142	79	23
2004	145	47	75	14	254	98	54	12
2005	251	99	124	20	157	77	77	20

Table 2: Average total phosphorus levels and flow rates for storm-treatment areas (STA) in the Comprehensive Everglades Restoration Plan prior to and after hurricane events of 2004/2005.

	January 2004			February 2006		
	Total phosphorus			Total phosphorus		
STA	Inflow (ppb)	Outflow (ppb)	Flow ($ft^3\ s^{-1}$)	Inflow (ppb)	Outflow (ppb)	Flow ($ft^3\ s^{-1}$)
1	149	48	95	222	112	153
2	79	15	456	118	51	529
5	275	121	45	197	91	75

almost 1,850,000 million l of inflow and removed 189 metric tons of total phosphorus from surface water. Total phosphorus concentrations were reduced from an average inflow of 179 ppb to and average outflow of 85 ppb. The Everglades marshes also generally showed little change in water quality from previous years despite the hurricanes in 2004 [40].

11 Future compliance

The Florida Department of Environmental protection proposed to set the first numeric ambient water-quality standard for phosphorus in the Everglades at 10 ppb in 2002. The revised rule for the phosphorus criterion from 100 to 10 ppb received final approval from the US Environmental Protection Agency in July 2005. To comply with the 10 ppb water-quality standard, The South Florida Water Management District has embarked on an ambitious research program for testing the feasibility of several advanced treatment technologies for the removal of phosphorus from waters leaving the STAs [40]. These technologies are being evaluated in pilot projects and are described below.

The first treatment technology evaluated was the use of indigenous submerged plants to remove P from the water column followed by limerock filtration at the downstream end of the STA system. Removal of P is accomplished by plant uptake as well as by coprecipitation with calcium carbonate that precipitates from the water column due to photsynthesis-related pH elevations. The limerock filter further removes a small amount of particulate P and dissolved organic P. Another treatment alternative was the periphyton-based stormwater-treatment area takes advantage of natural processes to sequester phosphorus. Treated post-STA water flows over a substrate colonized with calcareous periphyton (attached submerged algae) and macrophytes (floating aquatic plants). The macrophytes function as additional substrate and a stabilizing mechanism for the algal mats. Phosphorus is removed form the water column through biological uptake, chemical adsorption, and algal mediated coprecipitation with calcium carbonate within the water column. The third technology is the low-intensity chemical dosing treatment method incorporates small doses of aluminum salts directly into the STA influent. No apparent detrimental affect of Al accumulation has been noted at low dose rates. The chemical precipitation not only provides a mechanism for phosphorus removal and improved particulate removal, but also may enhance the phosphorus retention capacity of the sediments. No mechanical mixing or flocculation is used. This low-tech approach uses the STAs to provide the biological treatment as well as filtration and setting of the precipitate.

12 Conclusions

The actions taken by the water-management district in South Florida have improved the water quality of the Okeechobee and Everglades watersheds. Phosphorus-source control programs mandated by the 1994 Everglades Forever Act reduced water total phosphorus concentrations below the required 100 ppb well ahead of the 2003 deadline, and are continuing to exceed expectations. The use of BMPs and STAs have prevented more than 2200 metric tons of phosphorus from entering the Everglades over the past nine years. BMPs continue to be an effective tool for reducing phosphorus at its source north of Lake Okeechobee and the EAA. In 2005, the EAA reported a 59 per cent total phosphorus load reduction with its BMP program marking a strong continued performance at reducing nutrient inputs to the Everglades. One agricultural production area, the C-139 basin, continued to be out of compliance in 2005, the third year of BMP program implementation. However, a reduction in total phosphorus concentrations was observed during the water year suggesting that the program is having positive effects and moving the basin toward compliance with regulatory requirements. Lowering water phosphorus concentrations to 10 ppb established in 2005 will require the implementation of additional water-treatment measures.

The restoration of surface water flow in certain areas of the Okeechobee and Everglades watersheds have improved the surface-water flow and subsurface hydrology throughout both watersheds. Meanwhile, agricultural BMPs and water-treatment areas have improved water quality in these impacted watersheds. These restoration actions should reverse environmental impacts of increased P loading while maintaining the original goals of supporting agricultural production and urban development. The program of hydrological restoration and BMP implementation created in response to Everglades Forever Act should be a model for similarly impacted costal-plain ecologies.

References

[1] Bancroft, G.T., Status and conservation of wading birds in the Everglades. *American Birds,* **43**, pp. 1258–1265, 1989.

[2] Boodle, M.J., Ferreter, A.P. & Thayer, D.D., The biology, distribution, and ecological consequences of Melaleuca quinquenervia in the Everglades. *Everglades: The Ecosystem and Its Restoration*, eds S.M. Davis & J.C. Ogden, St. Lucie Press, Delray Beach, FL, pp. 341–356, 1994.

[3] Browder, J.A., Relationship between pink shrimp production on the Tortugas grounds and water flow patterns in the Florida Everglades. *Bulletin of Marine Science*, **37**, pp. 839–56, 1985.

[4] Brown, M.J., Boundurant, J.A. & Brockway, C.E., Ponding surface drainage water for sediment and phosphorus removal. *Transactions of the American Society of Agricultural Engineers,* **24**, pp.1478–1481, 1981.

[5] Brown, B., Weaver, J., Browder, J., Kitchens, W., Glaz, B., Armentano, T., Goodyear, C., Burns, L., Morrison, D., Thompson, N., Richards, P., Ogden, J.C., Hilton, R., Ambrose, R., Araujo, R., Barber, M.C., Bullock, R. & Loux, N., South Florida ecosystem restoration: scientific information needs. Science Subgroup. Management and Coordination Working Group, Interagency Task Force on the South Florida Ecosystem. South Florida Water Management District, West Palm Beach, FL, 1994.

[6] Daroub, S.H., Struck, J.D., Lang, T.A., Diaz, O.A. & Chen, M., Implementation and verification of BMPs for reduced P loading in the EAA. Final project report submitted to the Everglades Agricultural Area Environmental Protection District and The Florida Department of Environmental Protection. Tallahassee, FL, 2003.

[7] Davis, S.M., Phosphorus inputs and vegetation sensitivity in the Everglades. *Everglades: The Ecosystem and Its Restoration*, eds S.M. Davis, & J.C. Ogden, St. Lucie Press, Delray Beach, FL, pp. 357–378, 1994.

[8] DeBush, W.F., Newman, S., & Reddy, K.R., Spatio-temperal patterns of soil phosphorus enrichment in Everglades water conservation area 2A. *Journal of Environmental Quality,* **30**, pp.1438–1446, 2001.

[9] Diaz, O.A., Anderson, D.L. & Hanlon, E.A., Phosphorus mineralization from Histosols of the Everglades Agricultural Area. *Soil Science,* **56**, pp. 178–185, 1993.

[10] Duever, M.J., Meeder, J.F., Meeder, L.C. & McCollom, J.M., The climate in South Florida and its role in shaping the Everglades ecosystem. *Everglades: The Ecosystem and Its Restoration*, eds S.M. Davis & J.C. Ogden, St. Lucie Press: Delray Beach, FL, pp. 225–248, 1994.

[11] Fennema, R.J., Neidrauer, C.J., Johnson, R.A., MacVicar, T.K. & Perkins, W.A., A computer model to simulate natural Everglades hydrology. *Everglades: The Ecosystem and Its Restoration*, eds S.M. Davis & J.C. Ogden, St. Lucie Press: Delray Beach, FL, pp. 249–289, 1994.

[12] Fisher, M.M. & Reddy, K.R., Phosphorus flux from wetland soils affected by long term nutrient loading. *Journal of Environmental Quality*, **30**, pp. 261–271, 2001.

[13] Frohring, P.C., Voorhees, D.P. & Kushlan, J.A., History of wading bird populations in the Florida Everglades: A lesson in the use of historical information. *Waterbirds,* **11**, pp. 328–335, 1988.

[14] Harris, W.G., & Hollien, K.A., Changes across artificial E-Bh boundaries formed under simulated fluctuating water tables. *Soil Science Society of America Journal,* **64**, pp. 967–973, 2000.

[15] Gunderson, L.H., & Snyder, J.R., Fire patterns in the southern Everglades. *Everglades: The Ecosystem and Its Restoration*, eds S.M. Davis & J.C. Ogden, St. Lucie Press, Delray Beach, FL, pp. 291–306, 1994.

[16] Holling, C.S., Gunderson, L.H., & Walter, C.J., The structure and dynamics of the Everglades system: Guidelines for ecosystem restoration. *Everglades: The Ecosystem and Its Restoration*, eds S.M. Davis & J.C. Ogden, St. Lucie Press: Delray Beach, FL, pp. 741–756, 1994.

[17] Izuno, F.T., Sanchez, C.A., Coale, F.J., Bottcher, A.B. & Jones, D.B., Phosphorus concentrations in drainage waters in the Everglades Agricultural Area. *Journal of Environmental Quality,* **20**, pp. 608–619, 1991.

[18] Izuno, F.T., & Rice, R.W., Implementation and verification of BMPs for reducing P loading in the EAA. Final project report submitted to The Florida Department of Environmental Protection and The Everglades Agricultural Area Environmental Protection District, Tallahassee, FL, 1999.

[19] Light, S.S., Wodraska, J.R. & Sabina, S., The southern Everglades: The evolution of water management. *National Forum,* **69**, pp. 11–14, 1989.

[20] Light, S.S. & Dineen, J.W., Water control in the Everglades: A historical perspective. *Everglades: The Ecosystem and Its Restoration*, eds S.M. Davis & J.C. Ogden, St. Lucie Press, Delray Beach, FL, pp. 47–84, 1994.

[21] MacVicar, T.K., & Lin, S.T., Historical rainfall activity in central and southern Florida: Average, return period estimates and selected extremes. *Environments of South Florida: Present and Past II*, ed. P.J. Gleason, Miami Geological Society, Coral Gables, FL, pp. 477–509, 1984.

[22] Maltby, E. & Dugan, P.J., Wetland ecosystem protection, management, and restoration: An international perspective. *Everglades: The Ecosystem and Its*

Restoration, eds S.M. Davis & J.C. Ogden, St. Lucie Press, Delray Beach, FL, pp. 29–46, 1994.

[23] McIvor, C.C., Ley, J.A. & Bjork, R.D., Changes in freshwater inflow from the Everglades to Florida Bay including effects on biota and biotic processes: A review. *Everglades: The Ecosystem and Its Restoration*, eds S.M. Davis & J.C. Ogden. St. Lucie Press, Delray Beach, FL, pp. 117–148, 1994.

[24] Morris, D.R. & Gilbert, R.A., Inventory, crop use and soil subsidence in Florida. *Journal of Food, Agriculture & Environment,* **3**, pp. 190–193, 2005.

[25] Morris, D.R., Glaz, B. & Daroub, S.H., Organic matter oxidation potential determination in a periodically flooded Histosol under sugarcane. *Soil Science Society of America Journal,* **68**, pp. 994–1001, 2004.

[26] National Agricultural Statistics Service (NASS), Florida Agricultural Commodity Statistics, Washington, DC, 2005.

[27] Newman, S., Grace, J.B., & Koebel, J.W., Effects of nutrients and hydroperiod on Typha, Cladium and Eleocharis: Implications for Everglades restoration. *Applied Ecology*. **6(3)**, pp. 774–783, 1996.

[28] Nue, G.B., Childers, D.L. & Jones, R.D., Phosphorus biochemistry and impact of Phosphorus enrichment: why is the Everglades so unique? *Biosystems,* **4**, pp. 603–624, 2001.

[29] Ogden, J.C., Recent population trends of colonial wading birds on the Atlantic and Gulf coastal plains. *Wading Birds*, eds. A. Sprunt IV, J.C. Ogden & S. Winckler, New York, NY, pp. 137–153, 1978.

[30] Ogden, J.C., A comparison of wading bird nesting colony dynamics (1931–1946 and 1974–1989) as an indication of ecosystem conditions in the southern Everglades. *Everglades: The Ecosystem and Its Restoration*, eds S.M. Davis & J.C. Ogden, St. Lucie Press: Delray Beach, FL, pp. 533–570, 1994.

[31] Robbins, C.W. & Carter, D.L., Conservation of sediment in irrigation runoff. *Journal of Soil Water Conservation,* **30**, pp. 134–135, 1975.

[32] Robertson, W.B. & Kushlan, J.A., The South Florida fauna. *Environments of South Florida: Present and past, memoir No. 2.*, ed. P.J. Gleason, Miami Geological Society, Coral Gables, FL, pp. 414–452. 1974.

[33] Robertson, W.B. & Frederick, P.C., The faunal chapters: Contexts, synthesis, and departures. *Everglades: The Ecosystem and Its Restoration*, eds S.M. Davis & J.C. Ogden, St. Lucie Press: Delray Beach, FL, pp. 709–37, 1994.

[34] Ross, M.R., O'Brien, J.J. & Flynn, L.J., Ecological site classification of Florida Keys terrestrial ecosystems. *Biotropica,* **24**, pp. 488–502, 1992.

[35] Schuman, G.E., Spomer, R.G. & Priest, R.F., Phosphorus losses from four agricultural watersheds on Missouri Valley loess. *Soil Science Society of America Journal,* **37**, pp. 424–427, 1973.

[36] Sharpley, A.N., Daniel, T.C. & Edwards, D.R., Phosphorus movement in the landscape. *Journal of Production Agriculture,* **6**, pp.492–500, 1993.

[37] Shih, S.F., Glaz, B. & Barnes, Jr. R.E., Subsidence of organic soils in the Everglades Agricultural Area during the past 19 years. *Soil and Crop Science Society of Florida Proceedings,* **57**, pp. 20–29, 1998.

[38] Sievers, P., Pescatore, D., Daroub, S., Stuck, J.D., Vega, J., McGinnes, P. & Van Horn, S., Performance and optimization of agricultural best management practices. Everglades BMP program annual report, South Florida Water Management District, West Palm Beach, FL, 2003.

[39] South Florida Water Management District (SFWMD), Lower East Coast Regional Water Supply Plan, West Palm Beach, FL, 1994.

[40] South Florida Water Management District (SFWMD), Comprehensive Everglades Restoration Plan, 2005 Report to Congress, West Palm Beach, FL, 2005.

[41] Stuck, J.D., Izuno, F.T., Campbell, K.L., Bottcher, A.B. & Rice, R.W., Farm-level studies of particulate phosphorus transport in the Everglades Agricultural Area. *Transactions of the American Society of Agricultural Engineers,* **44**, pp. 1105–1116, 2001.

[42] Terry, R.E., Nitrogen mineralization in Florida Histosols. *Soil Science Society of America Journal,* **44**, pp. 747–750, 1980.

[43] Thomas, T.M., A detailed analysis of climatological and hydrological records of South Florida with reference to man's influence upon ecosystem evolution. *Environments of South Florida: Present and past, memoir No. 2.*, ed. P.J. Gleason. Miami Geological Society, Coral Gables, FL, pp, 82–122, 1974.

[44] Tomlinson, P.B., The Biology of Trees Native to Tropical Florida. Harvard: Harvard University Printing Office: Boston, MA, 1980.

[45] Qualls, R.G. & Richardson, C.S., Phosphorus enrichment affects: Litter decomposition, immobilization, and soil microbial phosphorus in wetland mesocosms. *Soil Science Society of America Journal,* **64**, pp. 799–808, 2000.

[46] US Army Corps of Engineers (USCOE), Comprehensive Review Study. Central and Southern Florida Project, Jacksonville, FL, 1994.

[47] US Census Report (USCR), *Florida Population Statistics*, Washington, DC, 2002.

[48] Vaithiyanathan, P. & Richardson, C.J., Macrophyle species changes in the Everglades: Examination along a eutrophication gradient. *Journal of Environmental Quality,* **28(4)**, pp. 1347–1358, 1999.

[49] Walters, C., Gunderson, L. & Holling, C.S., Experimental policies for water management in the Everglades. *Applied Ecology,* **2**, pp. 189–202, 1992.

[50] Weaver, J. & Brown, B., Federal Objectives for the South Florida Restoration. Science Sub-Group, South Florida Management and Coordination Working Group. US Department of Interior, Washington, DC, 1993.

CHAPTER 13

Waterborne zoonoses and changes in hydrologic response due to watershed development

Mark Walker[1], Bruce Wilcox[2] & Mayee Wong[2]
[1]*University of Nevada, Reno, Department of Natural Resources and Environmental Sciences, Reno, NV, USA.*
[2]*Asia-Pacific Center for Infectious Disease Ecology, John A. Burns School of Medicine, University of Hawaii, Honolulu, HI, USA.*

Abstract

Hawaii's mountain-to-sea ecosystems provide unique opportunities to evaluate watershed-scale processes. In this chapter, we use a watershed on Oahu, Hawaii (Manoa Stream) to explore the link between watershed development, hydrologic response and increased risk of waterborne disease as a result of flooding and the presence of commensal rodents chronically infected with leptospirosis. Flooding from the watershed in 2004 led to an outbreak of leptospirosis among people exposed to flood waters during clean-up efforts. We present hydrologic analysis of a flood-producing storm and animal sampling results to illustrate the hypothesis that watershed development may increase the likelihood of flooding and the magnitude of sources of pathogenic microbes that could be mobilized during flood-producing rainfall. The analysis considers land-use changes in the Manoa stream watershed that took place between 1939 and 2005.

Hydrologic response to the type of rain storm observed in 2004 has changed significantly since 1939. Peak flow rates would have been approximately 180% lower in 1939 than those observed during the 2004 floods. Rodent trapping results indicate that 27/354 rodents trapped over the course of 14 years were infected with serovars of *Leptospira.* When carried out intensively, trapping results suggest a low (~6%) but consistent prevalence of infection with serovars found in confirmed cases of leptospirosis following the 2004 flood. We discuss these results in the context of conditions likely to lead to an outbreak of waterborne disease, and how the likelihood may increase in areas experiencing urban expansion and high-density residential growth.

1 Introduction

A strong link exists between extreme weather events, flooding and the outbreak of zoonotic diseases caused by microbial pathogens [1]. Illness results in part from exposure to microbial contaminants in flood waters [2]. Flooding is likely to create an extra burden of microbial contaminants by disrupting human-sewage-treatment facilities and by mobilizing pathogens that are resident in soils and animal populations, increasing the likelihood that people will be exposed to them.

Flooding is most often caused by extreme weather events, such as cyclones and hurricanes [2]. These are considered to be rare and unusual, especially in the case of very destructive and disruptive flooding. Anthropogenic changes in landscape may increase the likelihood of flooding [3, 4].

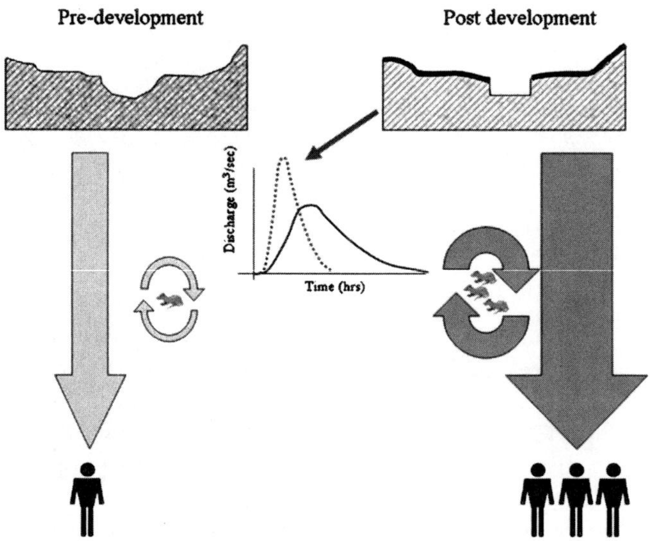

Figure 1: Illustration of hypothetical effects of watershed development on risk of infectious disease outbreaks from exposure to waterborne pathogens. Under predevelopment conditions, watershed response to storms is attenuated by abstraction losses in vegetated canopies and infiltration into soils. Resultant streamflow hydrographs caused by equivalent storms (center) have slower times to peak, and smaller quantities of peak flow and volume of storm flow than those observed postdevelopment. At the same time, commensal animal populations that are maintenance hosts for waterborne parasites may increase as food sources and habitat are introduced with development. Regardless of changes in prevalence, the magnitude of the source of pathogenic microbes in the watershed may also increase. At the same time numbers of downstream people increase with development as the risk of flooding also increases, because of changes in hydrologic response in the watershed.

Land-use changes may be an immediate cause for concern with respect to zoonotic diseases and transmission to susceptible populations by flooding. As discussed by Kaneshiro *et al.* [5], residential and commercial development in watersheds may synergistically increase the potential for waterborne, zoonotic diseases to emerge or re-emerge, in part as a consequence of change in factors that affect hydrologic response (Fig. 1). Kaneshiro *et al.* [5] propose that development of watersheds may create habitat and food sources for maintenance populations of zoonotic diseases and, concurrently, increase transport and exposure potential by changing watershed response to flood-producing rainfall.

Watershed development adds impervious surface to watersheds and can change hydrologic response because of several factors. These include a decrease of abstraction losses during storms because of loss of absorbing surfaces in watersheds (for example, vegetation and soil), increase in runoff volume because of added impervious surfaces (streets, parking lots and rooftops) and a decrease in the hydrologic response time to runoff-producing precipitation due to extensive networks of stormwater-management infrastructure, including conduits and paved portions of stream channels.

Emerging infectious diseases generally have increased globally in association with urbanization, agricultural intensification, and habitat alteration [1, 6, 7]. Emerging and re-emerging infectious diseases have been discussed in the context of environmental change, particularly with respect to anticipated effects of global warming [8, 9]. This may be the result of increased contact between individuals within host animal populations via specific media, such as contaminated water. Several factors may contribute to this, including habitat loss, alteration or creation [9, 10] and increased contact between host populations and human populations, especially at the margins of and within newly developed areas [11].

Several recent reviews suggest that zoonoses, transmitted from animal hosts to humans by a variety of routes of exposure, are an emerging class of disease that will affect greater numbers of people as a consequence of population growth, urbanization and climate change [12–14]. Wastes from infected hosts, including feces and urine, may contain disease-causing microbial agents, which may be transported by flowing water away from where wastes are deposited. Environmentally transmissive forms of microbial pathogens may survive environmental stresses and infect humans. For example, oocysts of the parasitic protozoa *Cryptosporidium parvum* remain infective for days to months while in suspension in aqueous solution [15, 16], fecal suspensions [17] and in soils and stored animal wastes [18]. *Cryptosporidium*, *Giardia*, *Microsporidia* and *Toxoplasma* have been the causative agents of recent waterborne-disease outbreaks in developed nations [14].

Leptospirosis is considered to be a re-emerging disease [19] that infects people by contact with contaminated water, soil and the urine of infected animal hosts [20]. The disease is commonly associated with flooding [20–24] and is prevalent in flood-prone areas [25]. Leptospirosis is caused by *Leptospira interogans* spirochetes. Infection may result from contact with contaminated water or urine from infected animals, especially through skin abrasions and the mucosa. Symptoms of illness range from mild febrile reactions to sometimes fatal disease. Leptospirosis is thought to be substantially underreported, because symptoms are easily confused with those associated with common influenza, dengue fever and other viral infections. The incidence of disease among humans has a marked association with seasonal weather trends; for example, the number of new cases in regions with endemic leptospirosis increased during wet months [26, 27]. Researchers also have noted that the incidence of leptospirosis in host animal populations coincides directly with seasonal fluctuations in rainfall [28–30].

A recent review paper suggested that leptospirosis be considered as a model zoonosis for evaluating the link between ecosystem change and disease emergence, especially related to the influence that potential reservoir hosts may have on susceptible human populations [31]. Other authors have suggested that Hawaii's islands offer ideal sites for case studies to understand anthropogenic influences on watershed processes, in part because watersheds can be viewed as scale models of large continental mountain-to-sea ecosystems [5]. The suggestion acknowledges the influence of Hawaii's topography on hydrologic processes. Hawaii's islands have topograhic relief sufficient to create significant orographic effects. Many of the islands have distinct windward/leeward topographic divides that define trends in precipitation distribution. Orographic effects lead to very large differences in rainfall over short distances. For example, on Oahu, the average annual rainfall estimated from 30 years of record at the Lyon Arboretum, in Manoa Valley (152 m (500 ft) AMSL), is 3886 mm (153 in). Average annual rainfall at the University of Hawaii rainfall gaging station (24 m (80 ft) AMSL), approximately 3.1 km (1.9 mi) south southwest of the Lyon Arboretum gage, is 1016 mm (40 in), based on 54 years of observation (www.wrcc.dri.edu/Climsum.html, last accessed 7/2006).

In the steep valleys that fall away from both the leeward and windward sides of topographic divides, such as the Ko'olau range on Oahu (Fig. 2), storms may be of moderate duration (hours) but high intensity (mm (in)/hr). As a consequence, the relationship between changes in land use and various aspects of hydrologic response, including water quality, should be readily discernible in modeling analysis and long-term measurements of the physical system. In the case of the latter, the violence of occasional floods has disrupted data-collection efforts in several locations, including Manoa Stream. However, modeling approaches remain feasible, and can be supported by historical aerial photographic records, land-use analyses and maps and databases related to the physical characteristics of watersheds, including soils and topography.

In this chapter, we examine the change in hydrologic regime in a specific location in Hawaii that experienced unusual, destructive flooding in 2004 and confirmed and suspected cases of leptospirosis among people exposed to flood waters during clean-up efforts. We focus on leptospirosis as an example of a pathogen associated with commensal animal populations that may increase in numbers as watersheds are developed. We estimate changes in hydrologic response due to changes in watershed use over a 65-year period, with animal-sampling results to illustrate a hypothesis presented by Kaneshiro *et al.* [5] (Fig. 1). The data are drawn from several sources and applied to the Manoa Stream watershed on Oahu, Hawaii, but suggest that watershed development was a factor that led to flooding and also to a small outbreak of this zoonotic, waterborne disease.

1.1 Physical setting

1.1.1 The Manoa Stream watershed

The Manoa Stream watershed, on the island of Oahu, Hawaii (Fig. 2) has a total relief of approximately 945 m (3100 ft) from the uppermost portion of the watershed to a gaging station operated by the U.S. Geological Survey at Kanewai Field (gaging station number 16242500), located on the southeastern portion of the University of Hawaii, Manoa campus (Fig. 3). The average annual rainfall recorded at a gage in the upper reaches of the watershed is 3 886 mm (153 in), based on thirty years of record (1975–2005, see http://www.wrcc.dri.edu/summary/Climsmhi.html (Manoa Lyon Arboretum, last accessed 6/06)). Average annual streamflow reported for gaging station 16242500 ranged from 0.2–0.3 m^3/s (7–11 ft^3/s) in 2000–2003. The perennial stream rises from the confluence of three primary tributaries at the base of extremely steep headwalls formed by the Ko'olau range. The stream course of 7.1 km (4.4 mi) drains an area of 14.2 km^2 (5.5 mi^2). The stream course is posted with signs warning of the risk of contracting leptospirosis from contact with stream water, from the base of Manoa Falls (accessible by foot only), and at almost every bridge crossing and access point along its course to the confluence with the Palolo Stream, downstream of gaging station number 16242500.

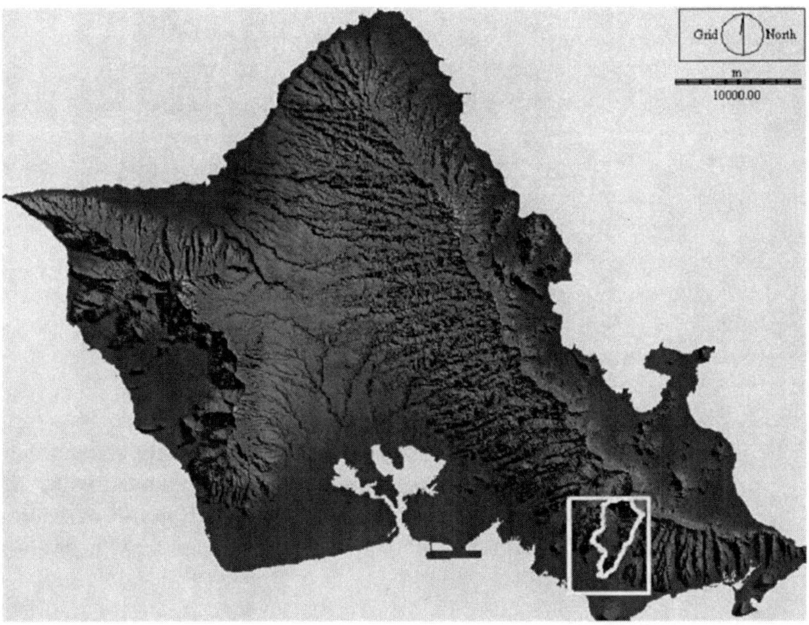

Figure 2: Study area (watershed boundaries shown within white box), Manoa Stream island on Oahu, Hawaii. The Ko'olau range trends from the northwest to the southeast on the eastern half of the island and causes significant precipitation gradients, including Mania Valley.

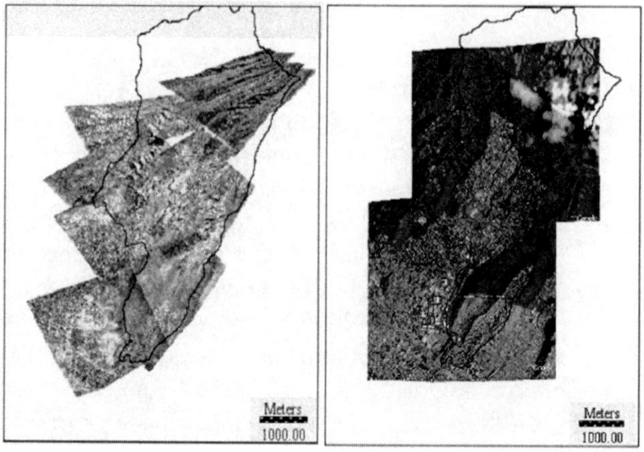

Figure 3: Photomosaics of land uses, 1939 (left, from Hawaii State Archives) and 2005 (right, from Google Earth®), with overlay of watershed bounds for Manoa Stream (black line). In 2005 nearly all land in the watershed with a slope of less than 22° was developed, while in 1939, most of the land was used for small-plot agriculture or covered in dense tropical forest.

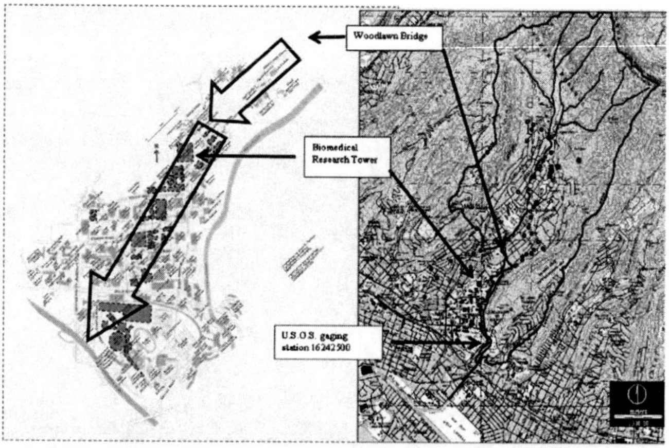

Figure 4: Manoa Stream watershed (right panel, outlined in blue), showing animal sampling points (red: collected animal positive for serovar ballum; yellow: positive for serovar icterohemorragiae; blue: positive for mixed serovars; black: negative results), and (left panel) detail of flood path (blue arrows) through University of Hawaii campus, with flood-damaged buildings depicted in red.

The watershed, with an overall area-weighted average slope of 22°, has two features that are common in the Hawaiian islands: (a) watershed relief is extreme, and (b) substantial portions of drainage areas are comprised of fractured volcanic rock formations with relatively thin soil covers. The latter feature is important with respect to hydrologic response. During dry periods dense canopies and thin soil covers with underlying fracture networks are thought to be the cause of initial abstraction losses during the early stages of storms [32]. However, soils and fracture networks are often saturated during periods of prolonged rain, which commonly occur from October through March (www.wrcc.dri.edu/narratives/HAWAII.htm, last accessed 4/06). Saturated soils accentuate watershed response to storms, especially the quantity of peak discharge and the time to peak discharge rates from the onset of storms [32].

The watershed has evolved from a largely undeveloped agricultural valley in the early part of the 20th century to a heavily developed residential and commercial area in the early part of the 21st century (Fig. 3). Land uses in the watershed are predominantly residential, with some open space (park lands, school yards and a cemetery) and commercial areas. The upper margins of the watershed are largely undeveloped. Much of the intensive building took place in the last quarter of the 20th century. Although the pace of development has slowed significantly, subdivision and new construction continue throughout the valley.

1.1.2 Manoa Stream flood, 2004

On October 30th, 2004 Manoa Stream swelled far beyond channel capacity at the Woodlawn Bridge (constructed in 1975), forcing debris- and sediment-filled waters into an adjacent subwatershed that contained a majority of the University of Hawaii, Manoa's buildings and facilities (Fig. 4). The rectangular opening in the stream channel defined by the Woodlawn bridge was designed to convey approximately 235 m^3/s (8300 ft^3/s). However, accumulated woody debris and sediment from upstream erosion and bank collapse reduced the capacity to carry flood waters. Flooding led to extensive damage of research, administrative and academic facilities in a swath across the campus (Fig. 4). In all, the flood damaged 32 buildings, 25 of which were sited and constructed before 1980. The flooding was especially destructive for researchers and librarians, who lost active experiments and digital and paper records. The flooding left standing water and sediment throughout affected areas, including inside buildings.

2 Methods

2.1 Estimated changes in hydrologic response associated with changes in land use

We carried out this analysis to evaluate the effects of increased proportion of impervious surface in the watershed during the past seven decades. We estimated the change in hydrologic response using the curve-number approach to predict peak discharge rates from observed rainfall amounts [33, 34]. We developed an area-weighted representation of key aspects of the watershed at four periods (1939, 1955, 1976 and 2005) using land-use and land-cover data and soil characteristics, as described below. The storm used for analysis was drawn from the National Oceanic and Atmospheric Agency's report of the 2004 Manoa Valley flood (www.prh.noaa.gov/hnl/pages/events/ManoaFlood20041030/, last accessed 1/06). Rain gages in the northeastern head of the valley recorded a cumulative total of 221 mm (8.7 in) of rain in six hours (*ibid.*). For the analysis, we assumed a storm duration of six hours and storm total of 221 mm (8.7 in). This duration and intensity of storm were estimated to have a 2% probability of occurring in a given year (*ibid.*). This is similar to the storm that caused flooding on Oahu in 1988 (51–102 mm/hr (2–4 in/hr) for 5–6 h [32]).

Excess rainfall was transformed to storm hydrographs using a dimensionless unit hydrograph and watershed physical characteristics (average slope, watershed area and watershed length [33]). The analysis relied on the following series of equations to derive time to peak and peak discharge.

$$t_p = \frac{\Delta D}{2} + L, \tag{1}$$

$$L = \frac{.00526 l^{0.8} (S+1)^{0.7}}{\sqrt{Y}}, \tag{2}$$

$$S = \frac{1000}{CN} - 10, \tag{3}$$

$$q_p = \frac{484AQ}{(\Delta D/2) + L}, \tag{4}$$

$$Q = \frac{(P - 0.2S)^2}{(P + 0.8S)}. \tag{5}$$

In these equations, t_p is the time to peak discharge (h), D is the duration of runoff-producing rainfall (h), L is the lag time (h) between the onset of runoff-producing rainfall and peak discharge rate, l is the hydraulic length of the watershed (ft), Y is the average slope of the watershed (ft/ft), CN is the curve number value typical of a combination of soil hydrologic group and land use, A is watershed area (square miles), Q is excess precipitation (inches, determined using the curve-number approach, eqn. (5)), and P is total storm precipitation (in). Application of these equations for simulation assumes uniform distribution of rainfall throughout a watershed. The average annual rainfall gradient between the head of Manoa Valley (at the Lyon Arboretum) and a rainfall gage in the lower portion of the stream course is approximately 2362 mm (93 in)/3.1 km (1.9 mi). For this storm the rainfall gradient was approximately 188.0 mm (7.4 in)/3.1 km (1.9 mi) (280.7 mm (11.1 in) at the head of Manoa Valley versus 61.0 mm (2.4 in) at the University of Hawaii's Experimental Farm [35]. However, simulations of peak flow rates recorded at a gaging station defined as the outlet of the watershed for these analyses corresponded well with observed annual peaks flows (see below), suggesting that the heterogeneous distribution of rainfall did not excessively violate the assumptions inherent in simulations.

Land uses in 1939 (Fig. 3), 1955, 1976 and 2005 were determined from aerial photographs obtained from the Hawaiian State Archives (1939, 1955), a land-use and land-cover analysis completed by the Earth Resource Observation and Science division of the U.S Geological Survey (1976) and satellite images obtained from Google Earth® v. 3.0.0762 (acquired in 2005). All digital files referenced NAD83, UTM zone 4N. Land-use and land-cover designations were based on the classification system proposed by Anderson *et al.* [36], for Level II analysis. Scanned aerial photos and satellite images were rectified to a digital line graph version of the 1: 24,000 Honolulu quadrangle, using the Global Mapper® v. 7.01 image-processing and analysis program.

Estimates of area-weighted curve numbers to represent the watershed during each selected year were derived from land-use/land-cover data and soil characteristics. The geographic information system Idrisi Kilimanjaro v. 14.01 was used for watershed delineation, digitizing land uses for the 1939, 1955 and 2005 data sets and developing a crosstabulation matrix with interpreted Soil Survey Geographic Database (SSURGO) soil types (land-use type × soil hydrologic group). Results of crosstabulations were used to develop area-weighted average curve numbers for the entire watershed for each year.

We evaluated area-weighted curve-number estimates for current conditions by comparing reported annual peak flows (see http://nwis.waterdata.usgs.gov/hi/nwis/peak/?site_no=16242500, last accessed 4/06) for 1999, 2001, 2002 and 2004. These flood peaks were selected for comparison because each occurred after periods of prolonged rainfall, which presumably led to saturated soils and approximately equivalent abstraction losses when peak rainfall occurred. Durations and intensities for these storms, reported by the National Oceanic and Atmospheric Administration's archived storm data for the upper portion of the watershed (see http://www.prh.noaa.gov/hnl/hydro/hydronet/hydronet-data.php (last accessed 4/06)) ranged from 1.5–5.25 h, with precipitation intensities of 60.2 mm (2.4 in)/1.5–224.3 mm (8.8 in)/5.25 h. The area-weighted curve-number estimate for this time period appeared to be an appropriate representation of the watershed under current conditions, with observed and estimated peak discharge rates correlated at 0.99.

2.2 Animal trapping: Manoa Stream Watershed, 1990–2003

The Hawaii Department of Health's (HDOH) Vector Control Branch (VCB) has statewide responsibilities to control disease outbreaks and monitor insect vectors and zoonotic diseases. The VCB traps rodents for several reasons, including requests from the Disease Investigation Division of HDOH related to potential exposure sites in human cases of leptospirosis or murine typhus, complaints of excessive rodent activity in an area, and special studies with intensive neighborhood surveillance.

Investigators select sites according to the reasons for trapping. In the case of sites of potential human exposure or in response to complaints, trap sites are specified by an investigator from the Disease Investigation Division or by the complainant and a limited number of traps is set. For intensive neighborhood surveillance, the VCB traps at multiple sites in a single neighborhood. Each trapping site is equipped with several rat traps and one mongoose and one mouse trap. Traps used include turtle-back cages for rats, 'tin cat' multiple catch traps for mice and Havahart wire cage traps for mongooses. Trapped animals were transported from the field to the HDOH's Zoonoses Laboratory and euthanized using carbon dioxide. Only live-trapped animals were tested.

To obtain the results reported in Table 2, kidneys were aseptically removed from animals and cultured for pathogenic leptospires. Sera from rats and mongooses were not tested for leptospiral antibodies. The results of serological surveys of rodent populations differ from those obtained by culturing kidney tissues, in part because colonized kidney tubules do not stimulate an immune response [37]. Accordingly, the VDC used cultures from kidneys to determine if trapped rodents were infected, using the microscopic agglutination test [38]. One antiserum from each of six or seven serogroups was allowed to react with the unknown isolate to attempt to place it in a serogroup.

3 Results

3.1 Peak-flow estimates

Table 1 presents, for each year considered, estimated flood characteristics associated with the design storm used for analysis. Figure 5 presents estimated hydrographs for each simulation. Table 1 and Fig. 5 indicate that hydrologic response to a storm equivalent to the October 2004 storm changed significantly with changes in land use. Watershed response was compressed and substantially increased. Estimated peak discharge rates increased by approximately 183% between 1939 and 2005.

Table 1: Estimated characteristics of hydrologic response to the design storm observed on October 30th, 2004.

Year	Excess precipitation mm (in)	Time to peak flow from onset of storm (h)	Peak discharge rate m^3/s (ft^3/s)
1939	104.1 (4.1)	4.0	86 (3051)
1955	115.3 (4.5)	3.9	101 (3583)
1976	116.8 (4.6)	3.9	99 (3513)
2005	165.4 (6.5)	3.6	158 (5591)
2004[a]	n.a.	<1h	166 (5870)

a. The time to peak and peak discharge rate are those reported for the Manoa Stream flood on the following web sites: http://hi.water.usgs.gov/projects/ data_manoa_peaks.html and www.prh.noaa.gov/hnl/pages/events/ ManoaFlood20041030/. Excess precipitation (rainfall that produces runoff and streamflow) estimates have not been made for this storm.

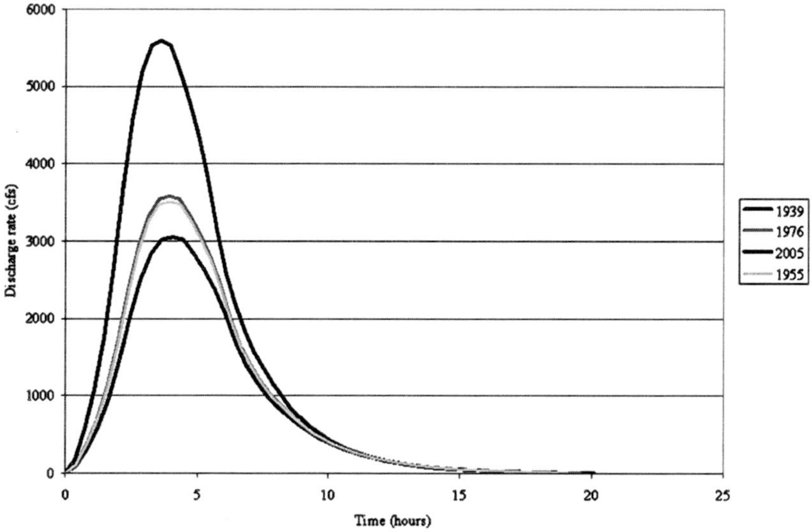

Figure 5: Estimated storm hydrographs for 1939, 1955, 1976 and 2005 from excess precipitation associated with a design storm of six hour duration and 8.71 inch volume.

3.2 Animal-trapping results

Trapping demonstrated that rodent species that could be maintenance hosts of pathogenic *Leptospira* serovars were found in the watershed (Table 2). The species *Rattus rattus* (the black rat) predominated, and *Rattus norvegicus* (the brown rat), *Mus musculis* (domestic mouse), and *Herpestes auropunctatus* (the Indian mongoose) were also present. Table 3 includes serovars associated with these and other potential maintenance hosts that are found in the watershed. It is important to note that host specificity does not appear to be strict by either maintenance species or location [20, 39, 40]. However, rodent-sampling results from the Manoa watershed (see below) correspond with those presented Table 3. When trapping was extensive (1990, 1995) results demonstrated that infected hosts were present with an overall prevalence of 21/131 and that serovar icterohemmoraghia was detected in cultures from 13/21 rodents that tested positive for infection (Table 2).

4 Discussion

Flooding of the University of Hawaii, Manoa, campus was an unexpected event that transferred water, sediment and debris between distinct hydrologic units. When the campus was partially inundated, sediment and water contaminated with pathogenic *Leptospira* spirochetes infected workers involved in the clean-up. In the several weeks after the flood, two cases of leptospirosis in flood clean-up workers were confirmed by diagnostic testing. Also, 90 of 271 respondents to

Table 2: Results of rodent sampling in the Manoa Stream watershed, carried out by the Hawaiian Department of Health from 1990–2003. Columns report aggregated annual results of trapping by species (R: *Rattus rattus* (the black rat); N: *Rattus norvegicus* (the brown rat); H: *Herpestes auropunctatus* (mongoose), E: *Rattus exulans* (the Polynesian rat); M: *Mus musculis* (domestic mouse). Cells report number of individuals trapped per species, with number of positives reported as n/N (for example 8/107 individual *Rattus rattus* trapped were positive for leptospirosis in 1990). The number of infected individual species, with specific serovars, (I: icterohemmoragaie, B: ballum) is reported in the last column (for example, in 1990, cultures from kidneys of 4 *R. rattus* individuals were positive for serovar icterohemmoragaie, 4 were positive for serovar ballum, 1 *R. exulans* was positive for serovar ballum, and 1 *R. norvegicus* was positive for serovar icterohemmoragaie). When listed as mixed, several serovars were present.

Year	*R*	*N*	*H*	*E*	*M*	Total	Serovars observed, by species
Intensive sampling efforts							
1990	*8/107*	*3/8*	0	*1/20*	3	*12/138*	R: I(4), B(4) N: I(3) E: B(1)
1995	*5/81*	*2/8*	1	1	*2/2*	*9/93*	R: I(3), B(2) N: I(2) M: B(2)
Total	*13/188*	*5/16*	1	*1/21*	*2/5*	*21/231*	R: I(7), B(6) N: I(5) E: B(1) M: B(2)
Surveillance, complaint response							
1991	4	0	0	0	0	4	
1992	0	0	1	0	0	1	
1993	20	2	0	0	0	22	
1994	*3/36*	0	1	0	2	*3/39*	R: I(3)
1997	*1/4*	0	0	0	0	*1/4*	Mixed serovars
1999	*1/4*	0	*1/1*	1	0	*2/6*	R: I(1), H: I(1)
2000	17	0	6	0	0	23	
2001	2	0	0	0	0	2	
2003	20	2	0	0	0	22	
Total	*5/107*	4	*1/9*	1	2	*6/123*	R: I(4), H: I(1)
Total	***18/295***	***5/20***	***1/10***	***1/22***	***2/7***	***27/354***	**R: I(11), B(6)** **N: I(5)** **E: B(1)** **M: B(2)** **H: I(1)**

Table 3: Observed associations between maintenance hosts and serovars of pathogenic leptospires.

Animal host	Potentially associated serovars
Rattus spp.	Icterohaemorrhagiae** [24], ballum [41]
Mus musculus	Ballum [24], autumnalis** [46]
Sus scrofa	Australis**, pomona and grippotyphosa** [47], Pomona [48]
Herpestes auropunctatus	Autumnalis** [46]
Dogs	Grippotyphosa** [49]

**Patients had high titers for these serovars following the 2004 Manoa Flood.

an e-mail survey limited to the University of Hawaii campus reported symptoms of leptospirosis [23]. Serovars that reacted in blood serum samples from the two confirmed cases showed elevated titres to several serogoups, including *Australis, Autumnalis*, *Icterohaemorrhagiae*, and *Grippotyphosa* [23].

Buildings affected by the flooding were constructed prior to 1980. This is significant because profound changes in land use in the Manoa watershed took place in the latter quarter of the 20th century. This suggests that although buildings were constructed in a swath that could be inundated by overflow waters from the Manoa watershed, flooding of this type had not been observed or anticipated at the time the buildings were constructed. The hydrologic analysis indicates that expected watershed response would have been significantly different when the buildings and Woodlawn bridge were sited and constructed, especially in terms of expected peak discharge quantities and flow energies. Prior to completion of the buildings and the bridge, even a storm with a low probability of occurrence may not have led to the type of flooding and overflow that occurred as a result of the intense storm of 2004. The Woodlawn bridge was especially critical because it had the potential to obstruct flow at a point that could cause overflow to the watershed containing the University of Hawaii. Although the amount of flow in the 2004 flood was less than the design capacity of the bridge, debris (including uprooted trees and accumulated sediment) reduced the capacity of the bridge, which led to flooding as high-velocity and -volume flows were forced out of the stream channel.

In order for pathogenic microbes to enter surface waters during flooding, at least four conditions must be met [16]:

1. a source of microbes must be present,
2. microbes must survive environmental stresses,
3. volumes of rainfall must be sufficient to generate drainage and
4. flows from areas that have a source of pathogens and the quantity of flow must have sufficient volume and energy to transport microbes to mainstream channels.

Each condition represents a series of micro- and macroprocesses about which little information is available for spirochetes, especially in the context of the living habits of rodent species. From the hydrologic analysis it appears that watershed development, especially addition of impervious surfaces, ensured that conditions 3 and 4 were met in the Manoa stream watershed. With regards to rodent populations, development created suitable habitat and food sources for rodent populations in the watershed, enhancing the likelihood that the first condition would be satisfied. In fact, research indicates that endemic infection rates are most related to commensal animal population density rather than environment [41]. The mode of transport of spirochetes in water has not been investigated (e.g. sorbed or not sorbed to sediment), nor is sufficient information available about condition 2 for spirochetes. However, several authors have speculated that wet environments are likely to favor survival of spirochetes and both vertical and horizontal transmission of disease within animal populations [37, 42].

Sporadic rodent trapping demonstrated that infected rodents were present in the watershed in five of the eleven years for which results are available prior to 2004. Several of the trapped rodents (*Rattus rattus* (the black rat), *Rattus norvegicus* (the brown rat), and *Mus musculis* (domestic mouse)) are closely associated with human presence. The results demonstrate that when sampling is intensive two species (*R. rattus* and *R. norvegicus*) were hosts of the same serovars found in patients with confirmed cases of leptospirosis following the 2004 flood. In addition, rodents infected with serovar icterohemmoraghia were found at several locations within 100 meters of the stream channel (Fig. 6). Although proximity is not the sole determinant of transport, runoff from almost all paved surfaces and roof tops in the watershed is conveyed directly to the mainstem of Manoa stream by gutters, culverts and concrete-lined channels. In addition, other potential maintenance hosts for pathogenic *Leptospira* were present in the watershed, but not tested. These include domestic dogs and feral pigs (*Sus scrofa*). Dogs may be regularly vaccinated as a prophylaxis against infection. However, the feral pig population potentially is a reservoir for leptospirosis. Feral pigs can be found throughout the upland portions of the watershed, though no studies have been conducted to determine numbers or rates of infection.

Research related to ecological factors associated with rat population dynamics suggests three characteristics likely to increase risk of waterborne disease occurrence, with respect to the four conditions listed earlier. First, both *R. rattus* and *norvegicus* have lifespans of 2–7 months, rarely exceeding a year [43, 44]. This suggests that sexually mature individuals will reproduce at least once in the course of a year, producing immunologically naïve newborns that will become reservoir hosts. Second, their range is limited, suggesting that short-term fluctuations in numbers due to reproduction and mortality will lead to very localized changes in density. Third, researchers have observed yearly density and reproductive cycles in populations of *R. rattus*, *exulans* and *norvegicus* [43, 44]. On the Hawaiian islands, these cycles correspond closely with annual cycles of precipitation, such that localized populations are likely to reach their maximum densities at the onset of the wet-weather months of October–May. This suggests that the source of pathogenic leptospires reaches a peak at approximately the same time that abstraction losses during rainstorms reach a minimum and transport potential is highest.

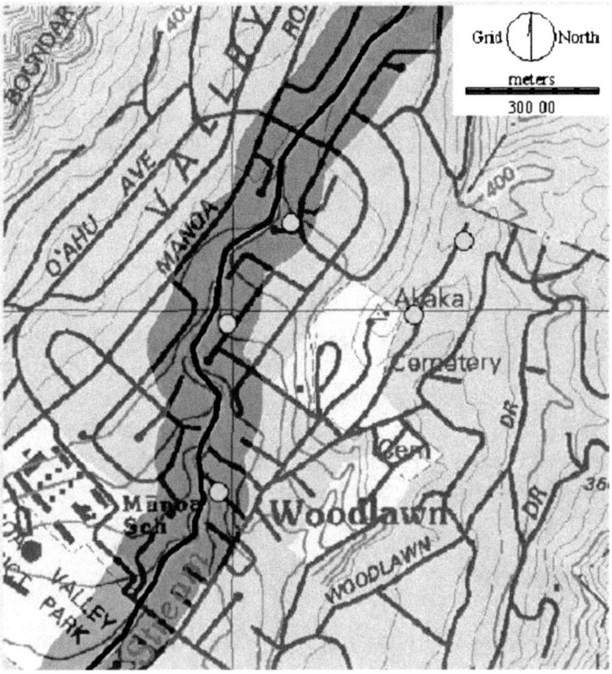

Figure 6: Detail of Manoa Stream watershed, showing 100 m buffer (shaded area adjacent to stream course, and selected sites where rodents that tested positively for serovar icterohemorraghia were captured.

The Manoa watershed offers an illustration of the model presented in Fig. 1, though data to support a functional version of the model are lacking. The available data indicate that development led to a change in hydrologic response that may have facilitiated transport of pathogens, resulting in an outbreak of leptospirosis caused by flooding. The enhanced hazard of flooding due to addition of impervious surface over the course of 65 years was likely accompanied by increased numbers of maintenance hosts of a microbial pathogen, which was important as an infectious agent during flooding in 2004. In this regard, the Manoa watershed may be a useful example for areas that are undergoing rapid expansion in previously undeveloped or lightly developed watersheds. As development progresses, populations of commensal animals, such as rodents, may increase significantly. This may be accompanied by proportional expansion of the types and amounts of microbial pathogens that may be transported by water. If addition of impervious surfaces significantly alters the hydrologic regime, increases in the magnitude of sources of pathogens may also increase the likelihood of zoonoses and horizontal transmission of disease to other populations, especially in downstream areas affected by flooding.

Significant anthropogenic change in watersheds can occur relatively quickly and have immediately noticeable effects. This is especially true in areas that are developing rapidly with influxes of industry and commerce, such as many of the nations in Asia and the Asian subcontinent. In such areas, significant alterations of hydrologic regime may take place within a span of decades. This suggests that localized changes in hydrologic regime associated with development may be important to anticipate and manage, especially in areas where diseases that may be transmitted by water are endemic in host-animal populations.

Although flood control and infrastructure management and protection are critical elements of landscape planning during the early phases of development, the potential for synergism between emergence of endemic, waterborne zoonoses and flooding is unlikely to be emphasized. However, it is important to anticipate such links by characterizing maintenance host populations in advance of development (including cyclical trends in host population densities and prevalence of disease). When established, reservoir host-animal populations may be difficult to control without social and behavioral changes coupled with aggressive eradication programs [45]. If the potential for zoonosis associated with flooding is recognized prior to development, such information could be a critical component of planning that emphasizes downstream public-health protection.

Acknowledgements

The Hawaii Department of Health's Vector Control Division generously provided access to trapping records. The Natural Resources and Environmental Management Department of the University of Hawaii's College of Tropical Agriculture and Human Resources provided support for hydrologic analysis. The University of Hawaii's John A. Burn's School of Medicine provided significant support for analysis of rodent sampling results.

References

[1] Rose, J.B., Daeschner, S. Easterling, D.R., Curriero, F.C., Lele, S. & Patz, J.A., *Climate and waterborne disease outbreaks. Journal American Water Works Association*, **92(9)**, pp. 77–87, 2000.

[2] Few, R. & Matthies, F., (eds) *Flood Hazards and Health.* 1st edn, Earthscan: London. p. 219. 2006.

[3] White, M.D. & Greer, K.A., *The effects of watershed urbanization on the stream hydrology and riparian vegetation of Los Penasquitos Creek, California. Landscape and Urban Planning*, **74(2)**, pp. 125–138, 2006.

[4] Florsheim, J.L., Mount, J.F. & Rutten, L.T., *Effect of baselevel change on floodplain and fan sediment storage and ephemeral tributary channel morphology, Navarro River, California. Earth Surface Processes and Landforms*, **26(2)**, pp. 219–232, 2001.

[5] Kaneshiro, K.Y., Chinn, P., Duin, K.N., Hood, A.P., Maly, K. & Wilcox, B.A., *Hawaii's mountain-to-sea ecosystems: social-ecological microcosms for sustainability science and practice. Ecohealth*, **2**, pp. 349–360, 2005.

[6] Gubler, D.J., *Resurgent vector-borne diseases as a global health problem. Emerging Infectious Diseases*, **4(3)**, pp. 442–450, 1998.

[7] Smolinski, M.S., Hamburg, M.A. & Lederberg, J., (eds) *Microbial Threats to Health: Emergence, Detection, and Response*, National Academy Press: Institute of Medicine, Washington, DC. 2003.

[8] Patz, J.A., Graczyk, T.K., Geller, N. & Vittor, A.Y., *Effects of environmental change on emerging parasitic diseases. International Journal for Parasitology*, **30(12–13)**, pp. 1395–1405, 2000.

[9] Patz, J.A., Epstein, P.R., Burke, T.A. & Balbus, J.M., *Global climate change and emerging infectious diseases. Journal of the American Medical Association*, **275(3)**, pp. 217–223, 1996.

[10] Bengis, R.H., Leighton, F.A., Fischer, J.R., Artois, M., Morner, T. & Tate, C.M., *The role of wildlife in emerging and re-emerging zoonoses. Revue Scientifique Et Technique De L Office International Des Epizooties*, **23(2)**, pp. 497–511, 2004.

[11] Mahy, B.W.J. & Brown, C.C., *Emerging zoonoses: crossing the species barrier. Revue Scientifique Et Technique De L Office International Des Epizooties*, **19(1)**, pp. 33–40, 2000.

[12] Macpherson, C.N.L., *Human behaviour and the epidemiology of parasitic zoonoses. International Journal for Parasitology*, **35(11–12)**, pp. 1319–1331, 2005.

[13] Gajadhar, A.A. & Allen, J.R. *Factors contributing to the public health and economic importance of waterborne zoonotic parasites. Veterinary Parasitology*, **126(1–2)**, pp. 3–14, 2004.

[14] Slifko, T.R., Smith, H.V. & Rose, J.B., *Emerging parasite zoonoses associated with water and food. International Journal for Parasitology*, **30(12–13)**, pp. 1379–1393, 2000.

[15] Walker, M., Leddy, K. & Hagar, E., *Effects of combined water potential and temperature stresses on Cryptosporidium parvum oocysts. Applied and Environmental Microbiology*, **67(12)**, pp. 5526–5529, 2001.

[16] Walker, M., Montemagno, C. & Jenkins, M., *Source water assessment and nonpoint sources of acutely toxic contaminants: A review of research related to survival and transport of Cryptosporidium parvum. Water Resources Research*, **34(12)**, pp. 3383–3392, 1998.

[17] Jenkins, M.B., Anguish, L.J., Walker, M.J., Bowman, D.D. & Ghiorse, W.C., *Assessment of a dye-permeability assay for determination of inactivation rates of Cryptosporidium parvum oocysts. Applied Environmental Microbiology*, **63(10)**, pp. 3844–3850, 1997.

[18] Jenkins, M., Walker, M.J., Anthony, L.C., Bowman, D.D. & Ghiorse, W.C., *Use of a Sentinel System for Field Measurements of Cryptosporidium parvum Inactivation in Soil and Animal Waste. Applied and Environmental Microbiology*, **65(5)**, pp. 1998–2005, 1999.

[19] Levett, P.N., *Leptospirosis: re-emerging or re-discovered disease? Journal of Medical Microbiology*, **48(5)**, p. 417–418, 1999.

[20] Levett, P., *Leptospirosis. Clinical Microbiology Reviews*, **14(2)**, pp. 296–326, 2001.

[21] Karande, S., Bhatt, M., Kelkar, A., Kulkarni, M., De, A. & Varaiya, A., *An observational study to detect leptospirosis in Mumbai, India, 2000. Archives Of Disease In Childhood*, **88(12)**, pp. 1070–1075, 2003.

[22] Sehgal, S.C., Sugunan, A.P. & Vijayachari, P., *Outbreak of leptospirosis after the cyclone in Orissa. National Medical Journal Of India*, **15(1)**, pp. 22–23, 2002.

[23] Park, S.Y., Effler, P.V., Nakata, M., Sasaki, D., Katz, A.R., Clark T.A. & Gaynor, K., *Brief Report: Leptospirosis After Flooding of a University Campus – Hawaii, 2004. Morbidity and Mortality Weekly*, **55(5)**, pp. 125–127, 2006.

[24] Bharti, A.R., Nally, J.E., Ricaldi, J.N., Matthias, M.A., Diaz, M.M., Lovett, M.A., Levett, P.N., Gilman, R.H., Willig, M.R., Gotuzzo, E. & Vinetz, J.M., *Leptospirosis: a zoonotic disease of global importance. Lancet Infectious Diseases*, **3(12)**, pp. 757–771, 2003.

[25] Morshed, M.G., Konishi, H., Terada, Y., Arimitsu, Y. & Nakazawa, T., *Seroprevalence of Leptospirosis in a rural flood prone District of Bangladesh. Epidemiology And Infection*, **112(3)**, pp. 527–531, 1994.

[26] Kuriakose, M., Eapen, C.K. & Paul, R., *Leptospirosis in Kolenchery, Kerala, India: Epidemiology, prevalent local serogroups and serovars and a new serovar. European Journal of Epidemiology*, **13(6)**, pp. 691–697, 1997.

[27] Sarkar, U., Nascimento, S.F., Barbosa, R., Martins, R., Nuevo, H., Kalafanos, I., Grunstein, I., Flannery, B., Dias, J., Riley, L.W., Reis, M.G. & Ko, A.I., *Population-based case-control investigation of risk factors for leptospirosis during an urban epidemic. American Journal of Tropical Medicine and Hygiene*, **66(5)**, pp. 605–610, 2002.

[28] Ward, M.P., *Seasonality of canine leptospirosis in the United States and Canada and its association with rainfall. Preventive Veterinary Medicine*, **56(3)**, pp. 203–213, 2002.

[29] Miller, D.A., Wilson, M.A. & Beran, G.W., *Relationships between Prevalence of Leptospira Interrogans in Cattle, and Regional, Climatic, and Seasonal Factors. American Journal of Veterinary Research*, **52(11)**, pp. 1766–1768, 1991.

[30] Shimizu, M.M., *Environmental and Biological Determinants for the Prevalence of Leptospirosis among Wild Small Mammal Hosts, Island of Hawaii. International Journal of Zoonoses*, **11(2)**, pp. 173–188, 1984.

[31] Vinetz, J.M., Wilcox, B.A., Aguirre, A., Gollen, L.X., Katz, A.R., Fujioka, R., Maly, K., Horwitz, P. & Chang, H. *Beyond Disciplinary Boundaries: Leptospirosis as a Model of Incorporating Transdisciplinary Approaches to Understand Infectious Disease Emergence. Ecohealth*, **2(4)**, pp. 291–306, 2005.

[32] Dracup, J., Cheng, E.D.H., Nigg, J.M. & Schroeder, T., *The New Year's Eve Flood on Oahu, Hawaii: December 31, 1987 – January 1, 1988.* Washington, D.C.: National Academy Press: National Research Council, Committee on Natural Disasters, 72, 1991.

[33] NRCS, *NRCS National Engineering Handbook – Part 630: Hydraulics and Hydrology* N.R.C. Service, ed. U.S. Department of Agriculture, 2001.

[34] McCuen, R.H., *Hydrologic Analysis and Design.* 2nd edn, Upper Saddle River, N.J.: Prentice Hall, 1998.

[35] NOAA, *Climatological data: Hawaii and Pacific*, D. of Commerce, ed. National Oceanic and Atmospheric Agency, National Climatic Data Center, Asheville, North Carolina, 2004.

[36] Anderson, J., Hardy, E., Roach, J.T. & Witmer, R.E., *A Land Use And Land Cover Classification System For Use With Remote Sensor Data*. U.S. Government Printing Office: Washington, DC, p. 41, 1976.

[37] Higa, H.H. & Fujinaka, I.T., *Prevalence of Rodent and Mongoose Leptospirosis on Island of Oahu. Public Health Reports*, **91(2)**, pp. 171–177. 1976.

[38] Sulzer, C. & Jones, L., *Leptospirosis: Methods in Laboratory Diagnosis*, E.A.W. Health, Public Health Service, ed. U.S. Department of Health, Education and Welfare: Washington, DC, 1976.

[39] Perry, G., *A Scientific Review of Leptospirosis and Implications for Quarantine Policy*. Biosecurity Australia: Agriculture, Fisheries and Forestry: Canberra, AU. p. 115. 2000.

[40] Daszak, P., Tabor, G.M., Kilpatrick, A.M., Epstein, J. & Plowright, R., *Conservation medicine and a new agenda for emerging diseases*, in *Impact of Ecological Changes on Tropical Animal Health and Disease Control*. pp. 1–11, 2004.

[41] Hathaway, S.C. & Blackmore, D.K., *Ecological Aspects of the Epidemiology of Infection with Leptospires of the Ballum Serogroup in the Black Rat (Rattus-Rattus) and the Brown-Rat (Rattus-Norvegicus) in New-Zealand. Journal of Hygiene*, **87(3)**, pp. 427–436, 1981.

[42] Barwick, R.S., Mohammed, H.O., McDonough, P.L. & White, M.E., *Epidemiologic features of equine Leptospira interrogans of human significance. Preventive Veterinary Medicine*, **36(2)**, pp. 153–165, 1998.

[43] Glass, G.E., Korch, G.W. & Childs, J.E., *Seasonal and Habitat Differences in Growth-Rates of Wild Rattus-Norvegicus. Journal of Mammalogy*, **69(3)**, pp. 587–592, 1988.

[44] Tamarin, R.H. & Malecha, S.R., *The population biology of Hawaiian rodents: demographic parameters Ecology*, **52(3)**, pp. 383–394, 1971.

[45] Lambropoulos, A.S., Fine, J.B., Perbeck, A., Torres, D., Glass, G.E., McHugh, P. & Dorsey, E.A., *Rodent Control in Urban Areas: An Interdisciplinary Approach. Environmental Health*, **61(6)**, pp. 12–17, 1999.

[46] Matthias, M.A. & Levett, P.N., *Leptospiral carriage by mice and mongooses on the island of Barbados. West Indian Medical Journal*, **51(1)**, pp. 10–13, 2002.

[47] Heise-Pavlov, P.M. & Heise-Pavlov, S.R., *Feral pigs in tropical lowland rainforest of northeastern Australia: Ecology, zoonoses and management.* Wildlife Biology, **9(Suppl. 1)**, pp. 21–27, 2003.

[48] Mason, R.J., Fleming, P.J.S., Smythe, L.D., Dohnt, M.F., Norris, M.A., & Symonds, M.L., *Leptospira interrogans antibodies in feral pigs from New South Wales. Journal of Wildlife Diseases*, **34(4)**, pp. 738–743, 1998.

[49] Brown, C.A., Roberts, A.W., Miller, M.A., Davis, D.A., Brown, S.A., Bolin, C.A., JareckiBlack, J., Greene, C.E. & MillerLiebl, D., *Leptospira interrogans serovar grippotyphosa infection in dogs. Journal of the American Veterinary Medical Association*, **209(7)**, pp. 1265–1272, 1996.

CHAPTER 14

The Waiāhole Ditch: a case study of the management and regulation of water resources in Hawai'i

L.H. Miike
Commission on Water Resource Management,
State of Hawaii, Honolulu, HI, USA.

Abstract

The Waiāhole Ditch Contested Case was the first opportunity the Hawaii Commission on Water Resource Management and the Hawaii Supreme Court had to rule on and review the 1987 State Water Code's approach to the complex interrelationships between preservation versus use of the state's freshwater resources. Historically, Hawaii's streams were the linchpin around which land was apportioned and managed. Streams were essential for the diversion of large amounts of water to grow the staple food, wetland taro, and were an important source for the traditional and customary gathering of fish, crustaceans, and mollusks for food and cultural purposes. This uniquely Hawaiian historical balancing between preservation and use found its modern counterpart in the Commission's and Court's struggles to quantify the Water Code's definition of "reasonable and beneficial" offstream uses, as well as how to establish priorities between stream preservation/restoration and offstream uses. The Waiāhole Ditch Contested Case not only revealed the intellectual challenges the Commission and Court had to and continues to face in performing this balancing act, but also provides a "behind-the-scenes" glimpse into how legal rulings are made that provide guidance for future decision making.

1 The Waiāhole Ditch

James Campbell was a carpenter who became known as "Kimo Ona Milliona" (James the Millionaire) when he made his initial fortune on the island of Maui by cofounding the Pioneer Mill sugar company. In 1877, he purchased 41,000 acres in Honouliuli, west of Pearl Harbor on the island of Oahu. It was arid ranch land, with five acres supporting only one head of cattle, but Campbell hired an experimental well-driller from California, who hit the Pearl Harbor aquifer at a depth of 240 feet in September 1879.

Campbell's Honouliuli lands were sub-leased in two parcels: 1) lands below 200-feet elevation, which became Ewa Plantation, with its first sugar crop harvested in 1892; and 2) the lands above 200-feet elevation, which became Oahu Sugar Company, Ltd. (hereinafter, "OSCO"), with its first crop harvested in 1899; James Campbell Company; LLC [1].

Pumping groundwater to OSCO's higher-elevation lands was costly, so a subsidiary, Waiāhole Water Company (hereinafter, "WWC") was established to transport surfacewaters from the windward side of the Koolau mountains to OSCO's leeward sugar cane fields:

> The general plan provided for collecting water from the many streams and gulches on the windward side of Oahu by means of tunnels through the ridges or spurs, and conveying the water, after collecting, through the mountain in the main tunnel to the leeward side of the Island, thence by tunnels, ditches and pipes, to the upper levels of Oahu Sugar Plantation. The tunnels connect up to the various streams on the North (windward) side, and take in the water at the adits in the gulches; Kluegel [2].

The greatest difficulty in constructing the Waiāhole Ditch system turned out to be drilling the Main Bore from the windward to the leeward side of the Koolau Mountains:

> While it was suspected at the outset that considerable water might be encountered in the main bore through the mountain, it was not anticipated at the beginning that enough water would be developed to materially interfere with the progress of the excavation. This hope was not realized, however, for the main bore had proceeded only about 200 feet from the North portal (on the windward side) when water to the extent of two million gallons daily was developed – this on breaking through the first dyke.
>
> These dykes are hard, impervious strata of rock lying approximately at an angle of 45° to the tunnel axis, and nearly vertical, and they occur at intervals of varying length. Between the dykes was the porous water-bearing rock, thoroughly saturated, and with the water pent up between the dykes often under considerable pressure, so that when a dyke was penetrated, the water would spout out from the drill holes and would gush forth from the openings blasted in the headings ...
>
> The texture and hardness of the rock varied considerably – some of it being particularly soft and porous and much of it hard and flinty – particularly at the dykes. The dykes varied in thickness from 14 feet down to about 4 feet, but all the dykes were composed of very hard, close-grained rock that was apparently waterproof; Kluegel [3].

Most of the dike-enclosed water was contained in or near the windward side of the Main Bore. While boring had proceeded only about 200 feet from the windward side when the first dike was breached, boring from the leeward side proceeded for 10,518 feet before water began to pore under pressure from the drill holes.

> Peak flow from the Main Bore was reached in April 1915, when the average daily discharge was 43 mgd (million gallons per day). It then declined rapidly, so that by the end of 1916 the flow averaged only 8 mgd. Between 1918 and 1927, flow from the Main Bore had decreased further and averaged 6 mgd. Between 1989 and 1993, flow from the Main Bore averaged 5–3.7 mgd from the portion of the Main Bore leeward of the crest of the Koolau Range and 1.3 mgd from the windward portion …
>
> The dikes in the area of the Waiāhole Ditch's tunnels are vertical or nearly vertical and run parallel to the Koolau Range. The Main Bore was driven perpendicular to the dikes, piercing the water-laden basalt between the hard rock of the dikes. The North Division tunnels were dug to channel water from the windward streams from Kahana Valley (at 790 feet elevation) south to Waiāhole Valley, where it met the Main Bore at the North Portal (at 753 feet elevation). This collection system runs parallel to the dikes and thus itself generates little if any dike-impounded groundwater; Kluegel [4].

After discovery of the large amounts of dike-contained water, six other tunnels were drilled perpendicular to the dikes. In 1915, two were drilled within 2000 feet of the Main Bore and on either side, but one was dry from the beginning and the other dried up in three months, the conclusion being that the Main Bore had drained the rocks for more than 2000 feet on either side.

Four other tunnels dug further away were successful in developing large amounts of water. In Waikāne Valley, two valleys north of Waiāhole Valley, Waikāne Tunnel 1 was started in 1925 and completed in 1927. It was 2635 feet long and pierced about 260 dikes from a few inches to 12 feet in width. Waikāne Tunnel 2 was started in 1927, completed in 1929, and was 2342 feet long. Kahana Tunnel in Kahana Valley was started in 1929, completed in 1931, and was 1975 feet long. In Uwau Valley, immediately north of Waiāhole Valley, Uwau Tunnel was started in 1932, completed in 1935, extended another 220 feet between 1964 and 1965 (177 feet of which extended leeward of the crest of the Koolau Mountains), and had a final length of 2500 feet; Miike [5].

As the Ditch system developed more dike-impounded water, the surfacewater collections were decreased. One of the last of these, 1 to 1.5 mgd pumped from Waiāhole Stream at 450 feet elevation, ceased in 1982 because of the cost of pumping, leaving only 2.1 mgd diverted from Kahana Stream as the last surfacewater diversion remaining. The Kahana Tunnel was also bulk-headed in 1992 to act as a dike-impounded reservoir to improve management of the water resource; Miike [6].

Currently, the Waiāhole Ditch flows average 27.0 mgd from the following sources; Miike [7]:

Kahana Stream:	2.1 mgd
Kahana Tunnel:	1.1 mgd (Originally 2.6 mgd, but bulk-heading toward the back of the tunnel decreased but did not completely cease flows)

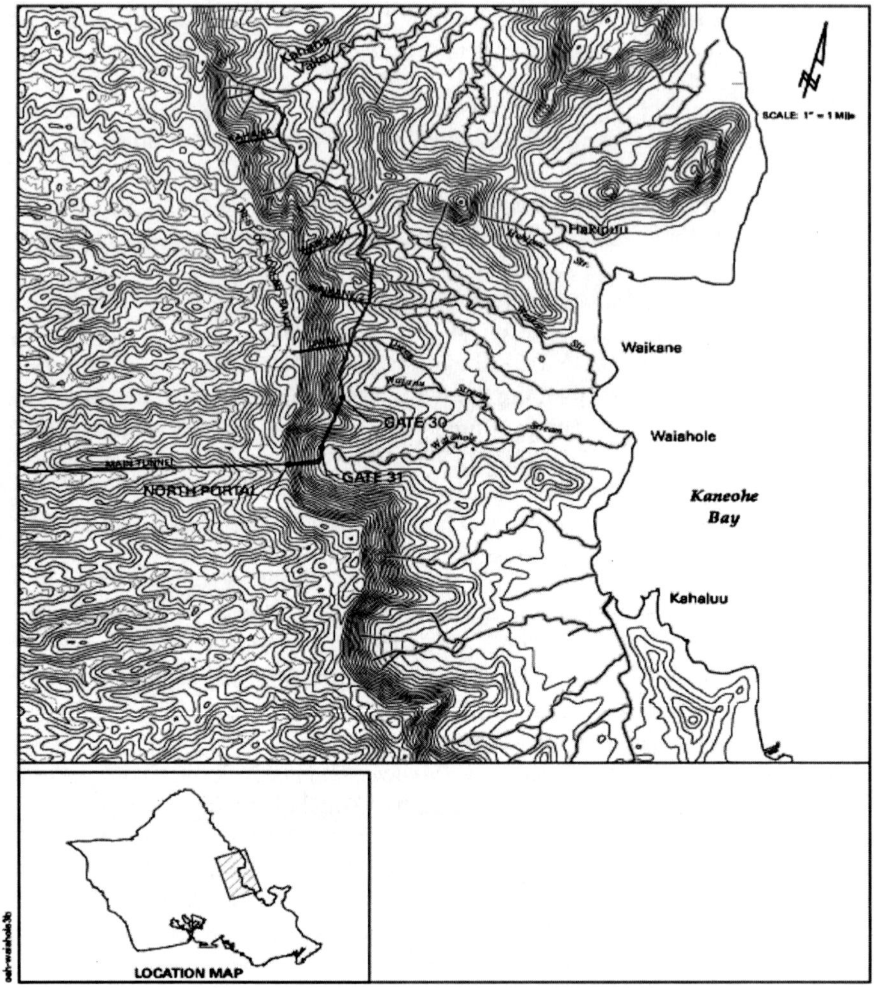

Figure 1 Windward side of the Waiāhole Ditch; Miike [9].

Waikāne Tunnel #1: 4.2 mgd

Waikāne Tunnel #2: 1.1 mgd

Uwau Tunnel: 13.5 mgd (8.7 mgd from the windward portion and 4.8 mgd from the leeward portion. The 1960s extension developed a net of only 2.8 mgd over the amount developed by the original tunnel, because about 2 mgd of the 4.8 mgd from the extension was previously already finding its way into the original tunnel.)

From Uwau to the crest of the Koolau: 1.3 mgd (windward portion of the Main Bore)

From the crest to the South Portal:	3.7 mgd (leeward portion of the Main Bore)
Total:	27.0 mg

The length of the original Waiāhole Ditch system before the four development tunnels were added was 26.53 miles, with the length of the open ditch where the tunnel emerges on the leeward side at 20,000 feet or 3.79 miles; Miike [8]. The windward side of the Waiāhole Ditch system is depicted in Fig. 1.

The Ditch was planned to deliver 125 mgd, but the resulting system could carry only a maximum of 100 mgd, and the actual amount averaged only 27–28 mgd of OSCO's average use of 140 mgd, the rest being provided by pumped groundwater from the Pearl Harbor and Ewa caprock aquifers; Miike [10].

2 Windward streams affected by the ditch system

2.1 Stream flows

The Ditch begins at 790 feet elevation in Kahana Valley and descends to 753 feet elevation in Waiāhole Valley, where the Main Bore transects the Koolau Mountains to the leeward side. There are four stream systems along this stretch: 1) Kahana; 2) Hakipuu; 3) Waikāne; and 4) Waiāhole and its tributary, Waianu (Uwau is in turn a tributary of Waianu). The Waiāhole Ditch decreased the flows of Kahana, Waikāne, Waianu (and its tributary, Uwau), and Waiāhole Streams. The four collection tunnels (Kahana, Waikāne 1 and 2, and Uwau) are at approximately 800 feet elevation, and the Ditch system has diverted all of the dike-impounded waters above 800 feet elevation that previously contributed to the flow of these streams. Leakage and overflow from dikes continue to contribute to stream flows at lower levels, with the exception of the lower parts of Hakipuu Stream; Takasaki & Mink [11].

The "base flow" is the groundwater contribution from springs and seeps and represents the stream's flow during prolonged dry periods, when contributions from rain and runoff are at a minimum (see Fig. 2). Average flows include rain and runoff.

Waiāhole, Waianu, and Waikāne Streams lie entirely in the dike complex, and their combined base flows are only about one-third of the flows coming out of Uwau, Waikāne 1 and Waikāne 2 Tunnels, which are above these streams. Kahana Stream lies only partly in the dike complex, and its base flow is about three times the flow that was developed by the Kahana Tunnel before it was bulk-headed; Takasaki & Mink [13]. Hakipuu Stream is not affected by the Ditch, as it originates well below the crest of the Koolau Mountains, and most of the stream is below 400 feet elevation (see Fig. 1).

The point of maximum base flow (below which there are no more groundwater contributions) for the streams affected by the Waiāhole Ditch is as follows; Takasaki *et al.* [14]:

Waiāhole Stream:	At 80 feet elevation, its confluence with its main tributary, Waianu Stream, where the long-term base flow is 3.9 mgd and the average flow is 6.9 mgd.

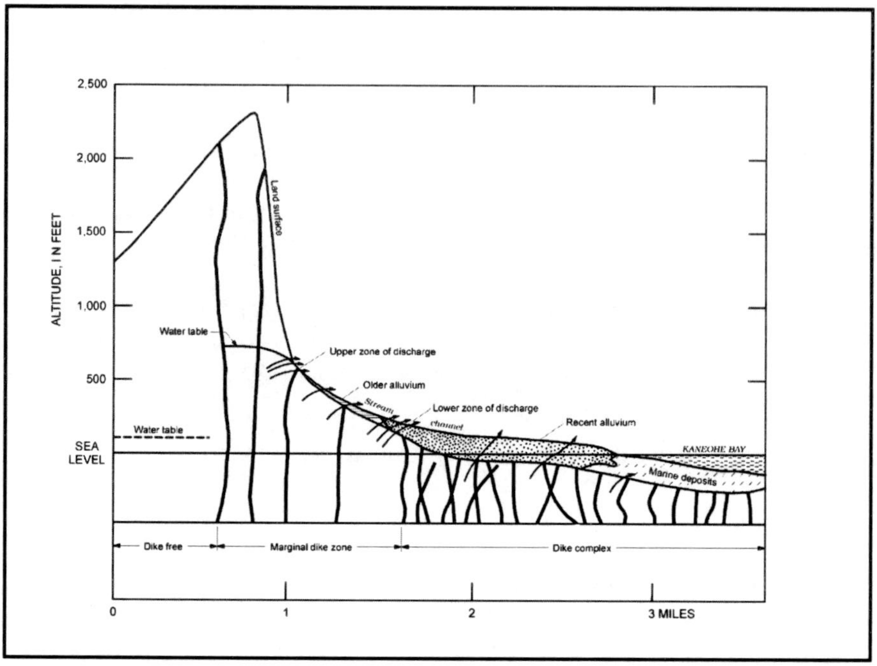

Figure 2: Base flow contribution to a windward stream; Takasaki & Mink [12].

Waianu Stream:	Also at 80 feet elevation at its confluence with Waiāhole Stream, where the long-term base flow is 0.5 mgd and average flow is 1.2 mgd.
Waikāne Stream:	At 75 feet elevation, where the long-term base flow is 0.5 mgd and average flow is 1.2 mgd.
Kahana Stream:	At 15 feet elevation, where the long-term base flow is 11.2 mgd and average flow is 29.5 mgd.

These base and average flows are from the U.S. Geological Survey, which uses multiple-year data and specifically, the period July 1, 1926, to June 30, 1960, Takasaki *et al.* [15]. The base flow, or index of reliable flow, is the flow equaled or exceeded 90 per cent of the time; Takasaki *et al.* [16].

On the windward side, the distance from the cliff faces to the shore is short, and streams are straight, steep, and flashy, rising and falling rapidly. Fig. 3 summarizes the flow of Waikāne Stream in the two-year period, 2000–2001. Daily mean flow varied from a low of 1.5 cubic feet per second (cfs) to a high of 114 cfs (1 cfs is approximately 0.65 mgd). Note that the lowest flow during this two-year period was twice the U.S. Geological Survey's base flow of 0.5 mgd. For the period 1960–2001, the base flow of Waikāne Stream was 1.37 mgd, compared to 0.5 mgd for the period 1926–1960; Taogoshi *et al.* [17].

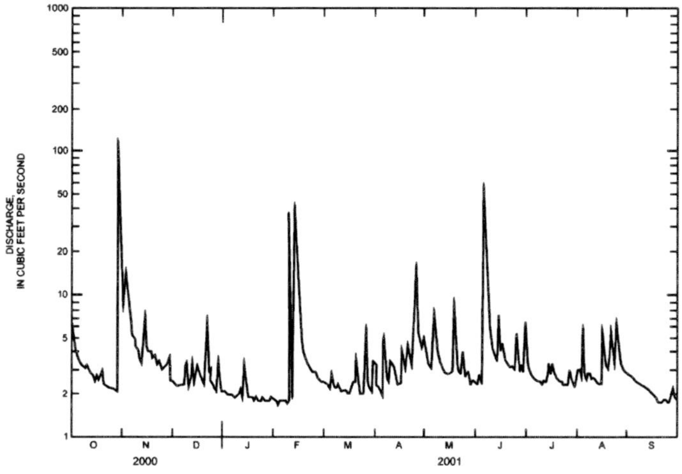

Figure 3: Waikāne stream at altitude 75 feet; Taogoshi *et al.* [18].

These streams' base flows before construction of the Waiāhole Ditch's development tunnels are not known, because no stream-flow gauging stations were in place prior to the Ditch's construction. However, geologists agree that in this windward area, the groundwater divide is shifted leeward, because the windward valleys penetrate deeply into the mountains and cut into the dikes. This causes proportionately more dike-impounded water to leak windward from the area underlying the crest; Takasaki & Mink [19]. While the location of this groundwater divide is not known, all of the development tunnels lie windward of the crest, except for parts of the Main Bore and the Uwau Tunnel extension. As summarized earlier, 3.7 mgd is developed by the part of the Main Bore leeward of the crest, and there was a net gain of 2.8 mgd when the Uwau Tunnel was extended into the leeward side. Note that the extension actually develops 4.8 mgd, but 2 mgd was already flowing windward into the original tunnel that was entirely on the windward side. Thus, the groundwater divide at the Uwau Tunnel is somewhere within the 177 feet of the extension into the leeward side. And of the 3.7 mgd developed in the Main Bore leeward of the crest, a portion probably would have flowed windward. Thus, of the 27 mgd developed by the Waiāhole development tunnels, between 20 to 25 mgd would have flowed windward prior to construction, and this would approximate the total decrease in flow of Waiāhole, Waianu, Waikāne, and Kahana Streams, with Kahana Stream being least impacted, because it is only marginally in the dike complex.

2.2 Stream ecology

> Little is known about the relationship between stream flows and the ecological systems of streams. Too many extraneous factors have been introduced to enable a one-to-one causal relationship between flow and stream viability. Watersheds have radically changed with the introduction of nonindigenous trees, shrubs, and grasses, altering the absorption characteristics of the watershed's soil and the amounts and patterns of water released into the streams. Introduced fishes and crustaceans not only compete for space and food with their indigenous counterparts, they also have brought with them parasites and other diseases. Yet some streams that have been severely degraded through reduced base flows, changes in their watersheds, and introduced aquatic species still support viable and thriving native species, while other comparable or even less degraded streams are nearly devoid of native life; Miike [20].

There are five fishes, two crustaceans, and two mollusks that are native to Hawaiian streams, and except for one fish, found nowhere else in the world (endemic). All are *amphidromous*:

> *Amphidromous* is illustrated by the spawning behavior of one of the fish, *Awaous guamensis*, known to the Hawaiians as *oopu nākea*. This fish inhabits the lower to middle reaches of the stream and migrates to estuarine areas nearer the ocean to spawn. After the eggs hatch, the fry are swept into the ocean, where they reside for about five months, at which time the juveniles return back to the stream, probably through cues associated with fresh water that cause the juveniles to orient shoreward, move into the stream mouth, and proceed upstream to the areas occupied by the adults. The return of the juveniles (which the Hawaiians called *hinana*) is not specific to the streams in which they were hatched, so streams devoid or severely depleted of *oopu* need not be stocked when they are restored (*citations omitted*); Miike [21].

Three of the fish, the two crustaceans (a shrimp and a prawn), and the two mollusks were important foods for the Hawaiians, and they had many sayings about them. For example the *hinana* were so numerous when they returned to the streams that they could be scooped up by hand, and the mollusks were referred to as "the fish of the stream that requires no bait (*ka ia maunu ole o ke kahawai*); Pukui [22]."

The impact of the streams probably also extends into their estuarine areas. Waiāhole (and its upstream tributary, Waianu) and Waikāne Streams empty into Kāneohe Bay (Kahana Stream empties into the ocean north of the Bay). These estuaries are nursery areas for a variety of ocean fish, and seaweeds, including edible varieties, thrive in the brackish waters. But, as with the impact of reduced stream flow on the ecological health of the streams, the effect of reduced stream flows on the health of Kāneohe Bay's estuaries is confounded by other factors such as "urbanization, overfishing, poaching, pollution, sediment run-off, dredging, sewage spills, algae blooms, growth of mangrove, development of culverts (cementing the sides of streams), and habitat degradation. All of these factors are interrelated and as such, efforts are needed to improve these conditions before fishing improves. The synergism of these factors is worse than the effects of any single factor (*citations omitted*); Miike [23]."

> Waiāhole and Waikāne Streams, situated at the northern end of Kāneohe Bay, nevertheless have a key advantage over the other streams that empty into the bay. Despite their reduced base flows and the altered flora and fauna of the streams and their watersheds, they are "pristine" compared to the streams farther to the south. Their watersheds are largely forested and have low population densities, devoid of large commercial districts and housing developments; the streambeds and banks are largely unaltered and have not been straightened and lined with concrete; and the shores of Kāneohe Bay at the northern end are cleaner and less densely populated than the shores to the south (*citation omitted*); Miike [24].

2.3 Historical and cultural significance

> Native Hawaians descend from a tradition and genealogy of nature deities: Wākea, Papa, Hoohoku-i-kalani, Hina, Kāne, Kanaloa, Lono and Pele, the sky, the earth, the stars, the moon, water, the sea, natural phenomena such as the rain and steam and from native plants and animals. Native Hawaiians today, inheritors of these genes and *mana*, are the *kino lau*, or alternative body forms of all their deities; D & O II [25].
>
> In Hawaiian mythology, the first *kanaka maoli* (Native Hawaiians), Hāloa, was the younger brother of the taro plant. The first child of Wākea, the sky father, and Hoohoku-i-kalani, his daughter by Papa, the earth mother, was stillborn and buried at a corner of Wākea's hut, where it sprouted into a taro plant. Wākea named the plant *hāloa-naka*, or "long-stalk-trembling," for the stirring of the taro leaf on its long slender stalk when the wind blew. A second child was born and named Hāloa, after the lengthy (*loa*) stalk (*hā*) of his older sibling; Miike [26].

These *moolelo*, or narratives about historical figures or events, must be taken from a Hawaiian and not western point of view, which focuses on whether or not the history is factually correct:

> It is rather in the intention than in the fact. Many a so-called *moolelo* that a foreigner would reject as fantastic nevertheless corresponds with the Hawaiian view of the relation between nature and man … Much that seems to us wildest fancy in Hawaiian story is to him a sober statement of fact as he interprets it through the interrelations of god with nature and with man; Beckwith [27].

But these differences between Hawaiian and western points of view do become problematic when the history of the origins of the Hawaiian people is introduced as evidence in legal proceedings. For example, in the litigation involving the Waiāhole Ditch (described below), witnesses for the Windward Parties, who were petitioning to restore the stream flows, had made the following claims:

> In Hawaiian history, the area of Windward Oahu from Kāneohe to Kualoa was the first land planned for creation by the gods Kāne, Kū and Lono. Kualoa was the land dedicated to Lono, god of fertility and agriculture. Waikāne Valley is located in this area, and therefore, restoration of the "Water of Kāne," Waikāne Stream, is vital to the restoration of the Hawaiians' spiritual and cultural image. Kāne is the chief deity among Hawaiian gods. The name Kāne is the male symbol for the procreative force. Native Hawaiians believe that the gods Kāne and Kanaloa especially looked for groundwater on Oahu in the region of Waikāne and Waiāhole Valleys in preparation for the coming of man. Waikāne is also considered a *puuhonua* (a place of refuge, asylum, place of peace and safety) for the district of Koolaupoko and worked in conjunction with the sacred land of Kualoa (*citations omitted*); D&O II [28].

But another version of the first lands planned by the gods identifies the island of Lānai; Beckwith [29], and there are at least five versions of the legend of creation involving the gods Kāne, Kū, Lono, and sometimes Kanaloa; Miike [30].

The mixture of Hawaiian and western viewpoints was nicely captured by the missionary William Ellis in 1823: "The general opinions entertained by the natives themselves, relative to their origins, are, either that the first inhabitants were created on the islands, descended from the gods, by whom they were first inhabited; or, that they came from a country which they called Tahiti ... The accounts they have of their ancestors having arrived in a canoe from Tahiti, are far more general and popular among the people; Ellis [31]."

Finally, the Hawaiian explanation for the origins of springs was that water, *wai*, was female and would gush out when Kāne, the male procreative force, thrust his spear into the ground. Kāne and Kanaloa were traveling companions and heavy *awa* (kava, a mild intoxicant) drinkers. When they stopped at places that had no water, Kāne would thrust his spear into the ground, and there are innumerable springs throughout the islands that are named after Kāne or Kanaloa, or both. Waikāne Valley is just one of the places on Oahu; Handy & Handy [32].

The historical/cultural aspect of the windward streams affected by the Waiāhole Ditch that is not confounded by "intention versus fact" is the gathering practices of Hawaiians. In the windward streams affected by the Ditch and the estuarine areas of Kāneohe Bay into which Waiāhole and Waikāne Streams emptied, an integral part of daily living was wetland-taro cultivation, gathering of freshwater fish, crustaceans and mollusks, various species of saltwater fish, crabs, and edible seaweed in the estuarine areas, as well as the use of the streams for drinking, bathing, and swimming; D&O II [33].

3 The Waiāhole Ditch contested case

3.1 Events leading to the contested case

In 1970, Ewa Plantation and OSCO merged into one company. In 1987, the State Water Code was enacted; Hawaii Revised Statutes [34]. Among its provisions were the establishment of Instream Flow Standards (IFS); Hawaii Revised Statutes [35] and triggers for designating ground; Hawaii Revised Statutes [36] and/or surface; Hawaii Revised Statutes [37], water-management areas, in which a water-use permit system would replace common-law water rights. The original bill would have designated the entire state as ground and/or surfacewater-management areas, but state-wide regulation of freshwater resources was resisted by the counties, so the designation system was devised as a compromise.

In April 1989, the Commission adopted "status quo" Interim Instream Flow Standards (IIFS) for all windward Oahu streams; i.e. the amount of water flowing in each stream on the effective date of this standard, which was May 4, 1992; Hawaii Administrative Rules [38].

On May 5, 1992, the Commission designated the five aquifer systems of windward Oahu as groundwater-management areas, effective July 15, 1992. After designation, existing users of groundwater would need a permit, but they would have superior status over new uses, because one of the conditions for a permit was that the proposed use of water would not interfere with any existing legal use; Hawaii Revised Statutes [39]. Applicants for existing-use permits have to apply within one year of the effective date of designation; Hawaii Revised Statutes [40].

OSCO's Waiāhole Ditch-management subsidiary, WWC, filed a combined water-use permit application (hereinafter, "WUPA") for all the existing users of the Ditch in June 1993. Then in August 1993, OSCO announced that it would cease its sugar operations by 1995.

In November 1993, the State Department of Agriculture (DOA) petitioned the Commission to reserve; Hawaii Revised Statutes [41], the present flow of the Ditch system for agricultural uses, to take effect upon OSCO's cessation of operations. The next month, three windward neighborhood associations, and later, the state Office of Hawaiian Affairs, petitioned to amend the IIFS of the windward streams affected by the Waiāhole Ditch. These four parties (as well as two other parties) had earlier petitioned to reserve water, but since their intention was to "reserve" water for the streams, the relevant section of the Water Code for this purpose was to petition to amend the IIFS.

Following an extensive series of meetings and attempts at mediation, in January 1995, the Commission ordered that a Contested Case Hearing (hereinafter, "CCH") be held on: 1) all related applications for water-use permits; 2) all related petitions to reserve water; 3) the petitions to amend the interim instream flow standards; and 4) any other matters related to the Waiāhole Ditch system.

Standing was granted to twenty-five parties and denied to nine parties. From May to November 1995, the Commission conducted six prehearing conferences, a field investigation, four hearings on existing uses, and six hearings on various motions.

3.2 Hawai'i water law prior to the Waiāhole decisions

3.2.1 Common-law rights

There are three types of common-law water rights in Hawaii: 1) appurtenant rights; 2) riparian rights; and 3) correlative rights. The former two are surfacewater rights, usually from streams, and the latter is a groundwater right and relatively more recently recognized because of its invisibility and inaccessibility until groundwater-pumping technologies were developed, as described previously on James Campbell's lands in 1879. Under the 1987 State Water Code, there is a process to designate surface- or groundwater-management areas when the water resources in an area may be threatened by existing or proposed withdrawals or diversions of water. A water-use permit system replaces common-law riparian and correlative rights in water-management areas, but appurtenant rights are preserved; Hawaii Revised Statutes [42]. Thus, once a water-management area is designated, unexercised riparian rights in a surfacewater-management area and unexercised correlative rights in a groundwater-management area are extinguished. However, riparian or correlative uses being exercised at the time of designation can qualify as "existing uses" if they meet the permit requirements. In nondesignated areas, all three types of common-law water rights are preserved.

3.2.1.1 Appurtenant rights "Appurtenant rights" was the label attached by westerners to water-management practices in the *ahupuaa* system of the Kingdom of Hawaii and its predecessors: "In the water law of Hawaii the term 'ancient appurtenant rights' include the right to drinking water and to water for other domestic purposes … Every portion of land, large or small, ahupuaa, ili, or kuleana, upon which people dwelt was, under this ancient Hawaiian system … entitled to drinking water for its human occupants and for their animals and was entitled to water for other domestic purposes; *Territory v Gay* [43]." (An *ahupuaa* was a division of land, typically extending from the seashore to the mountain, which was divided into strips called *ili*. A *kuleana* was a piece of land within an *ili* that the tenants lived on and cultivated for their own use.)

Because these ancient water uses were usually on lands not immediately adjacent to the streams that were their sources and were transported through ditches or *auwai*, the early Hawaii courts grouped "ancient appurtenant rights" with later, western concepts of water uses attaching to specific pieces of land that were not adjacent to the stream. Because western values that prevailed until relatively recently viewed freshwater as a commodity, exploitation of water was considered the highest and best use, and thus a superior right. This right was known as a "prescriptive" right: "A long or immemorial use of some right with respective to a thing so as to give a right to continue such use; Random House [44]."

The very first water-rights case before the newly constituted Hawaii Supreme Court set the tone that was to prevail for over 100 years that water could be owned through prescription, and that it was a superior right to the "natural" right to water from ownership of riparian lands (lands adjacent to the source of surfacewater): "(A) right to interfere with the natural right to make use of water belonging to another, when it is connected with the occupation of lands, constitutes an easement in favor of the latter, as the dominant estate; *Peck v Bailey* [45]."

Prescriptive rights no longer exist since 1973, when the Court ruled that the ownership of water in natural watercourses remained in the people of Hawai'i for their common good, and that title to water was not transferred with the land when private ownership of land was first authorized in the Kingdom of Hawaii; *Mcbryde v Robinson* [46]. The 1978 constitutional amendments also added "water" to the state's natural resources and proclaimed that "(a)ll public natural resources are held in trust by the State for the benefit of the people; Hawaii State Constitution [47]."

Prescriptive uses, however, including ground as well as surfacewater, may continue as long as the uses are reasonable and the rights of others (appurtenant, riparian and/or correlative rights) are not actually harmed; i.e. rightholders are not actually harmed until they try to exercise their rights and find that the remaining water is insufficient for their needs; Miike [48]. Such prescriptive (usually referred to now as "appropriated") uses are not based on any right but "are not prohibited as a matter of law. (D)iversions will be restrained only after a careful assessment of the interests and circumstances involved indicates a need for restraint; *Robinson v Ariyoshi* [49]." Note also that, if the area in which the prescriptive use is taking place is designated a (surface- or ground-) water-management area, it would qualify as an "existing use" under the Water Code and be issued a permit as long as the permitting conditions are met.

Appurtenant rights in the "ancient" sense, however, continue to exist and even enjoy superior status to riparian and correlative rights under the 1978 Hawaii constitutional amendments as reflected in the State Water Code. As noted earlier, appurtenant rights are preserved but riparian and correlative rights are not preserved in designated water-management areas.

An appurtenant right is a right to the amount of water used on a parcel of land at the time when private ownership was first obtained for that parcel of land – the Land Commission awards under the *Great Mahele* of 1848; *Territory v Gay* [50]. The right is to the amount of water being used at the time of the Land Commission awards, as shown by descriptions of the land at the time of the awards and approximation of cultivation methods at that time; *Reppun v Board of Water Supply* [51]. In designated surfacewater-management areas, appurtenant rights continue whether or not they are exercised, but when exercised for part or all of the water under those rights, they must apply for a water-use permit. Parties may petition the Commission to recognize their claims to appurtenant rights and the quantity of water under those rights; Hawaii Revised Statutes [52].

Prior to the enactment of the Water Code, which preserved appurtenant rights, such rights could be extinguished because it was based in the common law and did not have a statutory basis. For example, if the transferor of property that had appurtenant rights attempted to reserve the right in the deed, that reservation had been deemed invalid, because the right was appurtenant to the land. However, in severing the right from the land, the Court has ruled that the action extinguished the right; *Reppun v Board of Water Supply* [53]. Although no relevant cases have come before the Court since the 1987 enactment of the Water Code, appurtenant rights now have a statutory basis and probably cannot be extinguished, the right passing with the land regardless of the actions of the transferor. However, since statutes are rarely applied retroactively, any such attempts to sever the appurtenant right from the land prior to 1987 probably have extinguished the appurtenant right.

3.2.1.2 Riparian rights Lands adjacent to a surfacewater source such as a stream have riparian rights attached to them, in which owners of such lands have the right to the reasonable use of the streams' waters on those lands, but no one owns the water, and the rights of one riparian landowner are not superior to the rights of another riparian landowner. In the series of cases between 1867; *Peck v Bailey* [54], and 1930; *Territory v Gay* [55], that increasingly led to the legal doctrine that surface freshwater sources were private property, riparian rights clearly existed, although it was viewed as an inferior right to prescriptive rights. For example, in the 1867 case of *Peck v Bailey*, discussed earlier, the Court had explicitly referred to "the natural right to make use of water [56]" as being inferior to a prescriptive right. Yet in 1930, the Court had the temerity to conclude that "(r)iparian rights is not and never has been the law in Hawaii. It is utterly inconsistent with the system which from time immemorial has been recognized and enforced in these Islands; *Territory v Gay* [57]," referring to the ancient water practices that the Court equated with prescriptive rights.

In the 1973 case of *McBryde v Robinson*, cited earlier, the Court reinstated riparian rights by finding a statutory basis for it in an 1850 statute; Miike [58].

The 1973 *McBryde* court had identified the riparian right as "to the natural flow of the stream without substantial diminution and in the shape and size given it by nature," but in 1982, citing changing needs and circumstances, the Court endorsed riparian uses that affected the stream by holding that "in order to maintain an action for a diversion which diminishes the quantity or flow of a natural watercourse, a riparian owner must demonstrate actual harm to his own reasonable use of those waters; *Reppun v Board of Water Supply* [59]." Thus, the exercise of a riparian right must be reasonable and not actually harm the reasonable use of those waters by other riparian and appurtenant landowners.

Water transferred to other than riparian or appurtenant lands, whether or not conducted by the riparian landowner, is a prescriptive use – what the Court has now relabeled an "appropriated use" – that is allowed and is subject to the same restrictions described earlier.

3.2.1.3 Correlative rights Recall that in 1930, the Court in *Territory v Gay* had denied the existence of riparian rights, or the shared rights of riparian landowners, under which no one owned the waters. Curiously, the year before, the same Court had made an opposite ruling in a case involving claims of prescriptive rights over groundwater. The Court had established in two earlier cases in 1884; *Davis v Afong* [60], and 1896; *Wong Leong v Irwin* [61], that "(s)ubterranean waters, to be the subject of rights, must, like surfacewaters, in general flow in known and defined channels; *Wong Leong v Irwin* [62]." Thus, when faced with a groundwater dispute, the Court in 1929 considered three existing common-law doctrines and adopted what was known as "the rule of correlative rights": "all of the owners of lands under which lies an artesian basin have rights to the waters of that basin; that each may use water therefrom as long as he does not injure thereby the rights of others and that in times when there is not sufficient water for all each will be limited to a reasonable share of the water; *City Mill Co. v Hon. S&W Com.* [63]."

As with riparian rights: "(C)orrelative rights, however, extend only to uses on lands overlying the water source. Parties transporting water to distant lands are deemed mere 'appropriators,' subordinate in right to overlying landowners ... (T) he correlative rights rule grants overlying landowners a right only to such water as necessary for reasonable use. Until overlying landowners develop an actual need to use groundwater, nonoverlying parties may use any available 'surplus' (*citations omitted*); *Waiāhole I* [64].

3.2.2 The State Water Code

As explained earlier, the State Water Code's water-use permits replace the common-law's riparian rights in designated surfacewater-management areas and correlative rights in designated groundwater-management areas. Thus, in nondesignated areas, even though the Commission is the jurisdiction in which disputes involving riparian or correlative rights are adjudicated, the Commission has to follow the common law, not the regulatory, permitting conditions of the Code. However, the Commission has statewide jurisdiction, whether or not in designated water-management areas, over: 1) instream flow standards; 2) stream diversions and stream alterations through a permitting process; and 3) well constructions and pump installations, also through a permitting process. For example, in surfacewaters such as a stream, a new or expanded diversion for riparian or nonriparian uses would require a stream diversion permit and perhaps a stream alteration permit, and if the diversion would reduce the existing instream flow standard, a successful petition to amend the instream flow standard would have to be made. For groundwater, well-construction and pump-installation permits insure that the integrity of the aquifer being tapped is maintained. If a well is close to a stream, pumping tests may also include whether or not the stream flow is affected, because there might be a ground/surfacewater interaction.

3.3 The contested case and Hawai'i Supreme Court reviews

The CCH began on November 9, 1995, and ended on September 20, 1996, encompassing fifty-two days of hearings, including four evening sessions, and three days of closing arguments. There was written testimony from 161 witnesses, of which 140 orally testified. There were 567 exhibits introduced into evidence to accompany the witnesses' testimonies, and the parties submitted 2997 proposed Findings of Fact to be accepted or rejected by the; Commission D&O I [65].

The Commission issued its Decision and Order on December 24, 1997; D&O I [66]. It was immediately appealed by several parties, and the Hawaii Supreme Court issued its first decision in 2000, remanding several issues back to the Commission; *Waiāhole I* [67]. The Commission issued its second decision on December 28, 2001;

D&O II [68], which several parties again appealed. The Court issued its second decision in 2004; *Waiāhole II* [69], remanding some issues back to the Commission. In July 2006, the Commission issued its third decision; D&O III [70].

The dialogue between the Commission and the Court as presented in their respective decisions was unusual for what are usually dispassionate and detached documents. In no small part, the initial reaction of the Court, in its first decision, *Waiāhole I*, was influenced by two factors: 1) the public comments of the governor in support of leeward agricultural users while the Commission was conducting its hearing; and 2) the personal interest the Attorney General took in the case, first appearing in person to argue on behalf of the DOA at the Closing Arguments, and then firing the Assistant Attorney General serving the Commission before the final Decision and Order was issued and against the protests of the Commission. The Attorney General's actions, in particular, eventually led the Court to stop just short of accusing the Commission of being unduly influenced: "These eleventh hour developments, while falling short of a constitutional violation, strongly suggests that improper considerations tipped the scales in this difficult and hotly disputed case … Notwithstanding our feelings of unease regarding the circumstances under which the Commission rendered its final decision, our assessment of the totality of the circumstances prevents us from concluding that the aforementioned conduct constitutes a violation of (the windward parties') due process rights; *Waiāhole I* [71]."

Even if we set aside the contentious and controversial context in which the Waiāhole Ditch decisions was made, the legal doctrines at issue were themselves controversial. The Waiāhole decisions involved: 1) the first major Commission decision to interpret the application of the Water Code, the common law, and their possible interrelationships; and 2) the first major Court review of water law since the 1987 State Water Code partially but not totally replaced the common-law doctrines that had developed since 1867; *Peck v Bailey* [72]. In this summary review of the Waiāhole Ditch controversy, it would be distracting to discuss all of the issues addressed by the Commission and Court; Miike [73]. Thus, the focus will be on the key issues that have emerged and that have significant implications for how the Commission will address its future responsibilities and how the state's future water policies might be influenced.

3.3.1 Public-trust doctrine

> The public-trust doctrine is a court-created doctrine that arose from judicial review of executive or legislative management of the public's natural resources, particularly tidal or navigable waters, and now fresh waters … The purpose of the pubic trust has evolved with the changing public values and uses of water and navigable water ways. Originally applied to navigation, commerce, and fisheries, the public-trust doctrine now includes public values related to recreation, scenic value, the scientific study of the natural ecology, and environmental protection; D&O I [74].

In its initial Decision and Order, the Commission had relied on the public-trust doctrine as a legal authority in addition to the State Water Code, and on appeal, several parties argued that the Water Code "subsumes and supplants whatever common-law doctrine of public trust may previously have existed in Hawaii; *Waiāhole I* [75]." The Court disagreed, concluding that, while the Water Code displaced common-law rules on water law in water-management areas, the public-trust doctrine's basic premise is that the state has certain powers it cannot legislatively abdicate; *Waiāhole I* [76].

The Court then went on to identify the scope and purposes of the public-trust doctrine. First, it extended the scope to groundwater: "(J)ust as ancient Hawaiian usage reflected the perspectives of that era, the common law distinctions between groundwater and surfacewater developed without regard to the manner in which both categories represent no more than a single integrated source of water with each element dependent upon the other for its existence ... Few cases highlight more plainly its diminished meaning and utility than the present one, involving surface streams depleted by groundwater diversions and underground aquifers recharged by surfacewater applications (*quotation marks and citations omitted*); *Waiāhole I* [77]."

Second, the Court identified additional purposes of the public-trust doctrine: 1) resource protection, as a logical extension of the increasing number of public-trust uses of waters in their natural state; 2) domestic uses, particularly drinking, of the general public; and 3) what the Court deemed the trust's "original intent"–the exercise of Native Hawaiian and traditional and customary rights as a public-trust purpose, including appurtenant rights; *Waiāhole I* [78]. (In a subsequent 2004 decision, the Court identified a fourth purpose, reservations of water for Hawaiian homelands; *Wai ola o Molokai* [79].)

The Commission had identified the duty to protect public water resources as "a categorical imperative and the precondition to all subsequent considerations; D&O I [80]," but the Court disagreed. It first noted that the state water resources trust "embodies a dual mandate of 1) protection and 2) maximum reasonable and beneficial use; *Waiāhole I* [81]." It then concluded that the object is "the most equitable, reasonable, and beneficial allocation of state water resources, with full recognition that resource protection also constitutes 'use'... Given the diverse and not necessarily complementary range of water uses, even among public-trust uses alone, we consider it neither feasible nor prudent to designate absolute priorities between broad categories of uses under the water resources trust. Contrary to the Commission's conclusion ... we hold that the Commission inevitably must weigh competing public and private water uses on a case-by-case basis ... (A)ny balancing between public and private purposes begin(s) with a presumption in favor of public use, access, and enjoyment; *Waiāhole I* [82]."

Despite the Court's pronouncement, it is difficult to see how the Commission cannot treat resource protection as "a categorical imperative and the precondition to all subsequent considerations," because of the Water Code's provisions on aquifers and streams. In designated groundwater-management areas, the sustainable yield – or the amount that can be withdrawn without endangering the aquifer – has to be established before any water-use permits can be issued, and in both designated and nondesignated areas, withdrawal of water under the permit, correlative right, or appropriated use must meet well-construction and pump-installation permit requirements that protect the integrity of the aquifer. For streams, the instream flow standards must be established first before the amount of water available for offstream diversions can be identified. In fact, in the Waiāhole Contested Case, the Court had admonished the Commission when it mistakenly concluded that the Commission had issued water-use permits before setting the interim instream flow standards for the affected windward streams; *Waiāhole I*; D&O II [83].

The Court's inclusion of "domestic uses, particularly drinking, of the general public" also raises implementation questions. "Domestic uses" as defined in the State Water Code ("any use of water for individual personal needs and for household purposes such as drinking, bathing, heating, cooking, noncommercial gardening, and sanitation." HRS §174C-3) refers to uses of individual households and is exempt from the permit requirements [84]. The Court's reference is to municipal and other public water systems – in contrast to a private water system for a development, which is also for domestic uses but not of the "general public" – but such systems also serve industrial and commercial customers. As noted below, the Commission may prioritize among its applicants for water-use permits, and municipal use would be high on the list, but operationally, how would the Commission distinguish among the component uses of public water systems when applying the presumption test?

Finally, the remaining two public-trust purposes that the Court has dictated – traditional and customary rights, including appurtenant rights, and reservations for Hawaiian homelands – also raise implementation issues. The exercise of an appurtenant right presumably would have to be done in a traditional and customary manner if it is to be considered a public-trust purpose. Otherwise, commercial uses of appurtenant rights would be a protected public-trust use. But appurtenant rights are clearly superior to other rights under the State Water Code, and this superiority also applies to commercial uses of appurtenant rights as long as that use is reasonable and beneficial. As for reservations for Hawaiian homelands, the Water Code also subjects all water-use permits to the rights of the Department of Hawaiian homelands, and in nondesignated areas, arguably, such rights might be considered by the Court as falling under Native Hawaiian rights. Perhaps the potential impact of including traditional and customary appurtenant rights (i.e. manual cultivation of wetland taro) and water reservations for the Department of Hawaiian homelands is that resource protection enjoys no presumption against these two trust purposes. Realistically, however, rather than harm the resource to meet these other trust purposes, the Commission has the power under its permitting and instream flow standards authorities to reduce nontrust water uses.

3.3.2 Common-law rights

In its initial decision, the Commission had concluded that it had no regulatory authority for the 2.1 mgd of Kahana Stream water that was being diverted into the Ditch, because the area had only been designated a groundwater – but not surface-water – management area. Prior to enactment of the 1987 Water Code, the Court had ruled in the 1982 case of *Reppun v Board of Water Supply* that wetland-taro growers with riparian and appurtenant rights/uses could sue for damages and restoration of stream flows against the Board of Water Supply, which had dug development tunnels into the high-level, diked waters that contributed to the headwaters of the stream. The Board of Water Supply had argued that, regardless of the effect on the stream, it was diverting groundwater, not surfacewater. The Court disagreed, concluding that "where surfacewater and groundwater can be demonstrated to be physically interrelated as parts of a single system, established surfacewater rights may be protected against diversions that injure those rights, whether the diversion involved surfacewater or groundwater [85]." In *Waiāhole I*, the Court extended this reasoning by finding that where there is an "undisputed direct interrelationship between the surface and groundwaters ... designation of ... a groundwater-management area subjects both ground and (common law) surfacewater diversions from the designated area to the statutory permit requirement [86]." Note that in the 1982 case, the groundwater withdrawals were diminishing the stream's flow. But with the Waiāhole Ditch, the Kahana Stream diversions were not affecting the high-level, diked groundwater; in fact, it was the groundwater withdrawals that were affecting Kahana Stream. But the Court did not mention a cause-and-effect relationship; only that there was an undisputed direct interrelationship.

3.3.3 Water-use permit requirements

In *Waiāhole I*, the Court: 1) required greater scrutiny of the proposed uses under water-use permit applications; 2) added a "practical alternative" analysis requirement; and 3) concluded that the Commission could prioritize among different uses by imposing different standards and conditions. The Commission had been using county-based standards for irrigation water budgets, but the proposed water uses of large diversified agricultural farmers, small family plots, tree and house-plant nurseries varied widely. Thus, the Commission attempted to tailor the per-acre water budgets according to crop type, and the Court in *Waiāhole I* and *II* paid particular attention to the water practices of the different crops. As a consequence, the Commission is now attempting to develop more area- and crop-specific information, such as area-rainfall data and nearby comparable farming practices Commission on; Water Resource Management [87].

The Court based its requirement of the absence of practical alternatives as "intrinsic to the public trust, the statutory instream use protection scheme, and the definition of 'reasonable-beneficial' use (*citations and footnote omitted*); *Waiāhole I* [88]." However, the Court had identified the Pearl Harbor aquifer – a potable source – as possible alternatives for the Campbell Estate and another leeward water-use applicant, so the Commission noted that its "stated policy is to reserve potable groundwater for its highest and best use, domestic use, replacing it when appropriate for irrigation purposes with reclaimed or nonpotable groundwater," and concluded that "if water from the ... Pearl Harbor aquifer were to replace Ditch water ... windward public-trust resources that are available for nontrust purposes after measures have been taken to enhance those windward public-trust resources, would be given priority over a leeward public-trust resource; D&O II [89]." The Commission had reached this conclusion because it had interpreted the Court's order as requiring that potable water from the Pearl Harbor be used in place of nonpotable

Ditch water if it was practically available. However, in *Waiāhole II*, the Court clarified its order: "Considering whether alternative water resources are practicable innately requires prioritizing among public-trust resources [90]." The Commission therefore concluded in D&O III that the Waiāhole Ditch water and the Pearl Harbor aquifer water were practicable alternatives for each other and that "the highest use for Ditch water is for agricultural uses, while the highest use for (Pearl Harbor) Aquifer water is for potable purposes [91]."

Finally, the Commission had applied different standards and required different conditions when issuing a permit for a proposed golf course, whose owners contended that it was an agricultural use. The Court agreed with the Commission that "golf course irrigation raised different policy considerations than those uses typically associated with 'agricultural use'; *Waiāhole I* [92]" and that "such measures lay squarely within the Commission's appointed function of weighing and negotiating competing interests in regulating the water resources of this state; *Waiāhole I* [93]."

3.3.4 (Interim) instream flow standards

As anticipated, amending the instream flow standards proved to be the most difficult and contentious issue facing both the Commission and the Court. One of the interim orders that the Commission had issued prior to its first evidentiary hearings was to release 12 mgd into Waiāhole Stream and 2 mgd into Waianu Stream in order to avoid waste because of less use on the windward side as OSCO phased out of operations. The Commission had estimated that the 12 mgd released into Waiāhole Stream was about 150 per cent higher than preditch flow at the headwaters of the stream, and that the 2 mgd released into Waianu Stream was about 50 per cent of pre-Ditch flow at its headwaters. These additions allowed for preliminary studies of the impact of increased flows and instream studies of varying intensity, ranging from one day to monitoring over several months, were conducted by six scientists. The most that could have been expected of these studies would have been qualitative conclusions, and indeed, that is what most of the scientists reached:

> Font concluded that the water releases had reduced populations of exotic fishes and absolute numbers of fish parasites, and that it appeared likely that in time, abundances of parasites in native gobioid fishes would also increase. Devick concluded that substantial recruitment of all five native *oopu* species, along with the native *opae*, had occurred. Brasher concluded that the stream had a habitat suitable for native organisms, such as *oopu*, opae, and *hihiwai*. Kido concluded that the stream appeared to be in transition, which should translate into increases in existing biotic components. Hodges concluded that the increased flow would directly address the factors that have limited native macrofaunal abundance. And Englund concluded that the increased flow could improve habitat quality and displace introduced fish that serve as vectors for parasites; D&O II [94].

Three of these scientists, however, went on to reach quantitative conclusions on stream flows that they themselves contradicted on cross-examination:

> Two of these scientists, Hodges and Brasher, also stated that the added flows were the minimum required to improve the stream, while Englund recommended no further decreases in the releases. However, all three also made statements that directly contradicted these conclusions: 1) Hodges stated that there was no mathematical relationship yet developed for any stream in Hawaii between the density and/or abundance of native stream animals and the amount of water flowing through the stream; 2) Brasher stated that her one-day survey was not intended to be a comprehensive study but rather a reconnaissance survey, and that studies of at least two years, and perhaps up to five years, were needed to begin to evaluate the impact of changes in stream flow regimes; and 3) Englund stated that it didn't make biological sense to continue to put all the collected water of three major watersheds down just two separate stream channels, the Waiāhole and Waianu, and that flows should be adjusted once hydrologists have determined how much water the Waiahole Stream channel should normally hold during low base flow; D&O II [95].

A more realistic picture was painted by one of the experts:

> Besides the recommendation for long-term studies, Devick stated that restoration does not need to be an expansive effort to return a natural flow. Stream restoration is likely to be incremental through partial restoration of the original base flow. Restoration can take many forms, such as removal of a drainage pipe, replanting of riparian vegetation, removal of man-made alterations and the control or eradication of exotic species. Even small flow increases should be viewed as beneficial to the native biota, because those incremental improvements could not only become substantial with time but could also improve the knowledge base during the entire period, if appropriate simultaneous studies were undertaken; D&O II [96].

Personal testimonials had also been introduced into evidence by the windward parties, reflecting the opinions of long-time area residents of stream changes that had occurred in the early 1960s. The only construction that could have been related was the extension of the Uwau tunnel in 1964, which developed only an additional 2.77 mgd, all from the leeward side. And the cessation of pumping 1 to 1.5 mgd from the upper reaches of Waiāhole Stream in 1982 and the bulk-heading of the Kahana tunnel in 1992 would have increased the flows of these two streams compared to the 1960s. The testimonials also included Hakipuu Stream, which was not affected by the Ditch, and Punaluu Stream, north of Kahana Stream and outside the watersheds of the Ditch system:

> People in their fifties and sixties from Waiāhole, Waikāne, Hakipuu and Kahana recall that Waiāhole, Waikāne and Hakipuu Streams had clean, cold, year-round, swiftly flowing streams abundant with native stream life until there was a marked and qualitative decrease in the amount of water in the streams and in the *auwai*

> and aqueducts after 1962 and 1963. While all described abundance of stream life, stream flow and in some cases, stream watershed, during their childhood days, none used descriptions of the change that could be characterized as being gradual and imperceptible, and all ascribed the changes to multiple causes. The Hakipuu witnesses described changes in Hakipuu Stream and its watershed that were similar to those that had occurred in Waiāhole and Waikāne Streams and their watersheds, even though Hakipuu Stream is not hydrologically affected by the tunnel system. One witness described similar changes in Punaluu Stream, which also is not affected by the tunnel system; D&O II [97].

Expert and lay testimony on the impact on Kāneohe Bay from diminished flows in Waiāhole and Waikāne Streams led to similar findings on the impact of multiple factors – overfishing, population pressure, habitat destruction, etc. – and noticeable changes in the 1960s. But the scientists agreed that restoring both streams "would be particularly useful to factor out these influences and study the impact of increased flow in relative isolation, because of the relative absence of pollution and urbanization of the watersheds of these two streams; D&O II [98].

The lack of quantitative data to guide the Commission on amending the instream flow standards led it to look to history. It first rejected what it called "a minimalist approach to restoring stream flows (that) could look to the period of the 1960s and see what stream-flow-related changes occurred during that time that could have contributed to the decline in stream vitality; D&O II [99]." Under that scenario, the 2.77 mgd developed by the Uwau tunnel extension in 1964 would be added to Waiāhole and Waianu Streams, or 1.3 to 1.8 mgd to Waianu Stream, because pumping of 1 to 1.5 mgd from Waiāhole Stream had ceased in 1982. But the Commission determined that reasonable "margins of safety" should be adopted in amending the windward instream flow standards, and that it was "practicable to use increased stream flows to partially compensate for the other factors that have affected the vitality of the streams, as well as to increase the contribution that these stream flows may have on the vitality of Kāneohe Bay; D&O II [100]." The Commission then went on to conclude:

> One Hawaiian approach to diversion of stream waters … appears to limit diversions to no more than one-half of a stream's flow, although much more has been diverted on occasion … (T)here have been diversions limited to half the flow from a stream or place of diversion, and examples of other diversions taking up to or perhaps somewhat beyond the available water supply. However, it does not appear that there was any specific, quantified amount of water that should remain in the stream or be taken for off stream use. *Considering the specific facts of this case, not establishing a standard or generalized policy for future decisions, and in accordance with the precautionary principle, a reasonable and practicable approach would be to restore Waiāhole, Waianu, Waikāne, and Kahana Streams to one-half their pre-Ditch base flow levels which would also exceed their 1960 levels where testimony established the presence of aquatic biota at a higher level than today* (*emphasis added*); D&O II [101].

But on appeal, the Court concluded that "the Water Commission's decision to halve the possible stream flow, based solely on a quotation stating that 'no ditch was permitted to divert more than half the flow from the stream,' left unanswered the question whether instream values would be protected to the extent practicable. We, therefore, hold that the Water Commission's reliance on this approach was erroneous; *Waiāhole II* [102]." But the Court went on to state:

> If, on remand, the Water Commission is able to support its conclusion with findings quantifying the windward streams' flows during the 1960s, then the 1960s testimonials would be sufficient to set the IIFS at the levels established in the D&O II, inasmuch as: (1) more water would be added to the streams than that which adequately supported the streams' ecosystem in the 1960s; (2) the increase in stream flow over the 1960s stream flow would be beneficial in light of the Water Commission's finding that increasing a stream's flow results in stream habitat improvement; and (3) appurtenant rights, riparian uses, and existing uses would be accounted for by further increases in stream flow. The foregoing would then adequately establish that instream values would be protected to the extent practicable for interim purposes (*references and footnote omitted*); *Waiāhole II* [103].

The Commission had thought that its analysis had clearly shown that the base flows of the streams in the 1960s were unchanged from the 1930s to the present, except for the 1964 Uwau tunnel extension, cessation of pumping from upper Waiāhole Stream in 1982, and the bulk-heading of the Kahana tunnel in 1992, the latter two of which would have increased stream flows relative to the 1960s. So on remand, the Commission reiterated this analysis in greater detail and pointed out that the amended instream flow standards had resulted in increases in the base flows of the streams relative to the 1960s as follows; D&O III [104]:

1960s		Amended IIFS	% Increase
Waiāhole Stream	3.9 mgd	8.7 mgd	124%
Waianu Stream	0.5 mgd	3.5 mgd	600%
Waikāne Stream	1.4 mgd	3.5 mgd	150%
Kahana Stream	11.2 mgd	13.3 mgd	19%

4 Future water-resource issues

The Waiāhole Ditch case has lasted over a decade, introduced new procedures that are more use and area specific for determining how much water the Commission should award as a "reasonable and beneficial use," and confirmed that the water uses are governed by three separate legal doctrines: 1) the State Water Code; 2) common-law water rights and uses; and 3) the public-trust doctrine. Future decisions by the Commission – that will surely be appealed and hence result in further guidance by the Court – will address cases in which public-trust purposes may be in conflict with each other; i.e. appurtenant rights and/or domestic uses of the general public versus protection of the resource. In the initial phases, other nontrust private uses will have to give way to accommodate trust purposes, but trust purposes may eventually clash with each other. For example, there are two situations on Maui, where there are petitions to amend the instream flow standards of streams that have long been diverted, originally for sugar plantations. Now, as sugar phases out and diversified agriculture and housing developments take their place, past uses are coming under scrutiny either through the "reasonable and beneficial" criteria of the Water Code's permitting process or through the common-law criterion of reasonable use. As in the Waiāhole Ditch controversy, the competing interests of stream restoration and changing offstream uses are similarly converging and in need of resolution.

On Maui, withdrawals from groundwater resources are nearing the sustainable yield, so growing municipal water needs are requiring increased use of surfacewater resources. Thus, the Commission will eventually face the task of determining how high to prioritize municipal uses, which are currently part public-trust purpose (domestic use) and partly nontrust uses (industrial and commercial customers).

The most critical near-term issues to be faced by the Commission are the separate petitions to amend the instream flow standards of the streams: 1) in Iao Valley, where the Iao aquifer has recently been designated a groundwater-management area and the high-level, diked groundwaters have a direct relationship with the streams; and 2) in East Maui, a nondesignated area and hence subject to common-law rights and uses, where a sugar company's ditches divert (for appropriated uses) an average of 165 mgd from the streams, 12 mgd of which are provided to the; Maui Department of Water Supply Board of Land and Natural Resources [105]. The first scientific investigations of the streams in this area have only recently been initiated; U.S. Geological Survey [106]. Furthermore, there is a history of abundant wetland *loi* (taro patches), some of which are still being cultivated and others that some of the petitioners to amend the instream flow standards would like to reactivate. Some of these petitioners have refused so far to provide evidence of their appurtenant rights, claiming that they have no burden of proving anything in the contested case; Board of Land and Natural Resources [107]. However, appurtenant rights are rights attached to land, and Native Hawaiian and traditional and customary rights have to be asserted by parties who are "descendants of native Hawaiians who inhabited the Hawaiian Islands prior to 1778," and whose practices predate November 25, 1892, when a state statute recognized rights "established by Hawaiian usage"; Miike [108]. In either case, parties must prove that they are entitled to such rights. (Note, however, that refusal does not extinguish any such rights that they may have, and the parties can always return in the future to meet their evidentiary burden.)

Exercise of appurtenant rights for irrigating taro *loi* will pose one of the most difficult water-management issues for the Commission. Water budgets for such uses are based on the amount of water consumed; i.e. inflow minus outflow, which is site specific and will vary with the amount of leakage from the *loi* bottom and banks; the stage of the crop or crops; rainfall, elevation, average hours of sunshine and wind conditions; soil characteristics; and even whether or not it is necessary to conserve water and whether or not there is a profit incentive. Further, inflow varies widely; from over one million gallons per acre per day to keep the taro sufficiently cool to prevent rot, to very little or no water required for much of the fifteen-month crop cycle. Thus, one study in Waiāhole Valley concluded that "representative water use was 15,000 to 40,000 gallons per acre per day, allowing for sufficient outflow to assure good circulation. Since all stages of the crop were included in the study, the recommendation was that 40,000 gallons per acre per day be recognized as the fair requirement for an area of several taro patches in various stages of crop development, including patches requiring maximum irrigation and those requiring none; D&O II [109]." But the tremendous variation in daily irrigation requirements means that the contributing stream must have sufficient flow to accommodate periods of very high flow-through, even though the official withdrawal rate may be only several thousand gallons per acre per day and the flowthrough water is returned to the stream further downstream.

Thus, the most important situation facing the Commission in the near future will involve surfacewater resources, and specifically, pitting municipal, agricultural and commercial uses against traditional agricultural (wetland taro) and gathering (the stream's native fish, crustaceans, and mollusks) practices and the overall biological health and aesthetics of restored streams. The Waiāhole Ditch controversy will be played out again, only on a much larger scale.

References

[1] James Campbell, Esquire, James Campbell Company, LLC, www.jamescampbell.com.

[2] Kluegel, C.H., *Engineering Features of the Waiāhole Water Project of the Waiāhole Water, Co, Island of Oahu, Territory of Hawaii*, Hawaiian Engineering Association, Press Bulletin 53, Hawaiian Gazette Co.: Honolulu, pp. 5–6, 1916.

[3] Kluegel, C.H., *Engineering Features of the Waiāhole Water Project of the Waiāhole Water, Co, Island of Oahu, Territory of Hawaii*, Hawaiian Engineering Association, Press Bulletin 53, Hawaiian Gazette Co.: Honolulu, pp. 12–14, 1916.

[4] Kluegel, C.H., *Engineering Features of the Waiāhole Water Project of the Waiāhole Water, Co, Island of Oahu, Territory of Hawaii*, Hawaiian Engineering Association, Press Bulletin 53, Hawaiian Gazette Co.: Honolulu, pp. 12, 14, 1916.

[5] Miike, L.H., *Water and the Law in Hawaii*, University of Hawaii Press: Honolulu, pp. 142–144, 2004.

[6] *Ibid.*, p. 145. Miike, L.H., *Water and the Law in Hawaii*, University of Hawaii Press: Honolulu, pp. 142–144, 2004.

[7] Miike, L.H., *Water and the Law in Hawaii*, University of Hawaii Press: Honolulu, p. 146, 2004.

[8] Miike, L.H., *Water and the Law in Hawaii*, University of Hawaii Press: Honolulu, p. 141, 2004.

[9] Miike, L.H., *Water and the Law in Hawaii*, University of Hawaii Press: Honolulu, p. 144, Fig. 82004.

[10] Miike, L.H., *Water and the Law in Hawaii*, University of Hawaii Press: Honolulu, p. 140, 2004.

[11] Takasaki, K.J. & Mink, J.F., *Evaluation of Major Dike-Impounded Ground-Water Reservoirs, Island of Oahu*, prepared in cooperation with the Board of Water Supply, City and County of Honolulu, U.S. Geological Survey Water-Supply Paper 2217, U.S. Government Printing Office: Washington, DC, p. 35, 1985.

[12] Takasaki, K.J. & Mink, J.F., *Evaluation of Major Dike-Impounded Ground-Water Reservoirs, Island of Oahu*, prepared in cooperation with the Board of Water Supply, City and County of Honolulu, U.S. Geological Survey Water-Supply Paper 2217, U.S. Government Printing Office: Washington, DC, p. 35, 1985.

[13] Takasaki, K.J. & Mink, J.F., *Evaluation of Major Dike-Impounded Ground-Water Reservoirs, Island of Oahu*, prepared in cooperation with the Board of Water Supply, City and County of Honolulu, U.S. Geological Survey Water-Supply Paper 2217, U.S. Government Printing Office: Washington, DC, p. 35, 1985.

[14] Takasaki, K.J., Hirashima, G.T. & Lubke, E.R., *Water Resources of Windward Oahu, Hawaii*, Geological Survey Water-Supply Paper 1894, prepared in cooperation with the State of Hawaii Department of Land and Natural Resources, Division of Water and Land Development, U.S. Government Printing Office: Washington, DC, pp. 40, 75, 88, 1969.

[15] Takasaki, K.J., Hirashima, G.T. & Lubke, E.R., *Water Resources of Windward Oahu, Hawaii*, Geological Survey Water-Supply Paper 1894, prepared in cooperation with the State of Hawaii Department of Land and Natural Resources, Division of Water and Land Development, U.S. Government Printing Office: Washington, DC, p. 33, 1969.

[16] Takasaki, K.J., Hirashima, G.T. & Lubke, E.R., *Water Resources of Windward Oahu, Hawaii*, Geological Survey Water-Supply Paper 1894, prepared in cooperation with the State of Hawaii Department of Land and Natural Resources, Division of Water and Land Development, U.S. Government Printing Office: Washington, DC, p. 31, 1969.

[17] Taogoshi, R.I., Wong, M.F., Nishimoto, D.C. & Teeters, P.C., *Water Resources Data, Hawaii and other Pacific Areas, Water Year 2001, Volume* Hawaii Data Report HI-01-1, U.S. Department of the Interior, U.S. Geological Survey, prepared in cooperation with the State of Hawaii Department of Land and Natural Resources, Commission on Water Resource Management and with other agencies, U.S. Department of Commerce, NTIS: Springfield, VA, p. 155, 2001.

[18] Taogoshi, R.I., Wong, M.F., Nishimoto, D.C. & Teeters, P.C., *Water Resources Data, Hawaii and other Pacific Areas, Water Year 2001, Volume* Hawaii Data Report HI-01-1, U.S. Department of the Interior, U.S. Geological Survey, prepared in cooperation with the State of Hawaii Department of

Land and Natural Resources, Commission on Water Resource Management and with other agencies, U.S. Department of Commerce, NTIS: Springfield, VA, p. 155, 2001.

[19] Takasaki, K.J. & Mink, J.F., *Evaluation of Major Dike-Impounded Ground-Water Reservoirs, Island of Oahu*, prepared in cooperation with the Board of Water Supply, City and County of Honolulu, U.S. Geological Survey Water-Supply Paper 2217, U.S. Government Printing Office: Washington, DC, p. 18, 1985.

[20] Miike, L.H., *Water and the Law in Hawaii*, University of Hawaii Press: Honolulu, p. 13, 2004.

[21] Miike, L.H., *Water and the Law in Hawaii*, University of Hawaii Press: Honolulu, pp. 14–15, 2004.

[22] Pukui, M.K., *lelo Noeau: Hawaiian Proverbs & Poetical Sayings*, Bernice P. Bishop Museum Special Publication No. 71, Bishop Museum Press: Honolulu, pp. 145–146, 149, 1983.

[23] Miike, L.H., *Water and the Law in Hawaii*, University of Hawaii Press: Honolulu, p. 153, 2004.

[24] Miike, L.H., *Water and the Law in Hawaii*, University of Hawaii Press: Honolulu, p. 153, 2004.

[25] Commission on Water Resource Management, State of Hawaii, *Legal Framework, Findings of Fact, and Decision and Order, on Remand, in the Matter of Water Use Permit Applications, Petitions for Interim Instream Flow Standard Amendments, and Petitions for Water Reservations, for the Waiāhole Ditch Combined Contested Case Hearings: On Remand*, Case No. CCH-OA95-1, pp. 29–30, December 28, 2001.

[26] Miike, L.H., *Water and the Law in Hawaii*, University of Hawaii Press: Honolulu, p. 9, 2004.

[27] Beckwith, M.W., translator and editor, *The Kumuulipo: A Hawaiian Creation Chant*, University of Hawaii Press: Honolulu, pp. 1–3, 1972 (Originally published in 1951 by the University of Chicago Press).

[28] Commission on Water Resource Management, State of Hawaii, *Legal Framework, Findings of Fact, and Decision and Order, on Remand, in the Matter of Water Use Permit Applications, Petitions for Interim Instream Flow Standard Amendments, and Petitions for Water Reservations, for the Waiāhole Ditch Combined Contested Case Hearings: On Remand*, Case No. CCH-OA95-1, p. 30, December 28, 2001.

[29] Beckwith, M.W., translator and editor, *The Kumuulipo: A Hawaiian Creation Chant*, University of Hawaii Press: Honolulu, p. 11, 1972 (Originally published in 1951 by the University of Chicago Press).

[30] Miike, L.H., *Water and the Law in Hawaii*, University of Hawaii Press: Honolulu, pp. 25–28, 2004.

[31] Ellis, W., *Journal of William Ellis: Narrative of a Tour of Hawaii, or Owhyhee; with Remarks on the History, Traditions, Manners, Customs, and Language of the Inhabitants of the Sandwich Islands*, E. Charles Tuttle: Rutland, VT, p. 311, 1825; revised edition 1827.

[32] Handy, E.S.C. & Handy, E.G., with the collaboration of Mary Kawena Pukui, *Native Planters in Old Hawaii: Their Life, Lore, and Environment*, P. Bernice Bishop Museum Bulletin 233, Bishop Museum Press: Honolulu, pp. 64–65, 1972.
[33] Commission on Water Resource Management, State of Hawaii, *Legal Framework, Findings of Fact, and Decision and Order, on Remand, in the Matter of Water Use Permit Applications, Petitions for Interim Instream Flow Standard Amendments, and Petitions for Water Reservations, for the Waiāhole Ditch Combined Contested Case Hearings: On Remand*, Case No. CCH-OA95-1, p. 30, December 28, 2001.
[34] HRS Chapter 174C; HAR Chapters 13-167 to 1171.
[35] HRS §174C-71; HAR §§ 13-169-1 to 13-169-43.
[36] HRS §174C-44; HAR §13-171-7.
[37] HRS §174C-45; HAR §13-171-8.
[38] HAR §13-169-49.1.
[39] HRS §174C-49(a)(3).
[40] HRS §174C-50(c).
[41] HRS §174C-49(d); HAR §13-171-60(d).
[42] HRS §174C-41; HRS §174C-63.
[43] *Territory v Gay*, 31 Haw. 376, p. 395, 1930.
[44] *The Random House Dictionary of the English Language: The Unabridged Version*, Random House: New York, 1973.
[45] *Peck v Bailey*, 8 Haw. 658, p. 662, 1867.
[46] *McBryde v Robinson*, 54 Haw. 174; 504 P.2d 1330, (1973; *aff'd on rehearing*, 55 Haw. 260; 517 P.2d, 1973; *appeal dismissed for want of jurisdiction and cert. denied*, 417 U.S. 962, 1974.
[47] Hawaii State Constitution, Article 13, § 1.
[48] Miike, L.H., *Water and the Law in Hawaii*, University of Hawaii Press: Honolulu, pp. 99, 108, 2004.
[49] *Robinson v Ariyoshi*, 65 Haw. 641, pp. 649–650; 658 P.2d 287, 1982.
[50] *Territory v Gay*, p. 383.
[51] *Reppun v Board of Water Supply*, 65 Haw. 531, p. 554; 656 P.2d 57, 1982.
[52] HRS §174C-5(14).
[53] *Reppun v Board of Water Supply*, pp. 550–551.
[54] *Peck v Bailey*, 8 Haw. p. 658 (1867).
[55] *Territory v Gay*, 31 Haw. p. 376 (1930).
[56] *Peck v Bailey*, p. 652.
[57] *Territory v Gay*, p. 396.
[58] Miike, L.H., *Water and the Law in Hawaii*, University of Hawaii Press: Honolulu, pp. 82–96, 2004.
[59] *Reppun v Board of Water Supply*, p. 553.
[60] *Davis v Afong*, 5 Haw. 216, 1884.
[61] *Wong Leong v Irwin*, 10 Haw. p. 265, 1896.
[62] *Wong Leong v Irwin*, 10 Haw. p. 270, 1896.
[63] *City Mill Co. v Hon. S & W Com.*, 30 Haw. 912, p. 923, 1929.

[64] *In the Matter of Water Use Permit Applications, Petitions for Interim Instream Flow Standard Amendments, and Petitions for Water Reservations for the Waiāhole Ditch Combined Contested Case Hearing*, 94 Haw.97, p. 178; 9 P.3d 409, 2000.

[65] Commission on Water Resource Management, State of Hawaii, *In the Matter of Water Use Permit Applications, Petitions for Interim Instream Flow Standard Amendments, and Petitions for Water Resevations for the Waiāhole Ditch Combined Contested Case Hearings, Findings of Fact, Conclusions of Law, and Decision and Order*, Case No. CCH-OA95-1, pp. 2–5, December 24, 1997.

[66] Commission on Water Resource Management, State of Hawaii, *In the Matter of Water Use Permit Applications, Petitions for Interim Instream Flow Standard Amendments, and Petitions for Water Resevations for the Waiāhole Ditch Combined Contested Case Hearings, Findings of Fact, Conclusions of Law, and Decision and Order*, Case No. CCH-OA95-1, pp. 2–5, December 24, 1997.

[67] *In the Matter of Water Use Permit Applications, Petitions for Interim Instream Flow Standard Amendments, and Petitions for Water Reservations for the Waiāhole Ditch Combined Contested Case Hearing*, 94 Haw. 97, p. 178; 9 P.3d 409, 2000.

[68] Commission on Water Resource Management, State of Hawaii, *Legal Framework, Findings of Fact, and Decision and Order, on Remand, in the Matter of Water Use Permit Applications, Petitions for Interim Instream Flow Standard Amendments, and Petitions for Water Reservations, for the Waiāhole Ditch Combined Contested Case Hearings: On Remand*, Case No. CCH-OA95-1, p. 30, December 28, 2001.

[69] *In the Matter of Water Use Permit Applications, Petitions for Interim Instream Flow Standard Amendments, and Petitions for Water Reservations for the Waiāhole Ditch Combined Contested Case Hearing*, 105 Haw. 1; 93 P.3d 643, 2004.

[70] Commission on Water Resource Management, State of Hawaii, *In the Matter of Water Use Permit Applications, Petitions for Interim Instream Flow Standard Amendments, and Petitions for Water Reservations for the Waiāhole Ditch Combined Contested Case Hearings, Findings of Fact, Conclusions of Law, and Decision and Order*, Case No. CCH-OA95-1, July 13, 2006.

[71] *In the Matter of Water Use Permit Applications, Petitions for Interim Instream Flow Standard Amendments, and Petitions for Water Reservations for the Waiāhole Ditch Combined Contested Case Hearing*, 94 Haw. 97, p. 127, 2000.

[72] *Peck v Bailey*.

[73] Miike, L.H., *Water and the Law in Hawaii*, University of Hawaii Press: Honolulu, pp. 137–204, 2004.

[74] Commission on Water Resource Management, State of Hawaii, *In the Matter of Water Use Permit Applications, Petitions for Interim Instream Flow Standard Amendments, and Petitions for Water Resevations for the*

Waiāhole Ditch Combined Contested Case Hearings, Findings of Fact, Conclusions of Law, and Decision and Order, Case No. CCH-OA95-1, p. 7, December 24, 1997.

[75] *In the Matter of Water Use Permit Applications, Petitions for Interim Instream Flow Standard Amendments, and Petitions for Water Reservations for the Waiāhole Ditch Combined Contested Case Hearing*, 94 Haw. 97, p. 130, 2000.

[76] *In the Matter of Water Use Permit Applications, Petitions for Interim Instream Flow Standard Amendments, and Petitions for Water Reservations for the Waiāhole Ditch Combined Contested Case Hearing*, 94 Haw. 97, pp. 130–131, 2000.

[77] *In the Matter of Water Use Permit Applications, Petitions for Interim Instream Flow Standard Amendments, and Petitions for Water Reservations for the Waiāhole Ditch Combined Contested Case Hearing*, 94 Haw. 97, p. 135, 2000.

[78] *In the Matter of Water Use Permit Applications, Petitions for Interim Instream Flow Standard Amendments, and Petitions for Water Reservations for the Waiāhole Ditch Combined Contested Case Hearing*, 94 Haw. 97, pp. 136–137, 2000

[79] *In the Matter of the Contested Case Hearing on Water Use, Well Construction, and Pump Installation Permit Applications, Filed by Waiola o Molokai, Inc. and Molokai Ranch, Limited*, 105 Haw. 401, pp. 430–431; 83 P.3d 664, 2004.

[80] Commission on Water Resource Management, State of Hawaii, *In the Matter of Water Use Permit Applications, Petitions for Interim Instream Flow Standard Amendments, and Petitions for Water Resevations for the Waiāhole Ditch Combined Contested Case Hearings, Findings of Fact, Conclusions of Law, and Decision and Order*, Case No. CCH-OA95-1, p. 11, December 24, 1997

[81] *In the Matter of Water Use Permit Applications, Petitions for Interim Instream Flow Standard Amendments, and Petitions for Water Reservations for the Waiahole Ditch Combined Contested Case Hearing*, 94 Haw.97, p. 139, 2000.

[82] *In the Matter of Water Use Permit Applications, Petitions for Interim Instream Flow Standard Amendments, and Petitions for Water Reservations for the Waiahole Ditch Combined Contested Case Hearing*, 94 Haw. 97, pp. 140, 142, 2000.

[83] *In the Matter of Water Use Permit Applications, Petitions for Interim Instream Flow Standard Amendments, and Petitions for Water Reservations for the Waiāhole Ditch Combined Contested Case Hearing*, 94 Haw. 97, p. 153, 2000; and Commission on Water Resource Management, State of Hawaii, *Legal Framework, Findings of Fact, and Decision and Order, on Remand, in the Matter of Water Use Permit Applications, Petitions for Interim Instream Flow Standard Amendments, and Petitions for Water Reservations, for the Waiāhole Ditch Combined Contested Case Hearings: On Remand*, Case No. CCH-OA95-1, p. 99, December 28, 2001.

[84] HRS §174C-48(a).

[85] *Reppun v Board of Water Supply*, p. 555.

[86] *In the Matter of Water Use Permit Applications, Petitions for Interim Instream Flow Standard Amendments, and Petitions for Water Reservations for the Waiāhole Ditch Combined Contested Case Hearing*, 94 Haw. 97, p. 173, 2000.

[87] Commission on Water Resource Management, State of Hawaii, *Minutes for the Meeting of the Commission on Water Resource Management,* May 24, 2006.

[88] *In the Matter of Water Use Permit Applications, Petitions for Interim Instream Flow Standard Amendments, and Petitions for Water Reservations for the Waiāhole Ditch Combined Contested Case Hearing*, 94 Haw. 97, p. 161, 2000.

[89] Commission on Water Resource Management, State of Hawaii, *Legal Framework, Findings of Fact, and Decision and Order, on Remand, in the Matter of Water Use Permit Applications, Petitions for Interim Instream Flow Standard Amendments, and Petitions for Water Reservations, for the Waiāhole Ditch Combined Contested Case Hearings: On Remand*, Case No. CCH-OA95-1, pp. 127–128, December 28, 2001.

[90] *In the Matter of Water Use Permit Applications, Petitions for Interim Instream Flow Standard Amendments, and Petitions for Water Reservations for the Waiāhole Ditch Combined Contested Case Hearing*, 105 Haw. p. 20, 2004.

[91] Commission on Water Resource Management, State of Hawaii, *In the Matter of Water Use Permit Applications, Petitions for Interim Instream Flow Standard Amendments, and Petitions for Water Reservations for the Waiāhole Ditch Combined Contested Case Hearings, Findings of Fact, Conclusions of Law, and Decision and Order*, Case No. CCH-OA95-1, p. 58, July 13, 2006.

[92] *In the Matter of Water Use Permit Applications, Petitions for Interim Instream Flow Standard Amendments, and Petitions for Water Reservations for the Waiāhole Ditch Combined Contested Case Hearing*, 94 Haw.97, p. 168, 2000.

[93] *In the Matter of Water Use Permit Applications, Petitions for Interim Instream Flow Standard Amendments, and Petitions for Water Reservations for the Waiāhole Ditch Combined Contested Case Hearing*, 94 Haw.97, p. 169, 2000.

[94] Commission on Water Resource Management, State of Hawaii, *Legal Framework, Findings of Fact, and Decision and Order, on Remand, in the Matter of Water Use Permit Applications, Petitions for Interim Instream Flow Standard Amendments, and Petitions for Water Reservations, for the Waiāhole Ditch Combined Contested Case Hearings: On Remand*, Case No. CCH-OA95-1, p. 35, December 28, 2001.

[95] Commission on Water Resource Management, State of Hawaii, *Legal Framework, Findings of Fact, and Decision and Order, on Remand, in the Matter of Water Use Permit Applications, Petitions for Interim Instream*

Flow Standard Amendments, and Petitions for Water Reservations, for the Waiāhole Ditch Combined Contested Case Hearings: On Remand, Case No. CCH-OA95-1, pp. 35–36, December 28, 2001.

[96] Commission on Water Resource Management, State of Hawaii, *Legal Framework, Findings of Fact, and Decision and Order, on Remand, in the Matter of Water Use Permit Applications, Petitions for Interim Instream Flow Standard Amendments, and Petitions for Water Reservations, for the Waiāhole Ditch Combined Contested Case Hearings: On Remand*, Case No. CCH-OA95-1, p. 36, December 28, 2001.

[97] Commission on Water Resource Management, State of Hawaii, *Legal Framework, Findings of Fact, and Decision and Order, on Remand, in the Matter of Water Use Permit Applications, Petitions for Interim Instream Flow Standard Amendments, and Petitions for Water Reservations, for the Waiāhole Ditch Combined Contested Case Hearings: On Remand*, Case No. CCH-OA95-1, p. 29, December 28, 2001.

[98] Commission on Water Resource Management, State of Hawaii, *Legal Framework, Findings of Fact, and Decision and Order, on Remand, in the Matter of Water Use Permit Applications, Petitions for Interim Instream Flow Standard Amendments, and Petitions for Water Reservations, for the Waiāhole Ditch Combined Contested Case Hearings: On Remand*, Case No. CCH-OA95-1, pp. 48–49, December 28, 2001.

[99] Commission on Water Resource Management, State of Hawaii, *Legal Framework, Findings of Fact, and Decision and Order, on Remand, in the Matter of Water Use Permit Applications, Petitions for Interim Instream Flow Standard Amendments, and Petitions for Water Reservations, for the Waiāhole Ditch Combined Contested Case Hearings: On Remand*, Case No. CCH-OA95-1, p. 103, December 28, 2001.

[100] Commission on Water Resource Management, State of Hawaii, *Legal Framework, Findings of Fact, and Decision and Order, on Remand, in the Matter of Water Use Permit Applications, Petitions for Interim Instream Flow Standard Amendments, and Petitions for Water Reservations, for the Waiāhole Ditch Combined Contested Case Hearings: On Remand*, Case No. CCH-OA95-1, pp. 103–104, December 28, 2001.

[101] Commission on Water Resource Management, State of Hawaii, *Legal Framework, Findings of Fact, and Decision and Order, on Remand, in the Matter of Water Use Permit Applications, Petitions for Interim Instream Flow Standard Amendments, and Petitions for Water Reservations, for the Waiāhole Ditch Combined Contested Case Hearings: On Remand*, Case No. CCH-OA95-1, pp. 104–105, December 28, 2001.

[102] Commission on Water Resource Management, State of Hawaii, *Legal Framework, Findings of Fact, and Decision and Order, on Remand, in the Matter of Water Use Permit Applications, Petitions for Interim Instream Flow Standard Amendments, and Petitions for Water Reservations, for the Waiāhole Ditch Combined Contested Case Hearings: On Remand*, Case No. CCH-OA95-1, p. 11, December 28, 2001.

[103] Commission on Water Resource Management, State of Hawaii, *Legal Framework, Findings of Fact, and Decision and Order, on Remand, in the Matter of Water Use Permit Applications, Petitions for Interim Instream Flow Standard Amendments, and Petitions for Water Reservations, for the Waiāhole Ditch Combined Contested Case Hearings: On Remand*, Case No. CCH-OA95-1, p. 12, December 28, 2001.

[104] D&O III, p. 67.

[105] Board of Land and Natural Resources, State of Hawaii, DLNR File No. 01-05-MA, *In the Matter of the Contested Case Hearing Regarding Water Licenses at Honomanu, Keanae, Nahiku and Huelo, Maui; Hearings Officer's Recommended Findings of Fact, Conclusions of Law and Order Concerning Interim Relief*, p. 7, June 21, 2006.

[106] *Presentation by the U.S. Geological Survey on the Effects of Surface- Water Diversions on Habitat Availability for Native Macrofauna, Northeast Maui, Hawaii, Scientific Investigations Report 2005*-5213, before the Commission on Water Resource Management, June 21, 2006.

[107] Board of Land and Natural Resources, State of Hawaii, DLNR File No. 01-05-MA, *In the Matter of the Contested Case Hearing Regarding Water Licenses at Honomanu, Keanae, Nahiku and Huelo, Maui; Hearings Officer's Recommended Findings of Fact, Conclusions of Law and Order Concerning Interim Relief*, p. 31, June 21, 2006.

[108] Miike, L.H., *Water and the Law in Hawaii*, University of Hawaii Press: Honolulu, pp. 66–67, 2004.

[109] Commission on Water Resource Management, State of Hawaii, *Legal Framework, Findings of Fact, and Decision and Order, on Remand, in the Matter of Water Use Permit Applications, Petitions for Interim Instream Flow Standard Amendments, and Petitions for Water Reservations, for the Waiāhole Ditch Combined Contested Case Hearings: On Remand*, Case No. CCH-OA95-1, pp. 60–61, December 28, 2001.

Index

WIT*PRESS* ...*for scientists by scientists*

Groundwater Characterization

Edited by: ***M.CUNHA****, University of Coimbra, Portugal,* ***L.M. NUNES****, University of Algarve, Portugal and* ***L. RIBEIRO****, Technical University of Lisbon, Portugal*

This book addresses the theoretical background necessary to accomplish planning and management of groundwater systems, and presents up to date applications of the decision-aid techniques in this field.

Groundwater systems play an essential role in meeting the ever increasing use of water for different purposes. Proper design and management of such systems should therefore be a very important matter of concern, not only to ensure that water will be available in adequate quantity and quality to satisfy demands but also to guarantee that this would be done in an optimal manner in a IWRM perspective. There are many different decisions to be taken: where to locate wells, how much water is to be pumped, remedial strategies, to be adopted, water supply structures (especially pumping equipment and pipe networks) to be installed, monitoring networks to be defined, etc. These decisions must take many constraints into account, including drawdown limitations, flow gradients, quality standards. Given the uncertainty characterizing the groundwater flow and transport, risk issues have to be considered. Decision-aid techniques are methodological tools capable of handling simultaneously the various facets characterizing such problems (economic, social, technical, environmental, etc.) Therefore detailed simulation models have to be incorporated into the decision models. The application of simulation-optimization methods to planning and managing groundwater systems has become an area of active research.

ISBN: 978-1-84564-134-4 2008 apx 350pp apx £115.00/US$230.00/ €172.50

WIT*Press*
Ashurst Lodge, Ashurst,
Southampton,
SO40 7AA, UK.
Tel: 44 (0) 238 029 3223
Fax: 44 (0) 238 029 2853
E-Mail: witpress@witpress.com

WIT*PRESS* ...for scientists by scientists

Sustainable Irrigation Management, Technologies and Policies II

Edited by: ***Y. VILLACAMPA ESTEVE****, Universidad de Alicante, Spain*
C.A. BREBBIA*, Wessex Institute of Technology, UK and* ***D. PRATS RICO****,*
Universidad de Alicante, Spain

As the standard of living for a large section of the world population continues to increase, the strategic importance of fresh water resources also rises. It is essential to seek ways of achieving the most efficient and equitable use of these resources at the same time as making their use sustainable.

Even in those countries where fresh water is currently easily available, over-exploitation is leading to damaging long-lasting environmental effects, such as lowering of water tables or depletion of river flows. Adding to these effects, the problem of contamination effectively reduces the availability of sufficient clean water. Water is also essential for irrigation purposes, but its indiscriminate use can lead not only to shortages, but also to the deterioration of crop yields and soils. Hence it is vital to ensure that irrigation is applied as effectively as possible in order to achieve sustainability.

Attracting researchers in academia and industry, as well as professional practitioners and policy makers, the Second International Conference on Sustainable Irrigation provided a platform for a review of the state of the art from the scientific, technological, political and economic points of view. Papers from the Meeting are published in this volume, and cover the following topics: Irrigation Controls; Irrigation Systems and Planning; Irrigation Modelling; Irrigation Management; Remote Sensing Application; Irrigation in Arid and Semi-arid Regions; Re-use of Water; Water and Soil Pollution Control; Bio-economic Modelling.

WIT Transactions on Ecology and the Environment, Vol 112
ISBN: 978-1-84564-116-0 2008 apx 350pp apx £115.00/US$230.00/ €172.50

WIT*Press*
Ashurst Lodge, Ashurst,
Southampton,
SO40 7AA, UK.
Tel: 44 (0) 238 029 3223
Fax: 44 (0) 238 029 2853
E-Mail: witpress@witpress.com

WIT PRESS ...for scientists by scientists

Water Pollution IX

Edited by: ***D. PRATS RICO****, Universidad de Alicante, Spain,* ***C.A. BREBBIA****, Wessex Institute of Technology, UK and* ***Y. VILLACAMPA ESTEVE****, Universidad de Alicante, Spain*

Water pollution is a subject of growing public concern. The environmental problems caused by the increase of pollutant loads discharged into natural water bodies require the setting out of frameworks for regulation and control. The scientific community has responded to the need for studies capable of relating pollutant discharge with changes in water quality. The results of these studies are permitting industries to employ more efficient methods of controlling and treating waste loads, and water authorities to enforce stricter regulations regarding this matter.

Environmental problems are essentially interdisciplinary. Engineers and scientists working in this field must be familiar with a wide range of issues including the physical processes of mixing and dilution, chemical and biological processes, mathematical modelling, data acquisition and measurement to name but a few.

This book contains papers presented at the Ninth International Conference on Modelling, Monitoring and Management of Water Pollution and includes the following topics: Coastal Areas and Seas; Lakes and Rivers; Groundwater and Aquifer Issues; Oil Spills; Agricultural Contamination; Environmental Monitoring and Sensing; Experimental and Laboratory Work; Mathematical and Physical Modelling; Wastewater Treatment; Pollution Prevention; Remote Sensing Applications; Novel Techniques for Water Treatment; Low Cost Technologies; Pharmaceutical and Pesticides; Remediation; Bioaccumulation; Micropollutant Prevention in Drinking Water.

WIT Transactions on Ecology and the Environment, Vol 111
ISBN: 978-1-84564-115-3 2008 apx 600pp apx £198.00/US$396.00/ €297.00

Find us at
http://www.witpress.com

Save 10% when you order from our encrypted ordering service on the web using your credit card.

WIT PRESS ...for scientists by scientists

Water Resource Management IV

Edited by: ***C.A. BREBBIA****, Wessex Institute of Technology, UK and* ***A. KUNGOLOS****, University of Thessaly, Greece*

Water resources are under extreme pressure today all over the world. The resulting problems have given rise to many activities which reflect the growing concern about them and the importance of effective management.

As water increasingly becomes a precious resource on which the well-being of future generations depends, it is essential to discuss issues concerning quality, quantity, planning and other related topics.

Containing papers presented at the Fourth International Conference on Water Resources Management, this book examines the recent technological and scientific developments associated with the management of surface and sub-surface water resources. The wide variety of subjects covered are as follows: Water Resource Management and Planning; Waste Water Treatment and Management; Water Markets and Policies; Urban Water Management; Water Quality; Storm Water Management; Water Security Systems; Pollution Control; Irrigation Problems; Reservoirs and Lakes; River Basin Management; Hydrological Modelling; Flood Risk; Decision Support Systems; Groundwater Flow Problems and Remediation Technologies; Coastal and Estuarial Problems; Soil and Water Conservation and Risk Analysis.

WIT Transactions on Ecology and the Environment, Vol 103
ISBN: 978-1-84564-074-3 2007 720pp £235.00/US$465.00/€352.50

River Basin Management IV

Edited by: ***C.A. BREBBIA****, Wessex Institute of Technology, UK and* ***K.L KATSIFARAKIS****, Aristotle University of Thessaloniki, Greece*

In recent years, significant advances have been made in the overall management of riverine systems, including advances in hydraulic and hydrologic modelling, environmental protection and flood forecasting. Containing papers presented at the Fourth International Conference on River Basin Management this book addresses the latest developments in these fields.

Featured topics include: Hydraulics and Hydrology; Integrated Watershed Planning; River and Watershed Management; Water Quality Modelling; Flood Risk; Ecological Perspective; MIS, GIS and Remote Sensing; Sediment Transport; Environmental Impact; Hydrological Impact and Case Studies.

WIT Transactions on Ecology and the Environment, Vol 104
ISBN: 978-1-84564-075-0 2007 560pp £185.00/US$365.00/€277.50